Andriy M. Gusak

**Diffusion-controlled
Solid State Reactions**

Related Titles

Wetzig, K., Schneider, C. M. (eds.)

Metal Based Thin Films for Electronics

2006
ISBN: 978-3-527-40650-0

Skripov, V. P., Faizullin, M. Z.

Crystal-Liquid-Gas Phase Transitions and Thermodynamic Similarity

2006
ISBN: 978-3-527-40576-3

Jackson, K. A.

Kinetic Processes
Crystal Growth, Diffusion, and Phase Transitions in Materials

2010
ISBN: 978-3-527-32736-2

Andriy M. Gusak

Diffusion-controlled Solid State Reactions

in Alloys, Thin Films, and Nano Systems

With the collaboration of
Yuriy A. Lyashenko
Semen V. Kornienko
Mykola O. Pasichnyy
Aram S. Shirinyan
Tatyana V. Zaporozhets

WILEY-VCH Verlag GmbH & Co. KGaA

The Author

Prof. Andriy M. Gusak
Cherkasy State University
Department of Theoretical Physics
Cherkasy, Ukraine
amgusak@ukr.net

Consultant Editor

Dr. habil. Jürn W. P. Schmelzer
Physics Department
University of Rostock
Rostock, Germany
juern-w.schmelzer@uni-rostock.de

■ All books published by Wiley-VCH are carefully produced. Nevertheless, authors, editors, and publisher do not warrant the information contained in these books, including this book, to be free of errors. Readers are advised to keep in mind that statements, data, illustrations, procedural details or other items may inadvertently be inaccurate.

Library of Congress Card No.: applied for

British Library Cataloguing-in-Publication Data
A catalogue record for this book is available from the British Library.

Bibliographic information published by the Deutsche Nationalbibliothek
The Deutsche Nationalbibliothek lists this publication in the Deutsche Nationalbibliografie; detailed bibliographic data are available on the Internet at <http://dnb.d-nb.de>.

© 2010 WILEY-VCH Verlag GmbH & Co. KGaA, Boschstr. 12, 69469 Weinheim, Germany

All rights reserved (including those of translation into other languages). No part of this book may be reproduced in any form – by photoprinting, microfilm, or any other means – nor transmitted or translated into a machine language without written permission from the publishers. Registered names, trademarks, etc. used in this book, even when not specifically marked as such, are not to be considered unprotected by law.

Cover Design   Schulz Grafik-Design, Fußgönheim

Composition   Laserwords Private Limited, Chennai, India

Printing and Binding   Fabulous Printers Pte Ltd

Printed in Singapore
Printed on acid-free paper

ISBN: 978-3-527-40884-9

Foreword

I have known Prof. Andriy M. Gusak since 1991, and I visited him in the campus of Cherkassy National University twice. He has visited our campus at UCLA seven times and during each visit he has spent from one to four months in my research group to work together on diffusion-controlled solid-state reactions. It is a topic relevant to processing and reliability of modern microelectronic devices, especially thin film-based devices. For example, metal–silicon reactions are crucial in forming millions of electrical contacts and gates on a piece of Si chip. Another example is Pb-free solder joint reactions in flip-chip technology, which are the most important processing steps in the packaging of consumer electronic products. As we approach the end of Moore's law of miniaturization, a change of paradigm is being developed in order to focus on more applications of the existing CMOS technology besides pushing harder and harder to achieve smaller and smaller devices. When nanotechnology becomes mature and finds applications, diffusion-controlled solid-state reactions in nanoscale materials will be of major concern from the point of view of stability and reliability of the nanodevices.

Prof. Gusak is not only an outstanding physicist, but also a superb mathematician in applying mathematical analysis to new physical findings. He has published many highly cited papers, especially in kinetics of solid-state reactions. He has made significant contributions to reaction kinetics in bulk materials, thin films, and nanoscale microstructures. He has covered kinetic processes from nucleation, growth, to ripening. On reliability issues, he has applied the concept of irreversible processes to electromigration, thermomigration, and stress migration in interconnect and electronic packaging technology. I have benefited tremendously from working with him on the reliability issues, and I appreciate his deep insight and subtle understanding of some of the very salient nature of solid-state reactions.

In this book, Prof. Gusak and his former students have made a systematic presentation of the following topics:

(i) The initial stages of reactive reactions, including nucleation and lateral growth of nucleated islands. The nucleation of different phases is considered as a coupled process in open systems, under conditions of external fluxes and gradients of chemical potential (Chapters 3, 4, and 5).

(ii) Other flux-driven processes in open systems leading to specific morphology evolution – flux-drive ripening of intermetallic scallops during reaction between molten solder and metal (copper or nickel), flux-driven lateral grain growth in thin films during deposition, flux-driven bifurcations of Kirkendall planes and the related problem of dynamic stability or instability of these planes, electromigration-driven grain rotation, and electromigration and thermomigration-driven evolution of two-phase alloys (Chapter 6).
(iii) Void formation, growth, shrinkage, and migration during reactive diffusion in nanoparticles and during electromigration (Chapter 7).
(iv) Phase growth and competition of growth under stressing by direct electric current (Chapter 8).
(v) Interdiffusion and reactive diffusion (including the nucleation stage and the formation and growth of two-phase zones) in the ternary systems (Chapters 9 and 10).
(vi) Special emphasis on diffusion and phase transformations in nanosystems is made in Chapters 2 and 13 (although other chapters, especially Chapter 7, also include some analysis of nanosystems).
(vii) Since reactions can occur at various ways and proceed in different directions, there are unsolved (in general case) fundamental problems of choice between the evolution paths in closed and in open systems. These topics are intensively discussed in Chapters 11 and 12.

I highly recommend this book to students and to experienced researchers working in the field of materials science, nonequilibrium thermodynamics, and nucleation theory. I am sure it will become a source of new ideas and inspiration for anybody dealing with these topics.

Los Angeles, November 2009

King-Ning Tu
http://www.seas.ucla.edu/eThinFilm/

Contents

Editor's Preface *XVII*
List of Contributors *XIX*

1 **Introduction** *1*
Andriy M. Gusak

2 **Nonequilibrium Vacancies and Diffusion-Controlled Processes at Nanolevel** *11*
Andriy M. Gusak
2.1 Introduction *11*
2.2 Beyond Darken's Approximation *12*
2.3 The Model for Regular Chains of Ideal Vacancies Sinks/Sources *17*
2.4 Description of Interdiffusion in Alloys at Random Power of Distributed Vacancy Sinks *20*
2.5 Linear Phase Growth and Nonequilibrium Vacancies *22*
2.6 Intermetallic Layer Growth at Imposed Current and Nonequilibrium Vacancies Damping Effect *25*
2.7 Possible Role of Nonequilibrium Vacancies in Spinodal Decomposition *26*
2.8 Nanoshell Collapse *29*
2.9 The Role of Nonequilibrium Vacancies in Diffusion Coarsening *32*
2.10 Conclusions *34*
References *34*

3 **Diffusive Phase Competition: Fundamentals** *37*
Andriy M. Gusak
3.1 Introduction *37*
3.2 Standard Model and the Anomaly Problem *37*
3.3 Criteria of Phase Growth and Suppression: Approximation of Unlimited Nucleation *45*
3.4 Incubation Time *47*
3.5 Should We Rely Upon the Ingenuity of Nature? Nucleation Problems and Meta-Quasi-Equilibrium Concept *49*

3.6	Suppression of an Intermediate Phase by Solid Solutions	52
3.6.1	Unlimited Nucleation	53
3.6.2	Finite Rate of Nuclei Formation	54
3.7	Phase Competition in a Model of Divided Couple	55
	References	59

4 Nucleation in a Concentration Gradient 61
Andriy M. Gusak

4.1	Introduction	61
4.2	Nucleation in Nonhomogeneous Systems: General Approach	63
4.3	Thermodynamics of the Polymorphic Mode of Nucleation in a Concentration Gradient	65
4.3.1	Homogeneous Nucleation: General Relations	65
4.3.2	Spherical Nuclei	66
4.3.3	Ellipsoidal Nuclei	68
4.3.4	MC Simulations of the Shape of the Nucleus	70
4.3.5	Stress Effects	71
4.4	Thermodynamics of the Transversal Mode of Nucleation in a Concentration Gradient	74
4.4.1	Homogeneous Nucleation: General Relations	74
4.5	Thermodynamics of the Longitudinal Mode of Nucleation in a Concentration Gradient	79
4.6	Nucleation in Systems with Limited Metastable Solubility	81
4.6.1	Nucleation of a Line Compound at the Interface During Interdiffusion	82
4.6.2	Nucleation in between Dilute Solutions	86
4.6.3	Nucleation in between Two Growing Intermediate Phase Layers	86
4.6.4	Nucleation in between a Growing Intermediate Phase and a Dilute Solution	88
4.6.5	Application to Particular Systems	91
4.7	Conclusions	95
	References	97

5 Modeling of the Initial Stages of Reactive Diffusion 99
Mykola O. Pasichnyy and Andriy M. Gusak

5.1	Introduction	99
5.2	First Phase Nucleation Delay in Al–Co Thin Films	100
5.2.1	The Problem of Nucleation in a Concentration Gradient Field	101
5.2.2	Basic Model	102
5.2.3	Transversal Mode	105
5.2.4	Polymorphic Mode	107
5.2.4.1	Polymorphic Mode without Shape Optimization	108
5.2.4.2	Polymorphic Mode with Shape Optimization	109
5.2.5	Discussion and Conclusions	110

5.3	Kinetics of Lateral Growth of Intermediate Phase Islands at the Initial Stage of Reactive Diffusion *112*	
5.3.1	Problem Formulation *112*	
5.3.2	Physical Model *114*	
5.3.3	Numerical Results *116*	
5.3.4	Analytical Solution for the Steady State *117*	
5.3.5	Asymptotic Thickness of an Island *118*	
5.3.6	Estimates *119*	
5.3.7	Conclusions *121*	
5.4	MC-Scheme of Reactive Diffusion *121*	
5.4.1	Formulation of the Problem *121*	
5.4.2	The Model *122*	
5.4.3	Nucleation of Phase A_2B_1 at the Interface $A-A_1B_2$ *124*	
5.4.4	Competitive Nucleation of Phases A_1B_2 and A_2B_1 at the Interface $A-B$ *129*	
5.4.5	Lateral Competition *131*	
5.4.6	Conclusions *131*	
	References *132*	
	Further Reading *133*	

6 Flux-Driven Morphology Evolution *135*
Andriy M. Gusak

6.1	Introduction *135*	
6.2	Grain Growth and Ripening: Fundamentals *136*	
6.2.1	Main Approximations of the LSW Approach *136*	
6.2.2	Traditional Approaches to the Description of Grain Growth *138*	
6.3	Alternative Derivation of the Asymptotic Solution of the LSW Theory *139*	
6.4	Flux-Driven Ripening at Reactive Diffusion *142*	
6.4.1	Experimental Results *143*	
6.4.2	Basic Approximations *144*	
6.4.3	Basic Equations *145*	
6.5	Flux-Driven Grain Growth in Thin Films during Deposition *148*	
6.5.1	"Mushroom Effect" on the Surface of a Pair of Grains: Deterministic Approach *150*	
6.5.2	Analysis of Flux-Driven Grain Growth *151*	
6.5.3	Stochastic Approach *154*	
6.5.4	Monte Carlo Simulation of Flux-Driven Grain Growth *155*	
6.5.5	Lateral Grain Growth in Aluminum Nanofilm during Deposition *156*	
6.5.5.1	Hillert's Model *160*	
6.5.5.2	Models Leading to a Rayleigh Distribution *161*	
6.5.5.3	Pair Interaction Model (Di Nunzio) *161*	
6.6	Flux-Induced Instability and Bifurcations of Kirkendall Planes *163*	
6.6.1	Kirkendall Effect and Velocity Curve *164*	
6.6.2	Stable and Unstable K-Planes *165*	

6.6.3	Experimental Results	*166*
6.6.4	General Instability Criterion	*168*
6.6.5	Estimation of Markers' Distributions Near the Virtual K-Plane	*169*
6.6.6	Spatial Distribution of Markers	*170*
6.6.7	Possible Alternative to the Multilayer Method	*171*
6.7	Electromigration-Induced Grain Rotation in Anisotropic Conducting Beta Tin	*173*
6.8	Thermomigration in Eutectic Two-Phase Structures	*178*
6.8.1	Thermomigration Induced Back Stress in Two-Phase Mixtures	*183*
6.8.2	Thermomigration-Driven Kirkendall Effect in Binary Mixtures	*184*
6.8.3	Stochastic Tendencies in Thermomigration	*185*
	References	*186*
7	**Nanovoid Evolution** *189*	
	Tatyana V. Zaporozhets and Andriy M. Gusak	
7.1	Introduction	*189*
7.2	Kinetic Analysis of the Instability of Hollow Nanoparticles	*191*
7.2.1	Introduction	*191*
7.2.2	Mechanism of Nanoshell Shrinkage	*192*
7.2.3	Models of Nanovoid Shrinkage	*194*
7.2.3.1	Model 1: Shrinkage of Pure Element Nanoshells	*195*
7.2.3.2	Model 2: Shrinkage of a Binary Compound Nanoshell with Steady State Approximation for Both Vacancies and B Species	*197*
7.2.3.3	Model 3: Steady State and Non–Steady State Vacancies for Component B	*200*
7.2.3.4	Model 4: Non–Steady State Vacancies and Atoms	*204*
7.2.4	Segregation of Pure B at the Internal Surface	*205*
7.2.5	Kinetic Monte Carlo Simulation of Shrinkage of a Nanoshell	*206*
7.2.5.1	Model 1MC: Pure B-Shell in Vacuum	*207*
7.2.5.2	Model 2MC: Ordered IMC Nanoshell in Vacuum	*208*
7.2.6	Influence of Vacancy Segregation on Nanoshell Shrinkage	*208*
7.2.7	Summary	*215*
7.3	Formation of Compound Hollow Nanoshells	*216*
7.3.1	Introduction	*216*
7.3.2	Model of Nanoshell Formation	*216*
7.3.3	Simplified Analysis of the Competition Between "Kirkendall-Driven" and "Curvature-Driven" Effects	*218*
7.3.4	Rigorous Kinetic Analysis	*220*
7.3.5	Results and Discussion	*225*
7.3.6	Summary	*228*
7.4	Hollow Nanoshell Formation and Collapse in One Run: Model for a Solid Solution	*229*
7.4.1	Introduction	*229*
7.4.2	Shrinkage	*229*

7.4.3	Formation of a Hollow Nanoshell from Core–Shell Structure without the Influence of Ambient Atmosphere *233*
7.4.4	Results of the Phenomenological Model *234*
7.4.5	Monte Carlo Simulation of the Vacancy Subsystem Evolution in the Structure "Core–Shell" *238*
7.4.5.1	Formation of a NanoShell in a MC simulation *239*
7.4.5.2	Crossover from Formation to Collapse *239*
7.4.5.3	Shrinkage and Segregation Kinetics in an MC Simulation *241*
7.4.6	Summary *241*
7.5	Void Migration in Metallic Interconnects *245*
7.5.1	Hypotheses and Experiments *245*
7.5.2	The Model *248*
7.5.3	Results *249*
7.5.3.1	Migration of Voids in Bulk Cu and Determination of the Calibration Factor between MCS and Real Time *249*
7.5.3.2	Void Migration Along the Metal/Dielectric Interface *250*
7.5.4	Simplified Analytical Models of Trapping at the GBs and at the GB Junctions *253*
7.5.5	Summary *255*
	References *256*

8 Phase Formation via Electromigration *259*
Semen V. Kornienko and Andriy M. Gusak

8.1	Introduction *259*
8.2	Theory of Phase Formation and Growth in the Diffusion Zone at interdiffusion in an External Electric Field *260*
8.2.1	External Field Effects on Intermetallic Compounds Growth at Interdiffusion *260*
8.2.2	Criteria for Phase Suppression and Growth in an External Field *267*
8.2.3	Effect of an External Field on the Incubation Time of a Suppressed Phase *270*
8.2.4	Conclusions *271*
8.3	Effects of Electromigration on Compound Growth at the Interfaces *272*
8.4	Reactive Diffusion in a Binary System at an Imposed Electric Current at Nonequilibrium Vacancies *275*
8.4.1	Equation for the Growth of an Intermediate Phase taking into Account Nonequilibrium Vacancies *275*
8.4.2	Analysis of the Equation for the Rate of Intermediate Phase Growth in Limiting Cases *279*
8.4.3	Numerical Solution of the Equation for the Intermediate Phase Rate of Growth *281*
8.4.4	Conclusion *286*
	References *286*

9 Diffusion Phase Competition in Ternary Systems 289
Semen V. Kornienko, Yuriy A. Lyashenko, and Andriy M. Gusak

9.1 Introduction 289
9.2 Phase Competition in the Diffusion Zone of a Ternary System 289
9.2.1 Phase Competition in the Diffusion Zone of a Ternary System with Two Intermediate Phases 290
9.2.2 Influence of Pt on Phase Competition in the Diffusion Zone of the Ternary (NiPt)–Si System 295
9.2.2.1 Basic Considerations 295
9.2.2.2 Effect of Pt on Phase Competition in the Diffusion Zone of Ni–Si 297
9.2.2.3 Calculations and Discussion 300
9.3 Ambiguity and the Problem of Selection of the Diffusion Path 302
9.3.1 General Remarks 302
9.3.2 Analytical Solution of the Simplified Symmetric Model 304
9.3.3 Numerical Calculations for a Complex Model 309
9.3.4 Conclusions 320
9.4 Nucleation in the Diffusion Zone of a Ternary System 321
9.4.1 Model Description 321
9.4.2 Algorithm and Results for the Model System 325
9.4.3 Discussion 327
References 329
Further Reading 331

10 Interdiffusion with Formation and Growth of Two-Phase Zones 333
Yuriy A. Lyashenko and Andriy M. Gusak

10.1 Introduction 333
10.2 Peculiarities of the Diffusion Process in Ternary Systems 334
10.2.1 Notations 334
10.2.2 Thermodynamic Peculiarities 335
10.2.3 Diffusion Peculiarities 336
10.2.4 Types of Diffusion Zone Morphology in Three-Component Systems 337
10.3 Models of Diffusive Two-Phase Interaction 340
10.3.1 Model Systems 341
10.3.2 Phenomenological Approach to the Description of Interdiffusion in Two-Phase Zones 345
10.3.3 Choice of the Diffusion Interaction Mode 348
10.4 Results of Modeling and Discussion 350
10.4.1 One-Dimensional Model of Interdiffusion between Two-Phase Alloys 350
10.4.2 The Problem of Indefiniteness of the Final State 352
10.4.3 Diffusion Path Stochastization in the Two-Phase Region 353
10.4.4 Invariant Interdiffusion Coefficients in the Two-Phase Zone 354

10.4.5	Conclusions *356*	
	References *356*	
	Further Reading *358*	

11 The Problem of Choice of Reaction Path and Extremum Principles *359*
Andriy M. Gusak and Yuriy A. Lyashenko

11.1	Introduction *359*	
11.2	Principle of Maximal Entropy Production at Choosing the Evolution Path of Diffusion-Interactive Systems *359*	
11.3	Nonequilibrium Thermodynamics: General Relations *361*	
11.3.1	Isolated Systems *361*	
11.3.2	System in a Thermostat *363*	
11.3.3	Inhomogeneous Systems: Postulate of Quasi-Equilibrium for Physically Small Volumes *364*	
11.3.4	Extremum Principles *366*	
11.4	Application of the Principles of Thermodynamics of Irreversible Processes: Examples *368*	
11.4.1	Criterion of First Phase Choice at Reaction–Diffusion Processes *368*	
11.5	Conclusions *378*	
	References *379*	

12 Choice of Optimal Regimes in Cellular Decomposition, Diffusion-Induced Grain Boundary Migration, and the Inverse Diffusion Problem *381*
Yuriy A. Lyashenko

12.1	Introduction *381*	
12.2	Model of Self-Consistent Calculation of Discontinuous Precipitation Parameters in the Pb–Sn System *382*	
12.2.1	General Description of the Model Systems *384*	
12.2.2	Model Based on the Balance and Maximum Production of Entropy *387*	
12.2.2.1	Phase Transformations and Law of Conservation of Matter *388*	
12.2.2.2	Calculation of the Driving Force *389*	
12.2.2.3	Calculation of Energy Dissipation in the Transformation Front along the Precipitation Lamella *389*	
12.2.2.4	Calculation of Energy Dissipation Close to the Transformation Front *393*	
12.2.3	Calculation of Entropy Production Taking into Account Grain Boundary Diffusion and Atomic Jumps through the Grain Boundary *400*	
12.2.3.1	Optimization Procedure and Calculation Results *401*	
12.3	Model of Diffusion-Induced Grain Boundary Migration (DIGM) Based on the Extremal Principle of Entropy Production by the Example of Cu–Ni Thin Films *405*	
12.3.1	Model Description *406*	

12.3.1.1	Mass Conservation and Thermodynamic Description 408
12.3.1.2	Calculation of the Entropy Production Rate due to Grain Boundary Diffusion 409
12.3.1.3	Calculation of the Driving Force 410
12.3.2	Results of Model Calculations for the Cu–Ni System 411
12.3.2.1	Determination of the Curvature of the Gibbs Potential 411
12.3.2.2	Diffusion Parameters of the System 412
12.3.2.3	Grain Boundary Mobility 412
12.3.2.4	Results of the Model Calculation for the Cu/Ni/Cu-Like System 412
12.4	Entropy Production as a Regularization Factor in Solving the Inverse Diffusion Problem 416
12.4.1	Description of the Procedure of the Inverse Diffusion Problem Solution for a Binary System 416
12.4.2	Results of Model Calculations 418
12.5	Conclusions 421
	References 422
	Further Reading 424

13 Nucleation and Phase Separation in Nanovolumes 425
Aram S. Shirinyan and Andriy M. Gusak

13.1	Introduction 425
13.2	Physics of Small Particles and Dispersed Systems 427
13.2.1	Nano-Thermodynamics 427
13.2.2	Production of Dispersed Systems 428
13.2.3	Anomalous Structures and Phases in DSs and Thermodynamic Estimates 428
13.2.4	Influence of DSs on the Temperature of the Phase Transformation 430
13.2.5	State Diagrams of DSs 431
13.2.6	Shift of the Solubility Limits in DSs 431
13.2.6.1	Depletion 432
13.2.7	Concluding Remarks 432
13.3	Phase Transformations in Nanosystems 433
13.3.1	Solid–Solid First-Order Phase Transitions 433
13.3.1.1	Geometry of a Nanoparticle and Nucleation Modes 433
13.3.1.2	Depletion Effect 435
13.3.1.3	Regular Solution 435
13.3.1.4	Change of Gibbs Free Energy 436
13.3.1.5	Minimization Procedure 437
13.3.1.6	Probability Factor of the Phase Transformation 439
13.3.2	Phase Diagram Separation 439
13.3.2.1	Variation of Temperature T 439
13.3.2.2	Transition Criterion, Separation Criterion 440
13.3.2.3	Varying R 441

13.3.2.4	Varying C_0 441	
13.3.2.5	Phase Diagram 441	
13.3.2.6	Size-Dependent Diagram and Solubilities in Multicomponent Nanomaterials 442	
13.3.2.7	Critical Supersaturation 443	
13.3.2.8	Concluding Remarks 444	
13.4	Diagram Method of Phase Transition Analysis in Nanosystems 444	
13.4.1	Gibbs's Method of Geometrical Thermodynamics 445	
13.4.2	Nucleation of an Intermediate Phase 446	
13.4.2.1	Phase Transition Criterion 446	
13.4.2.2	Model of Intermediate Phase 446	
13.4.2.3	Separation in a Macroscopic Sample: Equilibrium State Diagram 447	
13.4.2.4	Separation in DSs: Size-Dependent Phase Diagram 448	
13.4.2.5	Influence of Size on Limiting Solubility 449	
13.4.2.6	Influence of Size of an Isolated Particle on the Phase Transition Temperature 449	
13.4.2.7	Concluding Remarks 450	
13.5	Competitive Nucleation and Growth of Two Intermediate Phases: Binary Systems 451	
	Case 1 454	
	Case 2 454	
	Case 3 or Crossover Regime 455	
13.5.1	Application to the Aluminum–Lithium system 456	
13.5.2	Concluding Remarks 458	
13.6	Phase Diagram Versus Diagram of Solubility: What is the Difference for Nanosystems? 458	
13.6.1	Some General Definitions 461	
13.6.1.1	What are the "solidus" and "liquidus"? 461	
13.6.1.2	What is the "Limit of Solubility"? 461	
13.6.2	Nanosized Solubility Diagram 462	
13.6.2.1	Solubility Limit 462	
13.6.2.2	Liquidus 462	
13.6.2.3	Solidus 462	
13.6.2.4	Nanosized Solubility Diagram 462	
13.6.3	Nanosized Phase Diagram 463	
13.6.3.1	Three Types of Diagrams 463	
13.6.3.2	T–C Diagram at Fixed R 464	
13.6.3.3	Varying R 465	
13.6.3.4	Concluding Remarks 465	
13.7	Some Further Developments 465	
13.7.1	Solubility Diagram of the Cu–Ni Nanosystem 465	
13.7.2	Size-Induced Hysteresis in the Process of Temperature Cycling of a Nanopowder 466	

13.7.2.1	Concluding Remarks *468*	
13.A	Appendix: The Rule of Parallel Tangent Construction for Optimal Points of Phase Transitions *469*	
13.A.1	Resume *470*	
	References *471*	

Index *475*

Editor's Preface

The present book is devoted to solid-state reactions covering the comprehensive analysis of birth (nucleation), growth, and competition of new solid phases as a result of interaction between parent solid phases. So, in order to cope with the theoretical description of such complex processes, the authors have to solve the whole spectrum of problems of description of nucleation and growth that one is confronted with in analyzing such kinds of problems. However, for the systems the authors analyze, additional problems occur, which are of less importance in a variety of other applications. One of them is the existence of a variety of possible reaction channels, the system may evolve at. So, the problem arises, regarding the choice of reaction channels by the systems at given initial conditions. The authors address this problem by connecting the choice of the reaction path with (heuristic) extremum principles for the rate of change of the appropriate thermodynamic functions. Another peculiarity of the analysis of solid-state reactions consists in the existence of well-expressed gradients of the concentration in the solids. These gradients in the concentration may affect significantly the process, and in particular, the work of critical cluster formation and thus the rate of nucleation of new phases.

Having obtained his scientific degrees at Moscow State University, the Institute of Metallurgy of USSR's Academy of Sciences, the Institute for Metal Physics of Ukrainian Academy of Sciences, and Kharkov State University, the leading author of this monograph, Professor Andriy M. Gusak, founded and now heads a highly productive group of applied theoreticians (at Cherkassy National University) involved in theoretical and computer modeling of diffusion, reactions, electromigration, void migration and microstructure change in metals, metal junctions, and nanosystems. Awarded with many prizes (among them a prize awarded by the American Physical Society) for his work, he has worked jointly as an investigator with several international teams and served on international advisory boards for major conferences in the field of nanoscience. Scientific visits have taken him to prestigious universities of Göttingen, Münster (Germany), UCLA (USA), Singapore, Grenoble, Marseille (France), Eindhoven (Netherlands), and Krakov (Poland). Written by an author with comprehensive and international expertise in cooperation with his young coworkers, the monograph coherently and comprehensively presents the approaches and results hitherto only available in various journal papers and at part only in Russian or Ukrainian language.

Diffusion-Controlled Solid State Reactions. Andriy M. Gusak
Copyright © 2010 WILEY-VCH Verlag GmbH & Co. KGaA, Weinheim
ISBN: 978-3-527-40884-9

The editor of the present monograph had the pleasure to discuss a variety of problems outlined in the present book on several research workshops in Dubna (Russia) devoted to the general topic "Nucleation Theory and Applications." In his research, Professor Gusak and his colleagues always combined profound theoretical analysis and applied research. In this way, the present book is not only of interest for people dealing with the theoretical concepts of phase formation but a must-have for all those involved with the public or corporate science of nanosystems, thin films and electrical engineering, and their applications.

Rostock and Dubna, December 2009 *Dr. habil. Jürn W. P. Schmelzer*

List of Contributors

This book is written by a group of applied theoreticians headed by **Prof. A. M. Gusak** and working at Cherkassy National University, Ukraine. The group is well-known, first of all, in the international diffusion community for their successful combination of deep insight into diffusion-controlled processes employing relatively simple mathematical tools for modeling these processes. The Cherkassy group has established long-standing cooperation links with diffusion and reactions centers in Debrecen, Eindhoven, Ekaterinburg, Göttingen, Grenoble, Kiev, Kharkov, Krakow, Los Angeles, Marseille, Moscow, Muenster, Rostock, and Singapore. The group is known also for organizing periodic international DIFTRANS conferences. The group and its members have obtained various international grants (International Science Foundation, CRDF, INTAS, etc.) The group includes:

Yuriy A. Lyashenko PhD (1992). Dean of Physics School. Having graduated from Cherkassy National University with the degree in physics, he defended his PhD thesis at G. V. Kurdyumov Institute for Metal Physics of the Ukrainian Academy of Sciences in Kiev. His field of expertise includes interdiffusion in ternary single-phase and two-phase alloys, cellular precipitation, diffusion induced grain-boundary migration, application of extremum principles to the description of complex systems.

Semen V. Kornienko PhD (1999). Chair of Theoretical Physics. Having graduated from Cherkassy National University with the degree in physics, he defended his PhD thesis at Kharkov National University. His field of expertise includes nucleation, inter- and reactive diffusion in ternary systems, phase growth under electromigration, diffusion with nonequilibrium vacancies.

Aram S. Shirinyan PhD (2001). Associate Professor in theoretical physics, now working on his habilitation thesis at T. Shevchenko National University in Kiev. He obtained his scientific degrees at Erevan State University, Armenia (1993), Cherkassy National University (1998), and Kharkov National University (2001) in solid-state physics. He has been awarded with the personal scientific grant of the Ukrainian Government, honorary diploma of the National Academy of Sciences of Ukraine (2002). He has profound experience in international cooperation (INTAS individual grant (2004), DAAD fellowship (2008), primary investigator in a joint Ukrainian–German project (2009) etc.), the main scientific contacts are established to colleagues in Bulgaria, Belgium, France, Germany, Hungary, Poland, Russia, USA. His field of expertise is in materials science, condensed matter physics, nanophysics, reaction–diffusion processes, and phase equilibria in low-dimensional and, in particular, nanosize systems, thermodynamics and kinetics of first-order phase transitions in multicomponent systems.

Mykola O. Pasichnyy PhD (2006). Chair of General Physics. Having graduated from Cherkassy National University with the degree in physics, he defended his PhD thesis in solid-state physics at the G. V. Kurdyumov Institute for Metal Physics of the Ukrainian Academy of Sciences in Kiev. The thesis was directed to the modeling of the initial stages of solid-state reactions. His field of expertise is in the theory and modeling of new phase nucleation, growth and competition during reactive diffusion or decomposition.

Tatyana V. Zaporozhets PhD (1999). Associate Professor at the Department of Theoretical Physics. Currently working on her habilitation thesis. Having graduated from Cherkassy National University with the degree in mathematics, she defended her PhD thesis in solid-state physics at the G. V. Kurdyumov Institute for Metal Physics of the Ukrainian Academy of Sciences in Kiev. Her field of expertise includes computer modeling of phase transformations and reactive diffusion in solids with preference to simulation at an atomic level by Monte Carlo methods, molecular dynamics, and molecular statics. She likes to confirm, visualize, and adorn the elegant but somewhat dull and long formulae derived by Prof. Gusak.

1
Introduction
Andriy M. Gusak

While writing the introduction to our book, I had in front of me the cover of the book by Slezov "Kinetics of first-order transitions," which was then almost ready for publication and has now been published by Wiley-VCH. This book is written by the "S" in the famous LSW (Lifshitz–Slezov–Wagner) theory of coarsening and will undoubtedly become a milestone in theoretical materials science. Our book is a logical (though, due to the complexity of the phenomena discussed, much less rigorous) continuation of Slezov's book, because it treats the kinetics of first-order transitions in open inhomogeneous systems, characterized by sharp gradients of chemical potentials and bypassing fluxes of matter, charge, and heat.

Solid state reactions (SSRs) are obviously the most interesting topic in the world since they mean birth, competition, and growth of new "worlds" (phases) as a result of interactions between parent phases. So, those who plan, realize, and analyze SSRs, are in fact the local gods with their six days (hours, seconds) of creation/annealing. SSRs are governed by two magic powers – thermodynamics and kinetics. Thermodynamics is something like a parliament, which decides how things should evolve, what is possible and impossible. Kinetics is something like a government or a local authority. It implements (or delays the implementation of) the laws given by the parliament. Common understanding is that kinetics determines only the rate of fulfilment (implementation) of thermodynamic laws. In this book we will see that life is more complicated.

For systems far from equilibrium, the problem of evolution path choice of an appropriate path of evolution appears frequently – which way to the equilibrium (among a variety of possibilities) should a system choose, and what factors influence the decision-making procedure. Well-known examples of such systems are the diffusion couples for reactive diffusion in binary and especially, ternary systems and supersaturated highly defectous alloys. For a rather long time, the phase spectrum of such nonequilibrium systems differs significantly from the equilibrium phase diagram. From the theoretical point of view, the best way of solving the problem of choice would be some kind of Prigogine-like principle. Yet, up to now there has been no simple algorithm that would allow one to predict, say, the sequence of phase formation by calculating, for example, the rate of free energy changes

Diffusion-Controlled Solid State Reactions. Andriy M. Gusak
Copyright © 2010 WILEY-VCH Verlag GmbH & Co. KGaA, Weinheim
ISBN: 978-3-527-40884-9

for different evolution paths. Moreover, in Chapter 11 we will see that Prigogine's principle cannot help us in this situation. So, to make some predictions, one should take into account the initial stage of the system's evolution including nucleation of the new phase.

It has been well known for several decades that reactive diffusion in thin films usually demonstrates "one-by-one" (sequential) phase formation. Despite the existence of several stable intermediate phases in the phase equilibrium diagram, only one growing phase layer is usually observed. The next phase appears or at least becomes visible, after one of the terminal materials has been consumed, so that the first phase has no more material for growth, becoming a material for second phase formation itself, and so on. Such sequential growth has at least three possible explanations:

1) "Just Slow Growth".
   The first phase to grow usually has the maximum diffusivity and, hence, grows fast (according to d'Heurle, "fast is the first" or "first is fast"). Other phases, with lower diffusivities, grow even more slowly than they could do alone. It means that if, by some reason (for example, nucleation problem), the fast growing phase would be missing, other phases would grow faster. Thus, according to this first explanation, other phases exist and grow, but too slowly, and their layers are so thin, that it is just difficult to detect them.
2) Interfacial Barriers Plus Competition (Gösele and Tu, 1982).
   Interfacial barriers are believed to cause the initial linear phase growth. In the case of single-phase formation the interfacial barriers simply slow down the rate of formation, making it linear instead of parabolic. For two phases, the barriers can make the growth rate of a certain phase formally negative even for zero thickness, which means that this phase will be not be present at all.
3) Diffusive Suppression of Critical Nuclei.
   In 1981, one of the authors (AMG) jointly with his teacher, the late Professor Cyrill Gurov (Institute for Metallurgy, Moscow) presented a simple (even "naive," as we see it today) model of phase competition taking into account the nucleation stage of each phase. Actually, the only concept, which had been taken in this initial work from the nucleation theory, was the existence of critical nuclei. They appear due to some miracle called *heterophase fluctuations*, which are stochastic events and cannot be described by some deterministic model. The initial idea was just that each phase cannot start from zero thickness – it should start from a critical size particle (about a nanometer). Contrary to standard nucleation theory, the critical nuclei of intermediate phases during reactive diffusion are formed in a strongly inhomogeneous region – the interface between other phases. Therefore, from the very beginning they have to allow the diffusion fluxes pass through themselves. Evidently, fluxes change abruptly when passing across each new-formed boundary of the newly formed nucleus, and thus drive the boundary movement. This picture of interface movement due to flux steps is well known for diffusion couples under the name of *Stephan problem* and refers to diffusive interactions between neighboring phases. Yet,

the initial width of each phase is taken to be the critical nucleus size (instead of zero). The peculiarity of the initial stage is just the possibility that the width of some phase nucleus (distance between left and right boundaries) can decrease as well as increase. If it decreases, the nucleus becomes subcritical and should disappear. Usually it happens if the neighboring phases have larger diffusivity and comparative thickness. Then these neighbors ("vampires" or "sharks") will destroy and consume all of the newly forming nuclei, making the new phase to be present only virtually – in the form of constantly forming (due to heterophase fluctuations) and vanishing (due to diffusive suppression by the neighbors) embryos.

Both models (b and c) (published simultaneously) predict suppression/growth criteria and a certain critical thickness of the first growing phase under which growth of other phases is kinetically suppressed. A simple mathematical scheme was built predicting the sequence of phase formation and incubation periods, provided that one knows the integrated Wagner diffusivities and critical nucleus size for each phase. The most simple example of this scheme is presented in Chapter 3. This scheme was applied to the competition of intermediate phase with solid solutions (Chapter 3), to the phase growth under strong electric current (jointly with Semen Kornienko, Chapter 8), to reactive diffusion in ternary systems (jointly with Semen Kornienko and Yuriy Lyashenko, Chapter 9), and to phase competition in reacting powder systems. Applications to the electric field case demonstrated that the phase spectrum in the reaction zone can be influenced and even controlled by sufficiently strong current density. Large current densities become real due to miniaturization of integrated schemes and introduction of the flip-chip technology. Latter results on reactive diffusion in UBM-solder contacts (UBM-standard abbreviation for "under-bump metallization" in microelectronics) under strong current crowding confirm the mentioned idea: without current or under a weak current, the reaction between copper and a dilute solution of tin in lead at 150 °C demonstrates only Cu_3Sn_1 phase formation. A current density of $j > 10^8$ A m^{-2} leads to the formation and fast growth of the Cu_6Sn_5 phase.

In our naive model of phase competition, we had taken the inhomogeneity of the nucleation region into account only partially. For instance, we treated the diffusive interactions of the newly formed nuclei, but we had not considered the possible change of the nucleation barrier, size, and shape, caused by the very fact of the existence of a sharp concentration gradient. Thus, we had to reconsider the thermodynamics of nucleation in a concentration gradient. The very first version of such theory was presented at DD-89 "river Volga boat-conference" in Russia and first published in the Ukrainian Journal of Physics in May 1990. The main idea was as follows: if, prior to intermediate phase formation, a narrow layer of the metastable solid solution or amorphous alloy had been formed at the base of the initial interface, the sharp concentration gradient inside this layer provides the decrease of the total bulk driving force of nucleation, and correspondingly the increase of the nucleation barrier. The nuclei were taken to be spheres, appearing in the strongly inhomogeneous concentration profile of the parent phase, so that the local driving force of transformation could change significantly from the left to the right along the diameter of the nucleus. This effect appeared to be nonnegligible, since the intermediate phases usually have a very strong concentration dependence on the Gibbs free energy. The main result was a principally new size dependence of the Gibbs free energy it contained, in addition to the terms of second-order (surface energy, positive) and third-order (bulk driving force, negative), and a new term proportional to the fifth power of size and the square of the concentration gradient, that is,

$$\Delta G(R) = \alpha R^2 - \beta R^3 + \gamma (\nabla c)^2 R^5 \tag{1.1}$$

Here the parameter γ has positive values and is proportional to the second-order derivative of the new phase Gibbs energy with respect to concentration. Equation 1.1 implies that for rather large gradients (typically larger than 10^{+8} m^{-1} (one hundred millions reciprocal meters) which means diffusion zone of about 10^{-8} m thickness), the dependence becomes monotonically increasing (infinitely high nucleation barrier) meaning thermodynamic suppression of nucleation by very sharp concentration gradients. Thus, according to our model, at the very initial stage of reactive diffusion, nucleation can be suppressed even without diffusive competition, due to a too narrow space region, favorable for transformation. Similar results were independently published by Pierre Desre *et al.* in 1990 and 1991. This approach had been applied then to the description of solid-state amorphizing reactions, explaining why stable intermetallics appear in the diffusion zone only after the amorphous layer exceeds some critical thickness.

Despite the similarity of the results, our models of nucleation in a sharp concentration gradient treated quite different possible mechanisms (nucleation modes). We treated a polymorphous mode keeping in mind the following picture: The initial diffusion leads to the formation and growth of a metastable parent solution with a sharp concentration profile. When this profile becomes smooth enough to provide sufficient space for compositions favorable for a new intermediate phase, this very phase nucleates just by reconstruction of atomic order, without immediately changing the concentration profile (at "frozen" diffusion) – polymorphic transformation. Desre suggested the transversal nucleation mode bearing in mind the following case: each thin slice of the newly formed nucleus, perpendicular to the direction of the concentration gradient, is considered as a result of decomposition in a corresponding thin infinite slice of the parent solution, leading, of course, to a redistribution of atoms among new and old phases. In this transversal mode, the redistribution proceeds within each slice, independent of others.

In 1996, one more mechanism had been suggested by Hodaj – the total mixing (longitudinal) nucleation mode, when the redistribution of atoms proceeds during nucleation, but only inside the newly forming nucleus. Contrary to the two previous modes, in this case the concentration gradient assists nucleation – in Equation 1.1, the coefficient γ is negative. The above-mentioned approach was generalized taking into account shape optimization, stresses, ternary systems, heterogeneous nucleation at grain boundaries and at interphase boundaries. Readers will find an almost full description in Chapter 4. The latest developments of this approach including atomic tomography experiments by Guido Schmitz and Monte Carlo simulations of Mykola Pasichnyy are described in Chapter 5.

It is quite possible to expect that nature will commonly use the mechanism with the lowest nucleation barrier – the total mixing mode. Yet, nucleation is ruled not only by thermodynamics but by kinetics as well. Thermodynamics of nucleation with constraints indicates only some probable paths of evolution. The real path is chosen by kinetics, taking into account not only the free energy profit, but also different "mobilities" along each path. Mobilities often appear to be inverse to profit – a kind of compensation rule, analogous to the relation between activation enthalpy and frequency factor in diffusion.

To understand nucleation kinetics quantitatively, we used the Fokker–Planck approach, first applied to nucleation problems by Farkas and recognized after the classical work of Zeldovich. Our contribution to this approach was just taking into account that the driving force depends on the concentration gradient, which in turn depends on time according to diffusion laws. It has been shown that the relative contribution of each mechanism depends on the ratio of atomic mobilities in the parent to that in nucleating phases. If atomic mobility in the new phase is much lower than in the parent one, we can forget about total mixing mode. In the opposite case (high mobility inside the new phase), nucleation will proceed via total mixing, in the very fast ("fast is the first") mode. One of the "raisins" of the total mixing (assisting) mode was that the easily formed nuclei, if not growing too fast in comparison with the decrease of the concentration gradient, after some period can find themselves to be subcritical and be destroyed. Unfortunately, so far we could not check the mentioned "kinetic nucleation" results experimentally, so, we did not risk including them in our book.

After discovering that a concentration gradient can function as a constraint on nucleation, it was natural for us to start looking for other constraints. Sharp concentration gradient means narrow layers, suitable for nucleation, for example, limited volume. So, the most natural thing was to treat the nucleation in small (nano-size) particles and multiple simultaneous nucleations during formation of bulk nanocrystalline materials. We tried it with Aram Shirinyan (Chapter 13). The main result was as follows: depending on the volume of the parent particle (or, in bulk, on the volume of "responsibility region" around the nucleation site), the same three possibilities exist as in a sharp concentration gradient – nucleation and growth (large size), metastable state (medium size), and forbidden nucleation (small size). Later we learned that similar results had been published by Rusanov 40 years earlier and later developed by Schmelzer *et al*. Yet, we hope that our application of these ideas to the "traffic jam" effect in bulk glasses might be of some interest: when many persons try to pass through a narrow door simultaneously, the process stops. A similar effect can be responsible for long-living nanocrystalline states. Moreover, we tried to generalize this approach to the case of competitive nucleation of two intermediate phases in nanovolumes. Later, Aram Shirinyan and I, jointly with Michael Wautelet, concluded that the common notions of phase transitions (like phase diagrams) should be revised accordingly in the case of transformations in nanosystems. This approach is not commonly accepted yet, and is also discussed in Chapter 13.

Diffusion controlled SSRs are, in fact, phase transitions in a concentration gradient. It is very important to keep in mind that the controlling step is not just diffusion but interdiffusion. It means that in most SSRs the vacancy fluxes appear as a result of different mobilities of diffusing species. Divergence of vacancy fluxes leads to competing Frenkel (Kirkendall) voiding and Kirkendall shift, depending on the relative efficiency of voids and dislocation kinks as vacancy sinks/sources. Competition of Frenkel and Kirkendall effects recently (in 2004) found interesting applications in the new method of hollow nanoshell production by SSRs (by Paul Alivisatos group). In this method, metallic nanoparticles react with oxygen or sulfur forming a compound in which metal atoms diffuse outside faster than oxygen or sulfur diffuse inside. Thus, the vacancy flux directed inside eventually leads to a hollow nanostructure with a central symmetric void. Tu and Gösele immediately noticed that such structures should be unstable due to the Gibbs–Thomson effect: the vacancy concentration at the inner shell boundary should be higher than that at the external one. Therefore, a vacancy gradient should exist leading to vacancy outflux and corresponding void shrinkage. Together with King-Ning Tu and Tatyana Zaporozhets, we analyzed the process of formation and collapse of nanoshells taking into account not only Frenkel, Kirkendall, and Gibbs–Thomson effects, but the inverse Kirkendall effect as well (segregation in the vacancy flux and corresponding damping of shrinking rate). We also demonstrated that the Gibbs–Thomson effect is important not only at the shrinking stage but also at the stage of formation, and may suppress the formation itself. These results are discussed in Chapter 7. Very recent results of Nakamura *et al.* (2008) prove that the shrinkage process is very real. In the same Chapter 7, we present a model of nanovoid evolution under high-density electric current in copper interconnects of VLSI (Very Large Scale Integrated) circuits. This model explains the recent discovery of a new failure mechanism by electromigration, when failure happens due to drift and coalescence of nanovoids along copper/dielectric interface instead of migration of individual vacancies. The latter two mentioned problems are examples of the flux-driven morphology of evolution. This general concept was formulated jointly with Professor King-Ning Tu.

In 2001, I visited Professor Tu at the Department of Materials Science at UCLA (Los Angeles) for the first time. These two months of stay were rather dramatic (they overlapped with the 9/11 tragedy). On the other hand, the joint work with Professor Tu appeared to be really exciting – we tried to theoretically address a very complicated experimental problem – the simultaneous growth and ripening of intermetallic compounds (IMCs) with scallop-like morphology during reaction between solid copper (or nickel) substrate and a liquid bump of tin-based solder. Despite our best efforts, the general picture remained unclear. I remember the moment, when we realized a very simple idea – in the case of almost hemispherical scallops, their growth and ripening on the fixed substrate area proceeds with growing volume and almost constant solid/liquid interface area (contrary to classic ripening – constant volume and decreasing area). Everything else was not easy, but more a technical task. The results (theory of flux-driven ripening) are presented in Chapter 6. After this, Prof. Tu reminded me of a Chinese saying, "if you see the corner of the table, please keep in mind that this table has three more corners." We therefore tried to look for other processes with growing volume and almost constant interface area. We first (mainly during my second visit in 2002) developed the FDGG (flux-driven grain growth) theory for lateral grain growth during deposition of thin films (see also Chapter 6). The above-mentioned ideas were applied to the behavior of two-phase systems (namely, solders) under electromigration and thermomigration, as well as for grain rotation under electromigration.

There is one more very interesting problem at the junction of interdiffusion and flux-induced morphology evolution in SSRs – it is a problem of bifurcation and instability of Kirkendall planes (K-planes). The honor of discovery of multiple K-planes (Section 5 of Chapter 6) has to be assigned to the Eindhoven diffusion group (van Loo, Kodentsov, van Dal, Cherhati). I was just lucky to come to the right place (Eindhoven) at the right time (beginning of 2000) to understand that K-planes can be stable or unstable, and a situation may be possible when only a single unstable K-plane exists leading to a broadening crowd of markers instead of a plane. Later, such a situation was experimentally verified. Actually, the problem is fundamental and exciting: the initial concentration step in the standard diffusion couple contains all the future of this couple like the initial fireball of Big Bang contained the future of our universe. Moreover, similar to small scales in the modern nonhomogeneous structure of the universe, small fluctuations of inert Kirkendall markers' distribution in the vicinity of the initial interface can lead to bifurcations and instabilities of K-planes. From a naive, purely intuitive point of view, it seems strange that markers, at first "overlapping" the whole concentration range of the diffusion couple, subsequently gather into a single plane corresponding to only one fixed (constant in time) composition, this plane usually being called a *Kirkendall plane*. It is commonly adopted that the K-plane represents the initial contact plane moving according to the parabolic law, its velocity being proportional to the difference of partial diffusivities and to the concentration gradient in the vicinity of the fixed composition. The existence of such a K-plane implies that this plane is a kind of attractor for markers. It is natural to ask why a system should have only one attractor. Is it possible for a binary diffusion couple to have two or more attractors? Is there any possibility for a system to interdiffuse without any attractors at all (without a stable K-plane), with a broadening, in time, of the distribution of markers? In the latter case, a system should "forget" the initial contact interface, which should lead to an especially tight bonding of the starting materials. Thus, the problem of bifurcations and instabilities of K-planes arises. As we will see below, this problem has a strong experimental background.

Other interesting examples of the flux-driven morphology evolution (cellular precipitation, diffusion induced grain boundary migration) can be found in Chapter 12, written by Yuriy Lyashenko. Here, he uses some general thermodynamic considerations introduced in Chapter 11, adding the entropy balance for steady state regimes. Of course, we could not leave aside interdiffusion and SSRs in ternary systems (Chapters 9 and 10). As early as 1989, together with my first official PhD student, Yuriy Lyashenko, we modeled interdiffusion in two-phase zones of ternary systems. Then we found (by numerical means) a so-called "stochastization" of the diffusion path in a two-phase region, and, in 1990, I published (in the Soviet Journal of Physical Chemistry) an explanation of this phenomenon, related to the degeneracy of the interdiffusivity matrix in a two-phase region. Similar results were later obtained John Morral's group, who went much further than we did and made an experimental verification.

Yet, we start from Chapter 2. We had two reasons for this. The first reason is the magic word "nano" in its title. The second reason is that this chapter gives a good example of how academic issues discussed by "pure theoreticians" can become very fresh and practically important, due to progress in technology and experimental possibilities. From a theoretical viewpoint, this chapter supplies one with a "trace" of the long activity (published mainly in Russian), related to the role of nonequilibrium vacancies in interdiffusion processes.

Thirty years ago, in the series of papers by Andrei Nazarov and Cyrill Gurov, the theory of interdiffusion with nonequilibrium vacancies was proposed. They predicted a much less (than in Darken's scheme) Kirkendall shift in the middle of the diffusion zone, and an interdiffusion coefficient

$$D_{NG} = \frac{D_A D_B}{C_A D_A + C_B D_B} \tag{1.2}$$

controlled by the slowest species (like in ionic crystals) instead of Darken's expression

$$\tilde{D} = C_B D_A + C_A D_B \tag{1.3}$$

controlled by the more mobile species.

I liked the NG (Nazarov-Gurov) approach but realized that its predictions of the Kirkendall shift were far from experimental. So, my aim was to somehow combine Darken's scheme and the NG theory. My main idea was to introduce the hierarchy of time-space scales. In "nanoscale" (when the size of a "physically small volume" is less than the free-walk distance of a vacancy to the sink) and in the corresponding time scale, the NG theory holds. In macroscale we come back to Darken. Evidently, in nanosystems the characteristic size is often less than the typical mean free path length for vacancies. Thus, nonequilibrium vacancies generated by interdiffusion obtain a second life in application to nanoshell formation and collapse, initial stages of SSR, electromigration, coarsening of nanoalloys, spinodal decomposition and so on.

The present book was written jointly with some of my former PhD students – Lyashenko (PhD, 1992), Kornienko (PhD, 1999), Zaporozhets (PhD, 1999), Shirinyan (PhD, 2001), Pasichnyy (PhD, 2006). I was happy to work with them. Many results covered by this book, were obtained jointly with such brilliant scientists as Cyrill Gurov (my teacher), King-Ning Tu, Frans van Loo and

Alexandr Kodentsov, Pierre Desre and Fiqiri Hodaj, Andrei Nazarov, and Guido Schmitz. I was very glad to discuss in detail various issues of SSRs with Dezso Beke, Boris Bokstein, Marek Danielewski, Sergiy Divinski, Patrick Gas, Christian Herzig, Francois d'Heurle, Leonid Klinger, Leonid Larikov, George Martin, Helmut Mehrer, Subodh Mhaisalkar, Anatoliy Mokrov, Ludmila Paritskaya, Jean Philibert, Jürn W. P. Schmelzer, Vitaly V. Slezov, Ulo Ugaste, and Chen Zhong. My special thanks are to King-Ning Tu, who helped us to find ourselves as a group of applied theoreticians in the diffusion and reactions community.

This book would have never appeared without Anne Vasylevska, who translated our Ukrainian Ukrainian, Russian Ukrainian, and Ukrainian English into (I hope) American English. This process was – as it turned out – so exciting that it led to another phase transformation of Anne from Vasylevska to Koval'chuk. And, finally, the authors acknowledge support of their research by collective and individual grants of the Ministry of Education and Science of Ukraine, Fundamental Research State Fund of Ukraine, INTAS, CRDF, DAAD (Germany), ISF (Soros), and BMBF (Germany).

2
Nonequilibrium Vacancies and Diffusion-Controlled Processes at Nanolevel
Andriy M. Gusak

2.1
Introduction

Old well-known words when having been added a "nano" prefix (nanophysics, nanomaterials, nanolevel, nanoparticles, nanopowders, etc.) have become very popular among physicists, material scientists, and even common people. Putting this "magic" addition is so favorable that it increases the possibility of getting grants. As the Americans say, "Go nano!" In fact, nanophysics is a connecting bridge between macroscopic systems physics and physics of single separate atoms and molecules. If a particle is of a few nanometers size, the number of its surface atoms is approximately equal to its bulk atoms, that is, the standard representation of a body as a bulk and a thin near-surface layer becomes questionable when speaking about nanoparticles and nanopowders. One cannot go over to the so-called thermodynamic limit ($N \to \infty, V \to \infty$) for nanoparticles. This statement means that the classical interpretation of a phase transition looses its meaning. This is for nanometric dimensions that size effects become the most apparent ones, that is, we have to account for a strong dependence of various intensive parameters on the size.

Some of these peculiarities are discussed in the next chapters of the book. In this chapter, we dwell upon one feature – for the majority of real systems, the "mean free path" of vacancies (depending on density and effectiveness of vacancies' sources and sinks) ranges within hundreds of nanometers to tens of microns. This means that in a randomly chosen zone of nanometer size, the vacancies probably neither arise nor disappear. Nonequilibrium vacancies in two cases are radiation damage and quenching from high temperatures. Yet, even without any radiation influence or rapid temperature changes, there can be other factors that constantly distort the balance of vacancies in the alloy. First, it is the difference of alloy components mobilities, which leads to vacancy fluxes rise in nonhomogeneous alloys. This means that, generally speaking, in nanoscale we must take into account nonequilibrium vacancies. Strange though it may seem, this simple truth had been paid little attention until recent times. We present some physical cases where nonequilibrium vacancy distribution at nanolevel appears to be significant.

Diffusion-Controlled Solid State Reactions. Andriy M. Gusak
Copyright © 2010 WILEY-VCH Verlag GmbH & Co. KGaA, Weinheim
ISBN: 978-3-527-40884-9

2.2
Beyond Darken's Approximation

The understanding and description of interdiffusion in alloys is based for about 60 years on the commonly accepted Darken's scheme. This scheme uses the following simple logic. The migration rate of atoms is different (tracer diffusion coefficients D_i^* and the corresponding partial diffusion coefficients $D_i, i = 1, 2$ are different). This results in different magnitude (and opposite direction) densities of components' fluxes in a crystalline lattice reference frame:

$$j_1 = -D_1 \nabla n_1, \quad j_2 = -D_2 \nabla n_2 \tag{2.1}$$

Here n_1 and n_2 are the numbers of atoms of the corresponding component per unit volume.

On the other hand, according to the law of conservation of matter, the vector sum of components fluxes in the laboratory reference frame (connected with one of the ends of an infinite, on diffusion scale, diffusion couple) must be equal to zero. According to Darken, the alloy provides fluxes balancing at the expense of lattice movement as a single whole at some certain velocity, u, which is measured by inert markers ("frozen" in lattice) displacement. Correspondingly in the laboratory reference frame, components fluxes acquire a drift component (similar to Galilean velocity transformation equations) as

$$J_i = j_i + nc_i u \tag{2.2}$$

Here c_1 and c_2 are the molar fractions of atoms of the corresponding species obeying the condition $c_1 + c_2 = 1$. We restrict ourselves to the case, when volume changes in the diffusion process are negligible, so that

$$n = n_1 + n_2 = \text{const.} \quad \Rightarrow \quad \nabla n_2 = -\nabla n_1 \tag{2.3}$$

One may define the magnitude of the lattice drift rate without analyzing the diffusion micromechanisms, but just from the condition that the sum of fluxes equals zero in the laboratory reference frame:

$$\sum J_i = 0 \quad \Rightarrow \quad u = -\left(\frac{1}{n}\right) \sum j_i = (D_2 - D_1) \nabla c_2 = (D_1 - D_2) \nabla c_1 \tag{2.4}$$

Similarly,

$$\Omega J_1 = -D_1 \nabla c_1 + c_1 (D_1 - D_2) \nabla c_1 = -(c_2 D_1 + c_1 D_2) \nabla c_1 = -\tilde{D} \nabla c_1$$
$$\Omega J_2 = -D_2 \nabla c_2 + c_2 (D_2 - D_1) \nabla c_2 = -(c_2 D_1 + c_1 D_2) \nabla c_2 = -\tilde{D} \nabla c_2$$
$$J_2 = -J_1 \tag{2.5}$$

holds. Here $\Omega = 1/n$ is the average atomic volume in the alloy. This lattice drift is known as *Kirkendall effect*, inert markers showing the movement of the lattice. From general considerations, one can expect that nature is multifaceted enough to possess plenty of means for implementing the conservation laws realization. If the movement of the lattice is totally or partially suppressed for some reasons, other mechanisms providing the fluxes balancing must arise.

The standard interpretation of the Kirkendall effect micromechanism, for substitutional alloys at least, is this: the partial fluxes difference causes a vacancy flux toward the more mobile component of the diffusion couple. Vacancies dismantle extra planes on this component's side. It means dislocations climb and vacancy annihilation at sinks (dislocation kinks). On the slow component's side vacancies are, conversely, generated (to support the flux) bringing to extra planes building up. Thus, on the slow component's side, new lattice planes appear and, on the side of a mobile one, the old lattice planes disappear. New lattice formation is vividly displayed in Mark van Dal's and Frans van Loo's experiments, showing the Kirkendall plane bifurcation into two stable K-planes moving in the opposite directions: between these two K-planes new grains appear and none of the markers remain (Chapter 6).

The necessary condition for the realization of Darken's scheme is a local quasi-equilibrium of vacancies. This means that the system must have a sufficient number of effective vacancies' sinks or sources. As a quantitative characteristic for both density and effectiveness of vacancies sinks/sources, we may use the concept of "mean free path length" of vacancies L_V and the corresponding relaxation time for vacancies, $\tau_V = (L_V^2/D_V)$, where D_V is the self-diffusion coefficient of vacancies. Obviously, Darken's scheme is realized only in a coarsened spatial scale of diffusion processes, in a physically small volume having a size, dx, much larger than L_V. In this case, sinks/sources can be considered uniformly "smeared" out over the volume and quite effective. But if we investigate diffusion in multilayer with 100 nm spacing, diffusion zone thickness being 10 nm and "mean free path" $L_V \sim 100 - 1000$ nm, it is clear that we cannot apply Darken's scheme. A similar problem arises in the description of grain boundary interdiffusion at low temperatures when it is not easy for the mobile component's atoms to build the lattice, since they have to pull apart rather massive grains.

Hence, the question arises as to how the system can implement the law of conservation of matter if lattice motion and vacancy sinks/sources functioning encounter difficulties or become impossible. The answer is simple – internal forces will appear in the system that will level the fluxes without displacement of lattice. If one excludes ionic crystals, there are two, known to us, types of such balancing forces – stress gradient and nonequilibrium vacancy concentration gradient. For the result to be obtained, the type of such "unclear force" is of secondary importance, while the result is significant.

Let F be an arising force, acting on each atom. Then components fluxes contain, in addition to the ordinary Fick's term, the drift (Einsteinian) term:

$$J_1 = -D_1 \nabla n_1 + \frac{n_1 D_1^*}{kT} F, \quad J_2 = -D_2 \nabla n_2 + \frac{n_2 D_2^*}{kT} F \tag{2.6}$$

The magnitude of the force is determined from the fluxes balancing condition:

$$J_1 + J_2 = 0 \tag{2.7}$$

Note that we consider the case $n_1 + n_2 = \text{const} \Rightarrow \nabla n_2 = -\nabla n_1$. Then

$$F = kT \frac{(D_1 - D_2) \nabla n_1}{n_1 D_1^* + n_2 D_2^*} \qquad (2.8)$$

Taking into account that partial diffusion coefficients are connected with tracer diffusivities by the same thermodynamic factor (from Gibbs–Duhem relation)

$$\frac{D_1}{D_1^*} = \frac{D_2}{D_2^*} = \varphi = \frac{c_1}{kT} \frac{\partial \mu_1}{\partial c_1} = \frac{c_2}{kT} \frac{\partial \mu_2}{\partial c_2} \qquad (2.9)$$

then from Equations 2.6 and 2.7 one obtains a rather simple formula:

$$J_1 = -\left(\frac{n D_1^* D_2^* \varphi}{c_1 D_1^* + c_2 D_2^*} \right) \nabla c_1 = -J_2 \qquad (2.10)$$

Thus, although the process is not reduced to mere random mixing, full fluxes prove to be equally proportional to the concentration gradient that allows one for the introduction of an interdiffusion coefficient

$$D = \frac{D_1^* D_2^*}{c_1 D_1^* + c_2 D_2^*} \varphi = \frac{D_1 D_2}{c_1 D_1 + c_2 D_2} \qquad (2.11)$$

which is named after various authors (Nernst and Planck, Bokstein and Shvindlerman, Nazarov and Gurov), who found Equation 2.11, which is being guided by different ideas concerning the interpretation of the "unclear force" F, namely, electric field, "osmotic pressure", nonequilibrium vacancies. Now, we dwell upon the latter aspects in more detail.

Apparently, in the absence of any acting vacancy sinks/sources, the flux of vacancies toward a more mobile component will lead to their accumulation on this component's side and depletion on the side of a less active component. As a result, a nonuniform distribution of nonequilibrium vacancies will evolve. Correspondingly, a vacancy concentration gradient will appear, and it must influence both vacancies and atoms fluxes.

To analyze these processes mathematically, we need formulae for the components fluxes taking into account nonequilibrium vacancies distribution. Let us illustrate how these formulae are obtained for the simplest example of a cubic lattice. Let the concentration gradient of the main components and vacancies $\frac{\partial c_A}{\partial x}, \frac{\partial c_B}{\partial x}, \frac{\partial c_V}{\partial x}$ be created in a square lattice along the $\langle 100 \rangle$ (x axis) direction. Since the vacancy concentration is much less than that of the main components (extremely rarefied alloys being an exception), we have

$$\frac{\partial c_A}{\partial x} \approx -\frac{\partial c_B}{\partial x} \qquad (2.12)$$

Let us find, for instance, the flux density of A-atoms through the plane with x coordinate, located between parallel atomic planes, $x - d/2, x + d/2$, where d stands for the interplanar spacing, which is equal to the lattice parameter a in this very case. Suppose ν_{AV} is the frequency (probability per unit time) of A atoms jumps into neighboring vacant sites at the condition that it is vacant indeed. This

2.2 Beyond Darken's Approximation

frequency depends on coordinate via main components concentration:

$$\nu_{AV} = \nu_{AV}\left(c_A\left(x \pm \frac{d}{2}\right)\right) \tag{2.13}$$

For an atom to jump, there must be a vacancy. The *a priori* probability of this is equal to the vacant sites fraction, which depends on the coordinate as well:

$$c_V = c_V\left(x \mp \frac{d}{2}\right) \tag{2.14}$$

The number of A atoms in one plane (per unit of area) being able to jump through the saddle plane x is $n \cdot d \cdot 1 \cdot c_A(x \pm d/2)$. Hence, obviously, the resulting flux density, which is defined as the difference between the number of jumps (per unit time) from the left plane to the right and from the right plane to the left, equals

$$I_A = nd \cdot c_A(x - d/2) \cdot c_V(x + d/2) \cdot \nu_{AV}(c_A(x - d/2)) \tag{2.15}$$
$$- nd \cdot c_A(x + d/2) \cdot c_V(x - d/2) \cdot \nu_{AV}(c_A(x + d/2))$$

In case that all gradients being not too large, we may neglect higher orders of smallness in interplanar spacings at the Taylor expansion. Otherwise, we will arrive at a nonlocal theory of Cahn–Hilliard type. Then elementary mathematics leads to the following expression for the flux

$$j_A = -nd^2 \cdot \frac{\partial(\nu_{AV} c_A)}{\partial x} + nd^2 \cdot \nu_{AV} c_A \frac{\partial c_V}{\partial x} \tag{2.16}$$

If one keeps in mind that the exchange frequency depends on the coordinate implicitly, that is, through concentration, one gets

$$j_A = -nd^2 \nu_{AV} c_V \left(1 + c_A \frac{\partial(\ln \nu_{AV})}{\partial c_A}\right) \frac{\partial c_A}{\partial x} + nd^2 \nu_{AV} c_A \frac{\partial c_V}{\partial x} \tag{2.17}$$

The term $d^2 \nu_{AV}$ represents the tracer diffusion coefficient of the component A.

Sometimes one stops at this point and writes the final equation in the following form:

$$j_A = -nD_A^* \varphi \frac{\partial c_A}{\partial x} + nc_A D_A^* \frac{\partial c_V}{\partial x} \tag{2.18}$$

interpreting the term $(1 + c_A(\partial(\ln \nu_{AV})/\partial c_A))$ as a thermodynamic factor from Darken's theory, which is equal to

$$\varphi = \frac{c_1}{kT}\frac{\partial \mu_1}{\partial c_1} = \frac{c_2}{kT}\frac{\partial \mu_2}{\partial c_2} \tag{2.19}$$

However, generally speaking, this is not the case. In fact, the vacancy concentration may be represented as a sum of the equilibrium concentration and its nonequilibrium deviation:

$$c_V(x) = c_V^{eq}(c_A(x)) + \upsilon(x) \tag{2.20}$$

The equilibrium vacancy concentration depends on local composition, while the latter depends on the coordinates in the heterogeneous alloy. Therefore,

$$c_V(x) = c_V^{eq}(c_A(x)) + v(x), \quad \frac{\partial c_V(x)}{\partial x} = \frac{\partial c_V^{eq}(c_A)}{\partial c_A}\frac{\partial c_A}{\partial x} + \frac{\partial v}{\partial x} \quad (2.21)$$

Substituting Equation 2.21 in 2.17 one gets

$$j_A = -nd^2 \nu_{AV} c_V \left(1 + c_A \frac{\partial(\ln \nu_{AV})}{\partial c_A} - c_A \frac{\partial \ln c_V^{eq}}{\partial c_A}\right) \cdot \frac{\partial c_A}{\partial x} \quad (2.22)$$
$$+ nd^2 \nu_{AV} c_A \frac{\partial v}{\partial x}$$

In the latter formula, we have taken into account that the deviation from equilibrium vacancy concentration in typical cases is essentially less than the equilibrium concentration itself.

Employing the usual symbols, we arrive at the following expressions:

$$\Omega j_A = -D_A^* \varphi \frac{\partial c_A}{\partial x} + \frac{c_A D_A^*}{c_V}\frac{\partial v}{\partial x} \quad (2.23)$$

$$\Omega j_B = -D_B^* \varphi \frac{\partial c_B}{\partial x} + \frac{c_B D_B^*}{c_V}\frac{\partial v}{\partial x} \quad (2.24)$$

$$\Omega j_V = (D_A - D_B)\frac{\partial c_A}{\partial x} - D_V \frac{\partial v}{\partial x} \quad (2.25)$$

Among these formulae, Equation 2.25 is the most significant one, showing that the vacancy flux is caused not only by the difference of the partial diffusion coefficients but also by the nonequilibrium vacancy distribution.

The principal result of the Nazarov–Gurov (NG) theory, developed in the 1970s, is that owing to high mobility of vacancies, their distribution can be considered as to be of quasi–steady state type, and so (if vacancy sinks/sources are not functioning) $\partial j_V/\partial x \approx 0$ and, moreover, $j_V \approx 0$. This means that $j_A + j_B = -j_V = 0$, so the vacancy concentration gradient levels components fluxes.

The notion of nonequilibrium vacancies is often considered to be connected either with generation of Frenkel pairs under the influence of high-energy particle radiation or with vacancy fluxes toward sinks after quenching of the alloy. Still, nonequilibrium vacancies may prove to be significant without any radiation, but in the processes of interdiffusion and reaction diffusion. In this case, the nonequilibrium spatial vacancy distribution is caused by differences of components mobility and (or) interface boundary motion at intermediate phase growth. In 1973–1978, Nazarov and Gurov developed a rigorous microscopic theory of interdiffusion in alloys taking nonequilibrium vacancies into account but having neglected (at least, explicitly) the vacancies sinks/sources effect [1–3]. In the above theory, the rate of concentration profile evolution is determined by an alternative to Darken's diffusion coefficient expression

$$D_{NG} = \frac{D_A^* D_B^*}{c_A D_A^* + c_B D_B^*}\varphi \quad (2.26)$$

instead of

$$\tilde{D} = (c_A D_B^* + c_B D_A^*)\varphi \tag{2.27}$$

From the viewpoint of "diffusion conductance," the NG and Darken equations describe extreme, marginal cases. Darken's equation corresponds to parallel connection when the interdiffusion rate is usually determined by the more mobile component. The NG equation conforms to series (consecutive) connection at which the interdiffusion rate is mainly determined by the less mobile component – the more mobile one has to wait until slower atoms accomplish their migration. The discrepancy between equations becomes most apparent whenever the components mobility ratio is much larger or much less than one. In the general case, we may expect the correct description of interdiffusion to correspond to a certain combination of parallel and series (consecutive) connection, depending on vacancies sinks/sources effectiveness. We will see that the NG and Darken equations conform to different spatial and time scales.

In Section 2.3, we consider a quite simplified model for a regular chain of vacancies ideal sources/sinks. We see that the D_{NG} coefficient describes details of the concentration profile between sources (sinks) in a "mesoscopic" scale of space and time. The coefficient $\tilde{D}$ (Darken) provides the interdiffusion process in a "macroscopic" (coarsened) scale. We get similar result in Section 2.4 in the framework of a "smeared" vacancies sources/sinks model. At that, the whole diffusion process is divided into three stages – parabolic NG stage, nonparabolic (nonlocal) stage, and Darken's (the longest) parabolic stage. In Section 2.5, we analyze the role of nonequilibrium vacancies at intermediate phase growth (reaction diffusion) when moving interphase boundaries become a sort of nonideal vacancies sources/sinks. At that, interface layer growth may appear to have three stages, too. In Section 2.6, the change of the described situation at direct current imposed on the diffusion zone is briefly analyzed (this topic is considered in details in Chapter 8). In Section 2.7, we suggest the modification of classical spinodal decomposition Cahn–Hilliard theory by taking into account nonequilibrium vacancies and finite (nonzero) free length of vacancies to sinks. The influence of vacancy fluxes on the kinetics of nanoshell collapse is discussed in Section 2.8 (the detailed theory is given in Chapter 7). Finally, in Section 2.9, the main ideas of our recently proposed theory of coarsening with taking into account nonequilibrium vacancies are briefly discussed.

2.3
The Model for Regular Chains of Ideal Vacancies Sinks/Sources

Let us consider interdiffusion in a nonhomogeneous binary substitutional alloy within the following simplified model. Suppose that ideal vacancies sinks/sources are located in a periodic way, then one can imagine a chain of grains with parallel large angle grain boundaries and without working dislocation sources between them. Sources/sinks coordinates are $x_n = nl$ (l = distance between sinks – grain size). The equilibrium vacancy concentration c_V^{eq} is maintained on each of these planes. We introduce two spatial scales at once – the coarsened one averages all parameters over grain sizes, while the microscopic scale defines the coordinates inside intervals between sinks. Thus, all three concentrations are determined as the functions of two spatial parameters:

$$c_{A,B,V} = c_{A,B,V}(x_n + \xi, t), \quad 0 < \xi < l \tag{2.28}$$

The concentration changes within each grain are small, so that we are able to use the linearization procedure

$$c_{A,B}(x_n + \xi, t) = c_{A,B}(x_n, t) + u_{A,B}(\xi, t)$$
$$c_V(x_n + \xi) = c_V(x_n, t) + \upsilon(\xi, t) = c_V^{eq} + \upsilon(\xi, t)$$
$$|u| \ll c, \quad |\upsilon| \ll c_V^{eq} \tag{2.29}$$

Hereinafter, we neglect equilibrium concentration dependence on concentrations. We also neglect correlation effects connected with memory of previous exchanges with vacancy (Manning corrections) [4]. As we have shown before [20], these corrections do not affect the main features.

Substituting Equation 2.29 into the equations for the fluxes Equations 2.23–2.25 and using the continuity equation not only for A and B but also for the vacancies located in the intervals between sinks, we arrive at the following generalizations for Fick's second law:

$$\frac{\partial u_A}{\partial t} = D_{AA}\frac{\partial^2 u_A}{\partial \xi^2} + D_{AV}\frac{\partial^2 \upsilon}{\partial \xi^2}, \quad \frac{\partial \upsilon}{\partial t} = D_{VA}\frac{\partial^2 u_A}{\partial \xi^2} + D_V\frac{\partial^2 \upsilon}{\partial \xi^2} \tag{2.30}$$

where

$$D_V = \frac{c_A D_A^* + c_B D_B^*}{c_V}, \quad D_{AA} \cong D_A^* \varphi \tag{2.31}$$

$$D_{AV} = -\frac{c_A D_A^*}{c_V^{eq}}, \quad D_{VA} = -\varphi\left(D_A^* - D_B^*\right) \tag{2.32}$$

In matrix form, Equation 2.30 may be presented as

$$\frac{\partial}{\partial t}\begin{pmatrix} u_A \\ \upsilon \end{pmatrix} = \begin{pmatrix} D_{AA} & D_{AV} \\ D_{VA} & D_V \end{pmatrix}\frac{\partial^2}{\partial \xi^2}\begin{pmatrix} u_A \\ \upsilon \end{pmatrix} \tag{2.33}$$

Equations 2.30 can be solved for each interval between sinks $(x_n, x_n + l)$ with zero boundary conditions for υ, using a diagonalization procedure and introducing such linear combinations of atom and vacancy concentrations $w^{(1)}$, $w^{(2)}$, for which the equations become independent:

$$\begin{pmatrix} w^{(1)} \\ w^{(2)} \end{pmatrix} = \begin{pmatrix} 1 & a_{12} \\ a_{21} & 1 \end{pmatrix}\begin{pmatrix} u_A \\ \upsilon \end{pmatrix} \equiv \hat{a}\begin{pmatrix} u_A \\ \upsilon \end{pmatrix}$$

$$\hat{a}\hat{D}\hat{a}^{-1} = \hat{\tilde{D}} = \tilde{D}_i\delta_{ij}, \quad \frac{\partial w^{(i)}}{\partial t} = \tilde{D}_i\frac{\partial w^{2(i)}}{\partial \xi^2}, \quad i = 1, 2 \tag{2.34}$$

Eigenvalues of the diffusion coefficients matrix in Equation 2.33 can be easily found and represent interdiffusion coefficient and vacancy diffusion coefficient in microscale, respectively,

$$\dot{D}_1 \cong \frac{D_A^* D_B^* \varphi}{c_A D_A^* + c_B D_B^*} = D_{NG}, \quad \dot{D}_2 \cong \frac{c_A D_A^* + c_B D_B^*}{c_v^{eq}} \cong D_V$$

$$\frac{\dot{D}_1}{\dot{D}_2} \approx c_v^{eq} \ll 1 \tag{2.35}$$

Although Equation 2.34 for $w^{(i)}$ ($i = 1, 2$) are independent themselves, their boundary conditions are not independent. In fact,

$$v\left(\xi = \substack{0 \\ 1}\right) = 0 \quad \Rightarrow$$

$$w^{(2)}\left(\xi = \substack{0 \\ 1}\right) = a_{21} w^{(1)}\left(\xi = \substack{0 \\ 1}\right) = -c_{ve} \frac{\left(D_A^* - D_B^*\right)\varphi}{c_A D_A^* + c_B D_B^*} w^{(1)}\left(\xi = \substack{0 \\ 1}\right) \tag{2.36}$$

Because of fast vacancy diffusion (atoms require vacancies to have an opportunity to jump, while vacancies in the atoms company are always "self-sufficient"), the relations $\dot{D}_2 \cong D_V \gg \dot{D}_1$ hold. Hence, the evolution of the combination $w^{(2)}$ proceeds much faster than that of $w^{(1)}$, and so the solution of Equation 2.34 for $i = 2$ is of steady-state type:

$$w^{(2)}(t, \xi) \approx -\frac{\varphi\left(D_A^* - D_B^*\right)}{D_V} \cdot \left(w^{(1)}(t, 0) + \frac{w^{(1)}(t, l) - w^{(1)}(t, 0)}{l}\xi\right)$$

$$= -\frac{\varphi\left(D_A^* - D_B^*\right)}{D_V}\left(u_A(t, 0) + \frac{u_A(t, l) - u_A(t, 0)}{l}\xi\right) \tag{2.37}$$

$$\Omega j_V = -D_{vA}\frac{\partial u_A}{\partial \xi} - D_V \frac{\partial v}{\partial \xi} \cong -D_V \frac{\partial w_2}{\partial \xi}$$

$$= \left(D_A^* - D_B^*\right)\varphi \frac{u_A(t, l) - u_A(t, 0)}{l} = \left(D_A^* - D_B^*\right)\varphi \frac{\partial c_A}{\partial x} \tag{2.38}$$

Thus, the vacancy flux inside each interval is constant along this interval but exhibits jumps (steps) at interval boundaries (at sources/sinks). At the same time, the flux, for example, of A atoms, is described by the formula:

$$\Omega j_A = -D_{AA}\frac{\partial u_A}{\partial \xi} - D_{AV}\frac{\partial v}{\partial \xi} = -D_1 \frac{\partial w^{(1)}}{\partial \xi} - \frac{c_A D_A^*}{c_A D_A^* - c_B D_B^*}\Omega j_V \tag{2.39}$$

Because of almost constant (steady-state) vacancy flux, the derivative of the second summand in Equation 2.39 tends to zero, so while passing on to Fick's second law (through the continuity equation), one gets

$$\frac{\partial u_A}{\partial t} = D_1 \frac{\partial^2 w^{(1)}}{\partial \xi^2} + 0 = D_1 \frac{\partial^2 u_A}{\partial \xi^2} \tag{2.40}$$

So, the concentration profile evolution between vacancy sources/sinks (that is, in microscopic scale, $d\xi \ll l$) is determined by the NG coefficient, D_{NG}. On the other hand, the full mean value (along the whole interval l) of A atoms flux is brought to the known Darken's form:

$$\Omega j_A = -D_1 \frac{\partial \bar{u}_A}{\partial \xi} - \frac{c_A D_A^*}{c_A D_A^* + c_B D_B^*} \Omega \bar{j}_V$$

$$= -\left(\frac{D_A^* D_B^*}{c_A D_A^* + c_B D_B^*} + \frac{c_A D_A^* (D_A^* - D_B^*)}{c_A D_A^* + c_B D_B^*}\right) \varphi \frac{\partial \bar{c}_A}{\partial x} = -D_A^* \varphi \frac{\partial \bar{c}_A}{\partial x} \quad (2.41)$$

Thus, in macroscopic scale ($dx \gg l$) interdiffusion is indeed described by Darken's scheme. Similar results can be obtained in other models, regarding sources/sinks as "smeared in space."

2.4
Description of Interdiffusion in Alloys at Random Power of Distributed Vacancy Sinks

Let us consider the interdiffusion process in a coarsened spatial scale, regarding vacancies sources/sinks as continuously distributed in an alloy. We take account of their action as a sink/source term in the equation for vacancy redistribution:

$$\frac{\partial c_V}{\partial t} = -\frac{\partial \Omega j_V}{\partial x} - \frac{c_V - c_V^{eq}}{\tau_v} = -\frac{\partial \Omega j_V}{\partial x} - \frac{\upsilon}{\tau_v} \quad (2.42)$$

Here τ_v is the vacancy relaxation time.

The initial set of equations for diffusion fluxes in the lattice reference frame Equations 2.23–2.25, after having made some algebraic transformations, can be presented in the following form:

$$\Omega j_A = -D^{(1)} \frac{\partial c_A}{\partial x} - \frac{c_A D_A^*}{c_V D_V} \Omega j_V \quad (2.43)$$

$$\Omega j_B = -D^{(1)} \frac{\partial c_V}{\partial x} - \frac{c_V D_B^*}{c_V D_V} \Omega j_V \quad (2.44)$$

$$\Omega j_V = (D_A - D_B) \frac{\partial c_A}{\partial x} - D_V \frac{\partial \upsilon}{\partial x} \quad (2.45)$$

At $t \gg \tau_v$ the characteristic time of vacancy concentration change is much larger than τ_v; therefore,

$$\frac{\partial \upsilon}{\partial t} \ll \frac{\upsilon}{\tau_v}$$

Thus, one may use the steady-state approximation for vacancies

$$\frac{\partial \Omega j_V}{\partial x} = -\frac{\upsilon}{\tau_v} \quad (2.46)$$

Substituting Equation 2.45 for vacancy flux in Equation 2.46, one obtains

$$\left(\frac{\partial^2}{\partial x^2} - \frac{1}{D_v \tau_v}\right) \Omega j_V(x) = \frac{1}{D_v \tau_v}(D_A - D_B) \frac{\partial c_A}{\partial x} \quad (2.47)$$

Let $L_V = \sqrt{D_V \tau_v}$ be the "mean free path length" of vacancies. Equation 2.47 gives the nonlocal solution for the vacancy flux with nonlocality radius L_V

$$\Omega j_V(x) = \frac{1}{2L_V} \int_{-\infty}^{\infty} \exp\left(-\frac{|x'-x|}{L_V}\right)(D_A - D_B)\frac{\partial c_A}{\partial x'}dx' \quad (2.48)$$

The component fluxes turn out to be also nonlocally bound up with a concentration gradient

$$j_A = -D^{(1)}\frac{\partial c_1}{\partial x} - \frac{c_A D_A^*}{c_V D_V}j_V$$

$$= -D^{(1)}\frac{\partial c_A}{\partial x} - \frac{c_A D_A^*}{c_V D_V}\frac{1}{2L_V}\int_{-\infty}^{\infty}\exp\left(-\frac{|x'-x|}{L_V}\right)\frac{\partial c_A}{\partial x'}(D_A - D_B)dx' \quad (2.49)$$

In the case of quite densely spaced ideal sinks of infinite power ($\tau_V \to 0, L_V \to 0$), nonlocality disappears and the obtained expressions coincide fully with Darken's scheme:

$$\frac{1}{2L_V}\exp\left(-\frac{|x'-x|}{L_V}\right) \xrightarrow[L_V \to 0]{} \delta(x'-x)$$

$$\Omega j_V(x) = (D_A - D_B)\frac{\partial c_A}{\partial x}$$

$$\Omega j_A = -\left[D^{(1)} + \frac{c_A D_A^*}{c_V D_V}(D_A - D_B)\right]\frac{\partial c_A}{\partial x} = -D_A\frac{\partial c_A}{\partial x}$$

$$j_B = -D_B\frac{\partial c^B}{\partial x} \quad (2.50)$$

In the other marginal case, at nonworking sinks ($\tau_V \to \infty, L_V \to \infty$)

$$j_V \to 0, \quad j_A = -nD^{(1)}\frac{\partial c_A}{\partial x} = -j_B \quad (2.51)$$

in such a way one gets (neglecting higher powers of vacancy concentration) the result of the NG theory.

If one accepts the classical interpretation to consider vacancy flux to cause the lattice displacement at a velocity $u = \Omega j_V$, and resulting components fluxes in the laboratory reference frame to be defined according to the Galilean equation (Equation 2.2), we have

$$\Omega J_A = -\Omega J_B = -\tilde{D}_{Dark}\left[(1-p)\frac{\partial c_A}{\partial x} + \frac{p}{2L_V}\int_{-\infty}^{\infty}\exp\left(-\frac{|x'-x|}{L_V}\right)\frac{\partial c_A}{\partial x'}dx'\right] \quad (2.52)$$

where

$$p = c_A c_B \frac{D_A^* - D_B^*}{c_V D_V}\frac{D_A - D_B}{\tilde{D}_{Dark}} = 1 - \frac{D^{(1)}}{\tilde{D}_{Dark}} \quad (2.53)$$

Hence, even having written Fick's first law in the laboratory reference frame at finite power of sinks, we get a nonlocal term.

Nonlocality is equivalent to a higher order derivatives contribution. In fact, if the concentration gradient is expanded into a Taylor's series with respect to x,

$$\frac{\partial c_A}{\partial x'}(x + (x' - x)) = \frac{\partial c_A}{\partial x} + \sum_{m=1}^{\infty} \frac{(x' - x)^m}{m!} \frac{\partial^{(m+1)} c_A}{\partial x^{(m+1)}} \qquad (2.54)$$

after integration one easily obtains

$$\frac{1}{2L_V} \int_{-\infty}^{\infty} \exp\left(-\frac{|x' - x|}{L_V}\right) \frac{\partial c_A}{\partial x'} dx' = \frac{\partial c_A}{\partial x} + L_V^2 \frac{\partial^3 c_A}{\partial x^3} + L_V^4 \frac{\partial^5 c_A}{\partial x^5} + \cdots \qquad (2.55)$$

Nonlocality of the diffusion equations leads to the violation of the parabolic law for diffusion.

An analysis of Equation 2.52 together with the continuity equation [5] results in the following conclusions. At $t \gg \tau_v/c_V$, interdiffusion is described by the parabolic law with the usual Darken coefficient. The above-mentioned law describes interdiffusion at $\tau_v \ll t \ll \tau_v/c_V$ as well, but there the NG coefficient has to be used. At intermediate stage $t \sim \tau_v/c_V$, the transition from one parabolic dependence to another one takes place. At that, the parabolic law for motion of the front of constant concentration is violated.

2.5
Linear Phase Growth and Nonequilibrium Vacancies

Intermediate phase growth at reaction diffusion often begins with a linear stage of growth ($\Delta X = \beta t$), which passes to a parabolic one ($\Delta X = k t^{1/2}$) only at large annealing times. This linear stage is commonly explained by the "boundary kinetics" (overcoming of an interface potential barrier) or, in other words, by the reaction at the interphase boundary (Chapter 3). The phase growth is linear whenever the characteristic relaxation time considerably exceeds the time of diffusion transfer of atoms through the phase layer. We emphasize that the discussion concerns the linear growth of an already existent phase layer, that is, nucleation and nucleus oppression problems in the concentration gradient field are excluded (Chapters 4 and 5). In such case, at a first view, there are no reasons for a notable delay of atom transitions through the interface boundary unless a thin layer of one of the components oxide appears between the phases. On the other hand, the linear growth is observed for those phases where the mobility of one component is predominant. Notable vacancy fluxes emerge in such phases. The vacancy flux generally exhibits jumps at the interface. This means that interfaces must serve as sources and sinks of vacancies.

Let us consider the case when the power of vacancy sinks/sources at the boundary of a growing phase layer is not sufficient to provide the subsystem's quasi-equilibrium. Here the vacancy concentration near the boundary toward which the flux is directed will be larger than the equilibrium one, and near the other boundary less as compared with the equilibrium concentration. Thus, the concentration gradient of nonequilibrium vacancies between the right and

2.5 Linear Phase Growth and Nonequilibrium Vacancies

the left boundaries of the phase layer will emerge, and this will contribute to the components' flux.

Suppose that there grows the only intermediate phase 1 ($c_1 < c_B < c_1 + \Delta c$) between mutually insoluble components A and B. Let c_{VL} (c_{VR}) be the vacancy concentration at the left boundary A – 1 (the right boundary 1 – B), c_V^{eq}, the equilibrium vacancy concentration in the phase 1. Let the A component mobility be larger than that of the B component: $D_A > D_B$, such that in the phase 1 vacancy flux is directed from B to A, that is, from the right boundary to the left one:

$$\Omega j_V = (D_B - D_A)\frac{\Delta c}{\Delta X} - D_V \frac{(c_{VR} - c_{VL})}{\Delta X} \tag{2.56}$$

Considering the fluxes in A and B phases to be absent, we assume that vacancy fluxes jumps are determined only by phase 1. Thus, we can write the vacancy flux balance at interfaces as

$$\frac{dc_{VL}}{dt} = -\frac{(c_{VL} - c_V^{eq})}{\tau_v} - \frac{\Omega j_V}{\delta} \tag{2.57}$$

$$\frac{dc_{VR}}{dt} = -\frac{(c_{VR} - c_V^{eq})}{\tau_v} - \frac{\Omega j_V}{\delta} \tag{2.58}$$

where δ stands for the boundary layer thickness, τ_v is the vacancy relaxation time.

Let us consider the steady-state regime ($dc_{VL,R}/dt \cong 0$), which is established at $t \gg \tau_V$. Then using Equations 2.56–2.58, one gets the expression for the vacancy concentration change between the right and the left phase boundaries as

$$\Delta c_V = c_{VR} - c_{VL} = \left(\frac{2\tau_v}{\delta}\right)\frac{(D_B - D_A)\Delta c}{(\Delta X + 2D_V\tau_v/\delta)} \tag{2.59}$$

The above-mentioned vacancy concentration change makes a contribution into component fluxes in the crystal lattice reference frame [1,5] in the form:

$$\Omega j_B = \frac{-D_B \Delta c}{\Delta X} + \left(\frac{c_B D_B^*}{c_V}\right)\frac{\Delta c_V}{\Delta X} \tag{2.60}$$

From this relation, one obtains (going over to the laboratory reference frame) the expression for the average (over the phase layer) flux in the laboratory reference frame:

$$\Omega J_B = \Omega j_B + c_B \Omega j_V = -\Omega J_A = \frac{-D\Delta c \cdot \left(1 + \frac{l_0}{\Delta X}\right)}{(\Delta X + l_1)} \tag{2.61}$$

where

$$D = c_1 D_A + (1 - c_1)D_B \tag{2.62}$$

is the mean interdiffusion coefficient inside the phase,

$$l_1 = \frac{2D_V \tau_v}{\delta} = \frac{2L_V^2}{\delta} \tag{2.63}$$

where L_V is the 'vacancies mean free length',

$$l_0 = \frac{D_{NG}}{\tilde{D}}l_1$$

and the inequality $D_{NG} < \tilde{D}$ is always fulfilled. For the phases with narrow homogeneity zones, the flux remains almost the same at each point of the phase, so Equation 2.61 can be used for writing the flux balance equations at the interfaces. This procedure results in the following equation for the rate of phase layer growth:

$$d\Delta X/dt = -\left(\frac{1}{c_1} + \frac{1}{(1-c_1)}\right) \cdot \Omega J_B$$

$$= \frac{D\Delta c}{(c_1(1-c_1))} \frac{\left(1 + \frac{l_0}{\Delta X}\right)}{(\Delta X + l_1)} \quad (2.64)$$

In the general case, the layer growth according to Equation 2.48 has three stages, the latter one being determined by two characteristic thicknesses l_1 and l_0,

1) $\Delta X \ll l_0 (< l_1)$

$$2\frac{d\Delta X}{dt} \cong \frac{k_{NG}^2}{\Delta X}, \quad \Delta X \cong k_{NG} t^{1/2} \quad (2.65)$$

(primary parabolic stage), where

$$k_{NG} = \left[\frac{2 D_{NG} \Delta c}{(c_1(1-c_1))}\right]^{1/2} \quad (2.66)$$

2) $l_0 \ll \Delta X \ll l_1$

This relation holds provided that $D_{NG} \ll D$, that is, $D_A \ll D_B$, or $D_B \ll D_A$

$$\frac{d\Delta X}{dt} \cong \beta, \quad \Delta X \cong \beta t \quad (2.67)$$

(linear growth stage), where

$$\beta = \frac{(D \Delta c / l_1)}{(c_1(1-c_1))} \quad (2.68)$$

3) $\Delta X \gg l_1$

$$2\frac{d\Delta X}{dt} \cong \frac{k^2}{\Delta X}, \quad \Delta X \cong k t^{1/2} \quad (2.69)$$

(Darken's parabolic regime with quasi-equilibrium vacancies), where

$$k = \left[\frac{2 D \Delta c}{(c_1(1-c_1))}\right]^{1/2} \quad (2.70)$$

Thus, if one component is considerably more mobile than the other in a growing phase, we observe a linear stage of growth, the upper limit of it $l_1 = 2 L_V^2 / \delta$ being determined by the vacancies mean free path length.

If we take reasonable estimations

$$L_V \sim 10^{-7} \text{m}, \quad \delta \sim 5 \times 10^{-10} \text{m} \quad (2.71)$$

then $l_1 \sim 40\,\mu\text{m}$, that is, the value is rather large.

When the phase layer becomes substantially thicker than L_V, then one should take vacancy sinks/sources inside the growing layer into account. The corresponding theory is now under construction.

2.6
Intermetallic Layer Growth at Imposed Current and Nonequilibrium Vacancies Damping Effect

Let us modify the scheme described in the previous paragraph for the case of phase formation at a direct current imposed onto the diffusion zone, from A to B or in the reverse direction (the pattern of phase formation at direct current is reviewed in detail in Chapter 8, alternative approach discussed in [19]). Current-induced electron wind results in drift components in component and vacancy fluxes described by

$$\Omega j_A = D_A \frac{\Delta c}{\Delta x} + \frac{c_A D_A^*}{kT} E_x z_A e + \frac{c_A D_A^*}{c_V} \frac{(c_{VR} - c_{VL})}{\Delta x} \quad (2.72)$$

$$\Omega j_B = -D_B \frac{\Delta c}{\Delta x} + \frac{c_B D_B^*}{kT} E_x z_B e + \frac{c_B D_B^*}{c_V} \frac{(c_{VR} - c_{VL})}{\Delta x} \quad (2.73)$$

$$\Omega j_V = (D_B - D_A) \frac{\Delta c}{\Delta x} - \frac{E_x e}{kT} \left(c_A D_A^* z_A + c_B D_B^* z_B \right)$$
$$- \frac{\left(c_A D_A^* + c_B D_B^* \right)}{c_V} \frac{(c_{VR} - c_{VL})}{\Delta x} \quad (2.74)$$

Here z_A and z_B are the effective charges (in elementary charge units), e is the absolute value of electron charge, and E_X is the electric field intensity.

The growth kinetics of the intermetallic layer is then determined by

$$\frac{d(\Delta x)}{dt} = \frac{\tilde{D} \Delta c}{c_A c_B} \frac{\left(1 + l_0/\Delta x\right)}{(\Delta x + L_V)} - \left(\frac{\tilde{D}}{(\Delta x + L_V)} \frac{(L_V - l_0)}{(D_A - D_B)} \right.$$
$$\left. \times \frac{c_A D_A^* z_A + c_B D_B^* z_B}{c_A c_B} - \left(D_A^* z_A - D_B^* z_B \right) \right) \frac{eE_x}{kT} \quad (2.75)$$

The detailed derivation of Equation 2.75 and the formal analysis of the growth kinetics on the basis of this equation are given in Chapter 8. Here we direct the attention to two cases that present great interest from a physical point of view:

1) $eE_x \cdot (D_A^* z_A - D_B^* z_B) > 0$.
   The current favors the phase growth. In this case, nonequilibrium vacancies slow down the process of phase formation when compared to quasi-equilibrium of vacancies (Figure 2.1).
2) $eE_x \cdot (D_A^* z_A - D_B^* z_B) < 0$.
   The current tends to inhibit the growth of the phase, so that the latter gains maximum thickness with time and does not grow any more. At that nonequilibrium vacancies increase this maximum thickness as compared with the case of quasi-equilibrium vacancies (Figure 2.2). In such a way, nonequilibrium vacancies "damp" any outer effects (similar to Lenz's rule in electromagnetism or to the le Chatelier–Braun general principle).

To be exact, when the current stimulates the phase growth, nonequilibrium vacancies reduce this augmentation (opposing the growth); if the current inhibits the phase growth, nonequilibrium vacancies reduce this suppression (promoting the growth).

2.7
Possible Role of Nonequilibrium Vacancies in Spinodal Decomposition

The classical theory of spinodal decomposition implies that a lamellar structure period is stipulated by nonlocal interaction intensity. The latter is taken into account in Gibbs potential density as a summand, proportional to the squared concentration gradient, that is,

$$G = \int \frac{dV}{\Omega} \left(g(c) + g^{el} + \Omega K (\nabla c)^2 \right) \tag{2.76}$$

Separation of the components at the initial stage of spinodal decomposition implies nonzero components fluxes that are proportional to partial diffusion coefficients, and thus, their sum is not equal to zero. Therefore, the components separation must cause local vacancy fluxes in the lamellae.

Suppose, for instance, that the B component is considerably more mobile than the A component. Then the emergence of the R lamella (enriched with the B component) and the P lamella (depleted with the B component) in the unstable alloy causes a more intensive flux of the B component from P to R than the opposite A component flux from R to P. Thus, the vacancy flux from R to P (i.e., opposite to the direction of a more mobile component's (the B one) diffusion) must arise. This must result in vacancy supersaturation in P and vacancy depletion in R.

In Darken's approach, any vacancy concentration deviation from equilibrium is immediately compensated by the effective work of the vacancies' sources and sinks. It leads to the Kirkendall effect (lattice movement) and equalizes the components' fluxes in the laboratory reference frame. Still, typical periods for lamellar structures in spinodal decomposition are notably less than the free path length of vacancies to sinks. Decomposition time is small as well. Therefore, the assumption of vacancy equilibrium in spinodal decomposition seems to be highly disputable. On the contrary, one may expect the vacancy concentration wave to intensify along with the main components concentration wave. In the lattice reference frame, considering nonequilibrium vacancies and nonlocal Cahn–Hilliard summands (but neglecting Manning's correction, that is, nondiagonal Onsager coefficients in the lattice reference frame), one can write the fluxes as follows:

$$\Omega j_A = -L_A \nabla \mu_A = -c_B L_A \nabla (\mu_A - \mu_B) = c_B L_A \nabla (\mu_B - \mu_A)$$
$$L_A = \frac{c_A D_A^*}{k_B T}, \quad \Omega j_B = -L_B \nabla \mu_B = -c_A L_B (\mu_B - \mu_A), \quad L_B = \frac{c_B D_B^*}{k_B T} \tag{2.77}$$

Nonlocality, as it is usual for the Cahn–Hilliard scheme, is "hidden" in chemical potentials.

2.7 Possible Role of Nonequilibrium Vacancies in Spinodal Decomposition | 27

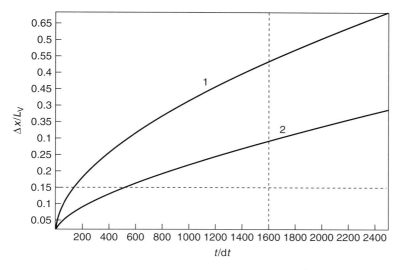

Figure 2.1 Current favors the growth, $eE_x \cdot (D_A^* z_A - D_B^* z_B) > 0$: 1) time dependence of the layer thickness in case of quasi-equilibrium vacancies; 2) the same dependence in case of nonequilibrium vacancies. The parameters are given by $j = 4 \times 10^8$ Am^{-2}, $D_{NG}/\tilde{D} = 0.209$, $l_1 = L_V^2/\delta = 1.6 \times 10^{-4}$ m.

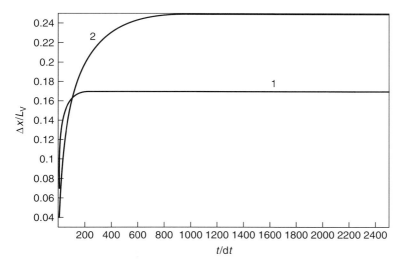

Figure 2.2 Current inhibits the growth, $eE_x \cdot (D_A^* z_A - D_B^* z_B) < 0$: 1) time dependence of the layer thickness in case of quasi-equilibrium vacancies; 2) the same dependence in case of nonequilibrium vacancies. The parameters are given by $j = 4 \times 10^8$ Am^{-2}, $D_{NG}/\tilde{D} = 0.209$, $l_1 = L_V^2/\delta = 1.6 \times 10^{-4}$ m.

In Equation 2.77, we have taken into account the Gibbs–Duhem relation. Having deciphered chemical potentials, taking into account gradient and elastic contributions, one gets

$$\Omega j_A = c_B L_A \nabla \left(\frac{\partial g}{\partial c_B} + \tilde{\mu}^{el} - 2\Omega K \nabla^2 c_B \right) + \frac{c_A D_A^*}{c_V} \nabla c_V \qquad (2.78)$$

$$\Omega j_B = -c_A L_B \nabla \left(\frac{\partial g}{\partial c_B} + \tilde{\mu}^{el} - 2\Omega K \nabla^2 c_B \right) + \frac{c_B D_B^*}{c_V} \nabla c_V \qquad (2.79)$$

$$\Omega j_V = (c_A L_B - c_B L_A) \nabla \left(\frac{\partial g}{\partial c} + \tilde{\mu}^{el} - 2\Omega K \nabla^2 c \right)$$
$$- \frac{c_A D_A^* + c_B D_B^*}{c_V} \nabla c_V = u \qquad (2.80)$$

where u stands for the drift rate.

The corresponding fluxes in the laboratory reference frame are

$$\Omega J_B = \Omega j_B + c_B u$$
$$= - \left(c_A D_B^* + c_B D_A^* \right) \frac{c_A c_B}{k_B T} \nabla \tilde{\mu}^{ef} + \frac{c_A c_B \left(D_B^* - D_A^* \right)}{c_V} \nabla c_V \qquad (2.81)$$

$$\Omega j_V = \left(D_B^* - D_A^* \right) \frac{c_A c_B}{k_B T} \nabla \tilde{\mu}^{ef} - \frac{c_A D_A^* + c_B D_B^*}{c_V} \nabla c_V \qquad (2.82)$$

$$\tilde{\mu}^{ef} \equiv \frac{\partial g}{\partial c} + \tilde{\mu}^{ef} - 2\Omega K \nabla^2 c \qquad (2.83)$$

Having applied the continuity equation (taking into consideration vacancies' sources and sinks), in the linear approximation one obtains

$$\frac{\partial c_B}{\partial t} \cong \left(c_A D_B^* + c_B D_A^* \right) \frac{c_A c_B}{k_B T} \left(g'' \nabla^2 c - 2\Omega K \nabla^4 c \right) \qquad (2.84)$$
$$- \frac{c_A c_B \left(D_B^* - D_A^* \right)}{c_V} \nabla^2 c_V$$

$$\frac{\partial c_V}{\partial t} \cong D_V \nabla^2 c_V - \left(D_B^* - D_A^* \right) \frac{c_A c_B}{k_B T} \left(g'' \nabla^2 c - 2\Omega K \nabla^4 c \right) \qquad (2.85)$$
$$- \frac{c_V - c_{Ve}}{\tau_V}$$

Let us consider the trial harmonic solution, according to the main scheme of stability analysis

$$c_B = A_B(t) \sin(kx), \quad c_V - c_{Ve} = A_V(t) \sin(kx) \qquad (2.86)$$

Since vacancies are considerably more mobile than the main components, we can reduce the equation for vacancy concentration to the quasi-stationary one at $t > \tau_V$. Then,

$$A_V \cong A_B k^2 \frac{\left(D_B^* - D_A^* \right) \frac{c_A c_B}{k_B T} \left(g'' + \frac{2\Omega K}{L_V^2} \left(kL_V \right)^2 \right)}{\left(kL_V \right)^2 + 1} \qquad (2.87)$$

$$\frac{dA_B}{dt} = R(k) A_B, \quad A_B = A_{B0} \exp(R(k) t) \tag{2.88}$$

$$R(k) = -k^2 \left(g'' + 2\Omega K k^2\right) \frac{c_A c_B}{k_B T}$$
$$\times \left\{ c_A D_B^* + c_B D_A^* - \frac{(D_B^* - D_A^*)^2}{c_V D_V} c_A c_B \frac{(kL_V)^2}{1 + (kL_V)^2} \right\} \tag{2.89}$$

Commonly, at $L_V \to 0$ (quasi-equilibrium case), the standard Cahn–Hilliard result is obtained.

Let us analyze another limiting case when $D_A^*/D_B^* \to 0$. Then,

$$R = -\left(\frac{c_A^2 c_B D_B^*}{k_B T}\right) \frac{k^2 \left(-|g''| + 2\Omega K k^2\right)}{1 + (kL_V)^2} \lim \tag{2.90}$$

The principal mathematical difference from the standard Cahn–Hilliard formula is connected with the second summand of the denominator. This summand can essentially change the optimal wave number for the lamellar structure. Indeed, having found the derivative of R and having set it equal to 0, one gets the optimal wave number (corresponding to the fastest growing harmonic):

$$(k^* L_V)^2 = -1 + \sqrt{1 + \frac{L_V^2 |g''|}{2\Omega K}} \tag{2.91}$$

The following limiting case is of particular interest:

$$\frac{L_V^2 |g''|}{2\Omega K} \gg 1 \to (k^* L_V)^2 \approx \sqrt{\frac{L_V^2 |g''|}{2\Omega K}}$$
$$\implies k^* \approx \frac{1}{\sqrt{L_V}} \left(\frac{|g''|}{2\Omega K}\right)^{1/4} = \left(\sqrt{2} \frac{k^{\text{Cahn}}}{L_V}\right)^{1/2} \ll k^{\text{Cahn}} \tag{2.92}$$

if $k^{\text{Cahn}} L_V \gg 1$. So, if the vacancies mean free path length to the sinks is larger than the optimal period for the lamellar structure at equilibrium vacancies, and the components' mobilities differ considerably, the real period for the structure will considerably exceed the theoretical Cahn–Hilliard prediction.

2.8
Nanoshell Collapse

The Kirkendall effect is commonly accompanied by the Frenkel effect, the void formation in the diffusion zone. In foreign literature, the Frenkel effect is often referred to as *Kirkendall voiding*, which is rather confusing, as Kirkendall and Frenkel effects are competitive: vacancies annihilating at the dislocation kinks and causing the Kirkendall shift, cannot be used for Kirkendall voiding, and vice versa.

The Kharkov research group on diffusion processes headed by Geguzin has invented the method for observation of the "pure" Kirkendall effect long ago. The method involves imposing comparatively low pressures up to 100 atm, having little influence upon activation barriers of atom jumps, but almost completely oppressing void formation. Kirkendall voiding (Frenkel effect) is a feature of a contact area between two substances, and it has always been considered as an undesirable effect, significantly worsening mechanical reliability of the diffusion contact, since the porous layer in the diffusion zone becomes a weak place of contact. However, recently, Kirkendall voiding has found a suitable place – it is appropriate for hollow nanoparticles. The method is based on the use of vacancy fluxes, caused by the difference of components' mobilities [6]. The reaction between Co nanoparticles and sulphur or oxygen leads to the formation of Co_3S_4 or CoO compounds (intermediate phases). The latter have a form of a nanoshell with a symmetrical void inside it. This is explained by a higher mobility of Co atoms in the newly formed compound when compared to S or O atoms. So, the reaction serves as a pump, Co atoms are drawn out from inside to continue the reaction, and vacancies are in the opposite direction (inside).

The authors of the method (the group headed by Alivisatos [6]) have not paid attention to the problem of thermal stability of the obtained nanoshells. This was done by Tu and Gösele [7], and we offered the corresponding rigorous theory at the end of the same year (Figure 2.3) [8]. The point is it is favorable for a nanoshell to turn into a compact nanosphere without internal surface (and, correspondingly, without a void inside the sphere), as the total surface energy of a system is reduced while the volume is preserved:

$$\gamma \left(4\pi r_{i0}^2 + 4\pi r_{e0}^2\right) > \gamma 4\pi r_f^2, \quad \frac{4\pi}{3}\left(r_{e0}^3 - r_{i0}^3\right) = \frac{4\pi}{3}r_f^3 \tag{2.93}$$

Here γ is the surface tension of a phase, r_{i0} and r_{e0} are the internal and external radii of a nanoshell right after the reaction, and r_f is the ultimate radius of the nanosphere after the nanoshell's collapse. The mechanism of such a collapse is stipulated by the vacancy flux from the internal surface to the outer one and corresponding atomic

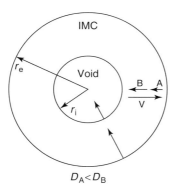

Figure 2.3 Scheme of the IMC shell shrinkage with the segregation of the B-component at the inner boundary.

fluxes in opposite direction. The driving force of the vacancy flux is a difference of vacancy concentrations at inner and external surfaces, $c_V(r_i) > c_V(r_e)$, due to the Gibbs–Thomson effect. In a linear approximation,

$$r_i \gg \beta, \quad \beta = \frac{2\gamma\Omega}{kT} \qquad (2.94)$$

we obtain

$$c_{V_e}(r_i) = c_{Ve}\left(1 + \frac{\beta}{r_i}\right), \quad c_{V_e}(r_e) = c_{Ve}\left(1 - \frac{\beta}{r_e}\right) \qquad (2.95)$$

Thus, the vacancy flux from inside must arise (from a higher value of concentration to a lower one). This vacancy flux will obviously generate the similar inside-directed flux of atoms, that is, the shell would collapse. However, this simple mechanism works only for one-component shells. Since typical nanoshells are binary, the collapsing process remains complicated because of the so-called inverse Kirkendall effect [9]. The factor, similar to that pertaining to a nanoshell formation, works here. It is the difference of components' mobilities. However, in the formation process, this factor augments the effect (actually, it causes voids), and at collapse, reduces its rate to a great extent (again an example of damping!). Whereas $(D_B \gg D_A)$, that is the vacancy chooses B among its neighbors with much higher probability, the vacancy flux must lead to a redistribution (segregation) of components, B atoms must segregate in the vicinity of the inner surface, and A atoms in that of the outer surface. This will create the additional concentration gradient that will reduce the vacancy flux (Equation 2.25), and slow down the collapse process. Therefore, the collapse rate is under control of a slower component's migration. The detailed analysis of the problem and results of modeling are presented in Chapter 7. Here, we give only the main idea.

The basic equations of the model are the following:

$$\frac{\partial c_B}{\partial t} = -\frac{1}{r^2}\frac{\partial}{\partial r}\left(r^2 \Omega J_B\right), \quad \frac{\partial c_V}{\partial t} = -\frac{1}{r^2}\frac{\partial}{\partial r}\left(r^2 \Omega J_V\right) + 0$$

$$r_i(t) < r < r_e(t)$$

$$\Omega J_B = -D_B \frac{\partial c_B}{\partial r} + \frac{c_B D_B^*}{c_V}\frac{\partial c_V}{\partial r}$$

$$\Omega J_V = (D_B - D_A)\frac{\partial c_B}{\partial r} - D_V \frac{\partial c_V}{\partial r} \qquad (2.96)$$

Inner and outer boundaries move due to the vacancy flux. To preserve the conservation of matter, the flux of B atoms through moving boundaries must be equal to zero:

$$\frac{dr_i}{dt} = -\Omega J_V(r_i), \quad \frac{dr_e}{dt} = -\Omega J_V(r_e)$$

$$\Omega J_B(r_i) - c_B(r_i)\frac{dr_i}{dt} = 0, \quad \Omega J_B(r_e) - c_B(r_e)\frac{dr_e}{dt} = 0 \qquad (2.97)$$

Equation 2.95 gives the boundary conditions for vacancy concentration for this boundary-value problem.

Let us apply the steady-state approximation both for vacancies and for atoms (the latter is not evident and is analyzed in Chapter 7). Then,

$$\frac{\partial}{\partial r}(r^2 J_V) \approx 0, \quad \frac{\partial}{\partial r}(r^2 J_B) \approx 0$$

such that

$$\Omega J_V \approx \frac{K_V}{r^2} (>0), \quad \Omega J_B \approx \frac{K_B}{r^2} (<0) \qquad (2.98)$$

Such identical dependencies of both fluxes on radius (at constant coefficients) allow one to apply similar dependencies to concentration gradients of atoms and vacancies:

$$\frac{\partial c_V}{\partial r} \approx -\frac{L_V}{r^2}, \quad \frac{\partial c_B}{\partial r} \approx -\frac{L_B}{r^2} \qquad (2.99)$$

If, say, $D_A^* \ll D_B^*$, the above-formulated equations, when having considered boundary conditions, lead to the following expressions for collapse rate (rate of movement for both nanoshell boundaries):

$$\frac{dr_i}{dt} = -\frac{D_A^*}{1 - c_1 + c_1 \frac{D_A^*}{D_B^*}} \cdot \frac{r_e + r_i}{r_e - r_i} \cdot \frac{\beta}{r_i^2} \approx -\frac{D_A^*}{1 - c_1} \cdot \frac{r_e + r_i}{r_e - r_i} \cdot \frac{\beta}{r_i^2}$$

$$\frac{dr_e}{dt} = \frac{r_i^2}{r_e^2} \frac{dr_i}{dt} \qquad (2.100)$$

In fact, it is apparent that the collapse rate is controlled by a slower component (A in our case), which increases the nanoshells' lifetime.

New developments of shrinkage and formation of nanoshells have been recently discussed in [10–12].

2.9
The Role of Nonequilibrium Vacancies in Diffusion Coarsening

In the analysis of diffusion coarsening, being a very slow process, it has always been assumed that nonequilibrium vacancies have enough time to "escape" to the sinks (or to appear at the sources), so that the nonequilibrium vacancy approximation is correct. However, the difference between A and B atoms' mobilities at their diffusion in the vicinity of a new phase's particles constantly generates local vacancy fluxes, and, accordingly, local deviations of vacancy concentrations from the equilibrium value. If the vacancies' mean free path length to the sinks (from sources) is commensurate (or larger) to the average distance between particles, one cannot neglect such deviations. If the distance between particles is about several tens or hundreds of nanometers, the vacancy may migrate from one particle to another without having come to the sinks. The greater the ratio of partial diffusion coefficients D_A/D_B differs from one, the larger deviations of vacancy concentrations from the respective equilibrium value should be expected. This statement especially concerns coarsening in ordered phases.

Thus, we have changed only two steps in the L(ifshitz)S(lezov)W(agner) – scheme [13], namely, the description of the quasi-stationary distribution of the concentration around a randomly chosen particle in the mean-field approximation, and the expression for flux in the flux balance at the moving precipitate boundaries. Instead of considering the standard steady-state equation $\tilde{D}\nabla^2 c_B = 0$, we have analyzed a system of two coupled steady-state equations, for B atoms and for vacancies [14]

$$\tilde{D}\nabla^2 c_B - (D_B^* - D_A^*)\frac{c_A c_B}{c_V}\nabla^2 c_V = 0$$

$$- (D_B^* - D_A^*)\varphi\nabla^2 c_B + D_V\nabla^2 c_V - \frac{c_V - c_V^{eq}}{\tau_v} = 0 \quad (2.101)$$

Having used these equations, we obtained the following expression for the growth (or shrinkage) rate of particles

$$\frac{dR}{dt} = \frac{\Delta - \frac{\alpha}{R}}{(c_B^\gamma - c_B^\alpha)R} D^{ef}(R) \quad (2.102)$$

where

$$D^{ef}(R) = \frac{D_{NG}\left(1 + \frac{R}{\lambda}\right)}{\left(1 + \frac{D_{NG}}{\tilde{D}}\frac{R}{\lambda}\right)} \quad (2.103)$$

Here R stands for the particle's radius, $\Delta = \bar{c} - c_B^{eq}$ is the supersaturation with respect to the B component, and

$$\lambda = L_V\sqrt{D_{NG}/\tilde{D}}$$

is a certain characteristic length determining the deviation from standard kinetics of coarsening.

Equation 2.102 looks like the growth equation in the LSW-theory, but here the introduced effective diffusion coefficient D^{ef} depends on the size of a particle: at $\lambda \gg R$ it approaches D_{NG} (the growth is controlled by the slower component), and if $\lambda \ll R$, then D^{ef} tends to Darken's value of $\tilde{D}$ (the more mobile component controls the growth). The analysis shows that a special coarsening regime may be expected in the following range of average sizes (and corresponding time interval typical for coarsening):

$$1 \ll (\langle R \rangle/\lambda) \ll \tilde{D}/D_{NG} \quad (2.104)$$

or

$$\sqrt{D_{NG}/\tilde{D}} \ll (\langle R \rangle/L_V) \ll \sqrt{\tilde{D}/D_{NG}} \quad (2.105)$$

For radii that are not far from the average value the effective coefficient D^{ef} is proportional to the size (in this interval of average values) that is,

$$D^{ef} \approx D_{NG} \cdot \frac{R}{\lambda} \quad (2.106)$$

so that

$$\frac{dR}{dt} = \frac{D_{NG}}{\lambda}\frac{\Delta - \alpha/R}{(c_B^\gamma - c_B^\alpha)} \quad (2.107)$$

In this case, the growth equation is quite similar to the Hillert equation for the normal grain growth, and it leads to a parabolic law $\langle R \rangle \sim t^{1/2}$ instead of the LSW-dependence $t^{1/3}$. At the initial stage, the average radius being small,

$$\langle R \rangle < \langle R_1 \rangle = L_V \left(D_{NG}/\tilde{D} \right)^{1/2} \tag{2.108}$$

the growth regime of LSW-type may be expected, but at a rate determined by the NG coefficient D_{NG} (i.e., the rate is controlled by the slow component) instead of Darken's one. At the latest stage,

$$\langle R \rangle \gg \langle R_2 \rangle = L_V \left(\tilde{D}/D_{NG} \right)^{1/2} \tag{2.109}$$

the system must pass to the LSW regime, again, yet the rate is determined by Darken's coefficient here. This scenario was realized in computer simulations and published recently in [14], and, in fact, coarsening proved to have three stages. At the transient stage (which may last quite long), the size distributions of particles are wider, more symmetrical, and closer to experimental observations [15].

2.10
Conclusions

Nonequilibrium vacancies must be taken into account in all diffusion-controlled processes with characteristic sizes (diffusion zone width, phase layer thickness, distance between precipitates, interlamellar distance, and so on) commensurate to the vacancies free length, and (or) with characteristic times commensurate to the relation of vacancies relaxation time to vacancy concentration. The effect of nonequilibrium vacancy distribution caused by the difference of components' mobilities or (and) external field is "damping." This refers to the linear case, when vacancy concentration change is not significant, such that partial diffusion coefficients (proportional to this concentration) experience little change. The nonlinear case was recently analyzed by Gapontsev [16]. Recently, Svoboda *et al.* develop similar approach (account of limited efficiency of vacancy sinks and sources) using the extremum principles of nonequilibrium thermodynamics [17, 18].

References

1. Nazarov, A.V. and Gurov, K.P. (**1974**) *Fizika Metallov I Metallovedenie*, **37**, 496 (in Russian).
2. Nazarov, A.V. and Gurov, K.P. (**1974**) *Fizika Metallov I Metallovedenie*, **38**, 689 (in Russian).
3. Nazarov, A.V. and Gurov, K.P. (**1978**) *Fizika Metallov I Metallovedenie*, **45**, 885 (in Russian).
4. Manning, J.R. (**1968**) *Diffusion Kinetics for Atoms in Crystals*, van Nostrand, Princeton, Toronto.
5. Gurov, K.P. and Gusak, A.M. (**1985**) *Fizika Metallov I Metallovedenie*, **59**, 1062 (in Russian).
6. Yin, Y., Rioux, R.M., Erdonmez, C.K. et al. (**2004**) *Science*, **304**, 711.
7. Tu, K.N. and Goesele, U. (**2005**) *Applied Physics Letters*, **86**, 093111.

8. Gusak, A.M., Zaporozhets, T.V., Tu, K.N., and Goesele, U. (2005) *Philosophical Magazine*, **85**, 4445.
9. Marwick, A.D. (1978) *Journal of Physics France*, **8**, 1849.
10. Nakamura, R., Tohozakura, D., Lee, L.-G. *et al.* (2008) *Acta Materialia*, **56**, 5276.
11. Gusak, A.M., Tu, K.N. (2009) *Acta Materialia*, **57**, 3367.
12. Gusak, A.M. and Zaporozhets, T.V. (2009) *Journal of Physics-Condensed Matter*, **21**, 5303.
13. Lifshitz, I.M. and Slezov, V.V. (1961) *Journal of Physics and Chemistry of Solids*, **19**, 35.
14. Gusak, A.M., Lutsenko, G.V. and Tu, K.N. (2006) *Acta Materialia*, **54**, 785.
15. Kim, D.M. and Ardell, A.J. (2003) *Acta Materialia*, **51**, 4073.
16. Gapontsev, V. (2008) Dr. Sciences thesis. Ekaterinburg, Russia.
17. Svoboda, J., Fischer, F.D., Fratzl, P. and Kroupa, A. (2002) *Acta Materialia*, **50**, 1369.
18. Svoboda, J., Fischer, F.D. and Fratzl, P. (2006) *Acta Materialia*, **54**, 3043.
19. Orchard, H.T. and Greer, A.L. (2005) *Applied Physics Letters*, **86**, 231906.
20. Gusak, A.M. (1994) *Materials Science Forum*, **155–156**, 55.

3
Diffusive Phase Competition: Fundamentals
Andriy M. Gusak

3.1
Introduction

Diffusion phase competition at interdiffusion became the "first love" of the author, after his greater involvement in the analysis of diffusion. It was the search for an answer to the phase competition point that made diffusion investigations attractive to him (in about 1980). Thus, diffusion turned from the "museum exhibit" into the master of nanoworlds fate (decision-maker) at solid-state reactions.

In 1981, the author, together with Gurov, proposed a quite naive (as for nowadays) competition model, where the possibility of diffusion suppression of some intermediate phase nuclei by their fast-growing neighbors was first emphasized. Sections 2.2–2.7 are dedicated to such a naive, still, physically clear approach. In particular, the obtained criteria for phase suppression/growth at the nucleation stage, the time for diffusion suppression of phases, are considered. In 1989, we were the first to propose a new geometrical model for diffusion interaction in powder mixtures, the model of a divided couple. The idea of diffusion competition was immediately put forward to the test for this model. Chapters 4 and 5 describe less naive approaches, which we developed later.

3.2
Standard Model and the Anomaly Problem

Reactive diffusion is the diffusion-controlled growth of intermediate phase layers between two reacting materials proceeding due to interdiffusion through the growing reaction products. Studies of reactive diffusion started in 1920s when the parabolic law of layer growth was discovered by Tammann. Evans found the possibility of a mixed linear–parabolic regime. Wagner, Frenkel, and Sergeev suggested theoretical explanations of the parabolic law. After World War II, the late stages of reactive diffusion were investigated by Pines, Kidson, Wagner, Geguzin, Kaganovski, Paritskaya, Gurov, Ugaste, van Loo, and others.

Diffusion-Controlled Solid State Reactions. Andriy M. Gusak
Copyright © 2010 WILEY-VCH Verlag GmbH & Co. KGaA, Weinheim
ISBN: 978-3-527-40884-9

Starting from 1980s, the main interest of the diffusion community steadily shifted from the problems of interdiffusion in solid solutions to the diffusive phase growth during interdiffusion (reactive diffusion) in the systems with limited solubility and the whole range of intermediate phases. In this chapter, we treat the history of some ideas that appeared in the early 1980s and became the basis of the new approach to the synergy of diffusion and reactions [1–32]. This chapter does not completely deal with reactive diffusion. It reflects the authors' personal view and is based on the traditions of the diffusion schools of the former Soviet Union. After 1990, when our group and simultaneously the group of Pierre Desre and Fiqiri Hodaj started to modify nucleation theory for the case of initial stages of reactive diffusion at nanoscale, much more interesting results were obtained [7, 33–43] (see also Chapters 4 and 5). Here, we almost omit these developments (which include rather intensive mathematical analysis) and concentrate the attention on rather naive but physically clear ideas about the interrelation between diffusion, reactions, and nucleation.

Reactive diffusion (the formation and growth of intermediate phase layers in the diffusion zone as a result of interdiffusion of the components through these layers) is a typical example of a solid-state reaction. It is notable for its reaction product – a crystal (generally, a polycrystal) of an intermediate phase rather than individual molecules (as in gas phase reactions) appears. The crystal remains at the site of the reaction and becomes a barrier to its further passing. For the reaction to proceed (when it is thermodynamically favorable), atoms have to diffuse through new-formed phase layers and react at one of the newly formed interfaces. The thicker the layers, the more time it takes to diffuse. And, therefore, the reaction takes place more slowly. Thus, even at a sufficient amount of reagents, the phase growth rate decreases with time. It has become known since the 1920s that in most cases the phase layer growth obeys a parabolic law. We briefly review the standard model for the phase growth kinetics [1, 2].

First, let us consider the case of phase 1 growing within a narrow concentration range $\Delta c = c_R - c_L \ll 1$ at the process of annealing of a sample couple containing almost mutually insoluble materials A, B (let the A component be on the left and the B component on the right; Figure 3.1). Thus, we neglect interdiffusion fluxes in the initial components of the diffusion couple. Thereby, we consider all A atoms, which "managed to diffuse" into the 1/B boundary, to react with B atoms and not to go further, growing the phase "on the right." In the same way, B atoms, which "managed to diffuse" into the A/1 boundary, react with A and grow the phase "on the left". Let us neglect the molar volume changes at intermetallic phase formation, thus disregarding those effects that are connected with stresses, arising at phase boundaries. Besides, we consider the condition of diffusion flux steadiness to be fulfilled over the phase thickness. In other words, the distribution of components in the phase layer is quasistationary. One can prove this assumption [44], but it would be easier to get it in an unsophisticated way.

Indeed, since the homogeneity interval of the intermediate phase is narrow (very large deviations from stoichiometric composition are energetically unfavorable), the concentration "has no way out" and it keeps itself almost constant (close

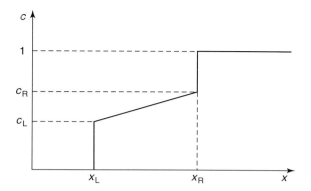

Figure 3.1 Concentration profile of one phase growing between insoluble components.

to stoichiometry) over the whole phase layer. That is why the time derivative of concentration is small, and so the flux divergence is close to zero. In the one-dimensional case (diffusion couple), the divergence is simply equal to the derivative with respect to the diffusion coordinate. The derivatives being equal to zero implies that the flux density almost does not change over the whole phase layer. This means that the product of interdiffusion coefficient and concentration gradient is nearly equal everywhere inside the phase. Still, this does not imply that the concentration gradient is equal everywhere, as the interdiffusion coefficient may considerably change even at small deviations from stoichiometry (especially in phases of B2 type, for example, NiAl). Hence, it is generally incorrect to simply say (as it is typically said and written) that the concentration profile inside the phase is linear.

Let us make some elementary transformations. If a certain parameter is constant, it can be put both inside and outside the integral sign. We take it inside the integral sign as

$$\Omega J = -\tilde{D}\frac{\partial c}{\partial x} \approx \text{const} \equiv \tilde{D}\frac{\partial c}{\partial x}\frac{\int_{x_L}^{x_R} dx}{\int_{x_L}^{x_R} dx} = -\frac{\int_{x_L}^{x_R} \tilde{D}\frac{\partial c}{\partial x} dx}{\int_{x_L}^{x_R} dx} = -\frac{\int_{c_L}^{c_R} \tilde{D} dc}{x_R - x_L} \qquad (3.1)$$

The term $\int_{c_L}^{c_R} \tilde{D} dc$ is called *Wagner's integrated coefficient* [45] and is often written in the following way:

$$\int_{c_L}^{c_R} \tilde{D} dc = D_1 \Delta c_1 \qquad (3.2)$$

where

$$D_1 \equiv \frac{\int_{c_L}^{c_R} \tilde{D} dc}{\Delta c_1} \qquad (3.3)$$

is the effective diffusion coefficient averaged over the phase.

We emphasize once more that, as a rule, the kinetics of solid-state reactions is not determined by the diffusion coefficient and the homogeneity region width separately but always by their product (i.e., the integrated coefficient). This is good for comparison with experiment since the concentration range in many phases is so narrow that it seems almost impossible to measure it experimentally. Wagner's integrated coefficient can be transformed using Darken's relation, expressing the interdiffusion coefficient in a binary system with tracer diffusion coefficients and the second derivative of Gibbs potential (per atom) g with respect to concentration

$$\tilde{D} = \left(cD_A^* + (1-c)D_B^*\right) \frac{c(1-c)}{kT} \frac{\partial^2 g}{\partial c^2} \tag{3.4}$$

Having substituted this expression into Wagner's integral coefficient, considering the narrow phase homogeneity range, one gets

$$\int_{c_L}^{c_R} \tilde{D}(c)dc = \overline{D}_1^* \frac{c_1(1-c_1)}{kT} \left(\frac{\partial g}{\partial c}\bigg|_{1,B} - \frac{\partial g}{\partial c}\bigg|_{A,1}\right)$$

$$\cong \overline{D}_1^* \frac{c_1(1-c_1)}{kT} \left(\frac{g_B - g_1}{1 - c_1} - \frac{g_1 - g_A}{c_i - 0}\right)$$

$$= \overline{D}_1^* \frac{\Delta g_1(A + B \to 1)}{kT} \tag{3.5}$$

Here, $\overline{D}_1^*$ is the combination of tracer diffusivities averaged over the phase

$$\overline{D}_1^* \equiv c_1 \overline{D}_A^* + (1-c_1)\overline{D}_B^* \tag{3.6}$$

and $\Delta g_1(A, B \to 1)$ is the thermodynamic driving force (per atom) of phase 1 formation (from A and B). Thus, the growth rate of the single intermediate phase is determined by the mobility of atoms in it (tracer diffusivities) and by the Gibbs energy of phase formation. As we can see, the "elusive" homogeneity interval itself is of little importance for us. In such a way, the product of the density of B fluxes through the phase layer and atomic volume (that is, "the flux density of the volume transferred by B atoms") is equal to

$$\Omega J_B = -\frac{\int_{c_L}^{c_R} \tilde{D} dc}{\Delta x_1} = -\frac{D_1 \Delta c_1}{\Delta x_1} \tag{3.7}$$

As it is well known, the rate of interface movement is equal to the ratio of the fluxes step at this boundary to the concentration step. We regard the fluxes in marginal phases of the diffusion couple to be zero (negligible solubility). So, the flux balance conditions at the interfaces yield the following two differential equations:

$$(1 - c_R)\frac{dx_R}{dt} = \frac{D_1 \Delta c_1}{\Delta x_1}, \quad c_L \frac{dx_L}{dt} = -\frac{D_1 \Delta c_1}{\Delta x_1} \tag{3.8}$$

From these equations, one can easily derive the differential equation for the phase thickness $\Delta x(t)$:

$$\frac{d\Delta x}{dt} = \frac{a}{\Delta x(t)} \tag{3.9}$$

where

$$a = \frac{1 - \Delta c_1}{(1 - c_R)c_L} D_1 \Delta c_1 \approx \frac{D_1 \Delta c_1}{c_1(1 - c_1)} \tag{3.10}$$

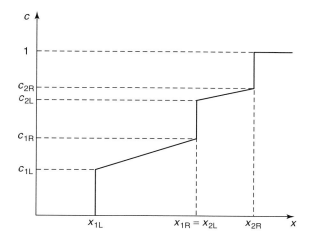

Figure 3.2 Concentration profile of two phases growing between insoluble components.

This gives the well-known parabolic law of phase growth:

$$(\Delta x)^2 = (\Delta x_0)^2 + \frac{2D_1 \Delta c_1}{c_1(1-c_1)} t \qquad (3.11)$$

If the thickness of the layer considerably exceeds the initial thickness (the latter is taken right after nucleation and lateral growth with formation of primary continuous layer), then

$$\Delta x \approx \sqrt{\frac{2D_1 \Delta c_1}{c_1(1-c_1)} t} \qquad (3.12)$$

Now, let us treat the case of two phases, 1 and 2, growing between two mutually insoluble metals (Figure 3.2). Here, we again make the assumption of constant fluxes over the thickness of each phase (steady state approximation)

$$\Omega J_B^{(1)} = -\frac{D_1 \Delta c_1}{\Delta x_1}, \qquad \Omega J_B^{(2)} = -\frac{D_2 \Delta c_2}{\Delta x_2} \qquad (3.13)$$

The equations of flux balance at the interfaces have the following form:

$$\begin{aligned}
\left(1 - c_R^{(2)}\right) \frac{dx_{2R}}{dt} &= \frac{D_2 \Delta c_2}{\Delta x_2} \\
\left(c_L^{(2)} - c_R^{(1)}\right) \frac{dx_{2L}}{dt} &= -\frac{D_2 \Delta c_2}{\Delta x_2} + \frac{D_1 \Delta c_1}{\Delta x_1} \\
\left(c_L^{(1)} - 0\right) \frac{dx_{1L}}{dt} &= -\frac{D_1 \Delta c_1}{\Delta x_1}
\end{aligned} \qquad (3.14)$$

It is quite easy to obtain the system of differential equations for phase thicknesses from Equation 3.14. We obtain

$$\frac{d\Delta x_1}{dt} = a_{11} \frac{D_1 \Delta c_1}{\Delta x_1} + a_{12} \frac{D_2 \Delta c_2}{\Delta x_2}$$

$$\frac{d\Delta x_2}{dt} = a_{21}\frac{D_1 \Delta c_1}{\Delta x_1} + a_{22}\frac{D_2 \Delta c_2}{\Delta x_2} \tag{3.15}$$

Here

$$a_{ik} = \begin{pmatrix} \frac{c^{(2)}}{c^{(1)}} & -1 \\ -1 & \frac{1-c^{(1)}}{1-c^{(2)}} \end{pmatrix} \frac{1}{c^{(2)} - c^{(1)}} \tag{3.16}$$

By direct substitution, one can check that Equation 3.15 has parabolic solutions

$$\Delta x_1 = k_1 t^{1/2}, \quad \Delta x_2 = k_2 t^{1/2} \tag{3.17}$$

as well; however, the expressions for the parabolic growth constants k_1 and k_2 are not so "elegant" as in Equation 3.12 for the growth of one phase

$$r_{2/1} = \frac{k_2}{k_1}$$

$$= \frac{a_{21} D_1 \Delta c_1 - a_{12} D_2 \Delta c_2 + \sqrt{(a_{21} D_1 \Delta c_1 - a_{12} D_2 \Delta c_2)^2 + a_{11} a_{22} D_1 \Delta c_1 D_2 \Delta c_2}}{2 a_{11} D_1 \Delta c_1}$$

$$k_2 = \sqrt{2 a_{22} D_2 \Delta c_2 + r_{2/1} a_{21} D_1 \Delta c_1}$$

The basic equations for any number of phase layers, growing simultaneously, are obtained in the same way. However, the mentioned simultaneity leads to some problems that have become crucial since the 1970s, with technological development of integrated chips manufacture and maintenance.

The standard model is quite sufficient to describe the late stages of reactive diffusion in a standard "infinite" diffusion couple (with a size much larger than the width of the diffusion zone). Yet, since the 1970s, the main field of reactive diffusion applications shifted to microelectronics, which uses thin film reactions. Researchers immediately found, in this case, that the standard model does not work, demonstrating various "anomalies" (sequential phase growth, deviations from parabolic law, etc.). Such problems led to a reconsideration of the standard model. This reconsideration was performed almost simultaneously by Gurov and Gusak [3], Gösele and Tu [4], and Dybkov [5].

According to [5, 6], diffusion theory proves to have such a principle drawback that it cannot explain (i) the disagreements between the observed phase composition in the diffusion zone and that of the phase diagram and (ii) breaking of the parabolic law for phase layer growth. Obviously, the author of [5, 6] might not have paid attention to the works [3, 4], which provide an appropriate explanation for the facts mentioned above. And, to our knowledge, a number of the author's [5, 6] statements and predictions appear to be disputable. In [5, 6], the diffusion theory is blamed for poor regard to "chemical phenomena" in the "chemical process." Author of [5, 6], obviously, applies the chemical concept of reactions to the processes proceeding at the moving interfaces of the growing phase layers. Let us treat these processes from the physical point of view.

Reactions in the diffusion zone during the formation and growth of phase layers can be divided into two types:

1) formation of new phase nuclei in the contact zone as a result of heterophase fluctuations due to chemical potential and concentration gradients;

2) "reactions" at the moving boundary of already existing phases, comprising three successive steps: (i) detachment of atoms from a lattice of a phase, (ii) transition through the interface, and (iii) attachment to the lattice of another phase.

The reactions of the second type are often meant, explicitly or implicitly, when "finite reaction rate" or "boundary kinetics" [4–6, 8–11] is taken into account. It makes sense to treat their influence on the process kinetics only when the "reaction" stage is limiting, that is, characteristic time of detachment, transition, and attachment of atoms are more or at least commensurate with that of diffusion delivery of both species of atoms at the interface.

The interface transition time is $\tau_1 \sim h^2/D_B$, where $h \sim 5 \times 10^{-10}$ m. The attachment time τ_2, at the worst, is reduced to the time required for searching the "appropriate place" (grain boundary dislocation, etc.) given by (d^2/D_B), d being equal to several atomic distances. The detachment time, obviously, is close to τ_2 by the order of magnitude. The characteristic time of transfer through the layer is

$$\tau_{\text{dif}} \sim \frac{\Delta x}{v} \sim \frac{\Delta x}{\frac{D\Delta c}{\Delta x}} \sim \frac{\Delta x^2}{D\Delta c} \tag{3.18}$$

Here, D is the effective interdiffusion coefficient in the layer.

The boundary kinetics appears to be limiting in the case of $\tau_1 + 2\tau_2 > \tau_{\text{dif}}$, that is, at

$$\Delta x < (h^2 + 4d^2)^{1/2} \left(\frac{D\Delta c}{D_B}\right)^{1/2} \tag{3.19}$$

According to Equation 3.5,

$$D\Delta c \approx D^* \frac{\Delta g}{kT} \tag{3.20}$$

where D^* is a combination of self-diffusion coefficients of the components in the phase. If there is no oxide film or other barrier layer at the interface, then, as it is known, $D^*/D_B \ll 1$. So, taking into account that h and d are of the order of a few atomic distances, the thickness of the phase layer Δx^*, at which the change from boundary kinetics to diffusion regime takes place, does not exceed the interatomic distance. Apparently, the situation must radically change when treating the reaction between a solid body and a gas. In this case, the boundary kinetics regime (and linear phase growth) can be reached by reducing the reagent's partial pressure in the gas medium, which leads the atoms to come at the surface.

In the contact zone of solid bodies, the boundary kinetics, connected with reactions of the second type, can be expected only in the presence of oxide films or other barrier layers at the interface. We developed an alternative idea for the late stage of linear growth [46, 47]. It presents certain interest, for it regards the limited power of interfaces as sinks/sources of vacancies and the corresponding contribution of nonequilibrium vacancies. In other cases, the regime of phase layer growth is a diffusion-controlled one from the very beginning, if the beginning implies the already formed layer of new phase nuclei which is able to undergo diffusion growth.

So far, we use the model that assumes the fact that critical nuclei of all phases, allowed by the phase diagram, appear at once (the unlimited nucleation model). It is known that the growth of a new phase from the nucleus is energetically favorable only in the case of nucleus size exceeding some critical value l_{cr}, determined from the extremum condition of Gibbs thermodynamic potential. In a one-component substance, the extremum condition is expressed simply by the derivative of G with respect to the nucleus size being equal to zero:

$$\left.\frac{\partial G}{\partial R}\right|_{l_{cr}} = 0 \tag{3.21}$$

In nucleation in a binary system, there are some problems concerning the difference between the composition of a nucleus and that of initial "parent" phases, occurring in general case [12]. Thus, the notions of the smallest (equilibrium), most probable and critical nuclei do not coincide. This matter is treated in more details below. At present, the actual existence of the critical nucleus size is important for us; this critical size becomes unstable against decomposition at $l < l_{cr}$.

We assume that at the initial period, successive layers of critical nuclei of all phases, allowed by phase diagram, appear as a result of heterophase fluctuations. It is significant that the nuclei arise in the chemical potential gradient field, so that finite differences $\Delta\mu$ over the thickness of each layer exist from the very beginning. Chemical potential gradients inside the nuclei cause diffusion fluxes through them. Owing to the difference in diffusivities, flux densities vary for different phases. The jumps of diffusion fluxes at interfaces make the boundaries move: if phase 1 provides more A atoms to the interface 1–2 than phase 2 takes, the interface will shift – phase 1 will grow at the expense of phase 2. At that nuclei layers come into diffusion interaction. The result of the interaction varies for different phases and depends on the characteristics of diffusion of all phases in the phase diagram. Those phases, for which $\left.\frac{d\Delta x}{dt}\right|_{l_{cr}} > 0$, start growing and reach the observed phase layers. The nuclei of those phases, for which $\left.\frac{d\Delta x}{dt}\right|_{l_{cr}} < 0$, shrink, become subcritical, and decay. The new ones arise at their place, and they suffer the same fate; they will be replaced by the fast-growing neighboring phases ("vampires").

The growth of such phases is suppressed; they exist in the contact zone only "virtually," in the form of nuclei that appear and decay straight away. This explains the disagreement between the phase composition of the zone and the phase diagram. Criteria of suppression and growth for the simplest cases are discussed in the following section. Yet, the phase suppression, as discussed in Section 3.4, lasts for a finite period (though it may be quite a long one). When growing phases amount to a certain thickness and the fluxes through them $\frac{-D\Delta c}{\Delta x}$ are reduced sufficiently, the value $\left.\frac{d\Delta x}{dt}\right|_{l_{cr}}$ for the previously suppressed phase gets positive, so that it starts growing as well. This behavior is confirmed by experiment [2]. The time of nuclei suppression actually represents the incubation time of a phase (if not taking into account the time of nuclei formation). Examples of calculations are given in Section 3.4.

Thus, our model shows that, when external influences are absent in a nonlimited diffusion couple, all phases allowed by the phase diagram must eventually grow (this does not concern thin films). In [5, 6], it is stated that at any time quite a small fraction of all phases, allowed by the phase diagram, grows. Generally speaking, this conclusion is wrong. In fact, during traditional investigation periods, involved in experiments, not all phases arise. And our model provides an explanation for this fact. However, if the periods are very long, other phases must also appear. As mentioned, it was experimentally proved that the phase that had been absent arose in the zone only after annealing for hundreds of hours [2].

When the previously suppressed phase starts growing, its own "building process" requires some material that in other cases would be used for the extension of other layers. This means that in the moment the suppressed phases start growing, the growth kinetics of other phases acquires some peculiarities. The latter processes are briefly analyzed in [3].

The above-described situation seems to be logically clear. However, nature is not obliged to keep to our logic. It may be quite a real case, at which for some reasons nuclei formation of some phases rather than phase growth from the nuclei is opposed, while the emergence of nuclei in metastable phases may appear to be easy, and these nuclei may start growing and suppressing the nuclei of "legal" stable phases, which had appeared subsequently. The conditions for the formation of these stable phases' nuclei depend on the type of neighboring phases and therefore are considered as the "parent" ones. Eventually, the solid-phase chemical interaction becomes highly complicated and turns to be strongly dependent on random factors arising at the initial stage of the contact.

3.3
Criteria of Phase Growth and Suppression: Approximation of Unlimited Nucleation

Let us treat the initial stage of phase formation in the process of annealing of the diffusion couple A–B, regarding the phase diagram to include two intermediate phases 1 ($c_1, c_1 + \Delta c_1$) and 2 ($c_2, c_2 + \Delta c_2$), and neglecting the solubilities of A in B and B in A. The concentration profile can be depicted from Figure 2.2 (see the previous chapter). Then the equations of diffusion interaction between phases are of the following form:

$$
\begin{aligned}
(c_1 - 0)\frac{dx_{A1}}{dt} &= -\frac{D_1 \Delta c_1}{\Delta x_1} \\
(c_2 - c_1)\frac{dx_{12}}{dt} &= \frac{D_1 \Delta c_1}{\Delta x_1} - \frac{D_2 \Delta c_2}{\Delta x_2} \\
(1 - c_2)\frac{dx_{2B}}{dt} &= \frac{D_2 \Delta c_2}{\Delta x_2}
\end{aligned}
\quad (3.22)
$$

According to the accepted model, the layers of critical nuclei of both phases appear in the contact zone at the very initial stage. From Equation 3.22, the expressions for phase thicknesses $\Delta x_1 = \Delta x_{12} - \Delta x_1$, $\Delta x_2 = \Delta x_{2B} - \Delta x_{12}$ are as follows:

$$\frac{d\Delta x_1}{dt} = \frac{1}{c_2 - c_1}\left(\frac{c_2}{c_1}\frac{D_1\Delta c_1}{\Delta x_1} - \frac{D_2\Delta c_2}{\Delta x_2}\right) \tag{3.23}$$

$$\frac{d\Delta x_2}{dt} = \frac{1}{c_2 - c_1}\left(-\frac{D_1\Delta c_1}{\Delta x_1} + \frac{1-c_1}{1-c_2}\frac{D_2\Delta c_2}{\Delta x_2}\right) \tag{3.24}$$

A simple analysis of Equation 3.24 shows that the phase behavior is determined by the value of the dimensionless parameter

$$r = \frac{D_1\Delta c_1 l_{cr}^{(2)}}{D_2\Delta c_2 l_{cr}^{(1)}} \tag{3.25}$$

Now, let us consider separately several different cases:

1) At
$$r < \frac{c_1}{c_2}$$
we get
$$\left.\frac{d\Delta x_1}{dt}\right|_{l_{cr}} < 0, \quad \left.\frac{d\Delta x_2}{dt}\right|_{l_{cr}} > 0$$
that is, phase layer 2 grows from the very beginning, suppressing the growth of phase 1 nuclei (phase 2 is a "vampire").

2) At
$$\frac{c_1}{c_2} < r < \frac{1-c_1}{1-c_2}$$
we get
$$\left.\frac{d\Delta x_1}{dt}\right|_{l_{cr}} > 0, \quad \left.\frac{d\Delta x_2}{dt}\right|_{l_{cr}} > 0$$
that is, both phase layers grow from the beginning.

3) At
$$r > \frac{1-c_1}{1-c_2}$$
we obtain
$$\left.\frac{d\Delta x_1}{dt}\right|_{l_{cr}} > 0, \quad \left.\frac{d\Delta x_2}{dt}\right|_{l_{cr}} < 0$$
that is, phase layer 1 grows from the beginning, suppressing the growth of phase 2 nuclei (phase 1 is a "vampire").

Thus, the criterion for the suppression and growth of the different phases at the initial stage for the system A-1-2-B is obtained; it is convenient to represent it on the diagram (Figure 3.3). The numbers denote the phases that grow from the very beginning. Similarly, we can find the criteria for the suppression and growth in the case of three intermediate phases between mutually insoluble A and B. The corresponding graphical representation of the criterion is given in Figure 3.4. The arrows indicate the sequence of phase composition change in the diffusion zone.

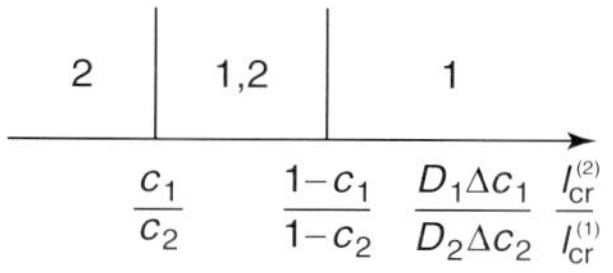

3.4 Incubation Time

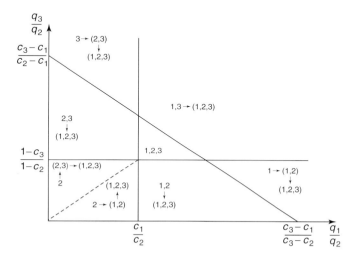

Figure 3.3 Suppression and growth criteria: case of two intermediate phases.

Figure 3.4 Suppression and growth criteria: case of three intermediate phases.

3.4
Incubation Time

As it was outlined in the first paragraph, phase suppression cannot be everlasting. Let us again consider the system B for the case of $r > (1 - c_1)/(1 - c_2)$, when at the initial moment of time the nuclei of phase 2 are "exhausted" by the growing phase 1. One can easily check that the growth of the single phase 1 between insoluble A and B takes place according to the following law:

$$\Delta x_1 = \left(\frac{2 D_1 \Delta c_1}{c_1 (1 - c_1)} \right)^{1/2} t^{1/2} \qquad (3.26)$$

At that critical nuclei of the phase 2 are constantly arising at the boundary of the growing phase 1 and B. Taking Equation 3.24 into account, the rate $d\Delta x_2/dt$ for these nuclei is determined by

$$\left. \frac{d \Delta x_2}{dt} \right|_{l_{cr}} = \frac{1}{c_2 - c_1} \left(-\frac{D_1 \Delta c_1}{\Delta x_1} + \frac{1 - c_1}{1 - c_2} \frac{D_2 \Delta c_2}{l_{cr}^{(2)}} \right) \qquad (3.27)$$

The thickness Δx_1 increases, such a moment comes when $\left. \frac{d\Delta x_2}{dt} \right|_{l_{cr}}$ passes through zero and becomes positive, that is, phase 2 will no longer be suppressed. This will occur at

$$\Delta x_1 = \frac{1-c_2}{1-c_1}\frac{D_1\Delta c_1}{D_2\Delta c_3}l_{cr}^{(2)} \tag{3.28}$$

that is, according to Equation 3.26, the time moment for this is

$$\tau_2 = \frac{c_1(1-c_2)^2}{2(1-c_1)}\frac{D_1\Delta c_1}{(D_2\Delta c_2)^2}\left(l_{cr}^{(2)}\right)^2 \tag{3.29}$$

It is quite natural for the time τ_2 of phase 2 suppression to be referred to as *incubation time*, if one disregards the time of first critical nuclei formation.

If $D_i\Delta C_i$ at least approximately fits the Arrhenius dependence ($(\exp(-Q/kT)$, for more details see [1]), then the exponential term is given by

$$\exp\frac{2Q_2 - Q_1}{kT} \tag{3.30}$$

and one may expect $Q_2 > Q_1/2$, so that the incubation time must reduce with an increase in temperature. If, by chance, $Q_2 < Q_1/2$, it should mean that the barriers for atom jumps in phase 2 are much lower than those in phase 1, so at commensurate preexponential factors one may expect the diffusivity of phase 2 to be much higher than that of phase 1, $D_2\Delta C_2 \gg D_1\Delta C_1$, and phase 2 will appear to be not the suppressed phase but the suppressive one (a "vampire"). The growth of phase 1 will be opposed and its incubation time will be reduced with the growth of T.

Let us specify the method of calculation of the critical nucleus size. Consider the formation of a phase 1 nucleus at the planar interface α and β. The formation involves an Gibbs bulk energy gain (decrement)

$$n(g_\alpha - g_1)V_\alpha + n(g_\beta - g_1)V_\beta \tag{3.31}$$

and unfavorable surface energy increment, $\sigma_{\alpha 1}S_{\alpha 1} + \sigma_{1\beta}S_{1\beta} - \sigma_{\alpha\beta}S_{\alpha\beta}$, connected with the change of $\alpha - \beta$ contact to two contacts $\alpha - 1$ and $1 - \beta$ ($\sigma_{\alpha 1}$, $\sigma_{1\beta}$, $\sigma_{\alpha\beta}$ are the interfacial tension coefficients). With regard to symmetry (disregarding anisotropy), the basis of the nucleus can be treated as a circle of a certain radius R ($S_{\alpha\beta} = \pi R^2$). Similarly to the nucleus at the boundary of two identical grains [13], it is quite easy to find boundary angles and dependencies of ΔG on R as

$$\cos\vartheta_\alpha = \frac{(\sigma_{\alpha\beta}^2 + \sigma_{\alpha 1}^2 - \sigma_{1\beta}^2)}{2\sigma_{\alpha\beta}\sigma_{\alpha 1}} \tag{3.32}$$

$$\cos\vartheta_\alpha = \frac{(\sigma_{\alpha\beta}^2 + \sigma_{1\beta}^2 - \sigma_{\alpha 1}^2)}{2\sigma_{\alpha\beta}\sigma_{\beta 1}} \tag{3.33}$$

$$\Delta G = -\frac{\pi R^3}{3}\left\{\frac{n(g_\alpha - g_1)}{\sin\vartheta_\alpha}\left(\frac{2}{1+\cos\vartheta_\alpha} - \cos\vartheta_\alpha\right)\right.$$
$$\left. + \frac{n(g_\beta - g_1)}{\sin\vartheta_\beta}\left(\frac{2}{1+\cos\vartheta_\beta} - \cos\vartheta_\beta\right)\right\}$$
$$+ \pi R^3\left(\frac{2\sigma_{\alpha 1}}{1+\cos\vartheta_\alpha} + \frac{2\sigma_{1\beta}}{1+\cos\vartheta_\chi} - \sigma_{\alpha\beta}\right) \tag{3.34}$$

The composition and components' chemical potentials of the appearing nucleus vary with X, which has a radical influence on its fate. Thus, in the strict sense, while calculating ΔG, one should take integrals of $\int n(g_\alpha - g_1)dV$ type instead of $n(g_\alpha - g_1)V_\alpha$. This fact was realized in [7, 33–40]. Now we consider the C gradient to be small and the function $g(C)$ to vary insignificantly inside the nucleus, such that g_1 implies the minimum of the Gibbs potential (per atom) for phase 1 (and g_α, g_β – Gibbs potentials of pure A and B).

From a thermodynamic viewpoint, the nucleus is able to grow at $R > R_{cr}$, $G(R)$ monotonically decreasing with the increase in R. However, it does not mean that neighboring phases, also able to grow, will give it such an opportunity. Investigating extremum properties of the function Equation 3.34, one can obtain

$$R_{cr} = \frac{2\left(\dfrac{2\sigma_{\alpha 1}}{1 + \cos \vartheta_\alpha} + \dfrac{2\sigma_{1\beta}}{1 + \cos \vartheta_\beta} - \sigma_{\alpha\beta}\right)}{\left\{\dfrac{n(g_\alpha - g_1)}{\sin \vartheta_\alpha}\left(\dfrac{2}{1 + \cos \vartheta_\alpha} - \cos \vartheta_\alpha\right) + \dfrac{n(g_\beta - g_1)}{\sin \vartheta_\beta}\left(\dfrac{2}{1 + \cos \vartheta_\beta} - \cos \vartheta_\phi\right)\right\}} \quad (3.35)$$

The transversal size of the critical nucleus is

$$L_{cr} = R_{cr}\left(\frac{1 - \cos \vartheta_\alpha}{\sin \vartheta_\alpha} - \frac{1 - \cos \vartheta_\alpha}{\sin \vartheta_\beta}\right) \quad (3.36)$$

3.5
Should We Rely Upon the Ingenuity of Nature? Nucleation Problems and Meta-Quasi-Equilibrium Concept

Until now we have been considering nature to be resourceful enough to always find the way (besides, a rather quick one) to perform heterogeneous fluctuations at intermediate phase nucleation. However, further we are going to consider such cases at which the time of intermediate phase nucleus formation may appear to be quite large. For the sake of simplicity, we will limit ourselves to the case of one intermediate phase. Note that two fundamentally different situations are possible here. They are illustrated in Figures 3.5 and 3.6.

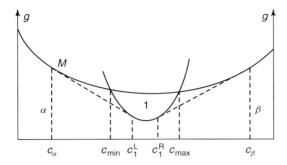

Figure 3.5 Dependence of Gibbs potential on the composition for the case of total mutual solubility.

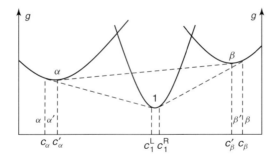

Figure 3.6 Dependence of Gibbs potential on the composition for the case of limited mutual solubility.

In Figure 3.5, there is a metastable solid solution M over the whole concentration interval, and an intermediate phase 1 (for instance, an ordered solution) is possible to appear, which is thermodynamically more favorable in a certain concentration interval. It is significant that at annealing of an A–B couple, the M phase lattice exists from the very beginning while phase 1 is absent. The system is not "aware" of a more favorable phase 1 up to fluctuation nucleation of a new lattice, so interdiffusion takes place within the single metastable phase M like any other quasiequilibrium process (each physically small volume has enough time to relax to an "equilibrium" state before its composition changes considerably). Since local relaxation reaches not truly equilibrium phases α, 1, β, but a metastable phase M, this process will be referred to as a *meta-quasi-equilibrium*.

As is discussed in the following section, even if we neglect the time required for the lattice building, successful nucleation needs time for concentration preparation $\tau_1 \sim R^2/D_M \Delta c_m$, where R stands for the size of a new phase viable nucleus, D_M being the interdiffusion coefficient in the "parent" phase M, and Δc_m is a concentration interval inside which phase 1 is more favorable than phase M. At small D_M or Δc_m, the time τ_1 may appear to be larger. Typical values are

$$R \sim 10^{-7} \text{ cm} \qquad \Delta c_m \sim 10^{-2} \tag{3.37}$$

If $D_M \sim 10^{-15}$ cm^2/s, then $\tau_1 \sim 10^3$ s $\sim$ 15 min.

Nucleation in case (b) appears to be even more complicated. Here, the lattices of α and β phases do not continuously go over one into another and "concentration preparation", that is, the formation of the interval Δx with concentrations (c_1^L, c_1^R) by the way of interdiffusion, in which only the lattice transformation remains to be done, is impossible.

Indeed, until the nuclei of phase 1 appear, the system is not "aware" of its profit and is supposed to establish quasiequilibrium between initially existing α and β phases. This means that quasiequilibrium boundary concentrations $\tilde{c}'_\alpha$ and $\tilde{c}'_\beta$ will set in locally at the moving interface after an initial kinetic period. These concentrations are determined by the common tangent to the curves $g_\alpha(\tilde{c})$ and $g_\beta(\tilde{c})$. This quasiequilibrium is not real (for it does not regard phase 1, which has not appeared yet); still it is not less steady. We again call it meta-quasi-equilibrium.

3.5 Nucleation Problems and Meta-Quasi-Equilibrium Concept

It is important that here the interdiffusion process will never make up the concentration intervals from $\tilde{c}'_\alpha$ to $\tilde{c}'_\beta$ including $(\tilde{c}_1, \tilde{c}_1 + \Delta\tilde{c}_1)$, corresponding to the intermediate phase. Thus, it turns out that we obtain, at a first sight, a paradoxical case – the legal phase 1 cannot arise despite its total thermodynamic favorability.

In the context of the "refined", free-of-drawbacks approach, only one way can be proposed: concentration zones from $\tilde{c}_\alpha$ to $\tilde{c}'_\alpha (\alpha')$ and from $\tilde{c}'_\beta$ to $\tilde{c}_\beta (\beta')$, formed in the process of meta-quasi-equilibrium diffusion, are metastable and must eventually decay: α' into α+1, β' into β+1. At that, the problem of intermediate phases rise at interdiffusion is reduced to the problem of supersaturated solid solutions decomposition. A thermodynamic analysis of this situation was made in [39]. The kinetic difficulties here are presented by concentration supersaturations $\tilde{c}'_\alpha$-$\tilde{c}_\alpha$ and $\tilde{c}_\beta$-$\tilde{c}'_\beta$, which may turn out to be quite small when compared to the differences required for new phase formation.

The probability of concentration fluctuations in a binary system is

$$W(\delta c) \sim \exp\left(-\frac{Ng''(\delta c)^2}{2kT}\right) \tag{3.38}$$

where g'' is the second-order derivative of Gibbs potential with respect to atomic concentration c for the phase in which the fluctuation occurs, and N is the number of atoms involved in the concentration fluctuation. For the systems with very low solubility $g''_\alpha \sim kT/c_\alpha$. At $\delta c \sim 1/2$, $N \sim 100$, and $c_\alpha \sim 10^{-2}$, we get $W \sim \exp(-1250)$. One may expect such fluctuation to occur for $\tau \sim (l^2/D_\alpha W)$. At $l \sim 10^{-7}$cm, $D_\alpha \sim 10^{-12}$ cm^2/s one gets $\tau_1 \sim 10^{400}$ s, which is absolutely unbelievable to be realistic. Hence, concentration preparation by the way of fluctuation in case (b) at low solubility of A and B is of very low probability. However, the phases still appear in the diffusion zone, so the concentration preparation is realized, though, not for all but for many phases.

We can presuppose the following possibilities:

1) At the intermediate stage, there occurs a metastable phase (amorphous, for instance) concentration range, where concentration preparation takes place, being rather wide. For this, the curves of Gibbs phase potentials must be arranged as indicated in Figure 3.7. The formation of the metastable phase itself is facilitated by, first, it's greater "overlapping" with α, β phases, and by the possibility of its nucleation as a result of segregation at grain boundaries or interfaces.
2) Formation of a zone with concentrations required for the intermediate phase resulting from the processes of such a type as "cold homogenization" or diffusion-induced grain boundary migration (DIGM).
3) Formation of concentration-prepared zones proceeding from segregation at grain boundaries or other defects.
4) High interfacial tension between α and β phases may lead to the following consequences: the size of the intermediate phase critical nucleus does not exceed the interatomic distance. This is correct anyway when $\sigma_{\alpha\beta} > \sigma_{\alpha 1} + \sigma_{1\beta}$ (see, for example, d'Heurle's review [14]). In this case, the Gibbs potential starts decreasing as the nucleus size increases from zero thickness. Here, the

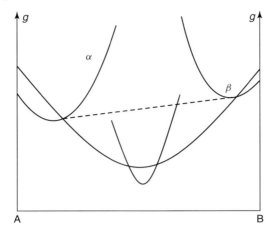

Figure 3.7 Illustration of the possibility for intermediate phase 1 formation from limited solid solutions via the intermediate stage of an amorphous phase with total solubility.

following situation seems to be quite appropriate: first, the two-dimensional nucleus of the surface intermediate phase appears at the interface, to which the atoms from the phase bulks attach, so that the two-dimensional nucleus continuously acquires a shape of a three-dimensional phase layer.

The author realizes that the proposed alternatives lack reasonable numerical estimations to obtain a firm basis; thus, we do not present any final explanations but simply present the formulation of the problem and some stimulating speculations.

3.6
Suppression of an Intermediate Phase by Solid Solutions

The incubation time for intermediate phases at interdiffusion can be stipulated not by only diffusion suppression of their nuclei by neighboring phases. The necessity of "concentration preparation" for the transformation of metastable formations can also influence this time. In solid-phase reactions, when intermediate layers form and grow in the contact zone of two metals, the phase composition of the diffusion zone in many cases appears to be incomplete when compared to the phase diagram [1, 15]. As shown above [3], this disagreement is temporary for massive samples, and it is connected with diffusion suppression of critical nuclei of some intermediate phases, realized by neighboring fast-growing phases. For the case of competition between two or three intermetallics, the suppression and growth criteria are already obtained. The suppression time for the nuclei of the phases being suppressed was also found. This time is considered as the phase's incubation time. It was assumed that the nuclei themselves appear in the contact zone almost immediately (as a result of heterophase fluctuations). This assumption does not always prove to be correct. Let us analyze the problem of incubation time taking into account the finite rate of nuclei formation by the example of a single intermediate phase 1 ($c_1 < c < \tilde{n}_1 + \Delta c_1$) between solid solutions α ($0 < c < \Delta c_\alpha$) and β ($1 - \Delta c_\beta < c < 1$).

3.6.1
Unlimited Nucleation

Suppose, because of heterophase fluctuations, that the nuclei of an intermediate phase 1 between α and β must appear. The appearance of an intermediate phase nucleus is thermodynamically favorable only if its size exceeds the critical value l_{cr}. Our case is complicated by the fact that the nucleus appears in the system that has not been homogeneous from the very beginning. So, the nucleus appears in the conditions of chemical potential and concentration gradients. And the driving force of the process is connected with concentration supersaturation rather than with overcooling (though there is a certain interrelation between them). This makes the composition of nuclei nonhomogeneous from the very beginning. Thus, when calculating points of extremum for Gibbs potential, one must use integrals of the $\int_{x_L}^{x_R} g(c(x)) S(x) dx$ type with varying limits ($S(x)$ – variable cross section area of a nucleus).

Besides, it should also be borne in mind that at essential overcooling (supersaturation) there is a higher possibility of polymorphic transformation without any composition change, diffusion of atoms being slow. Apparently, in the general case, the problem presents difficulties that will be considered in much more detail in Chapters 4 and 5. Here, we restrict ourselves to the case of narrow intermediate phases where Gibbs potential $g(c)$ almost does not change within a narrow homogeneity range Δc (though derivatives $\partial g / \partial c$ may change significantly [16]).

For a phase to grow, a critical nucleus must appear. As stated above, unlike phase transformations in a homogeneous system, we deal with the nuclei arising in a chemical potential gradient field: "on the left" we have a nucleus approaching equilibrium with phase α and "on the right", with phase β. Therefore, diffusion fluxes pass through it. Jumps of diffusion fluxes, occurring at the boundaries $\alpha - 1$ and $1 - \beta$, cause boundary motion, that is, a change in the nucleus' size. If the critical nucleus becomes larger (($d\Delta x/dt) > 0$ at $\Delta x = l_{cr}$), then the phase starts growing. If the critical nucleus becomes smaller (($d\Delta x/dt) < 0$ at $\Delta x = l_{cr}$), then the phase is not able to grow.

Suppose that D_α, D_l, and D_β are effective diffusion coefficients in phases α, l, and β. Let nuclei of phase 1 form a layer of thickness $\Delta x_1 = l_{cr}$ between α and β. Considering Δc_α, Δc_1, and $\Delta c_\beta \ll 1$, the equation of flux balance at moving phase boundaries is

$$(c_1 - \Delta c_\alpha) \frac{dx_{\alpha 1}}{dt} = \frac{D_\alpha \Delta c_\alpha}{\sqrt{\pi D_\alpha t}} - \frac{D_1 \Delta c_1}{l_{cr}}$$

$$(1 - \Delta c_\beta - c_1 - \Delta c_1) \frac{dx_{1\beta}}{dt} = \frac{D_1 \Delta c_1}{l_{cr}} - \frac{D_\beta \Delta c_\beta}{\sqrt{\pi D_\beta t}} \qquad (3.39)$$

For $\Delta x_1 = x_{\beta 1} - x_{\alpha 1}$, from Equation 3.39, we obtain

$$\left.\frac{d\Delta x_1}{dt}\right|_{l_{cr}} \cong \frac{1}{c_1(1-c_1)} \frac{D_1 \Delta c_1}{l_{cr}} - \frac{\left(\frac{\Delta c_\alpha}{c_1}\sqrt{D_\alpha} + \frac{\Delta c_\beta}{1-c_1}\sqrt{D_\beta}\right)}{\sqrt{\pi t}} \tag{3.40}$$

As it follows from Equation 3.40, $(d\Delta x/dt)_{l_{cr}}$ becomes positive, and growth of phase 1 is allowed at

$$t > \tau_1 = \left[\frac{\Delta c_\alpha(1-c_1)\sqrt{D_\alpha} + \Delta c_\beta c_1\sqrt{D_\beta}}{D_1 \Delta c_1} \frac{l_{cr}}{\sqrt{\pi}}\right]^2 \tag{3.41}$$

The suppression time τ_1 of phase 1 growth from the nuclei is called the *incubation time*, if one neglects the time required for nuclei formation.

3.6.2
Finite Rate of Nuclei Formation

Two fundamentally different cases, shown in Figures 3.5 and 3.6, are possible. Here, we examine the case as shown in Figure 3.5 in greater detail. Phases α and β have identical lattices and belong to a metastable form of a solid solutions. In this case before a nucleus of phase 1 appears, interdiffusion proceeds as a usual quasi-equilibrium process in a one-phase system. We denote the processes of this kind taking place in a metastable system as meta-quasi-equilibrium ones.

Analyzing the diagram in the figure, one can hypothesize that the concentration range of metastable phase formation between the points M and N (that is c_1' and $c_1' + \Delta c_1'$) serves as the "initial material" for phase 1 formation by the way of a polymorphic transformation. However, after the nucleus of phase 1 has appeared, concentration ranges $\Delta c_\alpha - c_1$ and $c_1 - 1 - \Delta c_\beta$ turn out to be unstable and correspond to two-phase regions on the phase diagram. However, according to Gibbs phase rule, these regions do not appear in the diffusion zone as they are used in the formation of the initial layer of phase 1. If the conditions of nonsuppression of phase growth are fulfilled for this initial layer, an ordinary quasiequilibrium process of phase layer growth proceeds due to diffusion interaction with solid solutions at its boundaries. Thus, we presuppose that there are three stages of the process:

1) appearance of a metastable solution, call this stage "concentration" preparation;
2) appearance of new phase nuclei, disappearance of unstable formations, and creation (at their expense) of a new stable phase layer (under certain conditions);
3) diffusive phase growth due to interaction with solid solutions.

Let us estimate the duration of the first stage.

The concentration profile $c(x)$ in metastable phase formation must become so flat that the length exceeding l_{cr}

$$x(c_1 + \Delta c_1) - x(c_1) > l_{cr} \quad \text{i.e.} \quad \Delta c_1/(\partial c/\nabla x) > l_{cr} \tag{3.42}$$

will be found within the interval of thermodynamically favorable (for phase 1) concentrations $(c_1, c_1 + \Delta c_1)$. Here l_{cr} indicates the thickness of the initial layer. Suppose D^M is the interdiffusion coefficient in the metastable solution. Let $c_1 \sim 1/2$. In the approximation, $D^M = \text{const}$

$$\frac{\partial c^M}{\partial x} \cong \frac{\exp\left(-\frac{x^2}{4D^M t}\right)}{\sqrt{\pi D^M t}}, \tag{3.43}$$

and $c \sim 1/2$ corresponds to $x \sim 0$, so that

$$\left(\frac{\partial c}{\partial x}\right)_{c \sim 1} \sim (\pi D^M t)^{-1/2} \tag{3.44}$$

So, the condition of concentration readiness for phase 1 growth becomes

$$\Delta c_1 (\pi D^M t)^{1/2} \geq l_{cr} \tag{3.45}$$

so that

$$\tau_{\text{prep}} \approx l_{cr}^2 / \left(\pi D^M (\Delta c_1)^2\right) \tag{3.46}$$

The value of D^M for the metastable phase can be estimated by the method, proposed in [17].

Here, we do not consider the time of lattice reconstruction. Then, the time of diffusion suppression of phase 1 nuclei can be considered as the incubation time Equation 3.41, provided it is larger than the time of concentration preparation for critical nucleus appearance τ_{prep}

$$\left[\frac{\Delta c_\alpha (1 - c_1)\sqrt{D_\alpha} + \Delta c_\beta c_1 \sqrt{D_\beta}}{D_1 \Delta c_1}\right]^2 \frac{l_{cr}^2}{\pi} > \frac{l_{cr}^2}{\pi D^M (\Delta c_1)^2} \tag{3.47}$$

If the condition, Equation 3.47 is not fulfilled, that is, diffusion permeability of phase 1 is high enough and its nuclei are competitive, the concentration preparation time becomes equal to the incubation time, Equation 3.46. A rigorous theory of nucleation in a concentration gradient and its influence on the incubation period and phase competition was developed by Gusak, Desre, and Hodaj in the series of papers [7, 33–40].

3.7
Phase Competition in a Model of Divided Couple

The problem of obtaining and suppressing intermediate phases is of particular interest when solid-phase reactions in powder mixtures are analyzed. Let us consider the initial stage of sintering of the simplest binary mixture when the particles of the initial components still preserve independence and one can speak of two connected surfaces with different total squares S_A, S_B, brought into contact via fast surface diffusion and diffusion through the gas phase. For simplicity, let us take a system of two almost mutually insoluble components, giving two

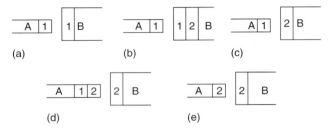

Figure 3.8 Possible variants of growth and suppression of intermediate phases 1 and 2 on free surfaces A and B being in ideal diffusion contact.

intermediate phases with concentrations c_1, c_2 and narrow homogeneity ranges Δc_1, $\Delta c_2 \ll 1$ on the phase diagram. We consider surface diffusion of both components to be fast enough to provide equal chemical potentials everywhere at both contacting surfaces at each moment of time. When both phases grow simultaneously, we may have the situations illustrated by schemes b, c, d in Figure 3.8. If one of the two phases is suppressed, cases a and e are implemented. Obviously, at the initial stage, the nuclei of phase 1 must appear on the surface of A and nuclei of phase 2 on the surface of B. Their further destiny is stipulated by diffusion phase competition, the theory of the latter being discussed above. Main principles of the theory are employed in this section.

We limit ourselves to small times of annealing, when the thickness of the growing phases is much less than the size of particles R, so that we could use planar geometry. As it is known [9], in the general case the rate of phase layer growth is determined by both diffusion transfer and the rate of reaction at interfaces (flux through the interface between ith and $(i+1)$-th phases is $j = k_{i,i+1}\delta c$, where δc is the boundary concentration deviation from the equilibrium value, and $k_{i,i+1}$ is the reaction rate constant). The flux through the ith phase may be expressed as

$$j = \frac{D_i \Delta c_i / \Omega_i}{\Delta x_i + D_i / k_i} \tag{3.48}$$

where D_i is the effective interdiffusion coefficient, k_i is the so-called effective constant of reaction rate obeying the condition $1/k_i = 1/k_{i,i+1} + 1/k_{i-1,i}$. So far, we analyze the case of diffusion kinetics, when phase thicknesses D_i/k_i typical for boundary kinetics do not exceed the sizes of critical nuclei. Thus, the effect of final reaction rate at interfaces can be neglected from the very beginning.

One can easily formulate the system of balance equations for interface movement in case c. In the approximation of the flux steadiness over the growing phase layer, the balance equations at moving interfaces are

$$\frac{dx_{A1}}{dt} = \frac{1}{c_1} \frac{D_1 \Delta c_1}{\Delta x_1} \tag{3.49}$$

$$\frac{1}{\Omega_1}\left(-\frac{D_1 \Delta c_1}{\Delta x_1} - u_1 c_1\right) S_A = \frac{1}{\Omega_2}\left(-\frac{D_2 \Delta c_2}{\Delta x_2} - u_2 c_2\right) S_B$$

$$\frac{1}{\Omega_1} u_1 S_A = \frac{1}{\Omega_2} u_2 S_B \qquad \frac{dx_{2B}}{dt} = \frac{1}{1-c_2} \frac{D_2 \Delta c_2}{\Delta x_2}$$

where Ω_1, Ω_2, and $\Omega_{B(A)}$ are the atomic volumes of the phases and the material B(A). Here we consider interface mobility between phases 1, 2 and a gas phase (u_1, u_2 are the boundary rates). The Kirkendall effect is neglected. Equations 3.49 easily result in the simple phase interaction picture:

$$\frac{d\Delta x_1}{dt} = \frac{1}{c_2 - c_1}\left(\frac{c_2}{c_1}\frac{D_1 \Delta c_1}{\Delta x_1} - \frac{\Omega_1}{\Omega_2}\frac{S_B}{S_A}\frac{D_2 \Delta c_2}{\Delta x_2}\right) \quad (3.50)$$

$$\frac{d\Delta x_2}{dt} = \frac{1}{c_2 - c_1}\left(-\frac{\Omega_2}{\Omega_1}\frac{S_A}{S_B}\frac{D_1 \Delta c_1}{\Delta x_1} + \frac{1-c_1}{1-c_2}\frac{D_2 \Delta c_2}{\Delta x_2}\right)$$

At the "initial moment of time", we may assume $\Delta x_1 = I_{cr}^{(1)}$, $\Delta x_2 = I_{cr}^{(2)}$, where I_{cr} stands for the critical size of the nucleus. In many cases, $I_{cr}^{(1)}$ and $I_{cr}^{(2)}$ can be considered to be equal by the order of magnitude. If the inequality

$$\frac{1-c_1}{1-c_2} > \frac{\Omega_2}{\Omega_1}\frac{S_A}{S_B}\frac{D_1 \Delta c_1}{D_2 \Delta c_2} > \frac{c_1}{c_2} \quad (3.51)$$

is fulfilled, both phases grow from the very beginning, obeying the parabolic law at $R \gg \Delta x \gg I_{cr}$. Having solved Equations 3.50 for this case, we find that the ratio of volumes for the growing phases is time constant and is equal to

$$\frac{V_2}{V_1} = \frac{S_B \Delta x_2}{S_A \Delta x_1} = \frac{r\varphi^2 - 1 + \sqrt{(r\varphi^2)^2 + 4r\varphi\frac{c_2}{c_1}\frac{1-c_1}{1-c_2}}}{2\frac{c_2}{c_1}} \quad (3.52)$$

where $r = (D_2 \Delta c_2)/(D_1 \Delta c_1)$, $\varphi = \Omega_1/\Omega_2 \cdot S_B/S_A$. Hence, the ratio of volumes is determined by $r\varphi^2$, that is, not only by diffusion coefficients but also by "relative dispersity" S_B/S_A. At $r\varphi^2 \ll 1$ and $r\varphi^2 \gg 1$, the dependence of V_2/V_1 on $r\varphi^2$ asymptotically approaches the lines with slopes $(1-c_1)/(1-c_2) > 1$ and $c_1/c_2 < 1$, respectively. The dependence Equation 3.52 was obtained within a quite rough model under the conditions $S_B/S_A = \text{const}$, $I_{cr} \ll \Delta x_i \ll R$, which are violated with time. Actually, we can hardly expect V_2/V_1 to be time constant, but the conclusion about the character of the V_2/V_1 dependence on S_B/S_A and $D_2 \Delta c_2/D_1 \Delta c_1$ must remain correct.

If inequality Equation 3.51 fails, and, for example,

$$\frac{\Omega_2}{\Omega_1}\frac{S_A}{S_B}\frac{D_1 \Delta c_1}{D_2 \Delta c_2} > \frac{1-c_1}{1-c_2} \quad (3.53)$$

then $\left.\frac{d\Delta x_2}{dt}\right|_{I_{cr}}$ in Equation 3.50 is negative. This means that critical nuclei of phase 2 on the surface B will decompose into phase 1 and pure B, so that phase 2 will be suppressed. This will inevitably lead to switching from case b to case a, that is, to the growth of phase 1 layer on surface B. At that, it can be shown that at boundaries of phase 1 and at interfaces between surfaces A and B with gas phase, we have the intermediate concentration from the range

$$c_1 < c_1' = c_1 + \frac{\Delta c_1}{1 + \left(\frac{S_B}{S_A}\right)^2 \frac{D_B}{D_A} \cdot \frac{1-c_1}{c_1}} < c_1 + \Delta c_1 \tag{3.54}$$

With some (incubation) time, the nuclei of phase 2 between 1 and B will stop being suppressed and a switch to case b will take place. Using the above given approximations and writing balance equations for cases a and b, taking into account the difference between partial diffusion coefficients D_1^A, D_1^B in phase 1, one can show that τ_2, that is, the moment of time at which the value becomes positive and the possibility of diffusion growth of phase 2 layer appears, is defined by the following formula:

$$\tau_2 = \frac{c_1(1-c_1) \Big/ \left(I_{cr}^{(2)}\right)^2}{2\Delta c_1 \left(c_1 D_1^A + \left(\frac{S_B}{S_A}\right)^2 (1-c_1) D_1^B\right)} \left(\frac{D_1 \Delta c_1}{D_2 \Delta c_2} \frac{1-c_2}{1-c_1}\right)^2 \tag{3.55}$$

that is, it strongly depends on the ratio $S_{B/A}$.

A detailed analysis of all possible variants is quite intricate; therefore, it is not given here (details can be found in [31, 32]). Instead, we would like to stress on the following. Varying the ratio S_B/S_A ("relative dispersity"), one may, first, transfer from suppression of one phase to growth of both phases or suppression of another phase; second, achieve the required relation of volumes of growing phases without changing the volumes of the initial components; and finally, change the incubation time of suppressed phases. For instance, if phase 2 has higher melting temperature, and therefore, $D_2 \Delta c_2 \ll D_1 \Delta c_1$, so that in a usual diffusion couple it is suppressed, then, enlarging the free surface of B particles by way of grinding, it is possible to make this phase appear and grow already at the initial stage of sintering (for this we must have $S_B/S_A = (D_1 \Delta c_1)/(D_2 \Delta c_2)$) or at least reduce the time of suppression.

If the growth of both phases is controlled by boundary kinetics at the initial stage, that is, phase thicknesses typical for this regime ($x_i^* = D_i/k_i$) exceed $I_{cr}^{(i)}$ considerably, the expression for fluxes through the phases, $D_i \Delta c_i / \Delta x_i$, should be changed to $D_i \Delta c_i / (\Delta x_i + D_i/k_i)$. This will alter the explicit expressions for the growth kinetics and cause deviations from the parabolic law. However, the general conclusions will remain unchanged. For example, the condition of simultaneous growth of two phases from the very beginning (Equation 3.51) will acquire the following form (at $x_i^* \gg I_{cr}^{(i)}$):

$$\frac{1-c_1}{1-c_2} > \frac{\Omega_2 S_A}{\Omega_1 S_B} \frac{k_1 \Delta c_1}{k_2 \Delta c_2} > \frac{c_1}{c_2} \tag{3.56}$$

that is, it strongly depends on S_B/S_A, again. A divided diffusion couple (for example, coaxial cylinders) can represent a model system for investigation. In this system α- and ε-brass stand for A and B elements, β-brass for intermediate element, and zinc atoms could realize diffusion contact in both directions. We deal about a divided couple in Chapter 7, discussing a hollow compound nanoshell formation.

References

1. Gurov, K.P., Kartashkin, B.A. and Ugaste, Y.E. (1981) *Inter-diffusion in Multiphase Metallic Systems*, Nauka, Moscow (in Russian).
2. van Loo, F.J.J. (1990) *Progress in Solid State Chemistry*, 20, 47.
3. Gusak, A.M. and Gurov, K.P. (1982) *Fizika Metallov I Metallovedenie*, 53, 842, 848 (in Russian).
4. Gösele, U. and Tu, K.N. (1982) *Journal of Applied Physics*, 53, 3552.
5. Dybkov, V.I. (1986) *Journal of Materials Science*, 21, 3078.
6. Dybkov, V.I. (1992) *Kinetics of Solid State Chemical Reactions*, Naukova Dumka, Kiev (in Russian).
7. Gusak, A.M. (1990) *Ukrainskii Fizicheskii Zhurnal (Ukrainian Journal of Physics)*, 35, 725 (in Russian).
8. Pines, B.Y. (1961) *Sketches on Metal Physics*, Kharkov State University Press, Kharkov (in Russian).
9. Geguzin, Y.E. (1979) *Diffusion Zone*, Nauka, Moscow (in Russian).
10. Mokrov, A.P. and Gusak, A.M. (1980) Diffusion in multiphase systems, *Diffusion Processes in Metals*, Tula Polytechnical Institute, Tula (in Russian).
11. Philibert, J. (1991) Atom Movements. Diffusion and Mass Transport in Solids. –Les Editions de Physique.
12. Kamenetskaya, D.S. (1967) *Crystallography (USSR)*, 12, 90.
13. Christian, J.W. (1975) *The Theory of Transformation in Metals and Alloys*, Oxford University Press, Oxford.
14. d'Heurle, F.M. (1988) *Journal of Materials Research*, 3, 167.
15. Poate, J.M., Tu, K.N. and Mayer, J.W. (1978) *Thin Films, Inter-diffusion and Reactions*, Wiley.
16. Gusak, A.M. and Gurov, K.P. (1988) *Metallofizika*, 10(4), 91 (in Russian).
17. Gurov, K.P. and Gusak, A.M. (1988) *Metallofizika*, 10(6), 55 (in Russian).
18. Gurov, K.P. and Gusak, A.M. (1990) *Izvestiya AN SSSR, Metally*, 1, 163 (in Russian).
19. Gusak, A.M. and Lyashenko, Y.A. (1990) *Fizika i Khimiya Obrabotki Materialov*, 5, 140 (in Russian).
20. Kalmykov, K.B. (1988) Interaction of aluminum and magnesium with cadmium based alloys. PhD thesis, Moscow (in Russian).
21. Gusak, A.M. (1982) Investigation of diffusion processes between bimetallic films. PhD thesis, Moscow, Institute of Metallurgy (in Russian).
22. Sokolovskaya, E.M. and Gusei, L.S. (1986) *Metallokhimiya*, Moscow State University Publication, Moscow (in Russian).
23. Yarmolenko, M.V. (1992) *Fizika i Khimiya Obrabotki Materialov*, 1, 103 (in Russian).
24. Gusak, A.M. and Gurov, K.P. (1992) *Solid State Phenomena*, 23&24, 117.
25. Gusak, A.M. and Dubiy, O.V. (1989) Diffusive competition in fine-structured systems, in *Diffusion Processes in Metals* (eds A.P. Mokrov and G.V. Shcherbedinski), Tula Polytechnical Institute, p. 37 (in Russian).
26. Bogdanov, V.V. and Paritskaya, L.N. (1991) International Conference on Diffusion and Defects in Solids: Abstracts Sverdlovsk, Institute of Metal Physics, June 26–July 4, p. 85.
27. Pierraggi, B., Rapp, R.A., van Loo, F.J.J. and Hirth, I.P. (1990) *Acta Metallurgica Et Materialia*, 38, 1781.
28. Nazarov, A.V. and Gurov, K.P. (1974) *Fizika Metallov I Metallovedenie*, 37, 496, 38, 486, 848, 38, 689 (in Russian).
29. Gurov, K.P. and Gusak, A.M. (1985) *Fizika Metallov I Metallovedenie*, 59, 1062 (in Russian).
30. Gurov, K.P., Gusak, A.M. and Yarmolenko, M.V. (1988) *Metallofizika*, 10(5), 91 (in Russian).
31. Gusak, A.M. (1989) *Poroshkovaya Metallurgiya*, 3, 39 (in Russian).
32. Gusak, A.M. and Lutsenko, G.V. (1998) *Acta Materialia*, 46, 3343.
33. Desre, P.J. and Yavari, A.P. (1990) *Physical Review Letters*, 64, 1553.
34. Desré, P.J. (1991) *Acta Metallurgica*, 39, 2309.
35. Hodaj, F., Gusak, A.M. and Desré, P.J. (1998) *Philosophical Magazine A*, 77, 1471.

36. Gusak, A.M., Dubiy, O.V. and Kornienko, S.V. (**1991**) *Ukrainskii Fizicheskii Zhurnal (Ukrainian Journal of Physics)*, **36**, 286 (in Russian).
37. Hoyt, J.J. and Brush, L.N. (**1995**) *Journal of Applied Physics*, **78**, 1589.
38. Gusak, A.M., Hodaj, F. and Bogatyrev, A.O. (**2001**) *Journal of Physics: Condensed Matter*, **13** 2767.
39. Hodaj, F. and Gusak, A.M. (**2004**) *Acta Materialia*, **52**, 4305.
40. Pasichnyy, M.O., Schmitz, G., Gusak, A.M. and Vovk, V. (**2005**) *Physical Review B*, **72**, 014118.
41. Pasichnyy, M.O. and Gusak, A.M. (**2005**) *Metallofizika I Noveishiye Technologii*, **27**, 1001.
42. Lucenko, G. and Gusak, A. (**2003**) *Microelectronic Engineering*, **70**, 529.
43. Shirinyan, A.S. and Gusak, A.M. (**2004**) *Philosophical Magazine A*, **84**, 579.
44. Gusak, A.M. and Yarmolenko, M.V. (**1993**) *Journal of Applied Physics*, **73**, 4881.
45. Wagner, C. (**1969**) *Acta Metallurgica*, **17**, 99.
46. Gusak, A.M. (**1994**) *Materials Science Forum*, **155–156**, 55.
47. Gusak, A.M. (**1992**) *Metallofizika*, **14**(9), 3 (in Russian).

4
Nucleation in a Concentration Gradient
Andriy M. Gusak

4.1
Introduction

The process of formation of intermediate phases at interdiffusion is accompanied by a competition between stable and metastable phases. Solid-state amorphizing reactions (SSAR) present a classical example, when at the initial stage a metastable amorphous layer is formed and grows rather than the stable intermetallides [1–3]. It is clear that the search for possible ways to establish control over phase competition is highly essential for technological processes. Another problem related to phase competition refers to the order of phases formed in thin films and multi-layers [4, 5].

What is "phase competition" in the process of a solid-state reaction? Consider a diffusion couple of two mutually soluble components A/B, which diffuse into each other forming intermediate phases and, possibly, a solid solution. Let a binary system A–B have three stable intermediate phases 1, 2, 3, a metastable compound 4, and an amorphous phase 5. The dependence of the Gibbs free energy on the composition of these phases is given in Figure 4.1.

At long times of annealing, according to the generally accepted theory, three stable flat phase layers must grow with the parabolic law $x_i = k_i t^{1/2}$. Quite often real experiment reveals, however, a different behavior – either one stable phase or a metastable crystal or an amorphous layer is formed. Thus, phases compete with each other like different regimes of reaction–diffusion.

The investigation of the initial stages of reaction–diffusion in multilayers, carried out during recent years, by differential scanning calorimetry (DSC), proved that the stage of intermediate phase nucleation at solid-state reaction does take place. DSC experiments [6–8] have shown that the formation of a new phase in multilayers can involve two stages. For example, the curve illustrating the dependence of heat flux on time at formation of $NbAl_3$ in multilayers Nb/Al (obtained by deposition) has two maxima. X-ray analysis and electron microscopy confirmed that both peaks correspond to the formation of the phase $NbAl_3$. Similar curves with two peaks are obtained for such systems as Co/Al, Ni/Al, Ti/Al, Ni/amorphous Si, and V/amorphous Si.

Diffusion-Controlled Solid State Reactions. Andriy M. Gusak
Copyright © 2010 WILEY-VCH Verlag GmbH & Co. KGaA, Weinheim
ISBN: 978-3-527-40884-9

4 Nucleation in a Concentration Gradient

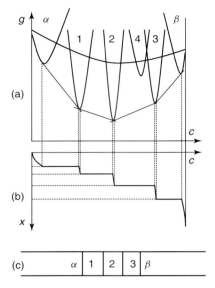

Figure 4.1 Phase spectrum of a binary system:
(a) schematic dependence of Gibbs free energy (per one atom) $g(c)$ on the composition of the intermediate phase;
(b) Concentration profile $c(x)$ in the diffusion zone at reaction–diffusion; (c) location of layers of the growing stable phases.

Coffee et al. [9] consider the shape of the curve to be the result of the following two-stage process. First, the nuclei of a new phase form and grow along the interface, subsequently generating a layer. Then the layer grows normally (perpendicular) to the interface. Divinski and Larikov [10] have analyzed the case of lateral (along the interface) and normal growth with rates being in the same order of magnitude. In such a case, we may expect the interface to have a fractal structure. As discussed later, the formation of pancakelike nuclei of an intermediate phase in a concentration gradient follows from our theory. So, our approach to the stage of nucleation is not simply a theoretical idea; now it has also got an experimental verification.

In general, here we treat the problem of selection between several different ways of relaxation to equilibrium. As far as we are concerned, there is no accepted thermodynamic approach (variation principles "a la Prigogine") to the choice of the particular path along which the system will develop (though there is an interesting idea using the extremum rate of Gibbs free energy release, see Chapters 11 and 12). Apparently, the fate of each phase is determined mainly by the nucleation stage. Thus, the analysis of the nucleation process for an intermediate phase at restrictions imposed by general features of the diffusion process is of great importance when treating phase competition. One should bear in mind that subcritical embryos and nuclei of intermediate phases appear in the diffusion zone under concentration gradients that change with time. Moreover, they are influenced by diffusion fluxes that change abruptly at newly formed boundaries and thus provide their displacement. So, new embryos undergo diffusion interaction with both the parent phase and the embryos of all neighboring phases.

The mentioned circumstances lead to the necessity of inevitable change of the classical theory of nucleation – both in its kinetic and thermodynamic aspects. Such modifications have been made during the last decades. These modifications are analyzed in this and the next chapters.

4.2
Nucleation in Nonhomogeneous Systems: General Approach

When an intermediate phase is formed in a diffusion zone, the nucleation barrier (saddle point) must be calculated taking into account the redistribution of the components outside the new embryo. A parent inhomogeneous phase (or two parent adjacent phases) is/are metastable with respect to the formation of a new phase, but at the same time it (they) is/are unstable with respect to subsequently occurring interdiffusion processes. Therefore, at fixed size and composition of the nucleus, the optimal distribution outside the parent phase(s) will be reached only after full homogenization. Obviously, the nucleus will not "wait" for this. This statement means that true minimization of the Gibbs potential for nucleation during diffusion is impossible. Therefore, the problem of nucleation in an inhomogeneous system should be solved under certain constraints, which are determined by the kinetics of the diffusion process.

Depending on the type of constraint, one can distinguish different nucleation modes. We can point out at least three possible nucleation modes:

1) *Polymorphic mode* (Gusak [11]): This mode can be realized if the parent metastable phase (for example, amorphous) can exist in the concentration range advantageous for a new intermediate phase. At first, the interdiffusion forms a concentration profile in the parent phase overlapping the concentration interval where the new intermediate phase has a lower Gibbs potential (Figure 4.2). Then, the polymorphic transformation in a limited region takes place, forming the lattice of the new phase at a frozen-in concentration gradient. After nucleation the diffusion proceeds, and the newly born nucleus interacts with the parent phase due to a steplike change of diffusion fluxes at its boundaries. The role of the parent phase can be played by the metastable solid solution or the previously formed amorphous layer. Obviously, to make this mode real, the rate of the lattice reconstruction must be much higher than the diffusion rate in the parent phase.

2) *Transversal mode* (Desre [12]): This mode is possible both for cases shown in Figures 4.2 and 4.3. In the case shown in Figure 4.3, interdiffusion before the formation of an i-phase nucleus leads to "meta-quasi-equilibrium" $\alpha - \beta$ with metastable concentration ranges $(c_{\alpha\gamma}, c_{\alpha\beta})$ and $(c_{\beta\alpha}, c_{\beta\gamma})$, unstable to decomposition into $\alpha + \gamma$ and $\gamma + \beta$.

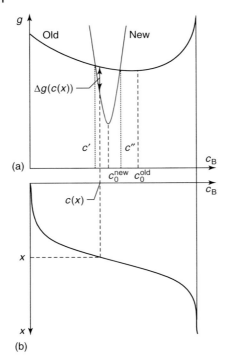

Figure 4.2 Nucleation at polymorphic mode and frozen-in concentration profile: (a) dependence of the Gibbs free energy per atom on the compositions of the old and new phases and (b) frozen-in profile in the diffusion couple, approximately linear in the nucleation region.

According to Desre [13], during the nucleus formation in a concentration gradient (in x-directions), each thin slice $(x, x + dx)$ of this nucleus is formed via unlimited redistribution of components inside this slice independently of other slices. The optimal concentration $c^{new}(x)$ in this nucleus slice is determined by the concentration $c^{old}(x)$ in the surrounding phase (for the same slice) according to the rule of parallel tangents (not the joint tangent!):

$$\frac{\partial g^{new}}{\partial c^{new}} = \frac{\partial g^{old}}{\partial c^{old}} \tag{4.1}$$

To make the transversal mode real, diffusivity in the parent phase(s) should be much larger than that in the new phase ($D_m \gg D_i$).

3) *Total mixing (longitudinal) mode* (Hodaj, Desre [14]): This mode means nucleus formation only at the cost of redistribution in the transformed region of the parent phase, the concentration distribution outside the nucleus being unchanged. To make this mode real, diffusivity in the forming phase should be much higher than that in the parent phase ($D_i \gg D_m$).

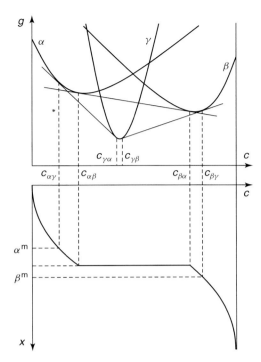

Figure 4.3 Stable and metastable equilibrium in a binary system with limited solubility on the base of solid solutions. The composition within the intervals $(c_{\alpha\gamma}, c_{\alpha\beta})$, $(c_{\beta\alpha}, c_{\beta\gamma})$ enables the formation of γ-phase nuclei.

In the general case, all these modes (and, maybe, some others) should operate simultaneously. Further, we discuss the thermodynamics for each mode in detail.

4.3
Thermodynamics of the Polymorphic Mode of Nucleation in a Concentration Gradient

Let a nucleus of an intermediate phase appear in the frozen-in concentration profile formed by interdiffusion in a homogeneous metastable phase (solid solution or amorphous phase) (Figure 4.2).

4.3.1
Homogeneous Nucleation: General Relations

The change of the Gibbs free energy caused by the formation of a nucleus is determined by the following expression:

$$\Delta G = n \int \left(g^{\text{new}}(c(x)) - g^{\text{old}}(c(x)) \right) S(x)\, dx + \sigma S \qquad (4.2)$$

4 Nucleation in a Concentration Gradient

Here, we neglect volume changes (atomic density $n_1 \approx n_2 = n$) and corresponding stresses (see below). S is the area of the newly formed boundary, σ is its surface energy per area unit, $S(x)$ is the cross-sectional area of a nucleus in the x-plane, which is perpendicular to concentration gradient, and g is the Gibbs free energy per atom. To simplify the mathematics, the following approximations are employed:

$$g^{\text{old}}(c) = g_0^{\text{old}} + \frac{\alpha^{\text{old}}}{2}\left(c - c_0^{\text{old}}\right)^2 \qquad (4.3)$$

$$g^{\text{new}}(c) = g_0^{\text{new}} + \frac{\alpha^{\text{new}}}{2}\left(c - c_0^{\text{new}}\right)^2$$

$$c(x) \cong c(0) + x\nabla c \qquad (4.4)$$

Taking into account Equations 4.3–4.4, we get

$$\Delta G = n \int \left(A_0 + A_1 \nabla c \cdot x + A_2 (\nabla c)^2 x^2\right) S(x)\, dx + \sigma S \qquad (4.5)$$

$$A_0 = -\left(g_0^{\text{old}} - g_0^{\text{new}} + \frac{\alpha^{\text{old}}\left(c(0) - c_0^{\text{old}}\right)^2}{2} - \frac{\alpha^{\text{new}}\left(c(0) - c_0^{\text{new}}\right)^2}{2}\right)$$

$$A_1 = \alpha^{\text{new}}\left(c(0) - c_0^{\text{new}}\right) - \alpha^{\text{old}}\left(c(0) - c_0^{\text{old}}\right) \qquad (4.6)$$

$$A_2 = \frac{\alpha^{\text{new}} - \alpha^{\text{old}}}{2}$$

4.3.2
Spherical Nuclei

Let the nuclei be a sphere with the center x_c so that

$$S(x) = \pi \left(R^2 - (x - x_c)^2\right) \qquad (4.7)$$

Then, simple algebraic transformations of Equation 4.2 give

$$\Delta G = \sigma \cdot 4\pi R^2 + n\pi \left(B_0 \frac{4}{3} R^3 + B_2 \frac{4}{15} R^5\right) \qquad (4.8)$$

where

$$\begin{cases} B_0 = A_0 + A_1 x_c \nabla c + A_2 x_c^2 (\nabla c)^2 \\ B_2 = A_2 (\nabla c)^2 \end{cases} \qquad (4.9)$$

First of all, it is necessary to find the optimal place for nucleation, proceeding from the condition

$$\frac{\partial \Delta G}{\partial x_c} = 0, \qquad \frac{\partial^2 \Delta G}{\partial x_c^2} > 0$$

$$x_c = -\frac{A_1}{2A_2 \nabla c} = \frac{\alpha^{\text{old}}\left(c(0) - c_0^{\text{old}}\right) - \alpha^{\text{new}}\left(c(0) - c_0^{\text{new}}\right)}{\left(\alpha^{\text{new}} - \alpha^{\text{old}}\right)\nabla c} \qquad (4.10)$$

These relations correspond to a minimum if $\alpha^{\text{new}} > \alpha^{\text{old}}$. Therefore, the optimal place for nucleation shifts with the change of the concentration gradient, but the corresponding concentration in the center remains unchanged, that is

$$c(0) + x_c \nabla c = c(0) - \frac{A_1}{2A_2} \qquad (4.11)$$

Hereinafter, we consider only nuclei appearing in the optimal place. In this case, the dependence of ΔG on the nucleus size is simple and is given by

$$\Delta G(R) = \alpha R^2 - \beta R^3 + \gamma (\nabla c)^2 R^5 \qquad (4.12)$$

where

$$\alpha = 4\pi\sigma, \qquad \gamma = 4\pi n/15 \cdot \left(\alpha^{new} - \alpha^{old}\right)/2$$

$$\beta = \frac{4\pi n}{3}\left(\frac{A_1^2}{4A_2} - A_0\right) \qquad (4.13)$$

$$= \frac{4\pi n}{3}\left(g_0^{old} - g_0^{new} + \frac{\alpha^{new} \cdot \alpha^{old}}{2\left(\alpha^{new} - \alpha^{old}\right)}\left(c_0^{old} - c_0^{new}\right)^2\right)$$

The coefficient $\beta > 0$ if the curves $g^{new}(c)$ and $g^{old}(c)$ intersect.

As it appears from Equation 4.12, the dependence $\Delta G(R)$ may be monotonic or nonmonotonic, depending on the concentration gradient (Figure 4.4). The case (a)

$$\nabla c > (\nabla c)_1^{crit} = \beta/\alpha\sqrt{\beta/5\gamma}$$

corresponds to a total suppression of nucleation at a sharp concentration gradient. Case (b)

$$(\nabla c)_1^{crit} > \nabla c > (\nabla c)_2^{crit} = \beta/\alpha\sqrt{4\beta/27\gamma}$$

presents the possibility of metastable nucleus formation. Case (c)

$$\nabla c < (\nabla c)_2^{crit}$$

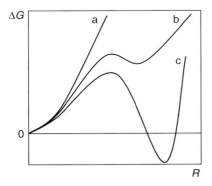

Figure 4.4 Dependence of the Gibbs thermodynamic potential change on the size of a spherical nucleus in the concentration gradient field: (a) nucleation is forbidden; (b) a metastable nucleus can be formed; and (c) nucleation is possible.

implies the possibility of formation of a new phase stable particle whose size increases with time as the concentration gradient (due to interdiffusion) decreases.

Simple transformations give the following expressions for $(\nabla c)_{1,2}^{\text{crit}}$, which correspond to transitions $a \leftrightarrow b$ and $b \leftrightarrow c$

$$(\nabla c)_1^{\text{crit}} = \frac{\beta}{\alpha}\sqrt{\frac{\beta}{5\gamma}}, \quad (\nabla c)_2^{\text{crit}} = \frac{\beta}{\alpha}\sqrt{\frac{4\beta}{27\gamma}}. \tag{4.14}$$

As it is seen, the values $(\nabla c)_1^{\text{crit}}$ and $(\nabla c)_2^{\text{crit}}$ are rather close, so the mode of metastable nucleation is difficult to reveal. Moreover, as discussed later, shape optimization excludes this mode (if we neglect stresses).

4.3.3
Ellipsoidal Nuclei

It is evident that, since concentration gradient suppresses nuclei growth along the longitudinal direction, nature will find the possibility to increase the nucleus's volume (and reduce Gibbs free energy) by transversal growth. This statement means that the nuclei formed in the diffusion zone must be nonspherical. Because of this, it is necessary to take into account the shape optimization for each fixed volume of a nucleus.

The first attempt in this direction was made in 1991 [15]. Nuclei (embryos) were taken as spheroidal (ellipsoids of rotation) with a symmetry axis being directed along ∇c, and the parameters $R_\parallel$ ($\parallel x$) and $R_\perp$ ($\perp x$). In this case, ΔG is a function of two arguments, the volume V and the shape parameter, $\varphi = R_\perp / R_\parallel$ at a fixed concentration gradient, $\nabla c = 1/L$. We get

$$\Delta G(R_\parallel, R_\perp) = 2\pi n \left\{ -\frac{2}{3}\Delta g_0 R_\parallel R_\perp^2 + \frac{g''}{15L^2} R_\parallel^3 R_\perp^2 + \frac{2\sigma}{n}\left[\frac{R_\perp^2}{2} + \frac{R_\perp R_\parallel^2}{2\sqrt{R_\perp^2 - R_\parallel^2}} \right. \right.$$

$$\times \left\{ \begin{array}{ll} \ln\left(\sqrt{\left(\frac{R_\perp}{R_\parallel}\right)^2 - 1} + \frac{R_\perp}{R_\parallel}\right) \right\}, & \frac{R_\perp}{R_\parallel} > 1, \\[1em] \arcsin\sqrt{1 - \left(\frac{R_\perp}{R_\parallel}\right)^2} \right\}, & \frac{R_\perp}{R_\parallel} < 1 \end{array} \right. \tag{4.15}$$

where

$$R_\parallel = \left(\frac{3V}{4\pi}\right)^{\frac{1}{3}} \varphi^{-\frac{2}{3}}, \quad R_\perp = \left(\frac{3V}{4\pi}\right)^{\frac{1}{3}} \varphi^{\frac{1}{3}} \tag{4.16}$$

For each fixed volume, having minimized the function $\Delta G(\varphi|V)$, the optimal shape $\varphi(V)$ was found. The function $\varphi_{\text{opt}}(V)$ approaches infinity at a certain value of V^* (Figure 4.5), which is determined by the concentration gradient (more than ∇c, less than V^*). The dependence $\Delta G(V, \varphi_{\text{opt}}(V))$ differs for large and

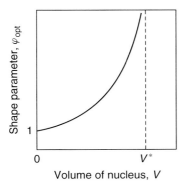

Figure 4.5 Dependence of the shape parameter on the volume of the nucleus.

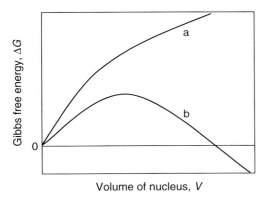

Figure 4.6 Dependence of the Gibbs free energy on the volume at shape optimization: (a) $\nabla c > (\nabla c)^{crit}$; (b) $\nabla c < (\nabla c)^{crit}$.

small concentration gradients (Figures 4.6 a and b). In such a way, we repeat the result of the previous paragraph: there is a concentration gradient above which the nucleation of the intermediate phase is forbidden. However, the possibility of metastable nuclei formation has disappeared as a result of shape optimization.

The new results obtained here are as follows: (i) the formation of pancakelike nuclei is observed (obviously, this is practically unattainable at critical ΔG values being much larger than 60 kT) and (ii) a decrease of the nucleation barrier and corresponding growth of the critical concentration gradient is found to be caused by shape optimization. Apparently, the assumption of a spherical shape of the nuclei is not rigorous. To check the appropriateness of the above results, we discuss below direct the Monte Carlo (MC) simulations of the nucleation process (the algorithm was suggested by the author; the simulations have been performed by O. O. Bogatyrev).

4.3.4
MC Simulations of the Shape of the Nucleus

Let interdiffusion in the binary compound A–B lead to the formation of a metastable parent phase (solid solution or amorphous phase) with a concentration gradient inversely proportional to $\sqrt{Dt}$, where D is the diffusion coefficient in the parent phase. Let us study the possibility for the nuclei of the stable intermediate phase to appear under the mentioned gradient. We approximate the concentration dependence of Gibbs potential for both phases by a parabola with a minimum at $c_0^{new} = c_0^{old} = 1/2$ (this approximation is essential only for comparison with the analytical solutions and is optional at MC simulations). The concentration profile of the parent phase near the forming nucleus will be approximated by a linear dependence.

In this case, we restrict ourselves to a polymorphic mode, assuming that fast nucleation takes place under a frozen-in concentration gradient and corresponding changes caused by diffusion start after the nucleus has been formed. The homogeneous alloy is divided into 'elementary' cells, each of them being able to transform from the old phase to the new one and conversely depending on thermodynamic favorability. The latter is determined by the bulk driving force and the number of neighboring cells with opposite state.

Consider the simulation procedure. Each cell may be found in either one of the two states, the old or the new one. The change of phase state leads to the change of bulk and surface energy. For example, if the cell changes the phase from the old to the new one, the change of the system's Gibbs potential equals

$$\Delta G = \left(g^{new}(c) - g^{old}(c)\right)(a^3 n) + \sigma \Delta N a^2 \tag{4.17}$$

Here, c is the concentration in the cell that depends on its location (the x coordinate if the concentration gradient is x-directed), n is the atomic density per unit volume, a is the cell's size, σ is the surface tension between the old and new phases, ΔN is the change of the number of neighboring cells with different phase state (even numbers from -6 to $+6$). If ΔG is negative, the transformation is accepted; Otherwise, the probability of acceptance is defined by $\exp(-\Delta G/kT)$ (Metropolis algorithm). To save time while implementing the MC procedure, we take the change of phase composition into account only for the cells located in the boundary layer of the forming nucleus.

The following algorithm was realized:

1) At first, all cells belong to the old (parent) phase. Randomly we choose a cell as that where the nucleation (new phase) takes place and try to transform it according to the Metropolis algorithm ($\Delta N = 6$).
2) Then, we randomly choose a boundary cell among the fixed number of cells located at the boundary (a cell belongs to a boundary set if it pertains to the nucleus and has at least one neighboring cell of the old phase).
3) Then we treat a cluster containing a boundary cell and its six neighbors. One of the seven cells is chosen at random. The choice is accepted if the chosen

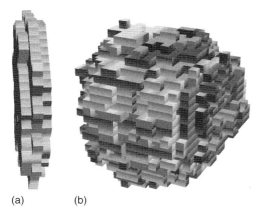

(a) (b)

Figure 4.7 Examples for simulation of the nucleus's shape for large (a) $\nabla c = 10^9 \mathrm{m}^{-1}$ and small (b) $\nabla c = 10^7 \mathrm{m}^{-1}$ concentration gradients. The simulation parameters are chosen as: $\sigma = 0.15 \mathrm{J/m^2}$, $a = 1.5 \cdot 10^{-10} \mathrm{m}$, $n = 10^{29} \mathrm{m}^{-3}$, $(\partial^2 \Delta g/\partial c^2) = 7.77 \cdot 10^{-19} \mathrm{J}$, $g_0^{\mathrm{old}} - g_0^{\mathrm{new}} = 7.48 \cdot 10^{-21} \mathrm{J}$.

cell is the central one (transformation new phase → old phase) or if it is a neighboring cell belonging to the old phase (transformation old phase → new phase). Otherwise, the attempt is repeated.

4) According to Equation 4.17, we calculate the change of Gibbs potential for the possible transformation and the decision concerning the confirmation of the change is made with the help of the Metropolis algorithm.

Step 2 of the given algorithm artificially increases the possibility of the nucleus's growth. Otherwise, the subcritical nucleus is most likely to be destroyed, and creating the supercritical nucleus will take much computing time. Simulation results for large and small concentration gradients are presented in Figure 4.7.

4.3.5
Stress Effects

In the previous analysis, we neglected the difference between molar volumes in the new and parent phases, leading eventually to stresses and additional free energy terms. The influence of stresses on the nucleation of intermediate phases was treated in detail in Ref. [16]. The change of the Gibbs potential at the formation of an intermediate phase nucleus in the field of a concentration gradient taking into account the elastic energy is

$$\Delta G = \Delta G_{\mathrm{bulk}} + \Delta G_{\mathrm{surf}} + \Delta G_{\mathrm{elastic}} \quad (4.18)$$

where the first two summands were described above, the elastic term is written with the assumption of a spheroidal nucleus and in accordance with the Nabarro

[17] model equals

$$\Delta G_{\text{elastic}} = \frac{2}{3}\mu\Gamma^2 nVE\left(\frac{R_\parallel}{R_\perp}\right), \quad \Gamma = \frac{\Omega^p - \Omega^i}{\Omega^p} \qquad (4.19)$$

Here Γ is the dilatation, Ω^i and Ω^p are the atomic volumes of the intermediate and parent phases, V is the volume of the nucleus, E is Nabarro's function (see below), $2R_\parallel$ is the longitudinal size, $2R_\perp$ is the transversal size, and μ is the shear modulus of the parent phase. Nabarro's function was approximated by a third-order polynomial as

$$E(x) = \frac{3}{4}\pi x - 3\left(\frac{\pi}{2} - 1\right)x^2 + \left(\frac{3}{4}\pi - 2\right)x^3 \qquad (4.20)$$

Straightforward transformations give the following dependence of Gibbs potential on the volume V and shape factor $\varphi = R_\perp/R_\parallel$

$$\Delta G(V, \varphi) = -n\Delta g_m V + c_3 \left(\frac{1}{\varphi}\right)^{\frac{4}{3}} V^{\frac{5}{3}} \qquad (4.21)$$

$$+ c_4 \left(\varphi^{\frac{2}{3}} + \frac{\ln\left(\sqrt{\varphi^2 - 1} + \varphi\right)}{\sqrt{\varphi^2 - 1}} \left(\frac{1}{\varphi}\right)^{\frac{1}{3}}\right) V^{\frac{2}{3}}$$

$$+ \frac{2}{3}\mu \frac{(\Omega^i - \Omega^p)^2}{\Omega^p} E\left(\frac{1}{\varphi}\right) nV$$

$$c_3 = \frac{\pi n g''}{10\left(\frac{4}{3}\pi\right)^{\frac{2}{3}}}, \quad c_4 = \frac{2\pi\sigma}{\left(\frac{4}{3}\pi\right)^{\frac{2}{3}}}$$

Further, the dependence $\Delta G(V)$ is computed under the condition of shape optimization for each value of volume. Introduction of the elastic energy changes the general form of the dependence only quantitatively. The main result of the previous theory remains the same: there exists a certain critical gradient above which nucleation is forbidden.

Dilatation created by the nucleus of the new phase varied from 0 to 0.10. The calculations were done for the following parameters:

$$g'' = 7.77 \cdot 10^{-19} \text{ J}, \quad \Delta g_m \equiv g_0^{\text{new}} - g_0^{\text{old}} = 7.48 \cdot 10^{-21} \text{ J}$$
$$\sigma = 0.15 \text{J m}^{-2}, \quad \mu = 5 \cdot 10^{10} \text{N m}^{-2}$$

The dependence of the nucleation barrier on the concentration gradient obtained at different dilatations is depicted in Figure 4.8. According to [14], nucleation can be considered as almost suppressed if $\Delta G_{cr} > 60kT$. In our case, the temperature was chosen as 600 K; at this temperature, diffusion amorphization in the Ni–Zr system occurs.

As it is seen from Figure 4.8, the critical gradient, taking into account the Stresses, changes insignificantly (decreases) and this corresponds to the results of [18, 19]. The actual barrier near the critical gradient may change quite considerably. For example, at a concentration gradient of $\nabla c = 3 \cdot 10^8 \text{ m}^{-1}$, the dilatation of 5% increases the nucleation barrier by 65%, and the possibility of nucleation at

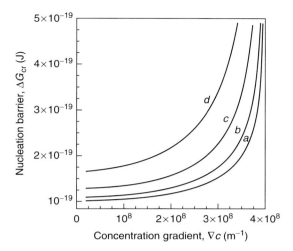

Figure 4.8 Dependence of the nucleation barrier, $G_{cr}(\Gamma)$, on the concentration gradient, ∇c (m^{-1}) at different dilatations: (a) $\Gamma = 0.0$, (b) $\Gamma = 0.01$, (c) $\Gamma = 0.03$, (d) $\Gamma = 0.05$, and (e) $\Gamma = 0.07$.

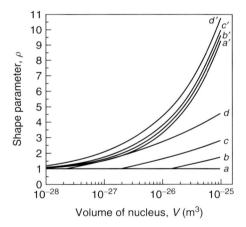

Figure 4.9 Dependence of the shape parameter $\varphi = (R_\perp/R_\parallel)$ on the volume of the nucleus V (m^3) at dilatations $\Gamma = 0.0, 0.05, 0.07, 0.10$ and a concentration gradient $\nabla c = 2.2 \cdot 10^8$ m^{-1} (curves a′, b′, c′, d′) and at its absence (curves a, b, c, d).

$T = 600$ K becomes approximately $3 \cdot 10^6$ times less. The stresses influence not only the barrier but also the nucleus's shape, which is demonstrated in Figure 4.9. Both factors – concentration gradient and stress – have equal influence upon the shape, making it disclike. So, the stresses that appear during nucleation of intermediate phases at the initial stage of reactive diffusion may have a significant influence on the probability of this process only near the critical value of the gradient, above which the nucleation is totally suppressed.

The basic results of the analysis for the polymorphic mode of nucleation are as follows:

1) Nucleation of the intermediate phase is thermodynamically forbidden if the diffusion zone is too narrow ($\partial c/\partial x > 1/L^*$).
2) In the case of possible nucleation ($\partial c/\partial x < 1/L^*$), the shape of the nucleus is not spherical and differs essentially from it. The shape of the nuclei is determined by the ratio V/L^2.
3) The nucleus grows mainly in the transversal to concentration gradient direction; the ratio $R_\perp/R_\parallel$ approaches infinity with the volume growth and can be limited only by the other nuclei, sample boundaries, or composition fluctuations.
4) Nucleation at grain boundaries makes only quantitative changes to the results [20].
5) Taking into account the stresses actually does not change the critical concentration gradient but does change the nucleation barrier.
6) Considering a nucleus as an arbitrary figure of rotation with an axis along the concentration gradient leads, qualitatively, to the same results [18, 19, 21].

4.4
Thermodynamics of the Transversal Mode of Nucleation in a Concentration Gradient

4.4.1
Homogeneous Nucleation: General Relations

This mode of nucleation was first suggested by Desre and Yavari [12] for a cubic nucleus without shape optimization. Latter generalization was performed by Hodaj et al. [22] for the simplest case of parallelepiped-shaped nuclei in 1998. Let an under- or overcritical parallelepiped nucleus $2h \times 2h \times 2r$ (Figure 4.10) appear in the metastable parent phase at the concentration gradient ∇c ($2r$ along ∇c). Each thin slice $2h \times 2h \times dx$ of the nucleus with a concentration $c^{new}(x)$ forms at the expense of the layer $\infty \times \infty \times dx$ with concentration $c^{old}(x)$ according to the parallel tangent rule (Figure 4.11). To avoid misunderstandings, we prove this rule.

Consider a small volume ΔV of the new phase with concentration c^{new} of B component forming in the large volume V_0 of the old phase with initial concentration c^{old} ($V_0 \gg \Delta V$). The appearance of the new phase slightly changes the concentration of the parent phase, so that $\Delta c = (c^{old} - c^{new}) \cdot (\Delta V/(V_0 - \Delta V))$, maintaining matter conservation. The change of the Gibbs free energy at formation of the new phase equals

$$\Delta G = \sigma S + n\left(g^{new}\left(c^{new}\right)\right) \Delta V + g^{old}\left(c^{old} + \Delta c\right)(V_0 - \Delta V)$$
$$- g^{old}\left(c^{old}\right) V_0\right) \tag{4.22}$$

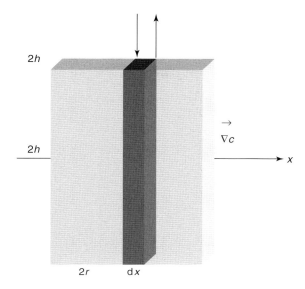

Figure 4.10 Schematic transversal mode of nucleation, vertical arrows indicating the direction of the components' redistribution in the layer dx.

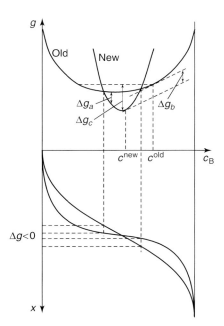

Figure 4.11 Schematic dependence of the driving force per atom of the nucleus for polymorphic (a), transversal (b), and longitudinal (c) modes.

4 Nucleation in a Concentration Gradient

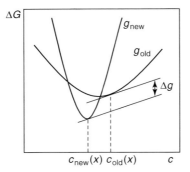

Figure 4.12 Parallel tangents rule.

where

$$g^{\text{old}}\left(c^{\text{old}} + \Delta c\right) \approx g^{\text{old}}\left(c^{\text{old}}\right) + \frac{\partial g^{\text{old}}}{\partial c}\Delta c \qquad (4.23)$$

Here S is the area of the boundary, and σ is the surface tension coefficient. Substituting Δc from the conservation condition, we get

$$\Delta G = \sigma S - \Delta V n \left(g^{\text{old}}\left(c^{\text{old}}\right) + \frac{\partial g^{\text{old}}}{\partial c^{\text{old}}}\left(c^{\text{new}} - c^{\text{old}}\right) - g^{\text{new}}\left(c^{\text{new}}\right) \right)$$

$$= \sigma S - N \Delta g \qquad (4.24)$$

The value of ΔG is minimal at such concentration c^{new} that

$$\frac{\partial \Delta G}{\partial c^{\text{new}}} = 0 = -\Delta V \left(\frac{\partial g^{\text{old}}}{\partial c^{\text{old}}} - \frac{\partial g^{\text{new}}}{\partial c^{\text{new}}} \right) \qquad (4.25)$$

holds. So, the minimal value of the Gibbs free energy (and, thus, the maximal probability of nucleation) is attained for such c^{new} that satisfies the parallel tangent rule. The driving force per atom of the nucleus (not the whole system) is equal to

$$\Delta g_b = g^{\text{old}}\left(c^{\text{old}}\right) + \frac{\partial g}{\partial c}\left(c^{\text{new}} - c^{\text{old}}\right) - g^{\text{new}}\left(c^{\text{new}}\right) \qquad (4.26)$$

and is determined by a vertical interval between the two parallel tangents, as shown in Figure 4.12. Note that the parallel tangents rule is reduced to the more widely known common tangents rule if the supersaturation tends to zero.

Applying the parallel tangents rule to each thin slice dx of the parallelepiped, we get

$$\Delta G = -n \int_{x_C - r}^{x_C + r} \Delta g \left(c^{\text{old}}(x) \to c^{\text{new}}(x) \right) \cdot 4h^2 dx + 2\sigma_1 \cdot 4h^2 + 4\sigma_2 \cdot 4hr \qquad (4.27)$$

where

$$c^{\text{old}}(x) = c(0) + x \cdot \nabla c, \quad c^{\text{new}}(x) = c_0^{\text{new}} + \frac{\alpha^{\text{old}}}{\alpha^{\text{new}}} \left(c^{\text{old}}(x) - c_0^{\text{new}}(x) \right) \qquad (4.28)$$

and σ_1, σ_2 are the surface tension coefficients for the side faces of the nucleus, perpendicular, and parallel to the concentration gradient. Having made some simple transformations, similar to those presented above, we find that the optimal nucleation center is determined by the following equation:

$$c_{opt} = x_C \nabla c + c(0) = c_0^{old} + \frac{c_0^{new} - c_0^{old}}{1 - \frac{\alpha^{old}}{\alpha^{new}}} \tag{4.29}$$

In this case,

$$\Delta G = -\alpha \cdot 8h^2 r + \gamma (\nabla c)^2 h^2 r^3 + 8(\sigma_1 h^2 + 2\sigma_2 hr) \tag{4.30}$$

$$\alpha = n \left(g_0^{old} - g_0^{new} + \frac{\alpha^{old}}{2\left(1 - \frac{\alpha^{old}}{\alpha^{new}}\right)} \left(c_0^{new} - c_0^{old}\right)^2 \right)$$

$$\gamma = \frac{4\pi n}{3} \alpha^{old} \left(1 - \frac{\alpha^{old}}{\alpha^{new}}\right) \tag{4.31}$$

If we express ΔG as a function of the volume $V = 8h^2 r$ and shape parameter $\varphi = (h/r)$ with

$$r = \frac{1}{2} V^{\frac{1}{3}} \varphi^{-\frac{2}{3}}, \quad h = \frac{1}{2} V^{\frac{1}{3}} \varphi^{\frac{1}{3}} \tag{4.32}$$

we get

$$\Delta G(V, \varphi) = -\alpha V^1 + \frac{(\nabla c)^2 \gamma}{32} \varphi^{-\frac{3}{2}} V^{\frac{5}{3}} + 2\sigma_1 \left(\varphi^{\frac{2}{3}} + 2s\varphi^{-\frac{1}{3}}\right) V^{\frac{2}{3}} \tag{4.33}$$

where $s = \sigma_2/\sigma_1$ is Wulf's parameter.

The function $\Delta G(V|\varphi)$ at each fixed volume has one minimum determined from the condition $\partial G/\partial \varphi = 0$ and gives the optimal Shape:

$$\varphi_{opt} = \left(\frac{h}{r}\right)_{opt} = \frac{s}{2} + \sqrt{\frac{s^2}{4} + \frac{\gamma (\nabla c)^2 V}{32\sigma_1}} \tag{4.34}$$

For small volumes, it is reduced to the Wulf rule. For $V \to 0$, we have

$$\varphi_{opt}(V \to \infty) = s = \frac{\sigma_2}{\sigma_1} \tag{4.35}$$

For larger volumes, the shape parameter approaches infinity as $V^{1/2}$:

$$\varphi_{opt}(V \to \infty) \approx (\nabla c^2 \cdot V)^{\frac{1}{2}} \cdot \left(\frac{\gamma}{32\sigma_1}\right)^{\frac{1}{2}} \tag{4.36}$$

which implies a flattened shape: the concentration gradient limits the longitudinal size,

$$r \underset{V \to \infty}{\to} r_{max} = \left(\frac{4\sigma_1}{\gamma (\nabla c)^2}\right)^{\frac{1}{3}} \sim (\nabla c)^{-\frac{2}{3}} \tag{4.37}$$

but does not limit the transversal growth,

$$h \underset{V \to \infty}{\sim} V^{\frac{1}{3}} \cdot \left(V^{\frac{1}{2}}\right)^{\frac{1}{3}} \sim V^{\frac{1}{2}} \to \infty \tag{4.38}$$

4 Nucleation in a Concentration Gradient

As it can be easily seen, the shape parameter depends on the product $(\nabla c)^2 V$, which presents a certain kind of scaling invariance. Thus, the dependence of ΔG on the volume with shape optimization has the following form:

$$\Delta G(V) = -\alpha V^1 + \frac{\gamma (\nabla c)^2}{32} \left(\frac{s}{2} + \left(\frac{s^2}{4} + \frac{\gamma (\nabla c)^2 V}{32\sigma_1} \right)^{\frac{1}{2}} \right)^{-\frac{4}{3}} V^{\frac{5}{3}}$$

$$+ 2\sigma_1 \left(\left(\frac{s}{2} + \left(\frac{s^2}{4} + \frac{\gamma (\nabla c)^2 V}{32\sigma_1} \right)^{\frac{1}{2}} \right) \right)^{\frac{2}{3}}$$

$$+ 2s \left(\left(\frac{s}{2} + \left(\frac{s^2}{4} + \frac{\gamma (\nabla c)^2 V}{32\sigma_1} \right)^{\frac{1}{2}} \right) \right)^{-\frac{1}{3}} V^{\frac{2}{3}} \quad (4.39)$$

Let us consider the limiting cases for small and large volumes

$$\Delta G(V \to 0) \approx 6 \left(\sigma_1 \sigma_2^2 \right)^{\frac{1}{3}} V^{\frac{2}{3}} \quad (4.40)$$

$$\Delta G(V \to \infty) \approx \left(-\alpha + (\nabla c)^{\frac{2}{3}} \frac{3}{2} \left(\frac{\sigma_1^2 \gamma}{4} \right)^{\frac{1}{3}} \right) V^1 \quad (4.41)$$

As evident, depending on the concentration gradient, ΔG can monotonically increase or have a maximum, corresponding to the nucleation barrier. Two cases are possible:

a)

$$\nabla c > (\nabla c)^{\text{crit}} = \frac{4\alpha}{3\sigma_1} \left(\frac{2\alpha}{3\gamma} \right)^{\frac{1}{2}}$$

Here nucleation is forbidden and

b)

$$\nabla c < (\nabla c)^{\text{crit}}$$

Here nucleation is possible.

Therefore, the transversal mode with shape optimization gives us the same qualitative results as the polymorphic mode:

1) The larger the volume and the concentration gradient, the flatter is the nuclei, the nucleus's shape being determined by the product $(\nabla c)^2 V$.
2) Nucleation is forbidden if the concentration gradient exceeds a certain critical value, close to 10^8 m^{-1}.

4.5
Thermodynamics of the Longitudinal Mode of Nucleation in a Concentration Gradient

At the longitudinal mode, the nucleus is formed at the expense of complete intermixing of the parent phases within this volume. In this case, the change of Gibbs free energy at the formation of the parallelepiped nucleus $2h \times 2h \times 2r$ is found from the following expression:

$$\Delta G = 4\left(2\sigma_1 h^2 + 4\sigma_2 hr\right)$$
$$+ n4h^2 \int_{x_C-r}^{x_C+r} \left(g^{new}\left(c^{new}(x)\right) - g^{old}\left(c^{old}(x)\right)\right) dx \quad (4.42)$$

where x_c is the coordinate of the nucleus's center,

$$c^{old}(x) = c(0) + x\nabla c, \quad c^{new}(x) = c(0) + x\nabla c = \text{const} \quad (4.43)$$

and g^{new} and g^{old} are parabolas as described in the previous chapters. Optimization of the nucleation place x_c gives

$$c(0) + x_c \nabla c = \frac{c_0^{new} - c_0^{old}}{1 - \frac{\alpha^{old}}{\alpha^{new}}} + c_0^{old} \quad (4.44)$$

For this optimal nucleation place, simple mathematical transformations lead to the following expression:

$$\Delta G = 4\left(2\sigma_1 h^2 + 4\sigma_2 hr\right) - \alpha 8h^2 r + \gamma_\parallel (\nabla c)^2 h^2 r^3, \quad (4.45)$$

where

$$\alpha = n\left(g_0^{old} - g_0^{new} + \frac{\alpha^{old}\left(c_0^{new} - c_0^{old}\right)^2}{2\left(1 - \frac{\alpha^{old}}{\alpha^{new}}\right)}\right) \quad (4.46)$$

$$\gamma_\parallel = -\frac{4n}{3}\alpha^{old} \quad (4.47)$$

The main peculiarity here is the negative sign of $\gamma_\parallel$. This particular feature means that unlike polymorphic and transversal modes at the formation of the nuclei at the longitudinal mode, the concentration gradient stimulates nucleation. Therefore, at any concentration gradient, nucleation at longitudinal mode is always possible (in the thermodynamic sense). However, the kinetics may suppress it.

The behavior of the shape of the critical nuclei appears to be the most interesting feature of the process in the case of longitudinal mode. The dependence $\Delta G(\varphi)$ at fixed volume is nonmonotonic with one minimum and one maximum for small volumes (Figure 4.13a) and monotonically increases for larger volumes (Figure 4.13b). In the first case (a), the relation

$$(\nabla C)^2 V < \frac{8S^2 \sigma_1}{|\gamma_\parallel|} \quad (4.48)$$

holds while the second case (b) is realized for

$$(\nabla C)^2 V > \frac{8S^2 \sigma_1}{|\gamma_\parallel|} \quad (4.49)$$

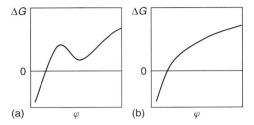

Figure 4.13 Dependence of ΔG on the shape at the longitudinal mode of nucleation for (a) small (Equation 4.48) and (b) large (Equation (4.49)) values $V(\nabla c)^2$.

The extremum condition $(\partial G/\partial \varphi)$ yields the following equation:

$$\varphi^2 - S\varphi + \frac{|\gamma_{||}|\,(\nabla c)^2\,V}{32\sigma_1} = 0 \tag{4.50}$$

with two solutions

$$\varphi_{1,2} = \frac{S}{2} \pm \sqrt{\frac{S^2}{4} - |\gamma_{||}|\frac{(\nabla c)^2\,V}{32\sigma_1}} \tag{4.51}$$

The first one corresponds to a metastable minimum and the second one to a maximum. These solutions disappear when

$$(\nabla c)^2\,V > \frac{8S^2 \sigma_1}{|\gamma_{||}|} \tag{4.52}$$

holds so that the nucleus must turn into a needle quickly (Figure 4.14). So, something similar to a phase transformation of the shape takes place. Apparently, all the speculations given above are correct only for $\nabla c = \text{const}$, since a needle cannot exceed the size of the diffusion zone.

Obviously, the longitudinal mode will be realized if the redistribution in the transversal direction is absent in the parent phase. To accomplish this, the diffusion coefficient of the new phase must essentially exceed that of the old phase. In fact, all nucleation modes are realized simultaneously. The description of their interference must be performed within a kinetic approach and it is described in Ref. [20].

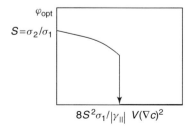

Figure 4.14 "Phase transformation of the shape" for the longitudinal mode.

4.6
Nucleation in Systems with Limited Metastable Solubility

Up to now, we have been investigating different ways (modes) of intermediate phase nucleation in the sharply inhomogeneous diffusion zone for the case when before nucleation the system can create a metastable parent phase with a wide concentration range including the interval of equilibrium concentrations of the new stable phase (Figure 4.15a). At such conditions, the sharp concentration gradient causes the change of the bulk driving force of nucleation and the respective change of the nucleation barrier:

$$\Delta G(r) = \Delta G^{\text{classic}} + \gamma \, (\nabla c)^2 \cdot r^5 \qquad (4.53)$$

Yet, in most cases, solid-state reactions proceed in diffusion couples with low mutual solubility even in the metastable state (Figure 4.15b). In the case of phase 1 nucleating between α and β, in the absence of phase 1, these phases have metastable solubilities c_α^m and $1 - c_\beta^m$. Interdiffusion must yield a steplike profile with a concentration gap (c_α^m, c_β^m), which does not contain the concentration interval of phase 1. Here the polymorphic mode cannot be realized.

Instead, the transversal mode is quite possible in this case and must take place as a simultaneous decomposition of inhomogeneous metastable alloys within the concentration ranges (c_α^e, c_α^m) and (C_β^m, C_β^e) on each side of the interface. Some aspects of the problem, regarding the shape of the nucleus, are considered in Refs [18, 19, 23]. The shape of the nucleus, being an important factor of nucleation

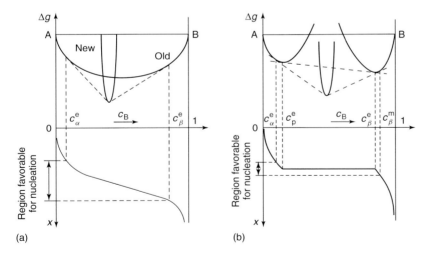

Figure 4.15 Schematic dependencies of Gibbs free energy on concentration and corresponding 'concentration on coordinate' for both cases: (a) the region of homogeneous intermediate phase inside the zone of metastable solution and (b) the region of homogeneous intermediate phase outside the zones of metastable solubilities of both parent phases.

under a concentration gradient, is not a decisive one (see the previous sections). Therefore, in the subsequent analysis, we disregard the problem of shape and focus on the influence of the gradient on the nucleation barrier.

The thermodynamic approach in the prediction of the sequence of occurrence of the different phases was also used by Lee et al. [24]. The authors have taken into account the change of the driving force due to the formation of a metastable solution. Still, they have disregarded the influence of concentration and chemical potential gradients on the driving force.

Now we will make sure that the influence of a concentration gradient (or, more precisely, chemical potential gradients) appears to be important during nucleation in systems with limited metastable solubility, especially when at least one intermediate phase is already growing. It is quite reasonable that the "gradient approach" can furnish the clue to an understanding of the phenomena of sequential phase formation in this case.

4.6.1
Nucleation of a Line Compound at the Interface During Interdiffusion

In this analysis, the phases L and R are assumed to have already developed some concentration profiles $c_L(x)$, $c_R(x)$ with a concentration gap (c_{LR}, c_{RL}) and metastable regions (c_{Li}, c_{RL}) and (c_{RL}, c_{Ri}), conductive to the nucleation of a line compound 'i' (Figure 4.16). The change in the Gibbs free energy due to nucleation (by the transversal mode) of a parallelepiped of the new phase $2h \times 2h \times 2r$ can be determined then, where $2h$ and $2r$ are, respectively, the size in the lateral (Y- and Z-axes) and the gradient of longitudinal (X-axis) directions. The longitudinal size $2r$ consists of parts overlapping the phases L and K and originating from these phases, that is, $2r = r_L + r_R$. Then, we have

$$\Delta G = 4h^2 (\sigma_{iL} + \sigma_{iR} - \sigma_{LR}) + 4 \cdot 2h (r_L \sigma_{iL} + r_R \sigma_{iR})$$
$$+ \frac{1}{\Omega} \int_{y-r_L}^{y} \Delta g (c_L(x) \to c_i) \cdot 4h^2 \, dx$$
$$+ \frac{1}{\Omega} \int_{y}^{y+r_R} \Delta g (c_R(x) \to c_i) \cdot 4h^2 \, dx \tag{4.54}$$

Here, Ω is the atomic volume of phase "i", σ_{iL}, σ_{iR}, σ_{LR} are the corresponding surface tensions, and y is the position of the L/R interface with respect to the x-axis ($x = 0$ is the position of the center of the embryo). Driving forces per atom of nucleus for precipitation of $L^{\text{metast}} \to i$ and $R^{\text{metast}} \to i$ are expressed by the "parallel tangent rule" (not to be confused with "common tangent rule") [10]; for $L^{\text{metast}} \to i$

$$-\Delta g (c_L(x) \to c_i) = g_L (c_L(x)) - g_i - (c_L(x) - c_i) \left. \frac{\partial g_L}{\partial c} \right|_{c=c_L(x)} \tag{4.55}$$

(for $R^{\text{metast}} \to i$ by analogy).

In addition, expansions into Taylor series are used both for concentrations and for Gibbs free energies per atom resulting in

$$c_L(x) \approx c_L(y) + (x - y) \cdot \nabla c|_{x=y-0} \quad \text{for} \quad x < y \tag{4.56}$$

4.6 Nucleation in Systems with Limited Metastable Solubility

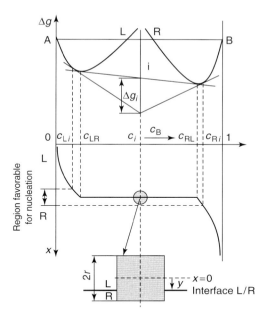

Figure 4.16 Schematic representation of intermediate phase nucleation at the L/R interface, with L and K being the parent phases. Δg_i is a driving force per mole of atoms for the $L + R \to i$ reaction. The distribution of nucleus volume among phases is determined by an additional optimization procedure (Equations 4.62 and 4.63).

$$c_R(x) \approx c_L(y) + (x - y) \cdot \nabla c|_{x=y-0} \quad \text{for} \quad x < y \tag{4.57}$$

leading to

$$g_L(c_L) \approx g_L(c_{LR}) + (c_L - c_{LR})g'_L + \frac{(c_L - c_{LR})^2}{2}g''_L \tag{4.58}$$

$$g_R(c_R) \approx g_R(c_{RL}) + (c_R - c_{RL})g'_R + \frac{(c_R - c_{RL})^2}{2}g''_R \tag{4.59}$$

Here the first- and second-order derivatives of the concentration are taken at the meta-equilibrium compositions (Figure 4.16).

Substitution of Equations 4.56–4.59 into Equation 4.55, and then of Equation 4.55 into Equation 4.54 gives, after simple but extended algebra, the following result:

$$\begin{aligned}\Delta G = {} & 4h^2(\sigma_{iL} + \sigma_{iR} - \sigma_{LR}) + 4 \cdot 2h(r_L\sigma_{iL} + r_R\sigma_{iR}) \\ & + \Delta g_i\frac{4h^2}{\Omega}(r_L + r_R) + \frac{2h^2}{\Omega}\left[(c_i - c_{LR})g''_L\nabla c_L r_L^2 + (c_{LR} - c_i)g''_R\nabla c_R r_R^2\right] \\ & + \frac{2h^2}{3\Omega}\left[g''_L(\nabla c_L)^2 r_L^3 + g''_R(\nabla c_R)^2 r_R^3\right]\end{aligned} \tag{4.60}$$

Here Δg_i is the driving force of the reaction L+R→i per atom of i. The first two terms in Equation 4.60 represent the classical model of heterogeneous nucleation $\Delta G^{\text{classic}}$ (without, however, taking into account Young's equilibrium conditions at three-phase junctions – otherwise a nonsymmetrical cap would be obtained with much less transparent mathematics for the gradient effect). The gradient effect is represented both by linear in ∇c and by quadratic $(\nabla c)^2$ terms, providing the fourth and fifth power size dependence, respectively.

In the case of total metastable solubility, the optimization of the nucleation place led to the elimination of the linear terms in the concentration gradient. In the case of limited solubility, these terms remain and, moreover, may play a decisive role. Indeed, $r_L \nabla c$ is evidently less than the metastable composition range $\Delta c_L = c_{LR} - c_{Li}$ of L-phase and $r_R \nabla c_R < \Delta c_R = c_{Ri} - c_{RL}$. If the parent material is considered to be only the phases with small mutual solubility, then the $(\nabla c)^2 \, r^3$ terms can be neglected in comparison with the $(\nabla c)^1 \, r^2$ terms, since $\nabla c \cdot r \ll 1$. In this case, we arrive at

$$\Delta G = \Delta G^{\text{classic}} + \frac{2h^2}{\Omega} \left[(c_i - c_{LR}) g_L'' \nabla c_L r_L^2 + (c_{LR} - c_i) g_R'' \nabla c_R r_R^2 \right] \quad (4.61)$$

Equation 4.61 is the basic equation for the further analysis.

Minimization of ΔG with respect to r_L or r_R (with fixed sum $2r$), in the particular case of equal surface tensions σ for all interfaces, gives

$$r_L = 2r \frac{\Lambda}{1 + \Lambda}, \qquad r_R = 2r \frac{1}{1 + \Lambda} \quad (4.62)$$

where

$$\Lambda = \frac{(c_{RL} - c_i) g_R'' \nabla c_R}{(c_i - c_{LR}) g_L'' \nabla c_L} \quad (4.63)$$

Substituting Equations 4.62 and 4.63 into Equation (4.61), the following expression is obtained for the change in the Gibbs free energy due to nucleation:

$$\Delta G = \Delta G^{\text{classic}} + \frac{2h^2 r^2}{\Omega} \cdot \frac{A_L A_R}{A_L + A_R} \quad (4.64)$$

with

$$A_L = (c_i - c_{LR}) g_L'' \nabla c_L, \qquad A_R = (c_{RL} - c_i) g_R'' \nabla c_R \quad (4.65)$$

The values of the products $g'' \nabla c$ will be most important for estimations. These products are assessed later, but for the time being, they are assumed to be known. From Equations 4.62 and 4.63, it is evident that the nucleus will prefer to overlap more with the smaller gradient term $g'' \nabla c$. In addition, from Equation 4.64, it may be concluded that the gradient contribution to the change in Gibbs free energy is controlled by the smaller of the two terms A_L, A_R determined by $g'' \nabla c$ (differences in composition are of the order of unity).

For the simplest case of cubic shape, $h = r$, and for limited solubility within the parent phase, the expression for $\Delta G (r)$ is

$$\Delta G (r) = \Delta G^{\text{classic}} + qr^4 = \alpha r^2 + \beta r^3 + qr^4 \quad (4.66)$$

with

$$\alpha = 20\sigma, \quad \beta = \frac{8\Delta g_i}{\Omega}, \quad q = \frac{8}{\Omega}\frac{A_L A_R}{A_L + A_R} \quad (4.67)$$

Depending on the value of q, the $\Delta G(r)$ dependence can be monotonically increasing (large q, nucleation forbidden), or with a metastable minimum (intermediate values of q), or with a stable minimum (small q, nucleation possible) as demonstrated in Figure 4.17. Crossover to possible nucleation (second minimum at zero level) means

$$\left.\frac{\partial \Delta G}{\partial r}\right|_{r*} - 0, \Delta G(r*) = 0 \quad (4.68)$$

This gives

$$r* = \frac{2\alpha}{\beta} = 5\frac{\sigma\Omega}{\Delta g_i}, \quad q^{crit} = \frac{\beta^2}{4\alpha} = \frac{4(\Delta g_i)^2}{5\sigma\Omega^2} \quad (4.69)$$

or

$$\left[\frac{A_L A_R}{A_L + A_R}\right]^{crit} = \frac{(\Delta g_i)}{10\sigma\Omega} \quad (4.70)$$

The intermediate phase suppression criterion is then

$$\frac{A_L A_R}{A_{lL} + A_L} > \frac{(\Delta g_i)^2}{10\sigma\Omega} \quad (4.71)$$

Remember that $(-\Delta g_i)$ is the driving force of the reaction $L + R \to i$ per atom of i.

The formal approach developed above will be applied now, further to different types of adjacent phases L and R.

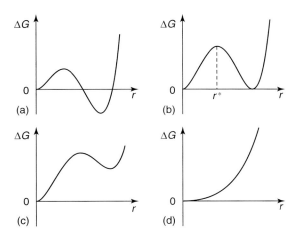

Figure 4.17 Schematic "Gibbs free energy change versus nucleus size" dependencies for different values of the 'gradient term' q: (d) nucleation forbidden ($q > q^{crit}$), (c) metastable nuclei possible ($q > q^{crit}$), (b) crossover to possible nucleation ($q = q^{crit}$), and (a) nucleation possible ($q < q^{crit}$) (if not suppressed kinetically).

4.6.2
Nucleation in between Dilute Solutions

Let α and β be dilute solutions of B in A and of A in B, respectively, with solubilities (prior to phase "i") $c_\alpha^m \ll 1$ and $(1 - c_\alpha^m) \ll 1$. Then $g''_\alpha \approx k_B T/c_\alpha^m$ and $g''_\beta \approx k_B T/(1 - c_\beta^m)$ can be estimated. The concentration gradients ∇c_α, ∇c_β at the α/β moving boundary ($\alpha = L$ and $\beta = B$) can be determined by the usual Stephan problem [25]. In general, the solution gives transcendent equations for determining the velocity and the parameters of the profiles. However, if the difference between $c_\alpha^m \sqrt{D_\alpha}$ and $(1 - c_\beta^m)\sqrt{D_\beta}$ (D_α and D_β being the diffusivities in the respective solutions) is not large, then the following approximate expressions can be employed:

$$\nabla c_\alpha \approx \frac{c_\alpha^m}{\sqrt{\pi D_\alpha t}}, \quad \nabla c_\beta \approx \frac{1 - c_\beta^m}{\sqrt{\pi D_\beta t}} \tag{4.72}$$

so that

$$g''_\alpha \nabla c_\alpha \approx \frac{k_B T}{\sqrt{\pi D_\alpha t}}, \quad g''_\beta \nabla c_\beta \approx \frac{k_B T}{\sqrt{\pi D_\beta t}} \tag{4.73}$$

Therefore,

$$\left(\frac{A_\alpha A_\beta}{A_\alpha + A_\beta}\right) \approx \frac{c_i (1 - c_i) k_B T}{\sqrt{\pi D_\alpha t}\left(c_i \sqrt{D_\beta/D_\alpha} + 1 - c_i\right)} \tag{4.74}$$

Then, according to Equation 4.71, nucleation becomes possible at

$$\frac{c_i (1 - c_i) k_B T}{\sqrt{\pi D_\alpha t}\left(c_i \sqrt{D_\beta/D_\alpha} + 1 - c_i\right)} < \frac{(\Delta g_i)^2}{10 \sigma \Omega} \tag{4.75}$$

or

$$c_i \sqrt{\pi D_\beta t} + (1 - c_i) \sqrt{\pi D_\alpha t} > 5 c_i \frac{k_B T}{\Delta g_i} l_i^{cr} \tag{4.76}$$

where $l_i = 2\sigma \Omega/\Delta g_i$ is a standard critical size of the nucleus, usually less than 1 nm. The ratio $k_B T/\Delta g_i$ is usually significantly less than unity. This means that nucleation becomes possible when at least one of two penetration depths reaches nanometric thickness. Thus, in this case nucleation is practically not suppressed.

4.6.3
Nucleation in between Two Growing Intermediate Phase Layers

Let L and R be the intermediate phases 1 and 3, growing simultaneously between almost mutually soluble materials A and B. Here, the nucleation of phase 2 at the interface 1/3 is studied (Figure 4.18). It is assumed that 1, 2, and 3 are line compounds. In this case, it is more convenient to treat the chemical potential gradients instead of the concentration gradient. Mathematically, it gives

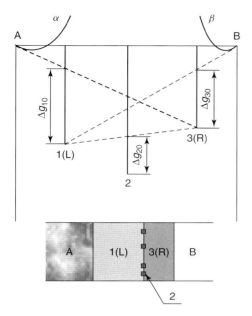

Figure 4.18 Schematic representation of phase 2 nucleation between phases 1 and 3, already growing in an A/B couple. Δg_{10}, Δg_{30}, and Δg_{20} are the driving forces per mole of atoms for the reactions $A + B \to 1$, $A + B \to 3$, and $1 + 3 \to 2$, respectively.

the following expression

$$g_1'' \nabla c_1 = \frac{\partial}{\partial x}\left(\frac{\partial g_1}{\partial c}\right) \approx \frac{\left.\frac{\partial g_1}{\partial c}\right|_{13} - \left.\frac{\partial g_1}{\partial c}\right|_{1\alpha}}{\Delta x_1}$$

$$= \frac{\frac{g_3 - g_1}{c_3 - c_1} - \frac{g_1 - g_\alpha^m}{c_1 - 0}}{\Delta x_1} = \frac{c_3}{c_1(c_3 - c_1)} \cdot \frac{(-\Delta g_{10})}{\Delta x_1} \quad (4.77)$$

where $(-\Delta g_{10})$ is the driving force of the reaction $A + 3 \to 1$ (Figure 4.18). Similarly, we find

$$g_3'' \nabla c_3 = \frac{1 - c_1}{(1 - c_3)(c_3 - c_1)} \cdot \frac{(-\Delta g_{30})}{\Delta x_3} \quad (4.78)$$

where $(-\Delta g_{30})$ is a driving force of reaction $1 + B \to 3$. Thus, in this case, we arrive at

$$A_L = A_1 = w_1 \frac{(-\Delta g_{10})}{\Delta x_1}, \quad A_R = A_3 = w_3 \frac{(-\Delta g_{30})}{\Delta x_3} \quad (4.79)$$

where

$$w_1 = \frac{(c_2 - c_1) c_3}{c_1 (c_3 - c_1)}, \quad w_3 = \frac{(1 - c_1)(c_3 - c_2)}{(1 - c_3)(c_3 - c_1)} \quad (4.80)$$

The phase 2 suppression criterion Equation 4.71 can then be easily reduced to the following form:

$$\frac{\Delta g_{10}}{\Delta g_{30}}\frac{\xi_3}{w_3} + \frac{\Delta g_{30}}{\Delta g_{10}}\frac{\xi_1}{w_1} < 5\frac{\Delta g_{10}\Delta g_{30}}{(\Delta g_{20})^2} \quad (4.81)$$

where

$$\xi_1 = \frac{\Delta X_1}{\left(\frac{2\sigma\Omega}{\Delta g_{10}}\right)} = \frac{\Delta X_1}{l_1^{cr}} \quad \xi_3 = \frac{\Delta X_3}{\left(\frac{2\sigma\Omega}{\Delta g_{30}}\right)} = \frac{\Delta X_3}{l_3^{cr}} \quad (4.82)$$

Note that w_1 and w_3 are of the order of unity. If phases 1 and 3 are mutually symmetrical ($\Delta g_{10} = \Delta g_{30}$, $w_1 = w_3$), then the criterion Equation 4.81 is reduced to

$$\frac{\Delta x_1}{l_1^{cr}} + \frac{\Delta x_3}{l_3^{cr}} < 5\frac{(\Delta g_{10})^2}{(\Delta g_{20})^2} \quad (4.83)$$

Here $(-\Delta g_{20})$ is the driving force of the reaction $1\,(L) + 3\,(R) \to 2$ (cf. Figure 4.18). If phase 1 is much wider than phase 3, that is, $\Delta x_1/\Delta x_3 \gg (\Delta g_{10}/\Delta g_{30})^2$, then it can be shown that the nucleation of phase 2 is preferable at the side of phase 1 ($r_R \approx 0$), and the criterion Equation (4.80) is reduced to

$$\frac{\Delta x_1}{l_1^{cr}} < 5\frac{(\Delta g_{10})^2}{(\Delta g_{20})^2} \quad (4.84)$$

The estimates given by Equations 4.83 and 4.84 are very important. They demonstrate that the critical thickness of the first (and fast-) growing phase may well be rather large, since the driving force of second phase formation (from the already-growing phases 1 and 3) is often significantly less than that of the first phase formation from one compound and one pure element (Figure 4.18).

4.6.4
Nucleation in between a Growing Intermediate Phase and a Dilute Solution

Let L and K be, respectively, the growing phase 1 and the dilute solution of A in B (β) (Figure 4.19). In this case, the criterion for suppression of phase 2 is given by

$$\left(\frac{A_1 A_\beta}{A_1 + A_\beta}\right) > \frac{(\Delta g_{20})^2}{10\sigma\Omega} \quad (4.85)$$

As before,

$$A_1 = (c_2 - c_1)\,g_1''\nabla c_1 = w_1\frac{\Delta g_{10}}{\Delta x_1} \quad (4.86)$$

and

$$A_\beta = (c_{\beta 1} - c_2)\,g_\beta''\nabla c_\beta \approx (1 - c_2)\frac{k_B T}{1 - c_\beta}\nabla c_\beta \quad (4.87)$$

where $(-\Delta g_{10})$ and $(-\Delta g_{20})$ are the driving forces of the reactions $A + B \to 1$ and $1\,(L) + B\,(R) \to 2$, respectively (Figure 4.19).

4.6 Nucleation in Systems with Limited Metastable Solubility | 89

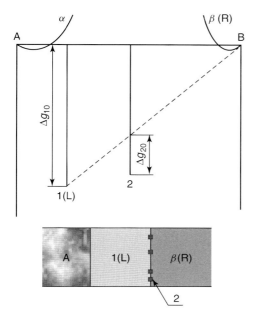

Figure 4.19 Schematic representation of phase 2 nucleation between intermediate phase 1 and dilute solution β, already growing in an A/B couple. Δg_{10} and Δg_{20} are the driving forces per mole of atoms for the reactions $A + B \to 1$ and $1 + \beta \to 2$, respectively.

To express the gradient in the dilute solution in the presence of a growing intermediate phase, it is necessary to solve the following set of algebraic equations for simultaneous parabolic growth of phase 1 layer and of β-solution with parabolic movement of interfaces A/1 and 1/β

$$c_1 \frac{k_{A1}}{2} = -\frac{D_1 \Delta c_1}{k_{1\beta} - k_{A1}} \tag{4.88}$$

$$(1 - c_1) \frac{k_{1\beta}}{2} = \frac{D_1 \Delta c_1}{k_{1\beta} - k_{A1}} - \sqrt{\frac{D_\beta}{\pi}} (1 - c_1) \frac{\exp\left(-\frac{k_{1\beta}^2}{4 D_\beta}\right)}{1 - \mathrm{erf}\left(-\frac{k_{1\beta}}{2\sqrt{D_\beta}}\right)} \tag{4.89}$$

where

$$y_{A1} = k_{A1} \sqrt{t}, \qquad y_{1\beta} = k_{1\beta} \sqrt{t}. \tag{4.90}$$

Analytically, two limiting cases can be treated (which are most often encountered): (i) $D_\beta \ll D_1 \Delta c_1$ (very low diffusivity in the solution, which is usual for high melting B) and (ii) $D_\beta \gg D_1 \Delta c_1$ (very high diffusivity in solution – low melting B). Omitting elementary algebra and expansions into Taylor series, the results are given hereafter:

1) $D_\beta \ll D_1 \Delta c_1$

In this case, we have

$$k_{1\beta} \approx \sqrt{\frac{c_1}{1-c_1}} \sqrt{2D_1 \Delta c_1}, \qquad \Delta x_1 \approx \sqrt{\frac{1}{c_1(1-c_1)}} \sqrt{2D_1 \Delta c_1 t}$$

$$\nabla c_\beta \approx (1 - c_\beta) \sqrt{\frac{c_1}{1-c_1}} \sqrt{\frac{2D_1 \Delta c_1}{2D_\beta^2 t}}$$

so that

$$A_\beta \approx (1 - c_2) \sqrt{\frac{c_1}{1-c_1}} \sqrt{2D_1 \Delta c_1} \frac{k_B T}{2 D_\beta \sqrt{t}}$$

and

$$\frac{A_1}{A_\beta} \approx \frac{c_2 - c_1}{c_1(1-c_2)} \frac{\Delta g_{10}}{k_B T} \frac{D_\beta}{D_1 \Delta c_1} \ll 1$$

Therefore, $A_1 A_\beta / (A_1 + A_\beta) \approx A_1$ and the suppression criterion is

$$\frac{\Delta x_1}{l_1^{cr}} < 5 \frac{c_2 - c_1}{c_1(1-c_1)} \left(\frac{\Delta g_{10}}{\Delta g_{20}} \right)^2 \tag{4.91}$$

Thus, in the case of a smaller driving force for the second phase and low diffusivity in B, the thermodynamic suppression of phase 2 may be quite significant.

2) $D_\beta \gg D_1 \Delta c_1$

In this case,

$$k_{1\beta} \cong -\sqrt{\frac{4}{\pi}} (1 - c_\beta) \sqrt{D_\beta} < 0$$

$$k_{1\beta} - k_{A1} = k = \frac{\Delta x_1}{\sqrt{t}} \approx \sqrt{\frac{\pi}{D_\beta (1-c_\beta)^2} \frac{D_1 \Delta c_1}{c_1}}$$

$$A_\beta \approx (1 - c_2) \frac{k_B T}{\sqrt{\pi D_\beta t}}$$

and

$$\frac{A_1}{A_\beta} \approx \frac{c_2 - c_1}{c_1(1-c_2)} \frac{\Delta g_{10}}{k_B T} \frac{D_\beta (1-c_\beta)}{D_1 \Delta c_1} \gg 1$$

so that

$$\frac{A_1 A_\beta}{A_1 + A_\beta} \approx A_\beta \approx (1 - c_2) \frac{k_B T}{\sqrt{\pi D_\beta t}}$$

This leads to the following suppression criterion:

$$\sqrt{\pi D_\beta t} < 5(1-c_2) \frac{k_B T}{\Delta g_{20}} l_2^{cr} < l_1^{cr} \tag{4.92}$$

So, in this case, suppression is actually absent.

The best candidates for second phase suppression by a chemical potential gradient are, therefore, systems with a considerable difference in melting temperature between constituents. More probably, the first phase to grow will be the phase that is closest to the low-melting-point element. Then, for the second phase, the situation will be as in case 1.

4.6.5
Application to Particular Systems

The reactions of metal layers with their silicon substrates resulting in the formation of various silicides are generally considered not only as phenomena common to all diffusion couples where new phases are formed [26]. The kinetics of silicide growth is classified into three different categories [26]: diffusion controlled, nucleation controlled, and others (reaction rate controlled). Many silicides such as $Mn_{11}Si_{19}$, $NiSi_2$, $ZrSi_2$, $PdSi$, $HfSi_2$, and so on, have been shown to occur via nucleation-controlled reactions. The above outlined approach will be applied, now, to (a) Ni–Si, (b) Co–Si, and (c) Co–Si, Ge thin film systems.

In the following, we do not discuss nucleation and growth of the first intermediate phase that is formed in these systems, that is, Ni_2Si and Co_2Si. These phenomena are discussed in detail elsewhere [26]. Nevertheless, we note that the influence of a sharp concentration gradient and/or chemical potentials on nucleation of the first intermediate phase in these systems is insignificant. We discuss only the second phase nucleation at the M_2Si/Si interface within the $M/M_2Si/Si$ system (M is here Ni or Si) in (a) and (b) and the $CoSi_2$ nucleation at the $CoSi/Si$ interface within the $Co_2Si/CoSi/Si$ system in (c).

(a) In the Ni–Si system, the reaction of nickel with silicon causes the successive formation of Ni_2Si and $NiSi$ at temperatures below 500 °C [24, 25]. Ni_2Si grows first at 200 °C and as long as Ni is not completely exhausted, $NiSi$ does not form at the Ni_2Si/Si interface. $NiSi$ grows at about 300 °C only after Ni is exhausted [27]. Further heating does not cause any other change until about 800 °C; then, one observes the sudden formation of $NiSi_2$. Nucleation of $NiSi_2$ is treated in Ref. [26]. Let us discuss $NiSi$ nucleation at the Ni_2Si/Si interface within the $Ni/Ni_2Si/Si$ system (Figure 4.20b). Note that, in the system, the first growing phase is Ni_2Si, which is the most stable and simultaneously the compound with the highest diffusivity of the three phases (Ni_2Si, $NiSi$, and $NiSi_2$) (Figure 4.20a).

For the $Ni/Ni2Si/Si$ configuration, Equations 4.56–4.57 can be used to determine the critical thickness of the Ni_2Si layer below which nucleation of the $NiSi$ compound at the Ni_2Si/Si interface is forbidden. The calculations are performed at $T = 600$ K (note that temperature has a very small influence on the values of the Gibbs free energy of phase formation (Table 4.1)).

With $c_1 = c_{Ni_2Si} = 1/3$, $c_2 = c_{NiSi} = 1/2$, $\Delta g_{10} = -48$ and $\Delta g_{20} = -4.7$ (-7.7) kJ/mol of atoms [28], the following value is obtained:

$$\frac{\Delta x_{Ni_2S}}{l^{cr}_{Ni_2Si}} < 5 \frac{c_{NiSi} - c_{Ni_2Si}}{c_{NiSi}\left(c_{Si} - c_{Ni_2Si}\right)} \left(\frac{\Delta g_{10}}{\Delta g_{20}}\right)^2 = \frac{10}{3}\left(\frac{\Delta g_{10}}{\Delta g_{20}}\right)^2 \approx 400 \ (150)$$

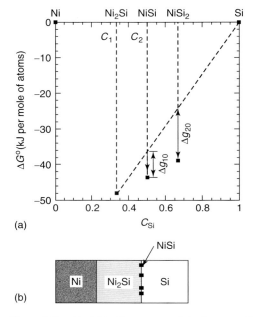

Figure 4.20 (a) Gibbs free energy of the intermetallic compounds in the Ni–Si system at $T = 600$ K. Δg_{10}, Δg_{20} are the driving forces per mole of atoms for the reactions $2Ni+Si \rightarrow Ni_2Si$ and $Ni_2Si + Si \rightarrow 2NiSi$, respectively. (b) Cross-sectional sketch of the Ni/Ni$_2$Si/Si system, showing localized nucleation of NiSi at the Ni$_2$Si/Si interface.

here $l^{cr}_{Ni_2Si}$ is the critical Ni$_2$Si nuclei size. A similar determination can be made in the case of NiSi$_2$ nucleation at the Ni$_2$Si/Si interface. This result indicates that, in the Ni/Ni$_2$Si/Si system, nucleation of NiSi at the Ni$_2$Si/Si interface is forbidden as long as Ni is not completely consumed and the Ni$_2$Si thickness remains submicronic. In other words,

- when the Ni layer is completely consumed, nucleation of NiSi becomes possible regardless of the Ni$_2$Si thickness (no concentration gradient effect), as observed in thin film experiments [25] and,
- when the Ni$_2$Si thickness becomes greater than a few micrometers, nucleation of NiSi becomes possible even in the presence of an Ni layer, as observed experimentally in bulk diffusion reaction experiments [29].

(b) In the Co–Si system, as in the Ni–Si system, Co$_2$Si grows first at temperatures greater than 200 °C but, in contrast to the Ni–Si system, CoSi$_2$, which forms above 350 °C, grows simultaneously with Co$_2$Si until all Co is exhausted. Once the Co layer is completely consumed, the Co$_2$Si phase dissociates, resulting in the growth of CoSi at the expense of Co$_2$Si [27]. Nevertheless, it may be noted that, in such case, contrary to the Ni–Si system, the first growing phase is Co$_2$Si, which is not

Table 4.1 Experimental values for the standard heat of formation (ΔH^0), entropy of formation (ΔS^0) and Gibbs free energy of formation (ΔG^0) of some Ni–Si and Co–Si compounds (from pure elements at 298 K) [27].

Compound	ΔH^0 (298 K) (kJ mol^{-1})	ΔS^0 (298 K) (J mol^{-1})	ΔG^0 (600 K) (kJ mol^{-1})	ΔG^0 (1073 K) (kJ mol^{-1})
Ni$_2$Si	−48		−48^a	
NiSi	−45; −42	−2.1	−43.7; −40,7	
NiSi$_2$	−31	−0.7	−38.9	
Co$_2$Si	−38		−38^a	−38^a
CoSi	−48	−3.1	−46.1	−44.7
CoSi$_2$	−33	−1.2	−32.3	−32.6

^{a} Because of the lack of data, $\Delta S^0 = 0$ is taken, "mole" always implies here "mole of atoms".

the most stable of the three phases (Co$_2$Si, CoSi and CoSi$_2$) reported in thin film systems (Figure 4.21a) but has the highest diffusivity.

The coexistence of Co, Co$_2$Si, CoSi, and Si phases and the simultaneous growth of the Co$_2$Si and CoSi phase are relatively unique features. We discuss here only this simultaneous growth or in other words why CoSi nucleation is not forbidden at the Co$_2$Si/Si interface within the Co/Co$_2$Si/Si system (Figure 4.21b). The same determination as in (a) gives the critical thickness of the first growing phase (Co$_2$Si) beyond which nucleation of CoSi at the Co$_2$Si/Si interface becomes possible. Calculations are performed here for $T = 600$ K. With $c_1 = 1/3$, $c_2 = 1/2$, $\Delta g_{10} = -38$ and $\Delta g_{20} = -17.6$ kJ/mol of atoms (Table 4.1) [26], $\Delta x_{Co_2Si}/l^{cr}_{Co_2Si} < 8$ is obtained. This result means that nucleation of CoSi at the Co$_2$Si/Si interface becomes possible at a very early stage of Co$_2$Si growth (Δx_{Co_2Si} is of the order of just a few nanometers), as observed experimentally.

(c) Film thickness effects in the Co − Si$_{1-x}$Ge$_x$ solid-phase reactions have recently been highlighted in Ref. [30]: The interfacial products of Co with Si$_{0.79}$Ge$_{0.21}$ after annealing at 800 °C depend on the thickness of the Co film. Complete conversion to CoSi$_2$ occurred only when the thickness of the Co layer exceeded 35 nm. Interface reactions with Co layers thinner than 5 nm resulted in CoSi formation. The threshold thickness for nucleation of CoSi$_2$ on Si$_{1-x}$Ge$_x$ was determined in the range $0 \leq x \leq 0.25$ and increases exponentially with x_{Ge}. The critical initial Co thickness is as high as 22 nm for $x_{Ge} = 0.25$ (Figure 4.21a).

To explain this critical thickness effect, it is assumed in this approach that, before the CoSi$_2$ nucleation could take place, the system is made up of Co$_2$Si/CoSi/Si (no Co remaining), as suggested by experimental results obtained by Lau et al. [27] at 300 °C for the Co–Si system. In such case, nucleation of CoSi$_2$ (phase 2) between the CoSi growing phase (1) and the right-hand phase Si (β) has to be considered. Co$_2$Si plays the role of the left-hand phase here (Figure 4.22b).

We exclusively discuss here CoSi$_2$ nucleation at the CoSi/Si interface within the Co$_2$Si/CoSi/Si system at $T = 800°$C (Figure 4.22b). Under these conditions, the

4 Nucleation in a Concentration Gradient

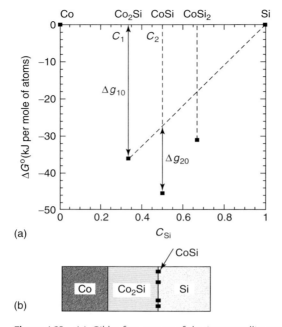

Figure 4.21 (a) Gibbs free energy of the intermetallic compounds in the Co–Si system at $T = 600$ K. Δg_{10}, Δg_{20} are the driving forces per mole of atoms for the reactions $2\text{Co} + \text{Si} \rightarrow \text{Co}_2\text{Si}$ and $\text{Co}_2\text{Si} + \text{Si} \rightarrow 2\text{CoSi}$, respectively; (b) cross-sectional sketch of the Co/Co$_2$Si/Si system showing localized nucleation of CoSi at the Co$_2$Si/Si interface.

suppression criterion of Co$_2$Si (phase 2) is given by the following equation:

$$\frac{\Delta x_{\text{CoSi}_2}}{l^{\text{cr}}_{\text{CoSi}_2}} < 5 \frac{c_{\text{CoSi}_2} - c_{\text{CoSi}}}{c_{\text{CoSi}}(c_{\text{Si}} - c_{\text{CoSi}})} \left(\frac{\Delta g_1}{\Delta g_2}\right)^2 = \frac{10}{3}\left(\frac{\Delta g_1}{\Delta g_2}\right)^2 \quad (4.93)$$

here Δg_1 and Δg_2 are the Gibbs free energy of reactions expressed via Equations 4.94 and 4.95, respectively:

$$\frac{3}{4}\text{Co}_{\frac{2}{3}}\text{Si}_{\frac{1}{3}} + \frac{1}{4}\langle\langle\text{Si}\rangle\rangle_{\text{Si,Ge}} \rightarrow \text{Co}_{\frac{1}{2}}\text{Si}_{\frac{1}{2}} \quad (4.94)$$

$$\frac{2}{3}\text{Co}_{\frac{1}{2}}\text{Si}_{\frac{1}{2}} + \frac{1}{3}\langle\langle\text{Si}\rangle\rangle_{\text{Si,Ge}} \rightarrow \text{Co}_{\frac{1}{3}}\text{Si}_{\frac{2}{3}} \quad (4.95)$$

and

$$\Delta g_1 = \Delta g_{10} - RT \ln a_{\text{Si}}^{1/4}, \quad \Delta g_2 = \Delta g_{20} - RT \ln a_{\text{Si}}^{1/3} \quad (4.96)$$

The standard Gibbs free energy of reactions given by Equations 4.94 and (4.95) at 800 °C are, respectively, $\Delta g_{10} = -16.2 \pm 1$ and $\Delta g_{20} = -1.9 \pm 1$ kJ per mol of atoms and can be easily calculated from the values of standard Gibbs free energy of formation of Co$_2$Si, CoSi, and CoSi$_2$ phases from pure silicon and cobalt at 800 °C

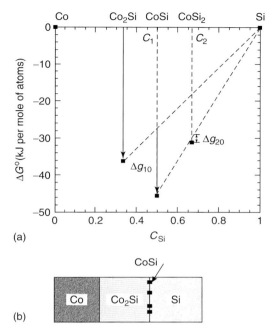

Figure 4.22 (a) Gibbs free energy of the formation of the intermetallic compounds in the Co–Si system at $T = 1073$ K. Δg_{10} and Δg_{20} are the driving forces per mole of atoms for the reactions $Co_2Si + Si \to$ 2CoSi and $CoSi + Si \to CoSi_2$, respectively. (b) Cross-sectional sketch of the $Co_2Si/CoSi/Si$ system showing localized nucleation of $CoSi_2$ at the CoSi/Si interface.

(Table 4.1). Here a_{Si} is the silicon activity in the Si, Ge solid solution, which may be considered as an ideal solution $a_{Si}=C_{Si}$ [31].

The determination, from Equation 4.93, of the critical thickness of the CoSi layer beyond which the nucleation of $CoSi_2$ between a CoSi layer and Si substrate may take place, shows that this thickness strongly depends on the composition of the Si–Ge solid solution, as observed experimentally (Figure 4.23b) in Ref. [28]. Even so, it was noted that these critical thickness values are very sensitive to the standard Gibbs free energy of formation of intermediate phases for which experimental errors are of the order of some kilojoules per mole of atoms. The presented here results were published first in Ref. [32].

4.7
Conclusions

A new possible explanation for phase suppression and for sequential phase growth in thin films is proposed in the context of the "nucleation in a concentration gradient" approach. Nucleation of intermediate phases at the initial stage of reactive diffusion is influenced by the sharp concentration and chemical potential gradients

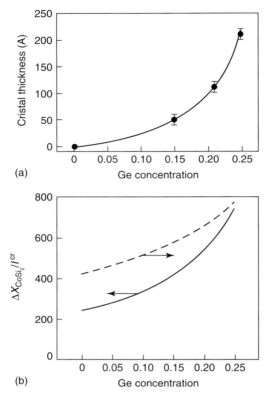

Figure 4.23 (a) Experimental values of the critical thickness for CoSi$_2$ nucleation in the Co/Si$_{1-x}$Ge$_x$ system ($\Delta X^{cr}_{CoSi_2}$) as a function of Ge concentration in a Si–Ge solid solution at $T = 800°C$ [29]; (b) Calculated values of $\Delta X^{cr}_{CoSi_2}$, from Equations 4.93–4.95 with $\Delta g_{10} = 16.2 \pm 1$; $\Delta g_{20} = -1.9$ (solid line) and $\Delta g_{20} = -2.9$ kJ/(mol of atoms) (dashed line) (Figure 4.22). l^{cr} is the critical size of the CoSi$_2$ nuclei.

in the vicinity of the contact interface. One should distinguish two major cases: case 1, full metastable solubility. In this case, the starting components have the same lattice and initially form a continuous diffusion profile. The new phase is trying to nucleate at the base of this continuous profile via one of the possible nucleation modes. Thermodynamics of nucleation is described in Equation 4.53 including the so-called gradient term, proportional to the fifth power of nucleus size. It leads to the possibility of additional suppression of nucleation by a sufficiently sharp concentration gradient. Case 2 refers to the case of limited solubility. If at least one phase with a narrow homogeneity range is already growing, the sharp chemical potential gradient in it strongly influences the nucleation barrier for the next phase to appear. In the expression of the Gibbs free energy change, the additional term, due to gradient concentration, appears to be proportional to the fourth power of size instead of the fifth power as obtained in previous models (developed for the case of a broad range of parent phase concentration). As a result, thermodynamic suppression of the new phase nucleation (in addition to kinetic suppression) may be effective as long as the thickness of the 'suppressing' phase remains less than a few tens or even hundreds of nanometers. Thus, the "gradient term effect" may well lead to the total absence of suppressed phases before consumption of the thin film by the "suppressing" growing phase. Comparison with available experimental data demonstrates that the approach presented fits at least reasonably the existing data.

References

1. Johnson, W.L. (1986) *Journal of Materials Science*, **30**, 81.
2. Vieregge, V. and Herzig, C. (1990) *Journal of Nuclear Materials*, **175**, 29.
3. Gusak, A.M. and Nazarov, A.V. (1992) *Journal of Physics-Condensed Matter*, **4**, 4753.
4. Philibert, J. (1989) *Defect and Diffusion Forum*, **66-69**, 995.
5. d'Heurle, F.M. (1988) *Journal of Materials Research*, **3**, 167.
6. Michaelsen, C., Barmak, K. and Weihs, T.P. (1997) *Journal of Physics D: Applied Physics*, **30**, 3167.
7. Emeric, E. (1998) *Etude des reactions a l'etat solide dans des multicouches Al/Co*. PhD thesis. Marseille.
8. Lucadamo, G., Barmak, K., Hyun, S. et al. (1990) *Materials Letters*, **39**, 268.
9. Coffee, K.R., Clevenger, L.A., Barmak, K. et al. (1989) *Applied Physics Letters*, **55**, 852.
10. Divinski, S.V. and Larikov, L.N. (1997) *Metal Physics and Advanced Technologies*, **6**, 40.
11. Gusak, A.M. (1990) *Ukrainskii Fizicheskii Zhurnal (Ukrainian Journal of Physics)*, **35**, 725.
12. Desre, P.J. and Yavari, A.P. (1990) *Physical Review Letters*, **64**, 1553.
13. Desre, P.J. (1991) *Acta Metallurgica*, **39**, 2309.
14. Johnson, W.C., White, C.L., Marth, P.E. et al. (1975) *Metallurgical Transactions A*, **6**, 911.
15. Gusak, A.M., Dubiy, O.V. and Kornienko, S.V. (1991) *Ukrainskii Fizicheskii Zhurnal (Ukrainian Journal of Physics)*, **36**, 286.
16. Bogatyrev, A.O. and Gusak, A.M. (2000) *Metal Physics and Advanced Technologies*, **18**, 897.
17. Christian, J.W. (1975) *The Theory of Transformation in Metals and Alloys*, Oxford University Press, Oxford.
18. Bogatyrev, O.O. (1999) *Nucleation at the initial stages of reactive diffusion*. PhD thesis. Institute for Metal Physics, Kiev, (in Ukrainian).
19. Gusak, A.M. and Bogatyrev, A.O. (1994) *Metallofizika I Noveishie Tekhnologii*, **16**, 28.
20. Gusak, A.M., Hodaj, F. and Bogatyrev, A.O. (2001) *Journal of Physics: Condensed Matter*, **13**, 2767.
21. Gusak, A.M., Lyashenko, Y.A. and Bogatyrev, A.O. (1995) *Defect and Diffusion Forum*, **129-130**, 95.
22. Hodaj, F., Gusak, A.M. and Desre, P.J. (1998) *Philosophical Magazine A*, **77**, 1471.
23. Hoyt, J.J. and Brush, L.N. (1998) *Journal of Applied Physics*, **78**, 1589.
24. Lee, B.-J., Hwang, N.M. and Lee, H.M. (1997) *Acta Materialia*, **45**, 1867.
25. Raichenko, A.I. (1981) *Mathematical Theory of Diffusion with Applications*, Naukova dumka Publishers, Kiev (in Russian).
26. Gas, P. and d'Heurle, F.M. (2000) Diffusion in silicides: basic approach and practical applications, in *Silicides: Fundamentals and Applications* (eds L. Miglio and F.M. d'Heurle), World Scientific, Singapore, pp. 34.
27. Lau, S.S., Mayer, J.W. and Tu, K.N. (1978) *Journal of Applied Physics*, **49**, 4005.

28. de Boer, F.R., Boom, R., Mattens, W.C.M. *et al.* (**1988**) Cohesion in metals: transition metal alloys, in *Cohesion and Structure* (eds F.R., de Boer and D.G. Pettifor), Elsevier Science Publishers, Amsterdam.
29. Gulpen, J.H., Kodentsov, A.A. and van Loo, F.J.J. (**1995**) *Zeitschrift Fur Metallkunde*, **86**, 530.
30. Boyanov, B.I., Goeller, P.T., Sayers, D.E. and Nemanich, R.J. (**1998**) *Journal of Applied Physics*, **84**, 4285.
31. Olesinski, R.W. and Abbaschian, G.I. (**1984**) *Bulletin of Alloy Phase Diagrams*, **5**, 180.
32. Hodaj, F. and Gusak, A.M. (**2004**) *Acta Materialia*, **52**, 4305.

5
Modeling of the Initial Stages of Reactive Diffusion
Mykola O. Pasichnyy and Andriy M. Gusak

5.1
Introduction

In the previous chapter, the general approach to problems of nucleation at the initial stages of diffusion zone evolution was outlined. When, in 1989–1990, the corresponding theory had been introduced, the direct experimental check of it was believed to be impossible. It was the theory that provided "post factum" the explanation for the reasons of the relative stability demonstrated by an amorphous layer growing instead of "legal" crystalline intermediate phases at the initial stages of reactive diffusion in such systems as nickel–zirconium or gold–lanthanum. Only new experimental techniques developed during the last decade (first, 3D atom probe – three-dimensional atomic tomography), which allow measuring the spatial distribution of atoms in solids with an accuracy of tenths of a nanometer, enabled a direct experimental investigation of solid state reactions in thin films with a thickness of several tens of a nanometer.

First studies that involved such methods have shown that previous theoretical ideas about the evolution of the processes in the initial stages of solid state reactions have to be thoroughly reconsidered, since the initial stage of a solid-state reaction is determined by the contact zone morphology, which, in turn, is influenced by diffusion and nucleation. At the same time, in the majority of cases, it is the initial stage during which the further path of the system's evolution is determined.

As a rule, there exist several possible paths of evolution. From the viewpoint of thermodynamics, one may treat the initial stage of reactive diffusion as a decomposition of a metastable solution, which takes place during interdiffusion, in the field of a variable concentration gradient. However, a purely analytical approach to the analysis of this topic presents certain difficulties resulting from the essential nonlinearity of the problem, size effects, the limitations of a deterministic approach, and the application of a hydrodynamic scale (diffusion approach) at nanoscales. In this connection, when treating the initial stages of reactions, computer modeling of diffusion processes at an atomic level is quite relevant. In this chapter, based on the theoretical interpretation of recent experimental

Diffusion-Controlled Solid State Reactions. Andriy M. Gusak
Copyright © 2010 WILEY-VCH Verlag GmbH & Co. KGaA, Weinheim
ISBN: 978-3-527-40884-9

observations of G. Schmitz et al. at the Institute of Materials Physics, Münster University, some new models of the initial stages of diffusion solid-state reactions are proposed.

In Section 5.2, phenomenological models of the first intermediate phase nucleation at reactive diffusion in a binary diffusion couple are presented, and methods of thermodynamic analysis of the process, based on the theory of nucleation in a concentration gradient field, are given. We have pioneered the explanation of experimental results using thermodynamic suppression of nucleation rather than kinetic factors [1].

In Section 5.3, computer and phenomenological models describing the mechanism, which is similar to diffusion-induced grain boundary migration (DIGM), for reactive diffusion at the lateral growth of intermediate phase islands are proposed. Analytical dependence of the layer thickness has been derived for the symmetrical case in the approximation of mere "chemical" driving force. Application of the model has been analyzed and it is shown that at lateral growth stage the "chemical" driving force of diffusion, caused by the concentration gradient along interfaces, is very important compared to that caused by the interface curvature gradient. The proposed model [2, 3] presents an alternative to the already existing Klinger–Brechet–Purdy model, since the latter is based on the interface curvature gradient.

In Section 5.4, the Monte Carlo method for the description of the initial stages of reactive diffusion at atomic level, with the formation of several ordered phases, is presented. The peculiarity of the model lies in taking the dependence of pair interaction energy upon the local environment into account [4]. On the basis of the model, the kinetics of the ordered-phase nucleation in multilayers was studied. This phase was actually diffusion nontransparent in its bulk. Using the above-mentioned modified Monte Carlo scheme, we found the same three stages that are usually distinguished in experiments – nucleation, lateral growth, normal growth. It was shown that the incubation period is caused by concentration preparation of the diffusion zone, which indeed confirms the theory of nucleation in the concentration gradient field. On the basis of the computer model, we have established a size effect which is rather interesting from the experimental point of view. Namely, it is the dependence of incubation time of phase formation on the multilayer period – beginning from a certain characteristic value, the decrease of the period leads to a decrease in the incubation time [5].

5.2
First Phase Nucleation Delay in Al–Co Thin Films

While investigating reactive diffusion in a thin film of Al–Co by using the high-resolution method of atomic tomography it was established that nucleation of the first phase Al_9Co_2 is possible only in the case of initial components interdiffusion to a depth of not less than 4 nm [6]. A numerical analysis of the nucleation possibility at the initial stage of reaction between Al and Co was carried out. Microscopic nucleation mechanisms in the diffusion zone were considered and it was realized that the polymorphic mode of nucleation provides a quantitative agreement with the experimental data.

5.2.1
The Problem of Nucleation in a Concentration Gradient Field

In Chapter 4, different modifications of nucleation theory in a concentration gradient field are described. Using the thermodynamic approach, we have introduced the notion of a critical concentration gradient above which nucleation becomes thermodynamically prohibited. Different microscopic schemes (nucleation modes) have been applied to the description of the nucleation mechanism.

Here, three modes (mechanisms) of the nucleation in a concentration gradient field are presented: the polymorphic mode, the transversal mode, and the longitudinal or total mixing mode. Only two of the above-mentioned modes [7–9] consider thermodynamic suppression of a phase in a concentration gradient field, while the third involves the increase of the thermodynamic driving force as a result of the gradient. Grounding upon a thermodynamic analysis, one may assume that nature will stick to the way with the lowest nucleation barrier; therefore, the total mixing mode would become the most probable one. However, if the kinetic aspects are taken into account, the situation becomes more complex, since the mobilities of the components appear too low to complete total mixing inside the nucleus. So far, there is no theoretical description of such an intricate process. Thus, the problem of realization of a particular mode in real experiments remains unsolved. The first attempt to take into account the superposition of different modes was performed by Gusak, Hodaj, and Bogatyrev [10].

Up to recent times, there have not been any experiments allowing confirmation a specific nucleation mode. This peculiarity is caused by concentration gradients (for the first phase) predicted by the theory, which are too high, and, thus, one needs to investigate diffusion zones where the characteristic thickness does not exceed several nanometers. In order to do so we must carry out the analysis with spatial resolution of about interatomic distance. In marginal cases, the critical thickness of the diffusion zone may be less than the lattice parameter of the expected phase. In such situations, thermodynamic suppression of nucleation due to the reduction of the driving force is not likely to take place at all.

One of the modern high-accuracy experimental methods is tomography atom probe (TAP). It enables the investigation of reactions in thin metallic diffusion couples by means of spatially resolved chemical analysis with a local accuracy up to interatomic distance [11, 12]. This technique was also applied to reactive diffusion in the thin film system Al–Co within the temperature range of 200–400 °C [6]. In accordance with the previous results obtained by the authors [13, 14] by DSC, the first phase to nucleate and grow in the temperature interval from 200 to 400 °C is Al_9Co_2. The authors of [6] have established that in the diffusion zone between Al and Co formation and growth of a solid solution layer occurs first, and nucleation of Al_9Co_2 particles is never observed before the diffusion zone reaches about 3–4 nm thickness. Thus, the appearance of the product phase is controlled by the width of the intermixing layer, that is, by a certain critical value of the concentration gradient in the diffusion zone. Usually, the first phase (stable or metastable) forms immediately and changes the conditions for the formation and growth of the second phase, so that the latter may be suppressed owing to kinetic [15] or thermodynamic [16] reasons.

According to d'Heurle [17], the fastest growing phase is also usually nucleated first, since this phase is distinguished by high atomic mobility. In our case Al_9Co_2 is clearly the first and the fastest growing phase. Therefore, it is difficult to explain the retardation of just this phase by kinetic arguments. As a consequence, thermodynamic suppression seems to be the most probable mechanism of inhibition. The total driving force for the formation of Al_9Co_2 from the pure elements is rather high. Furthermore, the specific interfacial energy between the nucleus and parent phase is quite moderate so that the theoretical limit of the nucleation thickness is estimated as merely $d = 2\sigma/\Delta g \approx 0.2$ nm; thus, one should not observe any barrier to nucleation in the homogeneous medium at all. However, according to the concept of critical gradient, the total driving force cannot be applied in full to the nucleation process. By a comparison of the behavior predicted by the suggested nucleation modes with the experimental one, we will figure out, which mode, if any, will provide a realistic description of the nucleation event. Since only the polymorphic and transversal modes predict a thermodynamic suppression by sharp concentration gradients, we present a numerical analysis of the nucleation of Al_9Co_2 via transversal and polymorphic modes.

5.2.2
Basic Model

According to the atom probe analysis, the concentration profile of the couple Al–Co, prior to nucleation of the first product phase, is continuous and can be well approximated by the error-function-shaped solution of Fick's laws (Figure 5.1), thus

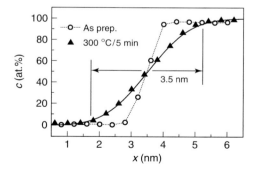

Figure 5.1 Composition profiles of Co determined normal to the Al–Co interface in the initially prepared state and after 5 min of annealing at 300 °C [6]. The solid line represents the approximated error function.

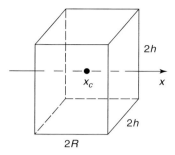

Figure 5.2 New phase nucleus with the shape of a rectangular parallelepiped with a size of $2R \times 2h \times 2h$ and its geometrical center at the point x_c. The concentration gradient in the parent phase is directed along the x axis.

$$c(x, L) = \frac{c_r + c_l}{2} + \frac{c_r - c_l}{2} \mathrm{erf}\left(\frac{x}{L}\right) \qquad (5.1)$$

where c_r and c_l denote the atomic fractions of Co on the "right" and "left" terminating layers. In order to avoid numerical problems with chemical potentials of very dilute solutions, the terminating compositions are chosen to be 0.01 and 0.99 instead of 0 and 1. L denotes the characteristic length of the diffusion zone. For comparison with the experimental data, we define the width K of the diffusion zone as the distance $x_2 - x_1$ between the positions with the concentrations $c(x_1, L) = 0.05$, $c(x_2, L) = 0.95$. L and K are connected via $K = 2.46 L$.

We consider the possibility of nucleation of Al_9Co_2 in an inhomogeneous concentration field. The nucleus of the intermetallic phase is supposed to be a rectangular parallelepiped with sizes of $2R \times 2h \times 2h$ (Figure 5.2). The thermodynamics of the diffusion zone before nucleation (inhomogeneous parent phase) is treated as an ideal solution with a Gibbs potential per one atom of the parent phase given by

$$g_0(c) = k_B T [c \ln(c) + (1 - c) \ln(1 - c)] \qquad (5.2)$$

where k_B and T are the Boltzmann constant and absolute temperature, respectively.

In a strict sense, one should model the solution at least as a regular one, but the formation enthalpy of the regular solution is much less than the driving force to form the intermetallic compound. Thus, neglecting the mixing enthalpy in the parent phase will not lead to any substantial error. For the intermetallic phase Al_9Co_2, we use the quadratic approximation of the Gibbs potential

$$g_n(c) = g_1 + \frac{\alpha}{2}(c - c_n)^2 \qquad (5.3)$$

where $c_n = 2/11$ denotes the ideal stoichiometric composition of the intermetallic; g_1 is the Gibbs free energy per atom for the stoichiometric composition, and the parameter α describes the curvature of the thermodynamic potential. Owing to the spatial concentration gradient, the Gibbs energy per atom as well as the local driving force of phase formation depends on the coordinates.

The formation of the nucleus within the concentration field leads to a change in the Gibbs free energy that can be written in general as

$$\Delta G(\varphi, R, x_c) = 4n\varphi^2 R^2 \int_{x_c-R}^{x_c+R} \Delta g(c(x))\, dx + 8\sigma \varphi R^2 (2+\varphi) \tag{5.4}$$

where φ is the shape factor, $\varphi = h/R$; n is the number of atoms per unit volume; σ is the energy of the interface (per unit area) between the nucleus and inhomogeneous matrix; $\Delta g(c)$ is the local driving force (per atom of the nucleus). The different nucleation modes are distinguished by different driving forces $\Delta g(c)$ and thus different nucleation barriers. However, the choice among these modes is controlled not only by the height of the nucleation barrier but by kinetic factors as well [10, 18]. In view of Cahn–Hilliard and modern phase field approaches, one may replace the last term on the right-hand side of Equation 5.4 by gradient energy terms restricting the expansion to a quadratic dependence on the gradients with respect to composition and of some order parameter describing the structural transformation.

As long ago as 1991, Desre [19] pointed out that in the transversal nucleation mode the influence of the concentration gradient energy on the nucleation barrier is negligible. Yet, in a polymorphic nucleation the composition gradient does not change at all. Thus, there is no first-order impact by the respective gradient energy too. The interface between the intermetallic and the parent lattice is reasonably assumed to be incoherent from the very beginning. Besides the incompatible lattice structures, this assumption is confirmed by the growth kinetics of the nuclei, which was observed in the TAP analysis and quantitatively discussed in [6]. As a consequence, the variation of the lattice structure is restricted to the narrow interphase boundary (IPB), so that a sharp interface characterized by a specific energy is a quite reasonable approximation. The change in molar volume by the lattice transformation is rather low, $\Delta V/V \approx 0.03$, and, as it is shown in [13], elastic contributions to the nucleation energy are negligible.

We treat the change in Gibbs free energy as a function of the shape parameter φ of the nucleus volume $v = 8\varphi^2 R^3$, and of the concentration c_x at the nucleus center x_c. The nucleation barrier ΔG_{cr} is defined by the saddle point of the surface, $\Delta G(\varphi, v, c_x)$, satisfying the set of the following conditions:

$$\frac{\partial \Delta G}{\partial \varphi} = 0, \quad \frac{\partial^2 \Delta G}{\partial \varphi^2} > 0, \quad \frac{\partial \Delta G}{\partial v} = 0 \tag{5.5}$$

$$\frac{\partial^2 \Delta G}{\partial v^2} < 0, \quad \frac{\partial \Delta G}{\partial c_x} = 0, \quad \frac{\partial^2 \Delta G}{\partial c_x^2} > 0$$

To determine the nucleation barrier ΔG_{cr} numerically, we proceed in the following way. The possibility of nucleation at each position of the parent solution is considered. At this position of the nucleus center we increase (step by step) the nucleus volume, optimizing the shape parameter at each step. Finally, the position corresponding to the minimum nucleation barrier is chosen.

Table 5.1 Total driving force per atom for the formation of intermetallic phases in the Al–Co system.

T (K)	$\Delta G_{Al_9Co_2}$ (J/atom)	$\Delta G_{Al_{13}Co_4}$ (J/atom)
298	-5.5×10^{-20}	-6.5×10^{-20}
573	-5.0×10^{-20}	-6.0×10^{-20}
773	-4.7×10^{-20}	-5.6×10^{-20}

Data at 298 K and 773 K were taken from [8]. The important value for 573 K was obtained by linear interpolation.

Table 5.2 Numerical parameters for the Al_9Co_2 phase at 573 K used for the calculation.

n (atom/m³)	T (K)	g_1 (J/atom)	σ (J/m²)
6.6×10^{28}	573	-5.0×10^{-20}	0.35

The thermodynamic parameters required for the calculation were taken from the data published in [8]. Since the nucleation of Al_9Co_2 was observed experimentally at 573 K, the Gibbs enthalpy of phase formation for this temperature was determined by a linear interpolation of literature data [13] (Table 5.1). The parameters used for the numerical evaluation are given in Table 5.2.

With the above explained algorithm, the composition and size dependence of the Gibbs nucleation free energy were calculated for the two nucleation modes under consideration and for various interdiffusion widths. As already mentioned, both polymorphic and transversal modes are distinguished by different driving forces and may take place only in restricted compositional ranges. A schematic overview indicating the driving forces and composition ranges is shown in Figure 5.3.

5.2.3
Transversal Mode

The transversal mode was introduced by Desre [8]. In this approach the nucleus of the new phase is constructed from thin transversal slices (perpendicular to gradient direction), each of which is formed as a result of redistribution of the components between new and parent phases only inside this slice. The composition of the parent phase varies from slice to slice due to the compositional gradient, and, thus, the local driving force varies as well. Therefore, the composition of the new phase and the driving force is determined individually for each slice from the composition of the respective parent phase according to the rule of parallel tangents.

Since the phase Al_9Co_2 has a very narrow homogeneity range, a new phase may be considered as a line compound, so it is assumed that $\alpha \to \infty$. Thus, we have

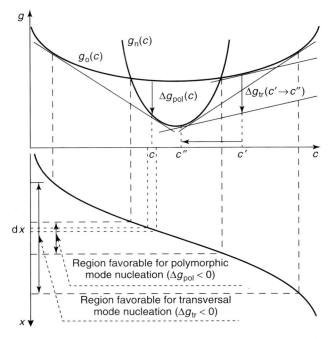

Figure 5.3 Schematic demonstrating the thermodynamic basis of the different possible nucleation modes. Here, $\Delta g_{pol}(c)$ is the driving force of the polymorphic transformation without a change of local composition $c(x)$ in a slice $dx = dc/(\partial c/\partial x)$. $\Delta g_{tr}(c' \to c'')$ represents the local driving force for the transversal mode, for which a slice of the new phase with composition c'' is formed by the transversal redistribution from a slice of the parent phase with composition c'.

for the driving force per atom of the nucleus

$$\Delta g(c) = g_1 - g_o(c) + (c - c_n)\left.\frac{\partial g_o}{\partial c}\right|_c \tag{5.6}$$

A numerical analysis of Equation 5.4 with $\Delta g(c)$ in the form of Equation 5.6 was done for different widths of the diffusion zone down to $K = 0.5$ nm. We considered two different cases: (i) without shape optimization (nucleus being a cube $2R \times 2R \times 2R$ and the shape parameter is fixed at $\varphi = 1$) and (ii) with shape optimization, that is, for each nucleus volume φ is determined to minimize ΔG. The calculated results demonstrate, in both cases, that even at a very narrow interdiffusion width the nucleation barrier is rather low, about $20 k_B T$ (Figure 5.4), and the critical size $2R_{cr}$ of the nucleus amounts to 0.45 nm, which nearly coincides with the value of classical theory and implies practically immediate nucleation.

This result obviously contradicts the experimental observation that the Al_9Co_2 phase appears only for a diffusion zone larger than 3 nm. Moreover, the mentioned critical size is less than the size of one structural unit cell, which actually includes 22 atoms, and its lattice parameters range from 0.6 to 0.8 nm. Therefore, we can

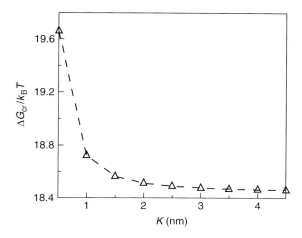

Figure 5.4 Nucleation barrier of Al_9Co_2 nucleus formation at the Al–Co interface calculated by the transversal mode without shape optimization for different diffusion zone widths.

conclude that the transversal mode is not relevant for the nucleation of Al_9Co_2 in Al–Co couples.

5.2.4
Polymorphic Mode

The polymorphic mode of nucleation is characterized by the conservation of the concentration profile, which is a reasonable assumption if the system has no time to redistribute atoms inside and outside the nucleus in the process of lattice transformation. For the polymorphic mode, we also consider the two cases discussed earlier: (i) without shape optimization ($\varphi = 1$) and (ii) with shape optimization.

In the case of polymorphic mode nucleation, unlike the transversal one, a new phase cannot be taken as linear, since for the given mode the homogeneity range determines the region of the diffusion zone where nucleation is possible. Thus, α in Equation 5.3 is a model parameter which characterizes the curvature of the thermodynamic potential for Al_9Co_2 and is a finite constant in the case of parabolic approximation, and equals $\partial^2 g_n / \partial c^2$. Regarding that the local concentrations do not change during nucleation, the driving force is represented by

$$\Delta g(c) = g_n(c) - g_o(c) \tag{5.7}$$

For each value of the curvature parameter α one finds a critical zone width K_{cr}, at which nucleation by the polymorphic mode becomes possible. We define this critical width by postulating that the nucleation barrier becomes lower than $60 k_B T$ for all $K > K_{cr}(\alpha)$ (nucleation allowed). No direct measurements of the curvature of the Gibbs potential of the Al_9Co_2 phase are known. Therefore, we first determine by our calculations the value of α so that the predicted critical width agrees with the experimental finding $K_{cr} = 3.5$ nm and we discuss the relevance of this numerical value later.

5.2.4.1 Polymorphic Mode without Shape Optimization

The numerical analysis at $\alpha = 8 \times 10^{-18}$ J/atom leads to the following results: At a zone width of $K_1 = 3$ nm, the surface $\Delta G(v, c_x)$ has no saddle point at all (Figure 5.5 a). For any c_x the dependence $\Delta G(v)$ is monotonically increasing.

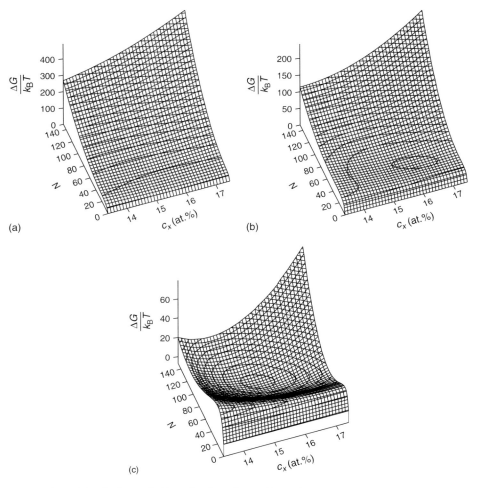

Figure 5.5 Surface $\Delta G(N, c_x)$ for the polymorphic mode without shape optimization ($\varphi = 1$) at $\alpha = 8 \times 10^{-18}$ J/atom for diffusion zone widths: (a) $K_1 = 3$ nm; (b) $K_2 = 3.5$ nm; (c) $K_3 = 4$ nm. Here, N is the number of atoms in the new phase nucleus ($N = nv$) and c_x is the atomic fraction of Co at the center of the nucleus.

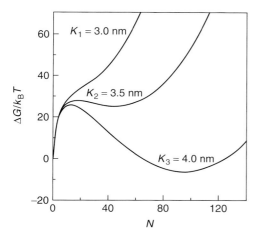

Figure 5.6 Section through the $\Delta G(N, c_x)$ surfaces for the polymorphic mode without shape optimization at a composition $c_x = 15$ at.% ($\alpha = 8 \times 10^{-18}$ J/atom, calculation for different diffusion zone widths: $K_1 = 3$ nm; $K_2 = 3.5$ nm; $K_3 = 4$ nm).

In other words, at this zone width, the nucleation of Al_9Co_2 is still thermodynamically suppressed.

Even a slight increment of the zone width to $K_2 = 3.5$ nm leads to the formation of a local minimum at the surface $\Delta G(v, c_x)$ (Figure 5.5 b), which is still of positive magnitude corresponding to a metastable state (Figure 5.6). A saddle point is found which allows determining a nucleation barrier. At $K_3 = 4$ nm, the minimum crosses the zero level and becomes negative (Figure 5.5 c), meaning that the intermediate phase becomes stable. The corresponding nucleation barrier ΔG_{cr} appears to be only $25 k_B T$ (Figure 5.6). The state of the local minimum (Figure 5.5 c) corresponds to a nucleus containing about 100 atoms (linear size about 1.2 nm). Interestingly, the composition at its geometrical center amounts to 15 at.% in significant deviation from the intermetallic's stoichiometry (18.2 at.%), which is due to the asymmetry of the error-function-shaped diffusion profile with respect to the nucleus center. The mentioned nucleation barrier is rather low, which means that immediately after the diffusion zone, in the parent phase, reaches a width of 3.5 nm, nucleation starts, and the intermediate phase appears.

5.2.4.2 Polymorphic Mode with Shape Optimization

If the system has the kinetic opportunity of shape optimization, the resulting nucleation barrier decreases, as expected. In the case of full optimization, the surface $\Delta G(v, c_x)$ does not reveal a minimum. After passing the saddle point, it is always favorable to increase the volume by transversal growth, but keeping limited the longitudinal size along the direction of the gradient. A critical diffusion zone width, close to the experimental value, is obtained at $\alpha = 4 \times 10^{-17}$ J/atom. This result implies that, for this parameter α and the diffusion zone width $K = 3.5$ nm,

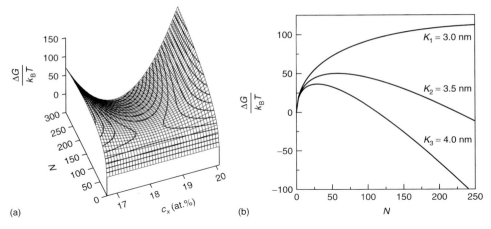

Figure 5.7 Polymorphic mode with shape optimization at $\alpha = 4 \times 10^{-17}$ J/atom: (a) surface $\Delta G(N, c_x)$ for a diffusion zone width 3.5 nm; (b) section $\Delta G(N, c_x)$ at $c_x = 18$ at.% for diffusion zone widths $K_1 = 3$ nm, $K_2 = 3.5$ nm, and $K_3 = 4$ nm (N is number of atoms in the new phase nucleus, c_x is the atomic fraction of Co at the center of the nucleus).

the nucleation barrier appeared to be $50k_B T$ (Figure 5.7), which corresponds to the upper limit for a barrier that allows nucleation to proceed at reasonable time scales. The calculated longitudinal size of the critical nucleus amounts to 0.4 nm. Since this value is still smaller than the dimension of a single unit cell of the Al_9Co_2 phase, it indicates that a full shape optimization during nucleation is a somewhat unrealistic assumption, because a viable nucleus should have a linear size not less than a single unit cell, along all three directions.

Full shape adaptation may be limited by kinetics and geometry of the process. The most probable is the case when the shape optimization is restricted to the following: the linear size of the supercritical nucleus must not be less than the size of a single unit cell (0.7 nm). In this case, the critical width of the diffusion zone which is close to the experimental result is obtained at $\alpha = 1.5 \times 10^{-17}$ J/atom. When the diffusion zone is 3 nm wide, the barrier amounts to $100k_B T$ (nucleation is prohibited), and at 4 nm to $30k_B T$ (nucleation is possible).

5.2.5
Discussion and Conclusions

The presented model calculations indicate that, indeed, the concentration gradient concept based on a polymorphic nucleation mode may quantitatively describe the observed retardation of the first product Al_9Co_2 in the case of Al–Co. However, this statement is bound to the curvature of the Gibbs potential of the intermetallic phase. It is important to check whether the figures used for the calculation, with α of about 10^{-17} J/atom, are reasonable. Since no direct measurements or explicit calculations of the respective Gibbs energy function are known, we have

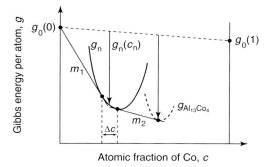

Figure 5.8 Schematic phase stabilities, Gibbs energies of phase formation, and equilibrium conditions of the terminating Al and the intermetallic phases Al_9Co_2 and $Al_{13}Co_{14}$.

to estimate this value from the known data of phase stabilities. The approximate value of $\partial^2 g_n/\partial c^2$ can be found from

$$\frac{\partial^2 g_n}{\partial c^2} \approx \frac{\left.\frac{\partial g_n}{\partial c}\right|_{c_R} - \left.\frac{\partial g_n}{\partial c}\right|_{c_L}}{\Delta c} \tag{5.8}$$

where $\Delta c = c_R - c_L$, c_R and c_L the compositions of the considered intermetallic compound in equilibrium to the "right-hand" neighbor phase $Al_{13}Co_4$, and to the "left-hand" terminating Al, respectively (Figure 5.8).

Furthermore, the values of the derivatives with respect to composition, $m_2 = \partial g_n/\partial c|_{c_R}$ and $m_1 = \partial g_n/\partial c|_{c_L}$, are fixed by the phase equilibrium according to the rule of common tangents. So, we have

$$\left.\frac{\partial g_n}{\partial c}\right|_{c_L} = \frac{g_n(c_n) - g_o(0)}{c_n} \tag{5.9}$$

and

$$\left.\frac{\partial g_n}{\partial c}\right|_{c_R} = \frac{g_{Al_{13}Co_4}(c_{Al_{13}Co_4}) - g_n(c_n)}{c_{Al_{13}Co_4} - c_n} \tag{5.10}$$

Since, Al_9Co_2 is a line compound, its range of existence in the composition space, Δc, is not exactly known, but it is hardly larger than 1 at.%, since otherwise this range would have been shown in the published phase diagrams. For the given marginal case, according to the approximations (Equations 5.8–5.10) and thermodynamic data (Table 5.1), the value of $\partial^2 g_n/\partial c^2$ is found to be

$$\alpha = 8.8 \times 10^{-18} \text{ J/atom} \tag{5.11}$$

Thus, it turns out that the curvature parameter required for the numerical calculation to obtain critical gradients in the order of the experiment is quite reasonable and corresponds to the thermodynamic situation of the Al–Co system. Especially, for the polymorphic nucleation mode without shape optimization, we find almost exact agreement. The value of $\alpha = 8 \times 10^{-18}$ J/atom for the polymorphic mode without optimization corresponds to a homogeneity range of $\Delta c = 1.1$ at.%. At full shape optimization ($\alpha = 4 \times 10^{-17}$ J/atom) the concentration range is $\Delta c = 0.22$ at.%. Thus, both marginal cases still seem reasonable.

Thus, we come to the following conclusions: Phenomenological models for nucleation of the first intermediate phase at reactive diffusion in a binary diffusion couple are presented. The methods of thermodynamic analysis of the process are described. They are based on the theory of nucleation in the concentration gradient field. The suggested models allowed us to study the delay in the occurrence of the first intermediate phase Al_9Co_2 in the system Al–Co. The numerical analysis of the process aimed at confirming the correctness of the concept of critical gradient (based only on the thermodynamic analysis) while explaining the delay of the first phase appearance and at establishing nucleation mechanisms.

Microscopic nucleation mechanisms (modes) in the diffusion zone were treated. It was found that the delay of the first product phase Al_9Co_2 in the thin film diffusion couple Al–Co can be the result of thermodynamic suppression of nucleation in the field of a concentration gradient. Experimental data on the delay of the first phase in Al–Co couple exclude the transversal mode from possible nucleation mechanisms of the system. However, this mechanism is quite possible for other systems and/or phases. Experimental data are explained by the realization of the polymorphic mode, provided the homogeneity range for Al_9Co_2 is located within the concentration range, $\Delta c = 0.22$–1.1 at.%. Thus, it was proven for the first time that the delay of the first intermediate phase Al_9Co_2 in the Al–Co system is stipulated by high concentration gradient in the contact zone, that is, nucleation occurs at a polymorphic transformation only in a concentration-prepared diffusion zone. The possibility of the polymorphic mode of nucleation in the field of a concentration gradient has been proven.

5.3
Kinetics of Lateral Growth of Intermediate Phase Islands at the Initial Stage of Reactive Diffusion

In this section, the interface of reacting components A–B overgrown by intermetallic A_1B_1 islands at frozen interdiffusion is analyzed. The growth occurs due to diffusion along moving interfaces A–A_1B_1 and B–A_1B_1. We demonstrate below that in most cases of diffusion, the "chemical" driving force along the interface boundaries contributes more than their curvature gradient. The asymptotic thickness of the islands is determined by the ratio of kinetic factors, the diffusivity along IPB versus the reaction constant.

5.3.1
Problem Formulation

At the initial stages of reactive diffusion a fast growth of intermediate phases is often observed. In particular, in DSC experiments, two peaks are distinctly registered

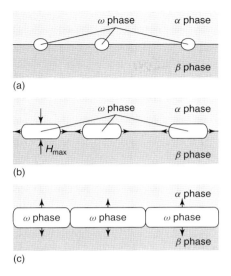

Figure 5.9 Schematic growth of ω-phase islands between α and β phases: (a) nucleation; (b) lateral growth; (c) normal growth.

on the diagram [14, 20, 21], indicating that the process must have two stages. The authors of [20, 21] explain the first peak as a stage of nucleation and lateral growth of an intermediate phase, and the second as the normal growth of a new phase layer.

In experimental works [22, 23], it was discovered that at the end of the lateral growth stage, the newly formed phase layer has a thickness of 7–8 nm, considerably exceeding the size of the critical nucleus. Lateral growth proceeds much faster than normal grain growth; thus, these two stages involve different mechanisms. Nucleation and lateral growth (Figure 5.9a,b) take place mainly due to diffusion of the components, required for new phase formation, along the moving boundaries between a new phase and the initial ones. This mechanism is similar to DIGM. Normal growth is limited by a slower mechanism, the diffusion through the new phase bulk, while at low temperatures, when bulk diffusion is frozen, the growth is possible only due to diffusion of the components along the grain boundaries of a newly formed phase (Figure 5.9c).

The detailed analysis of the lateral growth stage was realized in [24–26]. The authors of [24] have studied the process of intermetallic growth at a peritectoid reaction; however, the driving force for the new phase growth is considered to be stipulated by the curvature of the IPB, while the concentration profile along it is not taken into account. The authors argue that ordered phases have a very narrow homogeneity region and a concentration change at the very boundary can be neglected. But, it is the difference between chemical potentials that is known to be the real driving force of diffusion, not the concentration gradient, and the mentioned difference considerably changes for intermetallic phases in the homogeneity region (the product of a small concentration interval and a very large thermodynamic factor results in noticeable change of the chemical potential). So, our opinion in this regard is one must not neglect the "chemical" driving force along IPB. In [25, 26], both IPB curvature gradient and concentration gradient along IPB were taken into account, and the numerical analysis for the model of intermetallic island-like growth at reactive diffusion was performed.

In this section, we discuss a case of purely "chemical" driving force and analyze the limits within which the given model can be applied. Furthermore, we show that this driving force may appear to be the governing one in some cases.

5.3.2
Physical Model

Let us consider the process of intermetallic phase ω formation between pure components A (α phase) and B (β phase). Let the temperature be so low that bulk diffusion in α, β, and ω is frozen, but can take place along the moving IPBs $\alpha-\omega$ and $\beta-\omega$. The IPB is a layer of a certain thickness δ and concentration close to that of a new phase. We treat the diffusion along IPB (similar to Cahn's scheme for cellular precipitation) as "surface" diffusion on the ω-phase surface; that is, at IPB motion the concentration of components is "inherited" (frozen) by the intermediate phase.

Similar to the work carried out in [25, 26], we consider a symmetrical case when the ω phase has such a stoichiometry as $1:1$ ($c_\omega = 1/2$), the surface tensions $\sigma_{\alpha\omega}$ and $\sigma_{\beta\omega}$ are equal ($\sigma_{\alpha\omega} = \sigma_{\beta\omega} \equiv \sigma_\omega$), and the diffusion along $\alpha-\omega$ and $\beta-\omega$ IPB proceeds at equal rates. Owing to symmetry, at $\alpha-\beta-\omega$ triple junction, the B component concentration at IPB (from the ω-phase side) must be equal to $c_\omega = 1/2$. At planar regions of IPB, equilibrium concentrations are found with the help of the common tangent rule and are equal to $c_{\alpha\omega} = c_\omega - \Delta c_\omega/2$ between α and ω phases; $c_{\beta\omega} = c_\omega + \Delta c_\omega/2$ between β and ω phases, where Δc_ω is the homogeneity region for the ω phase (Figure 5.10). The indicated concentration change corresponds to

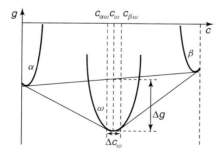

Figure 5.10 Dependences of Gibbs potential (per atom) for the intermetallic phase ω and the initial phases α and β. Common tangents specify the equilibrium concentrations between α, β, and ω phases, and the homogeneity region for the ω phase. The value Δg defines the driving force for the $\alpha + \beta \rightarrow \omega$ transformation.

5.3 Kinetics of Lateral Growth of Intermediate Phase Islands at the Initial Stage of Reactive Diffusion

the change of the reduced chemical potential

$$\tilde{\mu} = \mu_B - \mu_A = \partial g / \partial c$$

which is the diffusion driving force along IPB.

Owing to diffusion along IPB, concentrations will appear different from equilibrium, that is, supersaturation has to be accounted for. As a result of this effect, the motion of the boundary with "immuring" of atoms into the new phase will occur. From the problem specification, IPB will have the center of symmetry (center of nucleus), which we take as the origin of the coordinate system, and direct the x axis along the α–β planar interface. Now, it is sufficient to consider the part of the interface limited by the coordinate quarter. Let us take the IPB part between the α and ω-phases, located in the second quadrant, for example. From here onward we will consider only this part.

The boundary concentration can change only in two ways: first, via the diffusion fluxes along IPB, and second, via the boundary motion (Figure 5.11). The concentration profile $c(l, t)$ change along the moving boundary is described by Equation 5.12, formally corresponding to the law of conservation of matter

$$\frac{\partial c}{\partial t} = -\frac{1}{n} \frac{\partial j}{\partial l} - \frac{cU}{\delta} \tag{5.12}$$

where j is the flux of the B component along the boundary, l is the curvilinear coordinate, counted from the contact point of α–β–ω along IPB in the α–ω direction, n is the number of atoms per unit volume.

We will consider the rate of boundary motion $U(c)$ in the direction normal to IPB to be proportional to the difference between the local concentration at the boundary c and the equilibrium concentration in the ω phase (at α–ω contact) $c_{\alpha\omega}$ [25, 26]

$$U(c) = k(c - c_{\alpha\omega}) \tag{5.13}$$

Here, k is the reaction constant. In the case of a concentration gradient (proportional to the chemical potential gradient) being the diffusion driving force

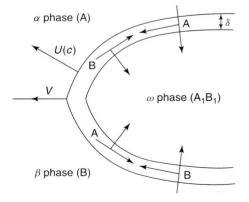

Figure 5.11 Schematic of interphase boundary and components' fluxes.

along IPB, the expression for the flux will be

$$j = j_{\text{chem}} = -n\tilde{D}\frac{\partial c}{\partial l} \qquad (5.14)$$

where $\tilde{D}$ is the reduced diffusivity. For the phase with a too narrow homogeneity region, the reduced diffusivity formally tends to infinity (because of the thermodynamic factor), and the concentration gradient tends to zero; yet, the flux remains finite [27, 28].

5.3.3
Numerical Results

The IPB evolution was investigated by using the computer model, the main features of which are given below. Consider IPB as a discrete set of sections, each of them being characterized by the sequence number, i, the average concentration, $c(i)$, and its center coordinates. The concentration change for each section is determined by the numerical solution of Equation 5.12 in finite-difference algorithm. The motion of the section in the direction normal to the tangent is realized simultaneously, at a velocity defined by Equation 5.13. For the numerical solution of the problem new variables were used: $\zeta = l/M$ is the dimensionless curvilinear coordinate, where $M = \sqrt{\tilde{D}\delta/k}$ holds, and $\tau = kt/\delta$ is the dimensionless time. Therefore, all spatial quantities must be treated within a new dimensionless coordinate system. The curvilinear coordinate ζ is counted from the point of triple junction ($\zeta = 0$) and is limited by the IPB length $\zeta_{\max}(\tau)$ in the given quadrant. Having employed new variables, we get Equation 5.12 as

$$\frac{\partial c}{\partial \tau} = \frac{\partial^2 c}{\partial \zeta^2} - c(c - c_{\alpha\omega}) \qquad (5.15)$$

The velocity in dimensionless form reads

$$u = \frac{\delta}{M}(c - c_{\alpha\omega}) \qquad (5.16)$$

The numerical solution of the problem, given by Equations 5.15 and 5.16, was accomplished at the boundary conditions

$$c(\zeta = 0) = 1/2, \qquad \theta(\zeta = 0) = \pi/4,$$

$$\left.\frac{\partial c}{\partial \zeta}\right|_{\zeta=\zeta_{\max}(\tau)} = 0, \qquad \theta(\zeta = \zeta_{\max}(\tau)) = 0$$

where θ is the angle between the x axis and the tangent to IPB. The section of $\zeta_{\max}(\tau = 0) < 1$ length, intercepting a certain distance on abscissa and ordinate axes, was taken as the initial shape of IPB. A linear dependence from $c(\zeta = 0) = c_\omega$ to $c(\zeta = \zeta_{\max}) = c_{\alpha\omega}$ with respect to ζ was chosen to be the initial concentration profile along IPB.

While carrying out the computer experiment, the following characteristics of IPB kinetics were discovered: Irrespective of the initial conditions a quasi-stationary

5.3 Kinetics of Lateral Growth of Intermediate Phase Islands at the Initial Stage of Reactive Diffusion

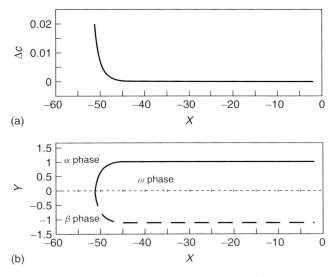

Figure 5.12 Results of the analysis of the numerical model: (a) dependence of supersaturation at IPB on coordinate; (b) IPB shape, solid line indicates IPB between α and ω (obtained numerically), dashed line indicates the IPB between β and ω (plotted symmetrically to α–ω IPB). The dependences are given for $\tau = 3.8 \times 10^4$ at $\Delta\xi = 0.1$, $\Delta\tau = 2 \times 10^{-4}$, $\delta/M = 0.05$.

regime of the process is established with time. During this process the concentration profile $c(\zeta)$ and the IPB shape $Y(\zeta)$ remain unchanged relative to the $\zeta = 0$ point. In the range $\zeta \gg 1$ the thickness of the ω-phase island becomes constant. The point of triple junction ($\zeta = 0$) moves at constant velocity, which leads to linear growth of the new phase bulk.

The IPB profile and the supersaturation obtained with the help of above-described model at $\tau = 3.8 \times 10^4$ for the parameters $\Delta\xi = 0.1$, $\Delta\tau = 2 \times 10^{-4}$, $\delta/M = 0.05$ are shown in Figure 5.12. We have managed to obtain the analytical solution for the steady-state case.

5.3.4
Analytical Solution for the Steady State

For the steady state, Equation 5.12 becomes a second-order differential equation for the composition profile $c(l, t)$ along the interface

$$\tilde{D}\frac{\partial^2 c}{\partial l^2} = \frac{k}{\delta} c(c - c_{\alpha\omega})$$

With respect to supersaturation, $\Delta c \equiv c - c_{\alpha\omega}$, we get

$$\frac{\partial^2(\Delta c)}{\partial l^2} = \frac{k}{\tilde{D}\delta}(c_{\alpha\omega} + \Delta c)\Delta c$$

A very narrow homogeneity range $\Delta c \ll c_{\alpha\omega}$ is a characteristic property of intermetallic phases; therefore, $c_{\alpha\omega} + \Delta c \approx c_\omega$ and we arrive at

$$\frac{\partial^2 (\Delta c)}{\partial l^2} = \frac{1}{L^2} \Delta c, \quad L = \sqrt{\frac{\tilde{D}\delta}{kc_\omega}} \quad (5.17)$$

where L is the characteristic length of the process. The differential equation Equation 5.17 with the boundary conditions

$$\Delta c|_{l=0} = c_\omega - c_{\alpha\omega} = \Delta c_\omega/2 \qquad \Delta c|_{l\to\infty} = 0$$

has the analytical solution

$$\Delta c = \frac{\Delta c_\omega}{2} e^{-\frac{l}{L}} \quad (5.18)$$

Strictly speaking, this solution is correct only for an island of infinite length, but, actually, we can apply it to finite islands with $l_{max} \gg L$ as well.

5.3.5
Asymptotic Thickness of an Island

We choose a reference system, whose zero coincides with the contact point of α–β–ω phases ($l = 0$). In the steady state, although the system moves constantly in the laboratory reference frame at certain velocity, V, the shape of the boundary $y(l)$ and concentration profile $c(l)$ along it remain unchanged. So, each physically small section of IPB gets an instantaneous velocity, $V'(l)$, directed along the tangent to the boundary (Figure 5.13).

According to the rule of velocity addition, we have

$$\sin \theta = \frac{U(l)}{V} = \frac{k\Delta c(l)}{V}.$$

At $l = 0$, we get

$$V = \frac{k\Delta c_\omega}{2 \sin \theta_0}$$

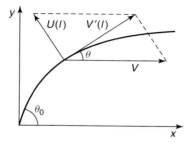

Figure 5.13 The rule of velocity addition in the reference frame of a moving IPB.

5.3 Kinetics of Lateral Growth of Intermediate Phase Islands at the Initial Stage of Reactive Diffusion

then

$$\sin\theta = \sin\theta_0 \exp\left(-\frac{l}{L}\right) \qquad (5.19)$$

The angle θ_0 is found from the condition of mechanical equilibrium for three phases in the triple junction point. In our case,

$$\sigma_{\alpha\beta} = \sigma_{\alpha\omega}\cos\theta_0 + \sigma_{\beta\omega}\cos\theta_0 = 2\sigma_\omega \cos\theta_0$$

correspondingly

$$\sin\theta_0 = \sqrt{1 - \frac{1}{4}\gamma^2} \qquad (5.20)$$

where $\gamma = \sigma_{\alpha\beta}/\sigma_\omega$.

Let us find now the $y(l)$ dependence integrating

$$y(l) = \int_0^l \sin\theta\, dl = L\sin\theta_0\left(1 - \exp\left(-\frac{l}{L}\right)\right)$$

then the asymptotic thickness of the α-phase layer overgrown by the new phase is

$$H_\alpha = \lim_{l\to\infty} y(l) = L\sin\theta_0 = \sqrt{\frac{\tilde{D}\delta}{kc_\omega}}\sin\theta_0$$

In the symmetrical case, the total thickness of the island becomes

$$H_{\max} = 2H_\alpha = \sqrt{\frac{\tilde{D}\delta}{kc_\omega}(4-\gamma^2)} \qquad (5.21)$$

Note that a "frozen-in" concentration profile in the new phase is linear in the direction perpendicular to the initial α–β contact

$$c(y) = c_\omega - \frac{\Delta c_\omega}{H_{\max}}y$$

Thus, in our model, the asymptotic thickness of the phase layer is determined by the characteristic size L, which, in turn, presents a combination of kinetic parameters $\tilde{D}$ and k. It is highly essential to analyze the behavior of this combination when the homogeneity region of the phase tends to zero. The interdiffusion coefficient at IPB may acquire different forms, but it is always proportional to the second-order derivative $\partial^2 g/\partial c^2$, which tends to infinity with Δc tending to zero [28]. The reaction constant k is also proportional to the same second-order derivative.

5.3.6
Estimates

Within the model we have assumed that the boundary curvature can be neglected. To check this assumption, let us analyze the ratio of the following fluxes: j_{curv} (caused by the curvature gradient) and j_{chem} (caused by the concentration gradient) in a quasi-stationary approximation. In the general case, we have

$$j_{\text{curv}} \approx \frac{cD^*}{k_B T}\frac{\partial(\sigma_\omega\Omega\rho)}{\partial l} = \frac{cD^*\sigma_\omega\Omega}{k_B T}\frac{\partial\rho}{\partial l} \qquad (5.22)$$

$$j_{\text{chem}} \approx \frac{cD^*}{k_B T} \frac{\partial \tilde{\mu}}{\partial l} = \frac{cD^*}{k_B T} \frac{\partial \tilde{\mu}}{\partial c} \frac{\partial c}{\partial l} = \frac{cD^*}{k_B T} \frac{\partial^2 g}{\partial c^2} \frac{\partial c}{\partial l} \quad (5.23)$$

To estimate the mentioned ratio we will use the shape of the boundary, obtained analytically without taking into account the gradient of curvature. In accordance with the geometry of the problem, the boundary curvature is $\rho = -d\theta/dl$.

Differentiation of Equation 5.19 gives

$$\cos\theta \, d\theta = -\frac{\sin\theta_0}{L} \exp\left(-\frac{l}{L}\right) dl = -\frac{\sin\theta}{L} dl$$

So, $\rho = \tan\theta / L$, and the gradient of curvature becomes

$$\frac{\partial \rho}{\partial l} = -\frac{\sin\theta}{L^2 \cos^3 \theta} \quad (5.24)$$

According to Equation 5.18 the concentration gradient will acquire the following form:

$$\frac{\partial c}{\partial l} = -\frac{\Delta c_\omega}{2L} \exp\left(-\frac{l}{L}\right) \quad (5.25)$$

Then

$$\frac{j_{\text{curv}}}{j_{\text{chem}}} \approx \frac{\sigma_\omega \Omega \frac{\partial \rho}{\partial l}}{\frac{\partial^2 g}{\partial c^2} \frac{\partial c}{\partial l}} = \frac{2\sigma_\omega \Omega \sin\theta_0}{\Delta c_\omega \frac{\partial^2 g}{\partial c^2} L \cos^3 \theta} \quad (5.26)$$

The ratio $j_{\text{curv}}/j_{\text{chem}}$ has a maximum at $\theta = \theta_0$. Thus, using Equation 5.21 and the almost obvious relation

$$\Delta c_\omega \left.\frac{\partial^2 g}{\partial c^2}\right|_{c_\omega} \approx \int_{\Delta c_\omega} \frac{\partial^2 g}{\partial c^2} dc = \frac{\Delta g}{c_\omega (1 - c_\omega)}$$

(see Chapter 4), we obtain

$$\frac{j_{\text{curv}}}{j_{\text{chem}}} = \frac{8\sigma_\omega \Omega c_\omega (1 - c_\omega)(4 - \gamma^2)}{\Delta g H_{\max} \gamma^3} \quad (5.27)$$

Experiments with multilayers give $H_{\max} \approx 10^{-8}$ m. After having reasonably estimated other parameters

$$\sigma_\omega \approx 10^{-1} \, \text{J/m}^2, \quad \Omega \approx 10^{-29} \, \text{m}^3, \quad \Delta g \approx 10^{-20} \, \text{J}, \quad \gamma \approx 1$$

we get

$$\frac{j_{\text{curv}}}{j_{\text{chem}}} = \frac{8 \times 10^{-1} \times 10^{-29} \times 0.5 (1 - 0.5)(4 - 1)}{10^{-20} \times 10^{-8}} \approx 6 \times 10^{-2} \ll 1 \quad (5.28)$$

Regarding Equation 5.21, one can formulate a general condition at which the contribution of the curvature gradient can be neglected (i.e., $j_{\text{curv}}/j_{\text{chem}} \ll 1$)

$$\frac{\tilde{D}(\Delta g)^2}{k} \gg \frac{(4 - \gamma^2)(8\sigma_\omega \Omega (1 - c_\omega))^2 c_\omega^3}{8\gamma^6} \quad (5.29)$$

Having estimated the fluxes ratio, we conclude that the IPB curvature gradient contribution into the driving force during the asymptotic stage can be neglected.

5.3.7
Conclusions

In Section 5.3, the phenomenological model for the evolution of the morphology of the IPB of intermediate phase islands at the lateral growth stage was proposed. The marginal case of the problem, involving intermetallic phase growth at reactive diffusion due to purely "chemical" driving force according to the mechanism similar to DIGM, was analyzed. It has been proved that the lateral growth of intermediate phase islands at reactive diffusion obeys a linear time dependence.

The thickness of the new phase layer depending on the kinetic parameters of the system was determined. The asymptotic thickness of the island is determined by the ratio of kinetic factors; it is proportional to the square root of the ratio of diffusivity along the IPB and the reaction constant at this boundary, which in the symmetrical case equals

$$H_{max} = \sqrt{\frac{\tilde{D}\delta}{kc_\omega}\left(4 - \left(\frac{\sigma_{\alpha\beta}}{\sigma_\omega}\right)^2\right)}$$

The curvature gradient contribution into the fluxes along IPBs and the corresponding effect on the growth kinetics depend on the particular mechanism of flux leveling employed. In most cases, at the stage of ordered-phase lateral growth, the "chemical" driving force along IPBs, caused by the concentration gradient, is more significant than that is caused by the gradient of IPB curvature. The presented analytical expression for the asymptotic thickness of the intermediate phase island allows forecasting the thickness of a new phase layer in a binary diffusion couple at the lateral growth stage, provided the estimation of their diffusion characteristics and reaction constants are performed.

5.4
MC-Scheme of Reactive Diffusion

A possible approach to solving the problem of atomistic Monte Carlo (MC) simulation of reactive diffusion is suggested. The approach is based on an MC model with pair interaction energies which strongly depend on the local atomic surrounding. The new MC model describes the competition of two intermediate phases appearing in the diffusion zone.

5.4.1
Formulation of the Problem

Many authors [15, 16, 29–33] addressed theoretical aspects of phase nucleation and growth at reactive diffusion. Still, the problem remains unsolved. The process of formation of several phases is of special significance as it is difficult to establish the criteria of phase growth and suppression as well as to find the incubation time of phase formation. The problem of competition and sequential phase formation at thin film solid-state reactions has been actively investigated since the 1970s when the microelectronics industry was boosted. However, it has not been solved yet. There are no unambiguous criteria for phase growth and suppression.

Modeling the processes of reactive diffusion by atomistic MC methods will give an opportunity to analyze the initial stage of the process, the nucleation of a new phase, and the kinetics of diffusion zone morphology. The realization of the MC simulation for such phenomena is complicated by IPBs that appear to and possess a certain surface energy. Another tricky task is to describe the thermodynamics of the new phase. It is quite clear that the formation of a new phase is caused by energetic profitability. The question arises whether using the same pair interaction energies for both new and parent phases is correct or not.

References [34–38] can be considered to be among the first to treat the simulation of the initial stages of reactive diffusion by atomistic MC methods. The authors of the mentioned papers have developed a new approach to modeling two-phase systems. The suggested method treats different pair interaction energies for an ordered phase and a solution. It is assumed that the new phase contains atoms whose local surrounding (within the first coordination shell) has an ideal order. For a two-dimensional quadratic lattice this criterion is formulated as follows: A atoms are surrounded only by B atoms and vice versa. The interaction energies for atoms belonging to different phases are introduced additionally. It is done so to keep the surface energy of interfaces positive. Model parameters are presented by six pair interaction energies: Φ_{AA}, Φ_{AB}, Φ_{BB} – for atoms of the solid solution; $\Phi_{A'B'}$ – for the intermetallic phase (neighborhood of A–A and B–B in the ordered phase is impossible; therefore, the corresponding energies are disregarded); $\Phi_{AB'}$, $\Phi_{A'B}$ – for atoms belonging to different phases ($\Phi_{AB'} = \Phi_{A'B}$).

Simulations using the "residence-time" algorithm for the vacancy mechanism demonstrated a two-stage growth of a single intermediate phase in multilayers. It was also shown that the first stage of reaction corresponds to the lateral growth of new phase islands and the second one to normal growth.

5.4.2
The Model

Consider a binary system as a two-dimensional rigid triangular lattice containing 40 000 sites. Each lattice site can be occupied by either A or B atom. Periodic boundary conditions are used. Any atom in this lattice has six neighboring atoms; there are also possibilities for vacant sites . In the simulations, one vacancy per whole simulation box is usually introduced. The energy of an atom in the site is determined as the sum of pair interaction energies with six atoms from the first coordination shell. The diffusion of atoms is accomplished through a vacancy mechanism employing the residence-time algorithm. In this mechanism, a vacancy always jumps into one of the six neighboring sites at each MC step. The jump

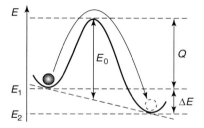

Figure 5.14 Energy profile along vacancy jump: E_1 and E_2 are the energies of the system before and after jump, respectively; Q the migration energy; and ΔE the energy change as a result of the jump.

probability in the ith direction is calculated as

$$p_i = \frac{\nu_{0i} \exp\left(-\dfrac{Q_i}{k_B T}\right)}{\sum_{j=1}^{6} \nu_{0j} \exp\left(-\dfrac{Q_j}{k_B T}\right)} \tag{5.30}$$

where ν_{0i} are the preexponential frequency factors related also to the mobilities of jumping atoms; k_B, the Boltzmann constant; and T, the absolute temperature. The migration activation energy Q is given as

$$Q = \frac{\Delta E}{2} + E_0 \tag{5.31}$$

where ΔE is the change of energy of the system as a result of the jump; E_0 is the parameter which characterizes the main barrier height (Figure 5.14).

To simplify the model we treat E_0 as a constant. It gives the following result for the probability:

$$p_i = \frac{f_i \exp\left(-\dfrac{\Delta E_i}{2k_B T}\right)}{\sum_{j=1}^{6} f_j \exp\left(-\dfrac{\Delta E_j}{2k_B T}\right)} \tag{5.32}$$

where $f_i = \nu_{0i}/\nu_{0B}$ characterizes the ratio of the mobilities of the components. The energy E of the system is given as the sum of pair interaction energies in the approximation of the first coordination shell. To simulate the first-order transformation, we use the approach suggested in [34, 36].

The chosen lattice type implies the existence of two ordered phases A_1B_2 and A_2B_1. The foundations of the suggested model are as follows. Each phase has its own pair interaction energy: Φ_0 – for the parent phase; Φ_1 – for the phase A_2B_1; Φ_2 – for the phase A_1B_2; Φ_s – for atoms belonging to different phases (any of them). Then the calculation of the jump probability needs only differences of these parameters (related to the formation energies of phases A_1B_2, A_2B_1, and to

Table 5.3 Identification criteria for phases A_1B_2 and A_2B_1.

Phase	A_1B_2					A_2B_1				
Types of atom	A			B		A			B	
Variants of neighborhood	6B	5B+1V	3A+3B	3A+2B+1V	2A+3B+1V	3A+3B	3A+2B+1V	2A+3B+1V	6A	5A+1V

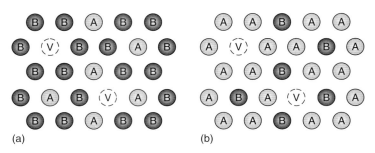

Figure 5.15 Illustrations of phase structures and possible vacancy locations that satisfy the phase identification criteria: structure of (a) A_1B_2 phase and (b) A_2B_1 phase.

surface tension, respectively)

$$\varphi_1 = \frac{\Phi_1 - \Phi_0}{k_B T}, \qquad \varphi_2 = \frac{\Phi_2 - \Phi_0}{k_B T}, \qquad \varphi_s = \frac{\Phi_s - \Phi_0}{k_B T} \qquad (5.33)$$

So, in our model we use three thermodynamic and one kinetic parameters: φ_1, φ_2 correspond to the formation energies of the ordered phases; φ_s is analogous to surface energy; f characterizes the ratio of atoms' mobilities.

To realize the identification criteria one needs to determine unambiguously to which of the two phases the atom belongs to. If the atom's neighbors meet the conditions given in Table 5.3, we will consider that the atom belongs to one of the ordered phases A_1B_2 or A_2B_1 (Figure 5.15), otherwise to the parent phase.

5.4.3
Nucleation of Phase A_2B_1 at the Interface $A-A_1B_2$

Initially, a layer of phase A_1B_2 was placed between the layers of pure component A (Figure 5.16). In the case of phase formation between pure A and phase A_1B_2, which already exists, three stages of phase formation are clearly distinguished: nucleation, lateral growth, and normal growth (Figure 5.17). Simultaneously, with the increase of the thickness of the phase layer, the transversal grain growth occurs. The latter is induced by the process of reactive diffusion itself. This growth leads to a decrease in the effective permeability of the layer.

The transition from one stage to another is characterized by an energy maximum due to overcoming the nucleation barrier (Figure 5.18). The time dependence of the average concentration of the solid solution is nonmonotonic (Figure 5.19). First, the concentration increases as a result of concentration preparation of the medium.

5.4 MC-Scheme of Reactive Diffusion | 125

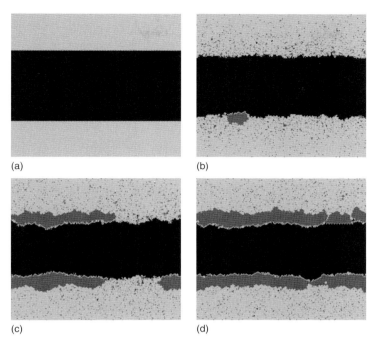

Figure 5.16 Formation of A_2B_1 at the interface $A-A_1B_2$: (a) initial state; (b) 1750 MC steps; (c) 3500 MC steps; (d) 5250 MC steps. Model parameters: $\varphi_1 = -0.25$, $\varphi_2 = -0.275$, $\varphi_s = 0.2$, and $f = 1$.

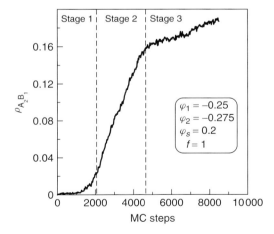

Figure 5.17 Time dependence of the "volume" of the A_2B_1 phase in MC steps at nucleation at the interface $A-A_1B_2$ for the following set of parameters: $\varphi_1 = -0.25$, $\varphi_2 = -0.275$, $\varphi_s = 0.2$, and $f = 1$.

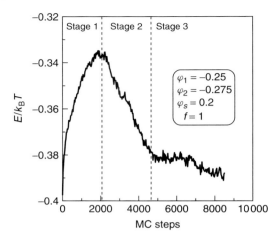

Figure 5.18 Dependence of the system's energy (per atom) on MC time at nucleation of A_2B_1 phase at the interface $A-A_1B_2$ for the following parameters: $\varphi_1 = -0.25$, $\varphi_2 = -0.275$, $\varphi_s = 0.2$, and $f = 1$.

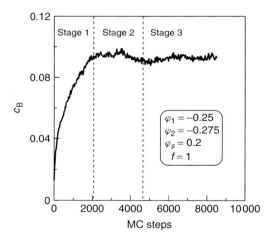

Figure 5.19 Dependence of the average concentration of component B in the solid solution on MC time at nucleation of A_2B_1 phase at the interface $A-A_1B_2$ for the following parameters: $\varphi_1 = -0.25$, $\varphi_2 = -0.275$, $\varphi_s = 0.2$, and $f = 1$.

At a certain moment which corresponds to nucleation of the intermediate phase A_2B_1, the concentration of the medium stops increasing. At the stage of lateral growth, the concentration slightly decreases. Thus, the formation of the new phase is preceded by the concentration preparation of the diffusion zone. Nonmonotonic dependence means concentration supersaturation of the solid solution with respect to the appearance of A_2B_1 phase.

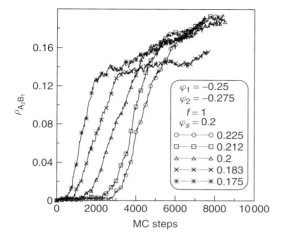

Figure 5.20 Dependences of the "volume" of the A_2B_1 phase on MC time at nucleation at the interface $A-A_1B_2$ for different values of φ_s. The values of the other parameters are $\varphi_1 = -0.25$, $\varphi_2 = -0.275$, and $f = 1$.

Stage-by-stage growth of the new phase allows us to define the incubation time for phase appearance and to determine its dependence on thermodynamic parameters of the system, the effective driving force and effective surface energy. On the basis of numerous simulations, it was confirmed that the increase of surface energy leads to the increase of incubation time (Figure 5.20). The dependence of incubation time of nucleation on thermodynamic parameters of the phase was studied as well. As it had been expected, the increase of the effective driving force led to a decrease of the incubation time. Since the phase A_2B_1 nucleates in a nonhomogeneous diffusion zone with sharp concentration gradient, its incubation time includes the concentration preparation, necessary for smoothing of the profile and corresponding increase of the nucleation driving force. According to nucleation theory in the field of a concentration gradient, phase nucleation requires a region to be formed where the concentration gradient is less than a certain critical value.

Due to periodic boundary conditions this model system is treated as a multilayer consisting of intermediate phase A_1B_2 layers and A-based solid solution. Since A_1B_2 is an ordered one, its diffusion coefficient is much lower than that of the solid solution. Thus, the thickness of the layer of the solid solution will determine the incubation time of phase appearance, while the thickness of the ordered-phase layer almost does not influence the process at all. In order to check the above sketched speculations, MC simulations were carried out. It was proved that the incubation time for phase A_2B_1 formation at the interface $A-A_1B_2$ increases with the growth of the solid solution layer, and reaches the asymptotic value (Figure 5.21).

To prove the fact that the dependence of incubation time is caused by the necessity to diminish the concentration gradient in the diffusion zone to a certain

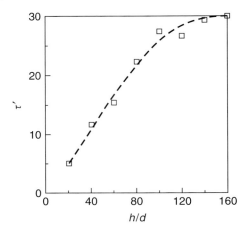

Figure 5.21 Dependence of the incubation time of phase A_2B_1 nucleation at the interface $A-A_1B_2$ on the thickness h of the solid solution layer (d is the lattice constant). The incubation time is taken in dimensionless units $\tau' = t\nu_0 \exp(-E_0/k_B T)$. MC simulations are performed for the following set of parameters: $\varphi_1 = -0.225$, $\varphi_2 = -0.25$, $\varphi_s = 0.2$, and $f = 1$.

critical value, we use the following simplified phenomenological model: Consider a two-dimensional diffusion couple which consists of the B component layer of h thickness, placed between semi-infinite layers of A component. The initial concentration profile is given as

$$c_0(x) = \begin{cases} 1, & |x| < h/2 \\ 0, & |x| > h/2 \end{cases} \quad (5.34)$$

For this problem, one may also obtain the concentration profile analytically. It is given by

$$c(t, x) = \frac{1}{2}\left\{\text{erf}\left(\frac{h/2 - x}{2\sqrt{Dt}}\right) + \text{erf}\left(\frac{h/2 + x}{2\sqrt{Dt}}\right)\right\} \quad (5.35)$$

where D is the diffusion coefficient, taken to be independent on concentration. The concentration gradient is evaluated from

$$\frac{\partial c}{\partial x} = -\frac{1}{2\sqrt{\pi Dt}}\left\{\exp\left(-\frac{(h/2 - x)^2}{4Dt}\right) + \exp\left(-\frac{(h/2 + x)^2}{4Dt}\right)\right\} \quad (5.36)$$

If we consider the incubation time, necessary for decreasing the gradient to a critical value, as ∇c_{cr} in the plane of initial contact, we obtain the equation which establishes the relation between incubation time τ and thickness of the layer, h

$$\left.\frac{\partial c}{\partial x}\right|_{x=h/2} = \nabla c_{cr} = \frac{1}{2\sqrt{\pi Dt}}\left\{1 - \exp\left(-\frac{h^2}{4Dt}\right)\right\} \quad (5.37)$$

This equation is transcendental with respect to τ, but allows us defining h via

$$h^2 = 4D\tau \ln \frac{1}{1 - 2\nabla c_{cr}\sqrt{\pi D\tau}} \quad (5.38)$$

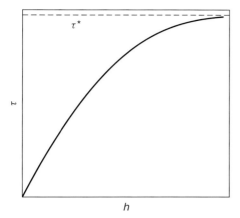

Figure 5.22 Characteristic dependence of incubation time τ on the thickness h of the solid solution layer, calculated from Equation 5.38. The asymptotic value of τ is equal to $\tau^* = \left(2\nabla c_{cr}\sqrt{\pi D}\right)^{-1}$.

It is easily seen that the thicker the layer, the larger the incubation time. At sufficient thickness of the layers, it becomes

$$\tau^* = \left(2\nabla c_{cr}\sqrt{\pi D}\right)^{-1}$$

The characteristic dependence of incubation time shown in Figure 5.22 qualitatively agrees with the results of MC simulations (Figure 5.21).

The above given analysis reveals a rather simple reason for the dependence of incubation time on thickness. The thinner the layer, the sooner the profile becomes different from that of an infinite couple. Thus, the gradient becomes more smooth and nucleation becomes thermodynamically possible.

5.4.4 Competitive Nucleation of Phases A_1B_2 and A_2B_1 at the Interface A–B

The initial configuration is presented by a layer of B component placed between two layers of A component (Figure 5.23). Periodic boundary conditions turn the system into a multilayer consisting of the layers of pure components with equal thicknesses.

Let us now analyze how the difference in thermodynamic driving forces influences the phase competition. As expected, the first growing phase is the one with the highest formation energy (Figure 5.24). The second phase forms either at the initial interface or between the first and parent phases. Both the first and the second phases demonstrate three stages: nucleation, lateral growth, normal growth. At that the second phase grows slower than the first one.

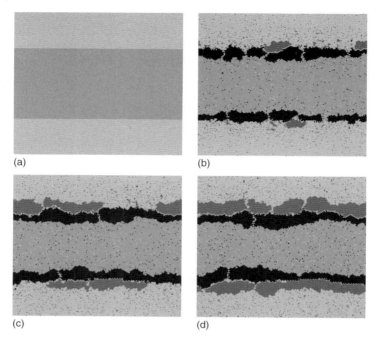

Figure 5.23 Formation of A_1B_2 and A_2B_1 at the interface A–B: (a) initial state; (b) 2500 MC steps; (c) 5000 MC steps; (d) 9250 MC steps. Model parameters: $\varphi_1 = -0.25$, $\varphi_2 = -0.275$, $\varphi_s = 0.2$, and $f = 1$.

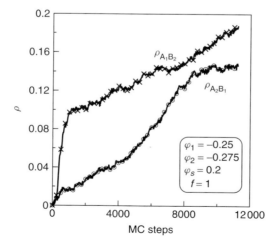

Figure 5.24 Dependences of "volumes" of phases A_1B_2 and A_2B_1 on MC time at nucleation at the interface A–B. The results of MC experiments are as follows: $\varphi_1 = -0.25$, $\varphi_2 = -0.275$, $\varphi_s = 0.2$, and $f = 1$.

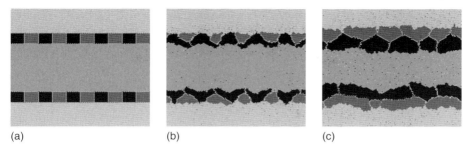

Figure 5.25 MC experiment for lateral competition: (a) initial state; (b) 625 MC steps; (c) 10 750 MC steps. Model parameters: $\varphi_1 = -0.325$, $\varphi_2 = -0.35$, $\varphi_s = 0.3$, and $f = 1$.

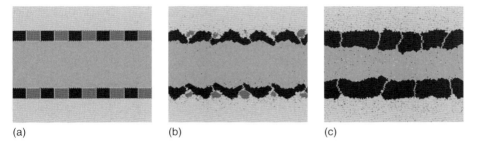

Figure 5.26 MC experiment for lateral competition: (a) initial state; (b) 625 MC steps; (c) 10 750 MC steps. Model parameters: $\varphi_1 = -0.25$, $\varphi_2 = -0.4$, $\varphi_s = 0.25$, and $f = 1$.

5.4.5
Lateral Competition

An interesting morphological evolution is observed when the contact zone is initially set as a parallel connection between the nuclei of both phases (Figures 5.25 and 5.26). When phase formation energies have similar values, lateral competition leads to consecutive connection in the diffusion zone (Figure 5.25). If phase formation energies differ considerably, a more favorable phase A_1B_2 destroys the nuclei of phase A_2B_1 (Figure 5.26).

5.4.6
Conclusions

The present section has treated one of the possible approaches to solve the problem of atomistic MC simulation of reactive diffusion. The main idea implies the use of the MC scheme with variable pair interaction energies when describing the initial stages of reactive diffusion. A new MC model involving the initial stages of reactive diffusion has been suggested. The model's peculiar property lies in the strong dependence of pair interaction energy on the local surrounding which enables clear distinction between the phases. The suggested model allows investigating the nucleation kinetics for ordered phases in multilayers. The obtained results prove that the model is a suitable tool for the analysis of phase formation and morphological evolution processes. In particular, it was shown that

- the formation of the first intermediate phase includes three stages: concentration preparation (incubation time), linear stage of lateral growth, normal growth;
- the incubation period is stipulated not only by the own nucleation barrier but by the concentration preparation (formation of a solid solution with a very low concentration gradient) as well;
- the incubation time increases with the increase in multilayer period and reaches the asymptotic value.

The suggested modification of the MC simulation for vacancy diffusion allows modeling the systems at an atomic level, which permits the description of first-order transformations with the formation of several ordered phases. The results of modeling the kinetics of solid-state reactions in multilayers with the formation of ordered phases can explain some experimental data because they enable the observation of the initial stages of the processes at an atomic level.

References

1. Pasichnyy, M.O., Schmitz, G., Gusak, A.M. and Vovk, V. (**2005**) *Physical Review B*, **72**, 014118.
2. Pasichnyy, M.O. and Gusak, A.M. (**2005**) *Metallofizika I Noveishie Tekhnologii-Metal Physics and Advanced Technologies*, **27**, 1001 (in Russian).
3. Pasichnyy, M. and Gusak, A. (**2008**) *Defect and Diffusion Forum*, **277**, 47.
4. Pasichnyy, M.O. and Gusak, A.M. (**2005**) *Defect and Diffusion Forum*, **237–240**, 1193.
5. Pasichnyy, M.O. (**2007**) *Bulletin of Cherkasy State University: Physics*, **114**, 65 (in Russian).
6. Vovk, V., Schmitz, G. and Kirchheim, R. (**2004**) *Physical Review B*, **69**, 104102.
7. Gusak, A.M. (**1990**) *Ukrainskii Fizicheskii Zhurnal (Ukrainian Journal of Physics)*, **35**, 725 (in Russian).
8. Desre, P.J. and Yavari, A.R. (**1990**) *Physical Review Letters*, **64**, 1533.
9. Hodaj, F., Gusak, A.M. and Desre, P.J. (**1998**) *Philosophical Magazine A*, **77**, 1471.
10. Gusak, A.M., Hodaj, F. and Bogatyrev, A.O. (**2001**) *Journal of Physics: Condensed Matter*, **13**, 2767.
11. Al-Kassab, T., Wollenberger, H., Schmitz, G. and Kirchheim, R. (**2002**) Tomography by atom probe field ion microscopy, in *High-resolution Imaging and Spectroscopy of Materials* (eds F. Ernst and M. Ruhle), Springer, p. 271.
12. Schleiwies, J. and Schmitz, G. (**2002**) *Materials Science and Engineering A*, **327**, 94.
13. Emeric, E. (**1998**) Etude des reactions a l'etat solide dans des multicouches Al/Co. PhD thesis. Marseille University, Marseille.
14. Emeric, E., Bergman, C., Glugnet, G. et al. (**1998**) *Philosophical Magazine Letters*, **78**, 77.
15. Gösele, U. and Tu, K.N. (**1982**) *Journal of Applied Physics*, **53**, 3252.
16. Hodaj, F. and Gusak, A.M. (**2004**) *Acta Materialia*, **52**, 4305.
17. d'Heurle, F.M., Gas, P., Philibert, J. and Zhang, S.L. (**2001**) *Defect and Diffusion Forum*, **194–199**, 1631.
18. Hodaj, F. and Desre, P.J. (**1996**) *Acta Materialia*, **44**, 4485.
19. Desre, P.J. (**1991**) *Acta Metallurgica et Materialia*, **39**, 2309.

20. Michaelsen, C., Barmak, K. and Weihs, T.P. (**1997**) *Journal of Physics D: Applied Physics*, **30**, 3167.
21. Rickman, J.M., Tong, W.S. and Barmak, K. (**1997**) *Acta Materialia*, **45**, 1153.
22. Lucadamo, G., Barmak, K., Carpenter, D.T. et al. (**1999**) *Materials Research Society Symposium Proceedings*, **562**, 159.
23. Lucadamo, G., Barmak, K., Hyun, S. et al. (**1990**) *Materials Letters*, **39**, 268.
24. Klinger, L., Brehet, Y. and Purdy, G. (**1999**) *Acta Materialia*, **46**, 2617.
25. Lucenko, G. and Gusak, A. (**2003**) *Microelectronic Engineering*, **70**, 529.
26. Lucenko, G.V. and Gusak, A.M. (**2002**) *Bulletin of Cherkasy State University: Physics*, **37–38**, 145.
27. Gusak, A.M. and Bogatyrev, A.O. (**1994**) *Metallofizika I Noveishie Tekhnologii-Metal Physics and Advanced Technologies*, **16**, 28 (in Russian).
28. Gusak, A.M. and Gurov, K.P. (**1988**) *Metallofizika I Noveishie Tekhnologii-Metal Physics and Advanced Technologies*, **10**, 116 (in Russian).
29. Gusak, A.M., Bogatyrev, A.O., Zaporozhets, T.V. et al. (**2004**) *Models of Solid State Reactions*, Cherkasy National University Press, Cherkasy (in Russian).
30. Kidson, G.V. (**1961**) *Journal of Nuclear Materials*, **3**, 21.
31. Gurov, K.P., Kartashkin, B.A. and Ugaste, Y.E. (**1981**) *Interdiffusion in Multiphase Metallic Systems*, Nauka, Moscow (in Russian).
32. Dybkov, V.I. (**1992**) *Kinetics of Solid State Chemical Reactions*, Naukova Dumka, Kiev (in Russian).
33. Pines, B.Y. (**1961**) *Sketches on Metal Physics*, Kharkov State University Press, Kharkov (in Russian).
34. Gusak, A.M., Kovalchuk, A.O. and Bogatyrev, A.O. (**1997**) *Defect and Diffusion Forum*, **143–147**, 661.
35. Gusak, A.M. and Kovalchuk, A.O. (**1998**) *Physical Review B*, **58**, 2551.
36. Gusak, A.M., Bogatyrev, A.O. and Kovalchuk, A.O. (**2001**) *Defect and Diffusion Forum*, **194–199**, 1625.
37. Gusak, A.M. and Kovalchuk, A.O. (**1997**) *Metallofizika I Noveishie Tekhnologii-Metal Physics and Advanced Technologies*, **19**, 39 (in Russian).
38. Kovalchuk, A.O. (**2001**) *Ukrainskii Fizicheskii Zhurnal (Ukrainian Journal of Physics)*, **46**, 1304 (in Russian).

Further Reading

Bokstein, B.S., Bokstein, S.Z. and Zhukhovitsky, A.A. (**1974**) *Thermodynamics and Kinetics of Diffusion in Solids*, Metallurgiya, Moscow (in Russian).

6
Flux-Driven Morphology Evolution
Andriy M. Gusak

6.1
Introduction

In the present chapter, we discuss some new approaches to structure evolution in open systems, developed jointly with our colleagues from the universities of Los Angeles and Eindhoven (King-Ning Tu, Frans van Loo, Alexander Kodentsov, and Mark van Dal).

Morphology evolution in closed systems, like ripening and grain growth, has been studied and modeled for many years. We discuss some peculiarities of standard models in Sections 6.2 and 6.3. In Section 6.4, the main ideas and models concerning reactive diffusion between copper and molten solder at simultaneous coarsening of the reaction products are presented. It has been shown that, unlike the classical case of LSW theory, here it is the gain in chemical energy that presents a driving force, while the total surface energy of the boundaries is nearly constant. Therefore, firstly, the kinetics of the process is controlled not by the surface tension but by the transmission capacity of liquid channels between the grains of the growing intermetallic. Section 6.5 briefly treats the idea of flux-driven grain coarsening as applied to the case of grain growth at thin film deposition. It is demonstrated that the deposition process itself must help large grains to consume (or, more precisely, to overlap) the small ones. This effect might result both in linear as well as parabolic growth of the average lateral grain size depending on temperature and on grain boundary (GB) tension.

Ripening and grain growth are used to demonstrate instability in the size space: any narrow size distribution of grains or precipitates broadens with time. Recently, a new type of instability was discovered, the possible instability of markers row (K-plane) during inhomogeneous Kirkendall shift during interdiffusion: Section 6.6 deals with the theory of Kirkendall planes bifurcation and instability at interdiffusion.

In Section 6.7, the influence of electric current on surface structure evolution and grain rotation in tin thin films is studied. Peculiarities of mass transfer in two-phase alloys at electro- and thermomigration are highlighted in Section 6.8.

Diffusion-Controlled Solid State Reactions. Andriy M. Gusak
Copyright © 2010 WILEY-VCH Verlag GmbH & Co. KGaA, Weinheim
ISBN: 978-3-527-40884-9

6.2
Grain Growth and Ripening: Fundamentals

During the last eight years, joint efforts of the research groups from the universities of California (Los Angeles) and Cherkassy have led to new approaches and results concerning the prediction of morphology evolution in open polycrystalline systems [1–6]. Here the concept "morphology evolution" firstly means the time evolution of sizes, shapes, and orientations of the grains in polycrystalline systems. The following open systems are treated here:

1) thin films in the process of deposition;
2) diffusion or reaction zones at inter- and/or reactive diffusion;
3) compounds of solders with metals and conductive thin films under electric current (and corresponding electromigration of atoms);
4) solders at a temperature gradient.

For comparison, classical examples of morphology evolution in "closed" systems are (i) ripening of precipitates at the last stage of decomposition of supersaturated alloys, (ii) void coarsening at the last stage of sintering, (iii) normal grain growth (the growth of large grains at the expense of small ones) in polycrystalline materials or soap films, (iv) colony growth during cellular decomposition, and (v) grain growth by grain rotation depending on misorientations between neighboring grains. All the processes listed above are caused by "internal" reasons (driving forces), that is, the tendency toward reduction of interface or GB surface energy.

The theory of both precipitates ripening and void coarsening has been established for about 50 years (LSW theory, [7–10]) and is generally accepted as a basis of material science, though it still remains in partial disagreement with experimental data [11–15]. Despite numerous attempts [5, 16–24], a universally accepted theory for grain growth has not yet been developed.

6.2.1
Main Approximations of the LSW Approach

The main approximations of the LSW approach are the following:

1) The asymptotic stage of decomposition is treated, assuming the supersaturation of the parent phase approaching zero, and the volume of the new phase is practically constant and only undergoes redistributions due to dissolution of small (undercritical) precipitates and growth of large (overcritical) ones, with a critical size being determined synergetically by the system itself.
2) The mean-field approximation is applied, that is, the supersaturation of the parent phase $\Delta = \langle C \rangle - C^{eq}$ in the vicinity of any precipitate is taken as one and the same, common for the whole system without considering correlations between the size and local supersaturation.
3) The steady state approximation is employed. This means that the Laplace equation $\nabla^2 C \cong 0$ is solved rather than the diffusion equation (second Fick's

law), with Gibbs–Thomson boundary conditions at spherical boundaries of the precipitate given by

$$C(t, r = R(t)) = C^{eq} + \frac{\alpha}{R} \tag{6.1}$$

Here α for the case of precipitation of an almost pure element B from a diluted solution of B in A equals

$$\alpha = C^{eq} \frac{2\gamma\Omega}{kT} \tag{6.2}$$

where γ is the surface tension and Ω is the atomic volume.

4) Taking into account the steady state and Gibbs–Thomson conditions, the growth (or dissolution) rate for a spherical precipitate is determined by the following balance equation at the moving spherical boundary:

$$\frac{dR}{dt} = D\frac{\langle C \rangle - C(t, r = R(t))}{R} = \frac{D}{R}\left(\Delta(t) - \frac{\alpha}{R}\right) \tag{6.3}$$

5) Equation 6.3 naturally introduces the critical radius:

$$R_{cr}(t) = \frac{\alpha}{\Delta(t)}, \quad \begin{array}{l} R < R_{cr} \Rightarrow dR/dt < 0 \\ R > R_{cr} \Rightarrow dR/dt > 0 \end{array} \tag{6.4}$$

The critical radius increases with time (as supersaturation diminishes), catching up with the growing particles one by one, which then start dissolving (ultimately, "the only one must remain" in a finite system, yet this falls beyond the statistical theory).

6) Approximate constancy of the volume of the new phase at the asymptotic stage results in a simple relation between critical and average sizes, namely:

$$0 \simeq \sum_{i=1}^{N} R_i^2 \frac{dR_i}{dt} = D\left(\Delta(t)\sum_{i=1}^{N} R_i - \alpha N\right) \tag{6.5}$$

$$\Rightarrow \Delta(t) = \frac{\alpha}{\langle R \rangle} \Rightarrow R_{cr}(t) = \langle R \rangle \tag{6.6}$$

Having combined Equation 6.3 and the condition of "almost constant volume," Equation 6.6 gives a simple-looking ultimate equation for growth/dissolution at the asymptotic stage:

$$\frac{dR}{dt} = \frac{D\alpha}{R}\left(\frac{1}{\langle R \rangle} - \frac{1}{R}\right) \tag{6.7}$$

7) The size (radius) distribution of precipitates obeys a continuity equation in the size space, provided there are no direct collisions of the particles and no fluctuations:

$$\frac{\partial f(t, R)}{\partial t} = -\frac{\partial}{\partial R}\left(f(t, R)\frac{dR}{dt}\right) = -D\alpha\frac{\partial}{\partial R}\left(\frac{f}{R}\left(\frac{\int_0^\infty f\,dR}{\int_0^\infty Rf\,dR} - \frac{1}{R}\right)\right) \tag{6.8}$$

8) A rather sophisticated analysis (performed by Slezov in 1958) of the integro-differential equation Equation 6.8 gives the following asymptotic solution:

$$\langle R \rangle^3 - \langle R_0 \rangle^3 = \frac{4}{9} D\alpha t \tag{6.9}$$

$$P(u) = \frac{3^4 e}{2^{5/3}} \frac{u^2 \exp(-1/(1-2u/3))}{(u+3)^{7/3} \left(\frac{3}{2} - u\right)^{11/3}}, \quad u < 3/2$$

$$P(u) = 0, \quad u > 3/2 \tag{6.10}$$

where the probability density as a function of reduced size $P(u)$ ($u = R/R_c$) is time independent and normalized to unity

$$\int_0^{3/2} P(u) \, du = 1 \tag{6.11}$$

The law as described by Equation 6.9 is generally in good agreement with experiment. The form of the distribution function almost never conforms to it.

6.2.2
Traditional Approaches to the Description of Grain Growth

Grain growth in a polycrystalline body or a soap film and is caused by the same thermodynamic origin as ripening (tendency toward reduction in surface energy by way of consumption of the small grains by the large ones), but is realized without long-range diffusion (if not segregation effects). Two marginal approaches to the description of the grain growth process are known: the deterministic and the stochastic ones (and, of course, a number of combined models employing the approximations of both approaches).

In the deterministic approach (the most well known are the Burke–Turnbull model [16], Hillert's theory [17], the topological model of Fradkov–Marder [18–20], the Di Nunzio model [21]), the behavior of each grain is unambiguously determined by several parameters: size (radius, cross-sectional area or volume), number of nearest neighbors, and so on. In the stochastic approximation, first suggested by Louat [22] and later modified by many authors, the grains are treated as "drunken sailors" who randomly walk within a semi-infinite size space with a boundary zero point working just as a sink (without the possibility of nucleation of new grains). Since a nonzero possibility of getting to the zero point (and disappearing) always exists, the number of "travelers" (grains) diminishes, while the average size of the grains increases.

The deterministic approach seems to be more physically founded since it deals with the explicit form of the driving force of grain growth process, namely the reduction in the grains' surface energy. However, this approach in the standard Hillert's form leads to a wrong prediction with respect to the size distribution of grains. In order to obtain a more precise form of the distribution, Fradkov and Marder extended the number of state parameters by introducing an additional parameter – the number of grain's sides, but this was done only for the two-dimensional case.

The stochastic approximation allows one to find a reliable size distribution. Already in his first work, Louat obtained the Rayleigh distribution, which is in rather good agreement with experiment. Later, Pande [23] introduced an additional drift term, inversely proportional to the size of the grain and of negative sign: this means that drift results in a decrease in each grain (which appears to be completely incorrect for the usual size space and works quite well as for the normalized space as discussed in [5]). Pande showed that the additional drift term in the equation for random walks transforms the Rayleigh distribution into the log-normal one. In spite of the indicated progress, the physical meaning of the stochastic approximation as well as the idea of "diffusion" of fairly massive grains in the size space remain unclear. Moreover, the role of grain growth driving force in this approach is not clear enough.

To adequately analyze numerous modifications of the LSW and LSW-like theories (those of Hillert's type), we need to simplify the mathematical method, probably by neglecting the analysis of uniqueness and stability of solutions. Such an attempt at simplified derivation is presented in Section 6.3.

6.3
Alternative Derivation of the Asymptotic Solution of the LSW Theory

If there is no physical reason to check whether or not the asymptotic solution is unique and stable, one may employ a simplified way used by the author first for the problem of flux-driven grain ripening in the zone of reaction copper-molten solder [1]. Here, we apply this approach to classical ripening in the 3D closed system (LSW case). The size distribution in classical LSW theory satisfies the well-known equation

$$\frac{\partial f}{\partial t} = -a_0 \frac{\partial}{\partial R}\left(\frac{f}{R}\left(\langle C\rangle - C^e - \frac{\alpha}{R}\right)\right), \quad a_0 = \frac{n}{n_i}\frac{D}{C_i} \quad (6.12)$$

The distribution function should also satisfy the constraint of matter conservation

$$n_i C_i V_i + n\langle C\rangle (V - V_i) = nC_0 V \quad (6.13)$$

$$V_i = \int_0^\infty \frac{4}{3}\pi R^3 f(t, R)\, dR \ll V, \quad \langle C\rangle \to C^e \quad (6.14)$$

where n_i, C_i, and V_i are respectively the atomic density, atomic fraction, and volume of the new phase undergoing coarsening in the parent phase matrix of n, $\langle C\rangle$ (initially C_0), and $V - V_i$ (initially V) respectively. For the late stage of coarsening, the limitation of nearly constant volume of the new phase is fairly correct

$$V_i = \int_0^\infty \frac{4}{3}\pi R^3 f(t, R)\, dR \cong \frac{n(C_0 - C^e)}{n_i C_i} V = \text{const} \quad (6.15)$$

Equation 6.15 is the only constraint on the LSW size-distribution function for the case of a closed system. Then

$$\frac{d}{dt}\int_0^\infty R^3 f \, dR = 0 = -a_0 \int_0^\infty R^3 \frac{\partial}{\partial R}\left(\frac{f}{R}\left(\langle C \rangle - C^e - \frac{\alpha}{R}\right)\right) dR \qquad (6.16)$$

Using the boundary conditions for $f(R)$, we obtain the supersaturation as

$$\Delta \equiv \langle C \rangle - C^e = \alpha \frac{\int f \, dR}{\int Rf \, dR} = \frac{\alpha}{\langle R \rangle} \qquad (6.17)$$

which means that in the LSW case, the critical radius is simply an average one.

The condition in Equation 6.17 transforms Equation 6.12 into

$$\frac{\partial f}{\partial t} = -A\frac{\partial}{\partial R}\left(\frac{f}{R}\left(\frac{1}{\langle R \rangle} - \frac{1}{R}\right)\right), \quad A = \frac{n}{n_i}\frac{D\alpha}{C_i} \qquad (6.18)$$

The structure of Equation 6.18 suggests new variables,

$$\tau = At, \quad \xi = R/(At)^{1/3} \qquad (6.19)$$

Solving with separation of variables ($f = \tilde{g}(\tau)\tilde{\varphi}(\xi)$), we get

$$\frac{d \ln \tilde{g}}{d \ln \tau} = \frac{d \ln \tilde{\varphi}}{d\xi}\left(\frac{\xi}{3} - \frac{1}{\langle \xi \rangle}\frac{1}{\xi} + \frac{1}{\xi^2}\right) - \left(\frac{2}{\xi^3} - \frac{1}{\langle \xi \rangle}\frac{1}{\xi^2}\right) = \lambda = \text{const} \qquad (6.20)$$

To find the parameter λ, we should use the constraint of constant volume:

$$\tilde{g}(\tau)\frac{4\pi}{3}\tau^{4/3}\int \xi^3 \tilde{\varphi}(\xi)\, d\xi = \frac{n}{n_i}\frac{\Delta_0}{C_i}V \qquad (6.21)$$

which means that

$$\tilde{g}(\tau) = \text{const} \cdot \tau^{-4/3}, \quad \lambda = -\frac{4}{3} \qquad (6.22)$$

Using Equation 6.22 in Equation 6.20, we have

$$\frac{d \ln \tilde{\varphi}}{d\xi} = \frac{2}{\xi} + 3\frac{1/\langle \xi \rangle - 2\xi^2}{\xi^3 - 3\xi/\langle \xi \rangle + 3} \qquad (6.23)$$

Thus,

$$\tilde{\varphi}(\eta) = \eta^2 \exp\left\{\int_0^\eta \frac{3\tilde{\Xi} - 6\xi^2}{\xi^3 - 3\tilde{\Xi}\xi + 3}\, d\xi\right\}, \quad \tilde{\Xi} \equiv \frac{1}{\langle \xi \rangle} \qquad (6.24)$$

The formal solution for the LSW distribution function is

$$f(t, R) = \frac{B_{\text{LSW}}}{(At)^{4/3}}\frac{R^2}{(At)^{2/3}}\exp\left\{\int_0^{R/(At)^{1/3}}\frac{3\tilde{\Xi} - 6\xi^2}{\xi^3 - 3\tilde{\Xi}\xi + 3}\, d\xi\right\}$$

$$= \frac{B_{\text{LSW}}}{\tau^{4/3}}\tilde{\varphi}(\eta) \qquad (6.25)$$

$$\tau = At, \quad \eta = \frac{R}{(At)^{1/3}}, \quad B_{\text{LSW}} = \frac{\frac{n}{n_i}\frac{\Delta_0}{C_i}V}{\frac{4}{3}\pi \int_0^\infty \xi^3 \tilde{\varphi}(\xi)\, d\xi} \qquad (6.26)$$

The parameter $\tilde{\Xi}$ should be found from the following self-consistency condition:

$$\frac{1}{\tilde{\Xi}} = \langle \xi \rangle = \frac{\int_0^\infty \xi \tilde{\varphi}(\xi, \tilde{\Xi}) \, d\xi}{\int_0^\infty \tilde{\varphi}(\xi, \tilde{\Xi}) \, d\xi} \quad (6.27)$$

To solve this equation, we need to know the explicit form of the function φ. We have three possibilities and three corresponding explicit forms found by standard integration:

1)
$$\tilde{\Xi} < \tilde{\Xi}^* \equiv (3/2)^{2/3}$$

This case refers to a single negative root of the denominator in Equation 6.24.

$$\xi_1 = -\sqrt{\Xi^*} \left\{ \left[1 + \sqrt{1 - (\tilde{\Xi}/\tilde{\Xi}^*)^3} \right]^{1/3} + \left[1 - \sqrt{1 - (\tilde{\Xi}/\tilde{\Xi}^*)^3} \right]^{1/3} \right\} \quad (6.28)$$

$$\varphi(\eta) = \eta^2 \left(1 + \frac{\eta}{|\xi_1|} \right)^{\frac{3\tilde{\Xi} - 6\xi_1^2}{2\xi_1^2 + 3/|\xi_1|}} \left(\frac{3/|\xi_1|}{\eta^2 + \xi_1 \eta + 3/|\xi_1|} \right)^{\frac{6\xi_1^2 + 18/|\xi_1| + 3\tilde{\Xi}}{2(2\xi_1^2 + 3/|\xi_1|)}}$$

$$\times \exp \left\{ \frac{3|\xi_1| \left(3/|\xi_1| - \xi_1^2 + \frac{3}{2}\tilde{\Xi} \right)}{(3/|\xi_1| + 2\xi_1^2) \sqrt{3/|\xi_1| - \xi_1^2/4}} \right.$$

$$\left. \times \left(\text{arctg} \frac{\eta - |\xi_1|/2}{\sqrt{3/|\xi_1| - \xi_1^2/4}} + \text{arctg} \frac{|\xi_1|/2}{\sqrt{3/|\xi_1| - \xi_1^2/4}} \right) \right\}$$

2) Degenerate case $\tilde{\Xi} = \tilde{\Xi}^* \equiv (3/2)^{2/3}$.
   Then

$$\xi_1 = \xi_2 = \sqrt{\tilde{\Xi}^*}, \quad \xi_3 = -2\sqrt{\tilde{\Xi}^*}$$

$$\varphi(\eta) = 0, \quad \eta > \left(\frac{3}{2} \right)^{1/3}$$

$$\varphi(\eta) = \frac{\eta^2 e^1 \exp\left(-\frac{1}{1 - \eta/\sqrt{\tilde{\Xi}^*}} \right)}{\left(1 + \frac{\eta}{2\sqrt{\tilde{\Xi}^*}} \right)^{7/3} \left(1 - \frac{\eta}{\sqrt{\tilde{\Xi}^*}} \right)^{11/3}}, \eta \langle \left(\frac{3}{2} \right)^{1/3} \quad (6.29)$$

3)
$$\tilde{\Xi} > \tilde{\Xi}^* \equiv (3/2)^{2/3}$$

In this case, we have two different positive roots and one negative root:

$$\xi_1 = 2\sqrt{\tilde{\Xi}} \cos(\phi/3)$$
$$\xi_2 = -2\sqrt{\tilde{\Xi}} \cos(\phi/3 + \pi/3)$$
$$\xi_3 = -2\sqrt{\tilde{\Xi}} \cos(\phi/3 - \pi/3), \quad \cos(\phi) = -\frac{3}{2\tilde{\Xi}^{3/2}}$$

$$\varphi(\eta) = 0, \quad \text{for} \quad \eta > \xi_2 \tag{6.30}$$

$$\varphi(\eta) = \eta^2 \left(1 - \frac{\eta}{\xi_1}\right)^{\frac{3\tilde{\Xi}-6\xi_1^2}{(\xi_1-\xi_2)(\xi_1-\xi_2)}} \left(1 - \frac{\eta}{\xi_2}\right)^{\frac{3\tilde{\Xi}-6\xi_2^2}{(\xi_2-\xi_3)(\xi_2-\xi_1)}}$$

$$\times \left(1 + \frac{\eta}{|\xi_3|}\right)^{\frac{3\tilde{\Xi}-6\xi_3^2}{(\xi_3-\xi_1)(\xi_3-\xi_2)}}, \quad \text{for} \quad \eta < \xi_2$$

Substituting Equations 6.28–6.30 in Equation 6.27 for a wide region of $\tilde{\Xi}$ values, we can check (by numerical calculation of integrals in Equation 6.27) that this equation is satisfied strictly for $\tilde{\Xi} = \tilde{\Xi}^* \equiv (9/4)^{1/3}$. For all other values, the product

$$\tilde{\Xi} \frac{\int \xi \tilde{\varphi}(\xi, \tilde{\Xi}) \, d\xi}{\int \tilde{\varphi}(\xi, \tilde{\Xi}) \, d\xi}$$

is less than unity. This means that only regime (2), with

$$\langle \xi \rangle = \frac{1}{\tilde{\Xi}^*} = \left(\frac{2}{3}\right)^{2/3}$$

and the distribution, given by Equation 6.29, are self-consistent. Thus, the critical size changes with time as

$$R_{cr} = \langle R \rangle = \langle \xi \rangle (At)^{1/3} = \left(\frac{4}{9} \frac{n}{n_i} \frac{D\alpha}{C_i} t\right)^{1/3}$$

which is the same as the classical LSW result. The size-distribution function presented in Equation 6.29 is the same as the asymptotic LSW expression.

To make this conclusion even more evident, let us express Equation 6.29 in terms of the reduced size

$$u = \eta/\langle \eta \rangle = \left(\frac{3}{2}\right)^{2/3} \eta \Leftrightarrow \eta/\sqrt{\tilde{\Xi}^*} = u/(3/2)$$

so that

$$\varphi(u) = \text{const} \frac{u^2 e^1 \exp\left(-\frac{1}{1 - u/(3/2)}\right)}{\left(1 + \frac{u}{3}\right)^{7/3} \left(1 - \frac{u}{3/2}\right)^{11/3}}, \quad u < \frac{3}{2}$$

Thus, the simplified method developed here gives the same asymptotic solution as the general LSW approach. However, the shortcoming of this method is the use of numeric (very simple, though) integration. Nevertheless, it seems to be less complicated and "more standard" as compared to the original LSW theory.

6.4
Flux-Driven Ripening at Reactive Diffusion

The investigation of reactions between Cu (or Ni) and molten solder have proven that the growth of intermetallics is accompanied by diffusion coarsening. A classical coarsening is a stage that follows nucleation and growth when the total volume of the new phase is practically maximal and the substance just undergoes diffusion redistributions between the particles [8–10]. The driving force of classical coarsening is the reduction in the interface area or surface energy. Therefore, the total area of the surface of new phase is reduced. The investigation of surface reactions between molten solder and metal has shown that ripening of the scallop-type intermetallide phase at the surface proceeds simultaneously with scallop growth. Since the volume of the intermediate phase increases with time, the type of coarsening is referred to as *nonconservative*. The driving force of this process is the release of free energy due to intermediate phase growth.

Assuming scallops to have a hemispherical shape (which appears to be realistic until the reaction zone becomes about 10 µm thick and if composition of liquid solder is close to eutectic), we consider the total area of the intermediate phase as not changing with time in the process of growth. The scallop growth is observed at the reaction between eutectic solder SnPb and Cu [25–27], and also between Pb-free solders (for example, SnAg) and Cu [28, 29], and between pure tin and copper [30, 31]. It was experimentally shown [32] that as the layers of reaction products grow, ripening of scallops takes place; the corresponding model that describes this phenomenon was proposed as well. In [1], we (Gusak and Tu) proposed the kinetic theory for the new type of coarsening grounding upon (i) conservation of the interface area at increasing volume and (ii) the dominating role of channels between the scallops. We have called this phenomenon *flux-driven ripening* (FDR).

6.4.1
Experimental Results

Solder reactions typically occur at 200 °C, lasting anywhere from a few seconds to several minutes. In case of pure Sn or Cu, the reaction temperature is typically around 250 °C. In these reactions, Cu_3Sn and Cu_6Sn_5 IMCs are formed at the interfaces. The phase Cu_6Sn_5 is the dominant growth phase and has scallop-type morphology. The size of the scallops can grow to several microns in diameter after a few minutes of reaction at 200 °C between eutectic SnPb and Cu. The radius of the scallops was found to obey an approximately $t^{1/3}$ dependence of growth. The number of scallops decreases at an approximately $t^{-2/3}$ dependence [25, 27]. The activation energy of the growth measured form the growth rates occurring from 200 to 240 °C during the reaction between eutectic SnPb and Cu, is about 0.2–0.3 eV/atom. On the other hand, Cu_3Sn is a thin layer and its growth is very slow. Liquid channels between the scallops remain during the whole process. The morphology of scallops and channels is thermodynamically stable in the presence of molten solder. Indeed, the molten eutectic solder SnPb quickly wets the GBs of Cu_6Sn_5 [33–35]. The channels serve as fast diffusion paths supplying copper to the molten solder and activating the scallop growth. According to the data of scanning electron microscopy [1], the

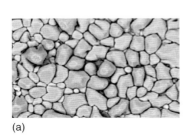
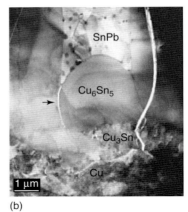

(a)   (b)

Figure 6.1 Scallop-like morphology of Cu_6Sn_5 during wetting reactions between molten eutectic SnPb solder and Cu: (a) top view of Cu_6Sn_5 scallops after 1 min reflow at 200 °C; (b) cross-sectional image of Cu_6Sn_5 scallops after 10 min reflow at 200 °C (channels indicated by arrows).

channels are less than 20 nm wide, which, however, exceeds the usual GB width (Figure 6.1).

6.4.2
Basic Approximations

The scallops have a shape of a hemisphere (Figure 6.2). Designating the area between the scallops and copper as S^{total}, take into account that the total surface area between hemispherical scallops and molten solder is equal to double S^{total}:

$$S^{scallops/melt} = \sum_{i=1}^{N} 2\pi R_i^2 = 2 \times \sum \pi R_i^2 = 2S^{total} = \text{const} \qquad (6.31)$$

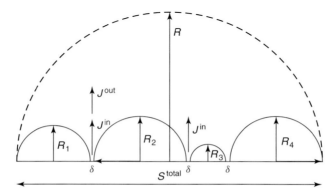

Figure 6.2 Schematic diagram of the cross section of an array of scallops on Cu.

J_{in} is an input flux of copper directed from the substrate into the melt via channels (serving for the growth of scallops) and J_{out} is an out-flux of copper that was not used for building up the IMC. For small solder bumps, this out-flux quickly tends to zero after their saturation. Thus, while the phase growth contributes to the total volume increase, the total area of scallops does not change.

Alternatively, Schmitz et al. suggested [36] that channels are not totally liquid but instead they are the grooves, transforming (near the substrate) into prewetted GBs of Cu_6Sn_5 phase (also with high permittivity for copper). Recent observations of Suh and Tu [37] demonstrate that actually the situation might be intermediate: the crystallographic orientation of the Cu_6Sn_5 scallops is not fully chaotic. There exist some agglomerates of scallops within which orientations are close, but the misorientation between agglomerates is far from being close. Since the wetting condition depends on the GB tension, and this tension depends on misorientation, one can expect that the GBs between scallops of the same agglomerate are not wetted, but boundaries between agglomerates (with large misorientations) are wetted by liquid solder and are transformed into liquid channels. All copper coming from the substrate is utilized for scallop growth, so one may neglect the out-flux of copper from the ripening region into the volume. It is fair for at least small solder bumps, since, say, a 100-μm bump becomes saturated with copper in 10 s. Interfacial diffusion of copper along the scallop/copper interface is not a rate-limiting factor (as shown in [1]).

Since the growth of a scallop must occur at the expense of its neighbors, it is a ripening process. In this process, there are two important constraints. The first (geometrical) constraint is that the interface of the reaction is constant. The second is conservation of mass, in which all the in-flux of Cu is consumed by scallop growth.

6.4.3
Basic Equations

Let $f(t, R)$ be the size-distribution function, so that the total number of grains and the average value of radius are equal to

$$N(t) = \int_0^\infty f(t, R)\, dR, \quad \langle R \rangle = \frac{1}{N} \int_0^\infty R f(t, R)\, dR \qquad (6.32)$$

respectively. The first constraint of constant interface takes the form

$$\sum_{i=1}^N \pi R_i^2 = \int_0^\infty \pi R^2 f(t, R)\, dR = S^{total} - S^{free} \cong S^{total} = const \qquad (6.33)$$

The surface area of channels for the copper supply is

$$S^{free} = \int_0^\infty \frac{\delta}{2} 2\pi R f(t, R)\, dR \qquad (6.34)$$

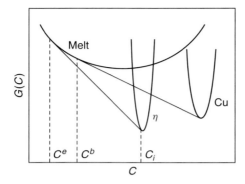

Figure 6.3 Schematic dependence of Gibbs free energy on composition. Quasiequilibrium concentration of Cu in the alloy near the substrate (C^b), in the intermetallic joint Cu_6Sn_5 or η phase (C_i) and in the alloy with stable equilibrium with the planar η phase (C^e).

The volume of the growing intermetallic scallop is

$$V_i = \sum_{i=1}^{N} \frac{2}{3}\pi R_i^3 = \int_0^\infty \frac{2}{3}\pi R^3 f(t, R)\, dR \qquad (6.35)$$

The major constraint of our system is the conservation of mass; that is, the "in-flux" is consumed by growing intermetallic scallops,

$$n_i C_i \frac{dV_i}{dt} = J^{in} S^{free} \qquad (6.36)$$

Here n_i and C_i are atomic density and mole fraction of copper in the intermetallic respectively. The "in-flux" density is determined by Fick's law with the average gradient being equal to the ratio of the difference $\Delta C = C^b - C^e$ (between the equilibrium concentration of Cu (in solder) with substrate and intermetallic in accord with Figure 6.3 and average width of the reaction zone (approximately equal to the average radius of a grain):

$$J^{in} \cong -nD\frac{C^e - C^b}{\langle R \rangle} \qquad (6.37)$$

(n standing for the number of atoms per unit volume). Since scallops must grow and shrink atom by atom, the distribution function should satisfy the usual continuity equation in size space

$$\frac{\partial f}{\partial t} = -\frac{\partial}{\partial R}(fu_R) \qquad (6.38)$$

where the velocity in the size space, u_R, is simply the growth rate of scallops, found from the equation for flux density:

$$u_R = \frac{dR}{dt} \cong \frac{-j(R)}{n_i C_i} \qquad (6.39)$$

In the classic ripening theory, when each grain is spherical and is surrounded by an infinite supersaturated solid solution, the expressions for $j(R)$ and u_R are found as a steady state solution of the diffusion problem in infinite space around a spherical grain with fixed supersaturation $\langle C \rangle - C^e$ at infinity. In our case, the scallops almost touch each other. Therefore, the classical expression is not applicable. In FDR, the diffusional transport of copper atoms through the channels in the reaction zone is the rate-controlling step, since ripening, which conserves the total surface area, would be impossible without growth and growth is impossible without incoming flux. Under the constraint of constant surface, the rate of ripening is controlled by the incoming flux, and this flux is redistributed among scallops. Flux in (out) of each individual scallop should be proportional to the difference between the average chemical potential of copper μ in the reaction zone (we take it to be the same everywhere – a mean-field approximation) and the chemical potential $\mu_\infty + (\beta/R)$ at the scallop-melt surface:

$$j(R) = -L\left(\mu - \mu_\infty - \frac{\beta}{R}\right) \tag{6.40}$$

where the parameters L, μ are determined self-consistently from the above-mentioned two constraints of constant surface and mass conservation.

Equations 6.32–6.40 lead to the following distribution function:

$$\frac{\partial f}{\partial t} = -\frac{k}{9} \frac{\langle R \rangle}{\langle R^2 \rangle - \langle R \rangle^2} \frac{\partial}{\partial R}\left(f\left(\frac{1}{\langle R \rangle} - \frac{1}{R}\right)\right) \tag{6.41}$$

where the growth rate k is determined by the incoming flux conditions rather than by surface tension (as in the LSW theory).

$$k = \frac{9}{2} \frac{n}{n_i} \frac{D(C^b - C^e)\delta}{C_i} \tag{6.42}$$

We find the analytical solution of this equation in the form of a time dependence for the average cubed radius

$$\langle R^3 \rangle = kt$$

$$f(t, R) = \frac{\text{const}}{t} \cdot \varphi\left(\frac{R}{(bt)^{1/3}}\right)$$

where

$$\varphi(\eta) = \frac{\eta}{\left(\frac{3}{2} - \eta\right)^4} \cdot \exp\left(-\frac{3}{\frac{3}{2} - \eta}\right), \quad 0 < \eta < \left(\frac{3}{2}\right) \tag{6.43}$$

$$b = \frac{k}{\langle \eta^3 \rangle} = \frac{k}{0.5535}$$

The form of the distribution is similar to Hillert's function for grain growth (Figure 6.4). Indeed, it is easy to check that the average value of η is

$$\langle \eta \rangle = \frac{\int_0^{3/2} \eta \varphi(\eta)\, d\eta}{\int_0^{3/2} \varphi(\eta)\, d\eta} = \frac{3}{4}$$

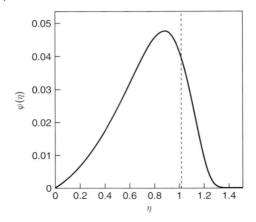

Figure 6.4 Scaling grain size-distribution function $\varphi(\eta)$, $\eta = R/(bt)^{1/3}$ at FDR in the reaction zone.

so that

$$\eta = \langle \eta \rangle u = \frac{3}{4}u$$

and

$$\varphi(u) = \frac{d\eta}{du}\varphi\left(\eta = \frac{3u}{4}\right) = \text{const} \frac{u}{(2-u)^4} \cdot \exp\left(-\frac{4}{2-u}\right), \quad 0 < u < 2$$

Thus, the mean size kinetics is "LSW-like" and the size distribution is "Hillert-like," demonstrating that FDR incorporates the features of both, ripening and grain growth.

For

$$\frac{n}{n_i} \approx 1, \quad C_i = \frac{6}{11}, \quad D \approx 10^{-5}\,\text{cm}^2\,\text{s}^{-1}$$

$$\delta \approx 5 \times 10^{-6}\,\text{cm}, \quad C^b - C^e \approx 0.001$$

where C^b is the equilibrium concentration of the melt and Cu_3Sn_1 aphase the growth rate

$$k \approx 4 \times 10^{-13}\,\text{cm}^3\,\text{s}^{-1}$$

For example, for a reflow time of $t = 300\,\text{s}$, we get $R \approx 5 \times 10^{-4}\,\text{cm}$, which is in good agreement with the experiment [25, 27].

6.5
Flux-Driven Grain Growth in Thin Films during Deposition

Numerous experiments show that at thin film deposition the size of a grain is comparable to the film thickness. This is especially true for vapor deposition of face-centered cubic metal films such as Al and noble metals on a fused quartz substrate at a few hundred degrees of centigrade. The thicker the film, the larger are the grains. It indicates that the very process of deposition somehow enables the grains to grow during deposition. The thickening of the film is accompanied by grain growth. It suggests a linear grain growth rate if the deposition rate is constant. It is very different from most of the known two-dimensional (2D) and three-dimensional (3D) models of grain growth providing the parabolic time dependence for the average grain size.

In this linear mode of grain growth (the model was developed jointly with Tu and Sobchenko), if we assume the grain radius to be the same as the film thickness, the total GB area remains constant during grain growth

$$S_{GB} = N \times \frac{1}{2} 2\pi r H = N \times \pi r^2 = S^{\text{substarte}} = \text{const}$$

This fact is illustrated in Figure 6.5, in which three sets of films of hexagonal grains with different grain size and thickness are shown on the same substrate area. If we assume the edge of a hexagonal grain to be the same as the film thickness, the total GB area in these three sets of film is the same. This is also true if we assume the grains to be cylindrical. Hence, we have a unique case of grain growth in which the total surface area is constant but the volume increases. We define it as flux-driven grain growth (FDGG); it depends on the incoming atomic flux of deposition [4]. This is quite different from the classical or normal grain growth in which the total volume of the grains is constant, but the grain boundary area decreases, and the grain growth rate is parabolic. In the previous section, we developed a kinetic theory of ripening, FDR, in which the ripening occurs under a constant surface area but growing volume. It is very different

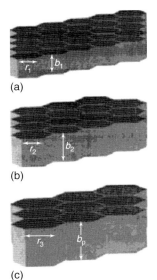

Figure 6.5 Schematic diagram to illustrate the conservation of total surface of grain boundaries during simultaneous deposition and lateral grain growth with proportionality between film thickness and lateral sizes of grains being maintained.

from the classic LSW ripening wherein the ripening phase, under the constraint of a constant volume, reduces its surface area. The driving force of FDR is the gain of bulk Gibbs free energy of IMC formation instead of the decrease in surface energy. Clearly, there is a strong similarity between FDR and FDGG; both are kinetic processes under the conditions of constant surface but growing volume.

6.5.1
"Mushroom Effect" on the Surface of a Pair of Grains: Deterministic Approach

In Figure 6.6, the cross section of a pair of large (radius r_1) and small (radius r_2) neighboring grains is shown. The top surfaces of the grains are assumed to be spherical with radii R_1 and R_2, respectively, and $R_1 > R_2$. We demonstrate below that the deposition flux enables the larger grain to grow by the so-called "edge" or "mushroom" effect and leads to a linear grain growth rate.

When atoms are being deposited onto the film surface, adatoms remain on the surface for some time, migrate on the surface, and are looking for a suitable place to join the lattice of one of the grains. We suppose that the migration length of adatoms is very short, and they tend to attach to the "home" grain surface at the place of their "landing." Furthermore, we suppose that the only place where a landing atom can choose its future "home" is located in a narrow band in the vicinity of a junction where a GB meets the surface. Denote the width of this band as $2d$, and take $a \leq d < d_{\text{sink}}$, where a is an interatomic distance, approximately equal to the spacing between atomic layers of the growing film, and d_{sink} is the distance to the nearest sink (surface step) for an adatom to join the lattice. We suppose that each atom, landing onto this narrow band, inevitably chooses the larger grain as its "home" and joins the lattice of this grain with corresponding orientation. There are atoms actually landing at the edge of the smaller grain

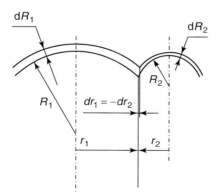

Figure 6.6 Schematic diagram of cross section of a pair of grains in thin film deposition. The edge of the large grain grows over the small grain due to the "mushroom" effect.

inside the 2d band, but they join the lattice of the larger neighbor. This means that the larger grain overlaps its neighbor by a distance d during the building up of a new atomic layer, leading to a linear grain growth rate in proportion to the thickening rate.

The reason for an adatom to choose the larger grain is simply a larger average radius of this grain, hence the lower potential energy of the corresponding sites. The difference in curvatures is not the only reason of choice. The difference in surface tensions of neighboring grains can well be a more powerful factor; it is known as a reason of abnormal grain growth in the absence of the deposition flux. Here, we do not take this factor into account explicitly. Yet, it is obvious that during the initial stage of film formation or islands growth, those grains with less surface tension should grow faster, and after the formation of the initial layer, they start as the larger grains. Thus, adatoms tend to choose them as their "home" grains.

In this simple model, the velocity of each GB is the same,

$$V = \pm \frac{d}{\Delta t} \quad (6.44)$$

where Δt is the time of one atomic layer deposition, which is determined by the flux density:

$$j S_t \Delta t = n S_t a \quad (6.45)$$

where n is atomic density of the film and S_t is substrate or film surface area, so that

$$\Delta t = \frac{an}{j}, \quad V = \pm \frac{d}{a} \frac{j}{n} \quad (6.46)$$

Hence, the rate of film thickening is

$$\frac{dh}{dt} = \frac{a}{\Delta t} = \frac{j}{n} = \frac{a}{d}|V| \quad (6.47)$$

Thus, the velocity of GB movement is approximately equal to the thickening rate and is independent of the absolute value of the differences between the curvatures of R_1 and R_2, as shown in Figure 6.6. The direction of motion is always from the larger grain to the smaller grain.

6.5.2
Analysis of Flux-Driven Grain Growth

In Figure 6.7, let S be the top surface area of a certain grain with Z neighbors. Then

$$\frac{dS}{dt} = \sum_{k=1}^{Z} L_k V_k \quad (6.48)$$

where L_k is the length of boundary with the kth neighboring grain and V_k is its velocity, positive if $S > S_k$ and negative if $S < S_k$. Naturally, the smaller the neighboring grain, the shorter (in average) should be the length of boundary with

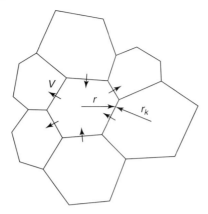

Figure 6.7 Schematic diagram of the top view of grain growth during deposition. Velocities of all boundaries are the same in absolute value and are directed from the larger grain to the smaller one.

it. We take simply that the grain size and the boundary length are proportional to each other:

$$L_k = qr_k, \quad r_k = \sqrt{\frac{S_k}{\pi}} \tag{6.49}$$

On the other hand, the perimeter of the grain $\sum_1^Z L_k$ can be approximated as $2\pi r$, $r = \sqrt{S/\pi}$. Thus,

$$q = \frac{2\pi r}{\sum r_k}, \quad L_k = 2\pi r \frac{r_k}{\sum_{i=1}^Z r_i} \tag{6.50}$$

Substituting Equation 6.50 in Equation 6.48, one obtains

$$2\pi r \frac{dr}{dt} = 2\pi r \frac{\sum_{k=1}^Z r_k V_k}{\sum_{k=1}^Z r_k} \tag{6.51}$$

$$\frac{dr}{dt} = \frac{\frac{1}{Z}\sum_{k=1}^Z r_k V_k}{\frac{1}{Z}\sum_{k=1}^Z r_k}, \quad V_k = \begin{cases} +|V|, r_k < r \\ -|V|, r_k > r \end{cases}$$

This is the basic equation for the growth or shrinkage of one arbitrary grain. It can be expressed in terms of a size distribution function if we neglect the "short-range order" effects (i.e., correlation between sizes of neighboring grains):

$$\frac{dr}{dt} = \frac{\frac{1}{Z}\sum_{k=1}^{Z} r_k V_k}{\frac{1}{Z}\sum_{k=1}^{Z} r_k} = \frac{\frac{1}{N}\left(\int_0^r r'|V|f(t,r')\,dr' + \int_r^\infty r'(-|V|)f(t,r')\,dr'\right)}{\frac{1}{N}\int_0^\infty r'f(t,r')\,dr'} \qquad (6.52)$$

Since a distribution function should obey a continuity equation, we have

$$\frac{\partial f}{\partial t} = -\frac{\partial}{\partial r}\left(f\frac{dr}{dt}\right) = -|V|\frac{\partial}{\partial r}\left(f\left(2\frac{\int_0^r r'f(t,r')\,dr'}{\int_0^\infty r'f(t,r')\,dr'} - 1\right)\right) \qquad (6.53)$$

Equation 6.53 indicates the possibility of (r/t)-scaling. Indeed, we introduce new variables:

$$\tau = |V|t, \quad \xi = \frac{r}{|V|t} = \frac{r}{\tau} \qquad (6.54)$$

$$\Rightarrow \frac{\partial}{\partial t} = |V|\left(\frac{\partial}{\partial \tau} - \frac{\xi}{\tau}\frac{\partial}{\partial \xi}\right), \quad \frac{\partial}{\partial r} = \frac{1}{\tau}\frac{\partial}{\partial \xi}$$

Then Equation 6.54 can be transformed into

$$\tau\frac{\partial f}{\partial \tau} = \xi\frac{\partial f}{\partial \xi} - \frac{\partial}{\partial \xi}\left(f\left(2\frac{\int_0^\xi \xi' f(\tau,\xi')\,d\xi'}{\int_0^\infty \xi' f(\tau,\xi')\,d\xi'} - 1\right)\right) \qquad (6.55)$$

A factorized solution is possible by taking

$$f(\tau,\xi) = g(\tau)\varphi(\xi) \qquad (6.56)$$

Then

$$\frac{d\ln g}{d\ln \tau} = \frac{d\ln \varphi}{d\xi}\left(1 + \xi - 2\frac{\int_0^\xi \xi'\varphi(\xi')\,d\xi'}{\int_0^\infty \xi'\varphi(\xi')\,d\xi'}\right) - \frac{2\xi\varphi}{\int_0^\infty \xi'\varphi(\xi')\,d\xi'} = \lambda \qquad (6.57)$$

We can determine the constant parameter λ from the constraint of conservation of surface area:

$$\text{const} = \int_0^\infty \pi R^2 f(t,R)\,dR = \pi g(\tau)\tau^3 \int_0^\infty \xi^2 \varphi(\xi)\,d\xi \qquad (6.58)$$

$$\Rightarrow g(\tau) = \text{const}\cdot \tau^{-3}$$

It leads to

$$\frac{d\ln g}{d\ln \tau} = \lambda = -3 \qquad (6.59)$$

Equation 6.59 makes it possible to find the type of time dependence for the average grain size and for the number of grains in the frame of our model:

$$\langle R \rangle = \frac{\int_0^\infty r' f(t, r') \, dr'}{\int_0^\infty f(t, r') \, dr'} = \frac{\tau^2 g \int_0^\infty \xi' \varphi(\xi') \, d\xi'}{\tau g \int_0^\infty \varphi(\xi') \, d\xi'} = \langle \xi \rangle \tau = \text{const} \times \tau \qquad (6.60)$$

$$N = \tau g \int_0^\infty \varphi(\xi') \, d\xi' = \frac{\text{const}}{\tau^2}$$

Thus, the average size in our model is indeed increasing linearly with time and, hence, it is proportional to the film thickness.

6.5.3
Stochastic Approach

So far, the choice of a "host" grain by adatoms within a narrow 2d band around the surface area of a GB was considered to be totally deterministic: adatoms choose the larger grain, independent of the difference in size of the two grains. It is physically evident that for a pair of grains having nearly the same size, the procedure could be stochastic. Therefore, we use a Boltzmann distribution of probabilities between two neighboring grains, taking into account the energy difference as $2\gamma\Omega(1/r_1 - 1/r_2)$. Then, from Equation 6.46 the velocity of GB is

$$V = \frac{d}{a} \frac{j}{n} \tanh\left[\frac{2\gamma\Omega}{kT}\left(\frac{1}{r_2} - \frac{1}{r_1}\right)\right] \qquad (6.61)$$

This equation resembles the well-known Potts model for normal grain growth, but deals only with adatoms arriving at the surface and choosing their host grain. In the mean-field approximation, the corresponding equation for size distribution has the following form:

$$\frac{\partial f}{\partial t} = -\frac{d}{a} \frac{j}{n} \frac{\partial}{\partial r} \left\{ f \tanh\left[\frac{2\gamma\Omega}{kT}\left(\frac{1}{\langle r \rangle} - \frac{1}{r}\right)\right]\right\} \qquad (6.62)$$

The numerical solution of Equation 6.62 has been performed for dimensionless time and space scales:

$$\rho = r/\lambda, \quad L = \frac{2\gamma\Omega}{kT}, \quad tt = t\,(d/a)\,j/nL$$

The typical evolution of the size distribution is shown in Figure 6.8. The time dependence of the average size depends on the ratio $\frac{\beta}{\langle r \rangle} = \frac{2\gamma\Omega}{kT\langle r \rangle}$. In the case $\frac{\beta}{\langle r \rangle} \ll 1$, the equations become Hillert-like

$$V = \frac{d}{a} \frac{j}{n} \frac{2\gamma\Omega}{kT}\left(\frac{1}{\langle r \rangle} - \frac{1}{r}\right) \qquad (6.63)$$

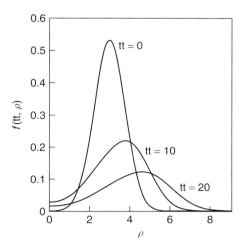

Figure 6.8 Typical size distributions, $f(tt, \rho)$, calculated according to Equation 6.62: (a) tt = 0; (b) tt = 10; (c) tt = 20; $\rho = r/L$, $L = 2\gamma\Omega/kT$, tt = $(d'/a)j'\ln L$.

and lead to a parabolic law for the mushroom effect instead of a linear one, but of course, the rate is, as before, determined by the flux density.

6.5.4
Monte Carlo Simulation of Flux-Driven Grain Growth

The above-given analytical and numerical results were checked by MC simulations. A full MC model, taking into account the rate of deposition, diffusion of adatoms, their distribution among sinks leading to grain growth, and possible changes of host grains at the free surface, was published in [38]. Here we present a simplified MC model, which almost directly corresponds to the deterministic analytical model, with the following assumptions:

1) Only those atoms arriving at the surface near a GB can choose their host grain.
2) If an atom arrived at the surface near a GB and if it is above the larger grain, it inevitably chooses this larger grain as its host.
3) If an atom arrived at the surface near a GB, but is above the smaller grain, it can nevertheless choose the larger grain as its host (edge effect) with probability $p = pn$ ($n = 0, 1, 2, 3$ in case of square lattice), depending on the number of nearest-neighbor atoms of the smaller grain excluding the atom just below.

Figure 6.9 demonstrates the typical top view of grain structure for the case $p3 = 0.1$, $p2 = 0.3$, and $p1 = p0 = 1$. Figure 6.10 shows the average grain size (in units of interatomic distance) versus the number of deposited atomic layers. In full correspondence with the analytical model, the relation appears to be linear. The initial nonmonotonic dependence is evidently due to random initial size distribution. Moreover, the proportionality coefficient, obtained by numeric

(a)   (b)   (c)

Figure 6.9 Top view: evolution of grain structure during deposition.

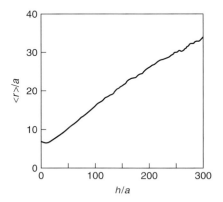

Figure 6.10 Average lateral size versus height in MC modeling.

solution of Equation 6.53 and calculation of $\langle R \rangle$ versus h, appears to be 0.406, which is rather close to the MC result (0.390).

6.5.5
Lateral Grain Growth in Aluminum Nanofilm during Deposition

Grain structure evolution of Al nanofilms during deposition at room temperature was investigated by transmission electron microscopy (the results outlined in the present section have been obtained in a common work with Ihor Sobchenko, Guido Schmitz, and Dietmar Baither and are published in [39]). Al thin films have been produced by ion-beam sputtering from Al targets of 99.99% purity. The typical background pressure of the vacuum chamber was 0.5×10^{-9} mbar, while during sputtering argon gas was introduced to a maximum pressure of 0.5×10^{-5} mbar. Silicon wafers or electron transparent carbon thin films were used as substrates for cross section or plane view images respectively. For the preparation of cross sections, a standard technique of grinding and ion-beam polishing was used. To be sure that aluminum films have the same structure on both, silicon and carbon substrates, the mean grain size was measured for layers deposited under identical conditions on both types of substrates. For both substrates, the mean grain sizes were found to be in close agreement. Specimens were investigated with a Zeiss TEM LIBRA 200-FE.

A fundamental question to the mechanism of growth is whether grains keep growing laterally inside the film, after being covered by subsequently deposited material. To check this, thin films were deposited with various thicknesses. First, a cross-sectional view of a very thin film ($h = 10$ nm) was investigated and the average lateral grain size was determined at the top surface. In the same manner, the lateral grain size was determined in cross sections of thicker films (25, 50, 100, and 250 nm) always (see the example in Figure 6.11). It was found that the average lateral grain size was almost identical for all of them. This means that there was no migration of grain boundaries at room temperature after the GB became covered by subsequent material.

To determine the dependence of mean grain size on the film thickness, we also used cross-sectional views as shown in Figure 6.11. Micrographs were acquired in dark and bright field mode to identify GBs reliably. The mean grain size was measured at different distances from the substrate in the range of 10–250 nm. It was found that for Al thin films the average lateral grain size fits a parabolic growth law (not linear!),

$$\langle R^2 \rangle = \langle R^2 \rangle_0 + \kappa h \tag{6.64}$$

in dependence on film thickness h (Figure 6.12). This experimental observation is in obvious contrast to the linear relation assumed or found in recent theoretical work (see first part of this section). An important consequence is that the grain shape, for example, the width-to-length aspect ratio varies with deposition time. At our experimental conditions, nucleation of new grains took place only at the interface to the substrate. At the initial stage, grains are quite flat. During further deposition, those grains that "survive" overlap the smaller ones and, hence, become more and more elongated.

Figure 6.11 Cross-sectional view, Libra 200 FE, dark field. Grain structure after deposition of a 250-nm film at room temperature.

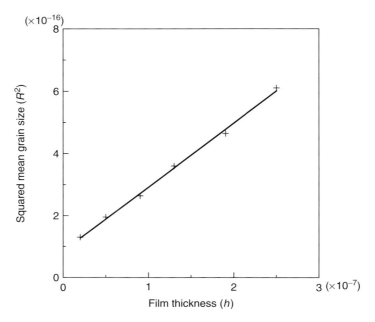

Figure 6.12 Squared mean grain size dependence on film thickness.

In the first paper on FDGG [4], the linear dependence $\bar{R} \propto h$ was taken for granted, and the model was adjusted to fit this dependence. The main idea was a so-called mushroom (edge) effect – the grain structure of a new atomic layer is basically inherited from the previous layer, but can slightly differ from the previous, in a very narrow band (of atomic thickness, 2δ) just above the GB. If an adatom is about to become attached in this region, it can choose any of two adjacent grains as host. In order to obtain the linear growth law, we had assumed in [4] that the above-mentioned choice is done in a deterministic way: an atom that lands within the 2δ band always chooses the larger grain as its host. So, the larger grain extends in each new deposited layer and gradually overlaps the neighboring smaller grain. This scheme has led authors, after doing some mathematics, to the conclusion that the average lateral diameter should develop according to $2\langle R \rangle \approx 0.8 \cdot (\delta/d) \cdot h$. Here d is the interplanar distance. In contrast, the presented experiment with Al deposition demonstrated that the growth of the lateral mean size is substantially slower and, moreover, it is better approximated by the parabolic dependence Equation 6.64. Thus, a revision of the FDGG model is required.

It is evident that in such reconsideration that it is sufficient to take into account thermodynamic probabilities when the choice of the host grain is made, the probability of choosing one of the two grains should be proportional to the tanh function, as in the Potts model. The probability of choosing the host grain by any atom at the GB is determined by the local curvature of GB. In our experiments, the surface of Al grains looked flat. So, from here we take into account only the curvature of the GB joint with the surface, but not the curvature of the surface

6.5 Flux-Driven Grain Growth in Thin Films during Deposition

itself. The local curvature, $1/r$, generates the difference in pressures across the GB (Laplace pressure/tension, depending on curvature sign): $\Delta p = \gamma/r$ (in 2D case). This means that the enthalpy changes steplike as $\Delta H = \Omega \Delta p = \gamma \Omega /r$. Evidently, the Laplace tension is a macroscopic parameter, so that r is a mean-field radius, averaged over at least several interatomic distances along the GB. Therefore, in our phenomenological scheme below, we neglect the change in local curvature after choosing of host grain by a single atom. Then the probabilities of belonging to one of the neighboring grains is equal to

$$\frac{e^{\pm \frac{\gamma \Omega}{2rkT}}}{e^{\frac{\gamma \Omega}{2rkT}} + e^{-\frac{\gamma \Omega}{2rkT}}}$$

Thus, the local lateral shift of GB after deposition of new atomic layer d is equal to $+\delta$ with the probability

$$\frac{e^{+\frac{\gamma \Omega}{2rkT}}}{e^{\frac{\gamma \Omega}{2rkT}} + e^{-\frac{\gamma \Omega}{2rkT}}}$$

and $-\delta$ with the probability,

$$\frac{e^{-\frac{\gamma \Omega}{2rkT}}}{e^{\frac{\gamma \Omega}{2rkT}} + e^{-\frac{\gamma \Omega}{2rkT}}}$$

so that average lateral displacement is

$$\langle \Delta x \rangle = \delta \cdot \tanh\left(\frac{\gamma \Omega}{2rkT}\right)$$

Accordingly, the velocity of GB lateral motion is also proportional to the tanh function

$$V = \frac{\langle \Delta x \rangle}{\Delta t_{\text{layer}}} = \frac{\langle \Delta x \rangle}{(d/(dh/dt))} = \frac{dh}{dt}\frac{\delta}{d} \cdot \tanh\left(\frac{\gamma \Omega}{2kT}\frac{1}{r}\right) \quad (6.65)$$

If the ratio $\gamma \Omega /(kTr)$ is large, then we obtain results of our previous FDGG model, with

$$V = \pm \frac{dh}{dt}\frac{\delta}{d}$$

depending on curvature direction. If the ratio $\gamma \Omega/(kTr)$ is small enough, that is, if

$$r = 20 \text{ nm}, \quad \gamma = 0.5 \text{ Jm}^{-2}, \quad \Omega = 2 \times 10^{-29} \text{ m}^3, \quad T = 300 \text{ K}$$

then this ratio is marginal (about 0.12) and one can approximately replace the tanh function by its argument, and we obtain the situation coinciding with the Neumann–Mullins approach for normal grain growth in the 2D case [40, 41], when the local velocity of any boundary is proportional to its local curvature:

$$\text{(a): } V = M\frac{\gamma}{r}, \quad \text{(b): } M = \frac{dh}{dt}\cdot\frac{\delta}{d}\cdot\frac{\Omega}{2kT} \quad (6.66)$$

As proven by many authors, the dependence of the type (a) definitely implies a parabolic law for the mean size, as in the case of normal grain growth. The basic peculiarity here is that the effective lateral mobility M of the boundary in our case (FDGG) is proportional to the deposition rate. It is quite different from the normal grain growth when the mobility M is determined by the kinetics of detachment of an atom from one grain structure and incorporation into another one. In our case, atoms have no need to detach – they just arrive and choose the "host" grain.

The following consequences (growth rate of the mean size and size distribution) depend on the choice of the model for grain growth. We treat three types of models: Hillert's (1965) [17] model, leading to the Rayleigh distribution [5, 18, 19, 22], and Di Nunzio's pair interaction model [42]. Each of these models leads to a parabolic growth law for the mean size of the form $\overline{R}^2 = \overline{R}_0^2 + kt$. In all these models, authors relate (in various manners) the local curvature of GB to the sizes of grains R_1, R_2 divided by this GB. Such interrelation is not straightforward, and not evident, but the transition from curvature radius to grain sizes is convenient for application of mean-field approaches. For example, in Hillert's model, $1/r$ is replaced by $\frac{1}{\langle R \rangle} - \frac{1}{R}$ (the local Laplace tension is replaced by the difference in effective Laplace tensions in neighboring grains).

In our case, the more convenient form of dependence is the following:

$$\langle R^2 \rangle = \langle R^2 \rangle_0 + \kappa h, \qquad \kappa = k/(dh/dt) \tag{6.67}$$

Interrelations between the growth rate k (or κ) and mobility M are all (for all mentioned kinds of models) linear,

$$k = p \cdot M\gamma, \qquad \kappa = p\frac{M\gamma}{dh/dt} \tag{6.68}$$

but with different proportionality coefficients p, and what is more important is that the size distributions are different.

We want to check now

1) whether our simplified model of flux-driven lateral grain growth due to the chosen procedure (without thermally activated motion of GBs) predicts a reasonable kinetic law for the average lateral sizes
2) and, if the answer is "yes," then what kind of model (after replacement of GB mobility by Equation 6.66) provides the lateral size distribution close to experiment.

6.5.5.1 Hillert's Model

In Hillert's model [17], Equation 6.65 for individual GB motion translates into an equation for growth/shrinkage of individual grains with some effective size R_i of the form

$$\frac{dR_i}{dt} = M\gamma \left(\frac{1}{\langle R \rangle} - \frac{1}{R_i} \right) \tag{6.69}$$

An LSW-type analysis of Equation 6.69 together with the continuity equation in the size space provides an asymptotic parabolic law and an asymptotic normalized distribution of reduced sizes

$$\langle R^2 \rangle - \langle R^2 \rangle_0 = \frac{1}{2} M\gamma t \tag{6.70}$$

Table 6.1 Parameters of distributions.

	$s = \langle(u-1)^2\rangle$	$sk = \frac{\langle(u-1)^3\rangle}{s^3}$	$K = \frac{\langle(u-1)^4\rangle}{s^4} - 3$	p
Hillert	0.33	−0.430	−0.400	0.50
Rayleigh	0.52	0.590	0.065	0.79
Di Nunzio	0.40	0.039	−0.490	2.70
Experiment	0.42	0.230	0.260	1.45

$$g(u) = \frac{8u}{(2-u)^4} \cdot \exp\left(-\frac{2u}{2-u}\right), \quad u \equiv \frac{R}{\overline{R}}$$

(so that the factor p in Equation 6.68 is equal to $p_{\text{Hillert}} = 0.5$). The Hillert size distribution is characterized by a standard deviation $s \approx 0.33$, skewness $sk \approx -0.43$, kurtosis $K \approx -0.40$ (see also Table 6.1).

6.5.5.2 Models Leading to a Rayleigh Distribution

For the first time, the Rayleigh distribution for normal grain growth

$$f(t, R) = C \cdot R \cdot \exp\left(\frac{-R^2}{\langle R^2 \rangle}\right) \tag{6.71}$$

with

$$s \approx 0.52, \quad sk \approx 0.59, \quad K \approx 0.065)$$

was obtained by Louat in 1974 [22], who treated grains as "drunken sailors" randomly migrating in the semi-infinite size space, and disappearing after reaching the zero point. Fradkov et al. [19] proposed a model using the Neumann–Mullins theorem plus account of switching events for ternary joints (vertexes). They obtained a distribution that almost coincided with the Rayleigh one. The same Rayleigh distribution was obtained by Gusak and Tu in the framework of an alternative approach – growth kinetics in the normalized size space [5]. In this approach, for the 2D case, the main growth/shrinkage equation is the following

$$\frac{dR_i}{dt} = \frac{M\gamma}{2}\left(\frac{R_i}{\langle R^2 \rangle} - \frac{1}{R_i}\right) \tag{6.72}$$

Standard analysis of Equation 6.72 leads to

$$\langle R^2 \rangle = M\gamma t, \quad \overline{R}^2 - \overline{R}_0^2 = \frac{\pi}{4} \cdot M\gamma t$$

so that $p_{\text{Rayleigh}} = \pi/4 \approx 0.79$ (see Table 6.1).

6.5.5.3 Pair Interaction Model (Di Nunzio)

Di Nunzio [42] recently proposed one more approach to the kinetics of grain growth taking pair interactions into account and using a more complicated interrelation between sizes and curvatures. We implemented his algorithm (as we understood it from [42]) and repeated the same results for the case of 2D growth with absent

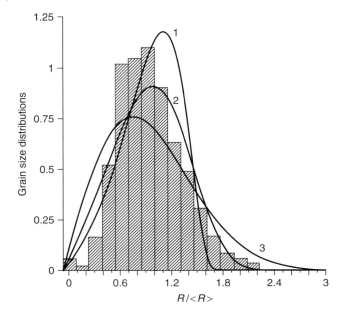

Figure 6.13 Grain size distributions: Hillert (dashed), Rayleigh (dashed-dotted), Di Nunzio (full curve), and histogram from our experiment (1325 grains processed, $R = \sqrt{\text{grain top area}/\pi}$).

inhibitors. It gave the following values for standard deviation, skewness, and kurtosis: $s \approx 0.40$, $\text{sk} \approx 0.039$, $K \approx -0.49$. For the mean size growth, we obtained $p_{\text{DiNunzio}} \approx 2.7$ (see Table 6.1).

The experimental grain size distribution is shown as a histogram in Figure 6.13. The main parameters are $s \approx 0.42$, $\text{sk} \approx 0.23$, and $K \approx 0.26$. Thus, experimental data are somewhere in between Di Nunzio and Rayleigh: the standard deviation almost coincides with Di Nunzio, skewness lies quite in the middle between Di Nunzio and Rayleigh, and kurtosis seems to be too sensitive (Table 6.1).

To obtain the experimental value for the factor p, we should, firstly, evaluate the value of κ in Equation (6.68),

$$\frac{M^{\text{FDGG}} \gamma}{dh/dt} = \frac{\delta}{d} \cdot \frac{\gamma \Omega}{2kT}$$

We take

$$d = d_{111} = \frac{a}{\sqrt{3}}$$

We roughly evaluate the maximal possible shift of GB in the next atomic layer as a distance from the atom to the nearest center of the triangle:

$$\delta \approx \frac{2\sqrt{3}}{3} \cdot 2r_0 = \frac{a}{\sqrt{6}}, \quad \frac{\delta}{d} = \frac{1}{\sqrt{2}}, \quad \Omega = a^3/4$$

Of course, the evaluation of the half-width δ is ambiguous, but gives the correct order of magnitude. So far, we have obtained

$$\frac{\delta}{d} \cdot \frac{\gamma \Omega}{2kT} \approx 1.42 \times 10^{-9}, \quad \kappa^{\text{experiment}} = 2.06 \times 10^{-9} \text{ m}$$

Thus, $p^{\text{experiment}} \approx 1.45$ (between Rayleigh-type and Di Nunzio models). Contrary to normal grain growth, when mobility is not well defined, in our case (FDGG) the lateral growth rate depends mainly on geometric (δ, d, Ω) and thermodynamic (γ) parameters.

One can see that at least the order of magnitude for the lateral grain growth constant, predicted by the modified FDGG approach, coincides with experimental data.

To conclude, the lateral grain growth during deposition can be described in terms of FDGG due to edge effects, when adatoms in the vicinity of GBs choose their host grain. The effective mobility of this lateral grain growth is proportional to the deposition rate. In case of Al deposition, the lateral grain size distribution is not far from the combination of FDGG with models of Di Nunzio or Gusak and Tu proposed for the usual thermally activated grain growth (more close to Di Nunzio).

6.6
Flux-Induced Instability and Bifurcations of Kirkendall Planes

The fundamental ideas and all the experiments presented in this section have been originated by Prof. Frans van Loo's group. The author was lucky to enter this group in Eindhoven Technical University practically at the moment of discovery (beginning of 2000) and to formulate the conditions of stable and unstable K-planes including the prediction of the only unstable K-plane. The initial concentration difference in the diffusion couple contains its entire future, similar to the fire ball of the Big Bang that contained the future of the Universe. Furthermore, just as a minor fluctuation in the Universe's fire ball may result in a the new inhomogeneous structure, a small fluctuation in Kirkendall markers distributed near the initial interface may lead to bifurcations and instabilities of K-planes.

From the intuitive viewpoint, it seems strange that the markers initially "smear" over the whole concentration range of the diffusion couple (located in quite a narrow region, though), and then gather into one plane which corresponds to only one fixed (constant) composition; this plane is generally referred to as the *Kirkendall plane*. It is accepted [31–35] that the (K-planes is a plane of initial contact moving at parabolic dependence

$$X_K(t) - X_K(0) = \text{const} \cdot \sqrt{t}$$

at a velocity proportional to the partial diffusion coefficients and concentration gradient in the region with fixed composition:

$$c_B(t, X_k) = c_B\left(\frac{X_K(t) - X_K(0)}{\sqrt{t}}\right) \equiv c_K \quad (6.73)$$

$$U_K = (D_B(N_K) - D_A(N_K))v_B \frac{\partial c_B}{\partial x}\bigg|_{x=X_k} \qquad (6.74)$$

(here c_B is the number of B atoms in the unit volume, v_B is a partial molar volume, $N \equiv N_B = v_m c_B$ is the molar fraction). The K-plane is supposed to be an attractor for markers.

Logically, a few questions arise: why should a system have only one attractor? Is it possible for a binary system to have two or more attractors? Can the diffusion proceed in the system without attractors (stable K-planes), with marker distributions broadening with time? In the latter case, a system may "forget" the initial contact plane that results in an especially tight connection between the materials. Thus, the problem of instabilities and bifurcations of K-planes appears. As shown below, this problem has a sound experimental basis.

6.6.1
Kirkendall Effect and Velocity Curve

According to Darken, the Kirkendall effect appears due to the inequality of own diffusion fluxes causing the vacancy flux [43–45]. If vacancy sources/sinks are effective enough, the vacancy concentration is in equilibrium everywhere. This means that the vacancy flux makes lattice planes disappear from the side of the faster diffusant and form from the side of a slower one. Thus, the lattice moves and it can be visualized with the help of inert markers (for example, tungsten wires or micron-size ThO_2 – *particles*). To obtain a clearer image of lattice movement, one may apply a so-called multilayer technique [36]. In this case, each half of the initial diffusion couple consists of thin stripes (10–20 µm), and the inert markers are placed at each interface between the layers. In the process of annealing, the diffusion zone widens, covering more and more layers, and, thus, more markers. Then one can measure the displacement of each plane of markers at a certain time, t. Having the displacement $\Delta x_i(t)$, it is possible to find a so-called Kirkendall velocity curve $U(t, x)$ (see Equation 6.94 below) [45–50].

Obviously, there is only one K-plane with the markers located at the initial contact surface; it moves at a parabolic dependence

$$X_K(t) = X_K(0) + \text{const}\sqrt{t}$$

$$U_K = \frac{dx_K}{dt} = \frac{\text{const}}{2\sqrt{t}} = \frac{\Delta X_K}{2t} \qquad (6.75)$$

Therefore, we may use a simple geometric presentation of a K-plane: it corresponds to an intersection of a straight line

$$U = \frac{1}{2t}(X_K - X_K(0))$$

and the marker velocity plot, $U(t, x)$.

6.6.2
Stable and Unstable K-Planes

It can be easily seen that for the case shown in Figure 6.14 the K-plane is a stable one. Indeed, if due to some incidental noise one of the markers is found to be slightly ahead of the K-plane (Figure 6.14), its velocity is less than that of the K-plane, so the marker comes back to the plane. If the marker is located behind the K-plane, its velocity increases and it also reaches the K-plane. Therefore, if a K-plane is fixed with an intersection in the region with negative slope of the velocity curve, it is a moving attractor for markers.

There are several systems for which the difference $D_A - D_B$ can change the sign depending on the compound: for example, for the β phase of Au–Zn, Ni–Al, Co–Ga, and so on. In such a case, the straight line $U = \frac{1}{2t}x$ and markers velocity plot may have three intersections rather than only one (Figure 6.15), and three possible K-planes.

We can say that K_1 and K_3 must be stable, while the K_2-plane must be "virtual" and unstable. Indeed, it corresponds to a positive slope of the velocity curve. So, a marker placed in front of the K-plane moves faster than it and does not return to it. The marker, located behind the K-plane, has a smaller velocity and also leaves it.

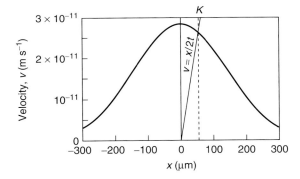

Figure 6.14 Velocity curve for a single stable K-plane.

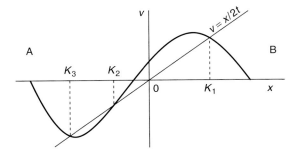

Figure 6.15 Velocity curve for the case of three K-planes: two are stable (1, 3) and one is unstable (2).

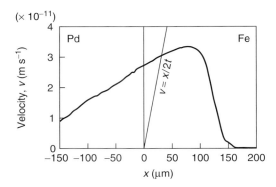

Figure 6.16 Velocity curve for a single unstable K-plane (real system Pd–Fe: annealing at 1100 °C for 144 h.

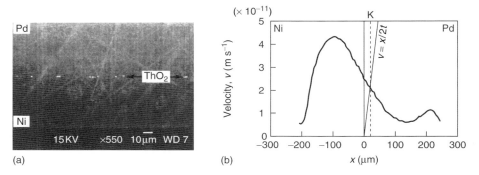

Figure 6.17 Location of the stable K-plane (left) and velocity curve (right) for the couple Ni–Pd annealed at 1100 °C for 121 h.

Thus, all markers that could be a part of K_2 would be captured by K_1 and K_3. It is also possible in case of one (unstable) K-plane (systems Fe–Pd and $Au_{40}Zn_{60}$-$Au_{70}Zn_{30}$; Figures 6.16 and 6.17). All the cases indicated above were found experimentally.

6.6.3
Experimental Results

1) A single, stable K-plane: ThO_2-particles, placed at the initial contact of Ni/Pd diffusion couple, join into one K-plane and are clearly observed (Figure 6.17).
2) A single, unstable K-plane: The initial row of ThO_2-particles located at the initial contact of Fe/Pd diffusion couple transform into a "smeared" set of markers (Figure 6.18). In this case, provided no other reasons for such spread are "to blame," the diffusion couple consisting of two previously made two-phase aurum–zinc alloys, $Au_{40}Zn_{60}$-$Au_{70}Zn_{30}$, presents no doubts (Figure 6.19), especially if we consider better enlargement (Figure 6.20).

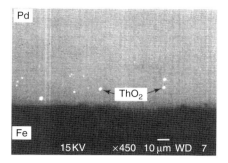

Figure 6.18 Markers spread for the system Fe–Pd (annealed at 1100 °C for 144 h).

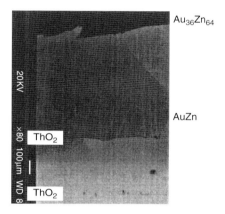

Figure 6.19 Two stable K-planes moving simultaneously in $Au_{36}Zn_{64}$-AuZn from the region of the new lattice formed in between them (annealed at 500 °C for 17.25 years).

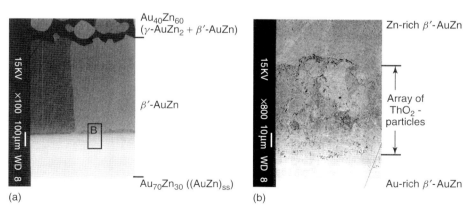

Figure 6.20 (a) Markers spread in the system $Au_{40}Zn_{60}$-$Au_{70}Zn_{30}$ (annealed at 500 °C for 6 h). (b) Detailed spread of markers for the same system.

3) Multiple K-planes: Experimental proof of the case given in Figure 6.15 was found for the system Au–Zn (Figure 6.19). Two K-planes appear to be stable (K_1 and K_3) and are observed in the interaction zone.

6.6.4
General Instability Criterion

As shown before, K-plane(s) appear(s) to be stable if it (they) Correspond(s) to a negative slope of the velocity curve. A K-plane is considered to be unstable (virtual) if it is found at a positive slope of the velocity curve. Let us try to derive the general criterion for K-instability taking into account the concentration dependence of diffusion coefficients, which makes it somewhat difficult to find the expression for the velocity curve.

The condition of K-plane instability,

$$\left.\frac{\partial U}{\partial x}\right|_{x_K} > 0 \tag{6.76}$$

can be written explicitly, having introduced the substitution $\xi = x/\sqrt{t}$ and using Fick's law:

$$\left(-\frac{\xi}{2}\frac{dc_B}{d\xi} = \frac{d}{d\xi}\left(\tilde{D}\frac{dc_B}{d\xi}\right)\right) : 0 < \frac{d}{d\xi}\left((D_B - D_A)\upsilon_B\frac{dc_B}{d\xi}\right)_{\xi_K}$$

$$= \frac{d}{d\xi}\left(\frac{D_B - D_A}{\tilde{D}}\upsilon_B\frac{dc_B}{d\xi}\right)_{\xi_K}$$

$$= \frac{d}{dc_B}\left(\frac{D_B - D_A}{\tilde{D}}\upsilon_B\right) \cdot \tilde{D}\frac{dc_B}{d\xi}\bigg|_{\xi_K} + \frac{D_B - D_A}{\tilde{D}}\upsilon_B\frac{d}{d\xi}\left(\tilde{D}\frac{dc_B}{d\xi}\right)$$

$$= \frac{d}{d\xi}\left(\frac{D_B - D_A}{\tilde{D}}\upsilon_B\right) \cdot \tilde{D}\left(\frac{dc_B}{d\xi}\right)^2 - \frac{\xi_K}{2}\frac{dc_B}{d\xi}\bigg|_{\xi_K}\frac{D_B - D_A}{\tilde{D}}\upsilon_B \tag{6.77}$$

The value of ξ_K for the K-plane is determined by the equation:

$$(D_B - D_A)\upsilon_B\frac{\partial c_B}{\partial x}\bigg|_{x_K} = U_K = \frac{dx_K}{dt} = \frac{d}{dt}\left(\xi_K\sqrt{t}\right) = \frac{\xi_K}{2\sqrt{t}} \tag{6.78}$$

resulting in

$$\frac{\xi_K}{2} = (D_B - D_A)\upsilon_B\frac{dc_B}{d\xi}\bigg|_{\xi_K}$$

Substituting Equation 6.77 in Equation 6.76, we get the criterion

$$\tilde{D}\left(\frac{dc_B}{d\xi}\right)^2 \cdot \left[\frac{d}{dc_B}\left(\frac{D_B - D_A}{\tilde{D}}\upsilon_B\right) - \left(\frac{D_B - D_A}{\tilde{D}}\upsilon_B\right)^2\right] > 0 \tag{6.79}$$

A transformation of the expression obtained in Equation 6.79 leads to the inequality

$$\frac{d}{dc_B}\left(\frac{\tilde{D}}{(D_B - D_A)\upsilon_B}\right) < -1 \tag{6.80}$$

The interdiffusion coefficient relates to partial diffusion coefficients according to Darken (remembering that partial volumes have different values):

$$\tilde{D} = c_A \upsilon_A D_B + c_B \upsilon_B D_A$$

Then,

$$\frac{d}{dc_B}\left(\frac{c_A \cdot \upsilon_A/\upsilon_B + c_B r_{AB}}{1 - r_{AB}}\right) < -1 \qquad (6.81)$$

where $r_{AB} \equiv D_A/D_B$. For simplicity, we further treat the case when $\upsilon_A = \upsilon_B = \upsilon =$ const. Then, the instability criterion is reduced to such inequality:

$$\frac{d}{dN_B}\left(\frac{D_A}{D_B}\right)\bigg|_{N_K} < 0 \qquad (6.82)$$

The above inequality is true when, for example, the value $D_A/D_B - 1$ turns from positive to negative as the B-content increases.

We note that the instability criterion in the form of Equations 6.79–6.81 can be applied only for the K-plane of special composition N_K. This is the additional condition, similar to the sign of the second-order derivative determining the type of extremum (maximum, minimum, or inflection point) for some function (the first derivative being equal to zero). The condition Equation 6.76 can be used for finding the prospective unstable "candidates."

6.6.5
Estimation of Markers' Distributions Near the Virtual K-Plane

Apparently, somewhere near the unstable K-plane, the system has one or two stable ones, the latter acting as attractors and gathering all markers initially located at the virtual K-plane. As shown above, a case when there is only one unstable K-plane may be observed. Then the initial distribution of markers simply sprawls out (the system gradually forgets the initial contact plane).

Let Δx be the distance between two markers near the virtual K-plane (for which $(\partial U/\partial x) > 0$). The relative velocity of these markers is roughly equal (for small Δx) to

$$\Delta U = \frac{d\Delta X}{dt} \approx \frac{\partial U}{\partial X}\bigg|_{X_K} \Delta X \qquad (6.83)$$

The composition of each (even virtual) K-plane must remain constant; therefore,

$$\frac{\partial U}{\partial x}\bigg|_{X_K} = \frac{1}{\sqrt{t}}\frac{\partial U}{\partial \xi} = \frac{1}{t}\frac{d}{d\xi}\left((D_B - D_A)\upsilon_B \frac{dc_B}{d\xi}\bigg|_{\xi_K}\right) \qquad (6.84)$$

Denoting this constant as

$$t\frac{\partial U}{\partial x}\bigg|_{X_K} = \frac{d}{d\xi}\left((D_B - D_A)\upsilon_B \frac{dc_B}{d\xi}\bigg|_{\xi_K}\right) \equiv \Upsilon \qquad (6.85)$$

Then Equation 6.82 is reduced to

$$\frac{d\Delta x}{dt} = \Upsilon \frac{\Delta x}{t} \tag{6.86}$$

so that

$$\ln \Delta x = \Upsilon \ln t + \text{const} \tag{6.87}$$

Let Δx_0 be the initial distance between the markers. Then, formally, we may write

$$\frac{\Delta x(t)}{\Delta x_0} = \left(\frac{t}{t_0}\right)^\Upsilon \tag{6.88}$$

where the value of the "initial moment" is found from the initial conditions.

If, say, one marker is placed at the contact plane ($x_1 = 0$) and the other is at some distance Δx_0 from it, then it takes some time to involve the other marker in the diffusion process. If we suppose that at the initial stage an ordinary bulk diffusion takes place, we may take

$$t_0 \sim (\Delta x_0)^2 / 2\tilde{D}$$

$\tilde{D}$ standing for maximal diffusion coefficient for the whole concentration interval. In this case, the time dependence becomes

$$\Delta x(t) \approx \Delta x_0 \left(2\tilde{D}t/\Delta x_0^2\right)^\Upsilon \tag{6.89}$$

The behavior of markers in the multilayered sample may be calculated numerically. This is what the next section features.

6.6.6
Spatial Distribution of Markers

The above-described experiment on the markers smearing in the diffusion couple is incorporated into the model describing the spatiotemporal evolution of the distribution of markers, $\rho(t, x)$. Here $\rho(t, x)\,dx$ is the number of markers in a thin layer $(x, x + dx)$. Obviously, from the condition of markers' conservation, it follows that the continuity equation must be fulfilled

$$\frac{\partial \rho}{\partial t} = -\frac{\partial j}{\partial x} \tag{6.90}$$

j standing for the flux of markers' density. For inert markers, it equals

$$j = \rho U = \rho(D_B - D_A)\,v_B\frac{\partial c_B}{\partial x} \tag{6.91}$$

Further, we treat (for simplicity) the case of constant and equal partial volumes for all constituents. Then, the time evolution of markers' distribution is described by the system of two equations

$$\frac{\partial N_B}{\partial t} = \frac{\partial}{\partial x}\left((N_B D_A + (1 - N_B) D_B)\frac{\partial N_B}{\partial x}\right) \tag{6.92}$$

$$\frac{\partial \rho}{\partial t} = -\frac{\partial}{\partial x}\left(\rho(D_B - D_A)\frac{\partial N_B}{\partial x}\right)$$

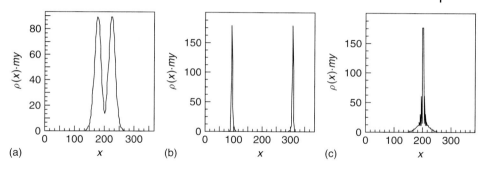

Figure 6.21 Markers redistribution during interdiffusion:
(a) one unstable K-plane, (b) two stable K-planes, and
(c) one stable K-plane.

The second of these relations is not very convenient for numerical calculations, since the initial distribution is Gaussian with very small dispersion, rather resembling the Dirac δ-function. So, we rewrite it taking $\ln \rho$ as the unknown function:

$$\frac{\partial \ln \rho}{\partial t} = (D_B - D_A) \frac{\partial N_B}{\partial x} \frac{\partial \ln \rho}{\partial x} - \frac{\partial}{\partial x}\left((D_B - D_A)\frac{\partial N_B}{\partial x}\right) \quad (6.93)$$

Typically, the partial diffusion coefficients D_B and D_A depend exponentially on composition with a linear dependence on N_B in the exponent. The coefficients in this dependence have been varied with the aim of getting different forms of the velocity curve. We have drawn several typical examples of numerical simulation for the single but unstable K-plane (Figure 6.21a, the peaks moving in opposite directions and eventually spreading out), two stable K-planes (Figure 6.21b, two peaks moving in opposite directions and approaching the Dirac delta-function), and one stable K-plane (Figure 6.21c, one peak becoming narrower and narrower).

6.6.7
Possible Alternative to the Multilayer Method

As proven, in order to find K-planes as well as the concentration dependencies for partial coefficients, one should have the velocity curve. The usual experimental method, providing this, is the multilayer method. The diffusion couple consists of several thin layers (generally, 10–20 μm thick) with inert markers placed between them. After annealing, the displacement of each marker plane is measured, and the velocity of each plane is determined by the equation [34, 35]

$$U(t, x) = \frac{1}{2t}\left(y - x_0 \frac{dy}{dx_0}\right) \quad (6.94)$$

Here the displacement is $y(t, x_0) = x(t, x_0) - x_0$, x being the marker's coordinate (at time t), and at the starting time the marker was at x_0 point. All coordinates are written in Matano's reference frame ($x = 0$ corresponding to the Matano plane).

Obviously, the calculation of the finite-difference derivative for markers, located in the 10–20 μm distance, leads to a great mistake and poor accuracy of the method viewed. Discussions with Prof. van Loo resulted in the formulation of the alternative method grounding upon the initial homogeneous markers' distribution. Here we present some algebra demonstrating the advantages of the new method.

Assume the inert markers to be homogeneously distributed over the diffusion couple before annealing. Actually, one may employ the steplike distribution (provided the markers' distribution is different but constant in the left and right halves of the diffusion couple). Let

$$\rho(t, x) = \frac{M(x, x + dx)}{dx} = \frac{dM}{dx}$$

be a markers' distribution after annealing for time t. Here $M(x, x + dx) = dM$ stands for the number of markers inside the layer $(x, x + dx)$ at time t. Markers neither disappear nor arise (the law of conservation of markers number). Moreover, they cannot mix, that is, the difference $x_2(t) - x_1(t)$ changes its value but not the sign with time. This means that the markers that belong to some thin layer $(x_0, x_0 + dx_0)$ will belong to it all the time, while the layer itself will undergo changes in its location and thickness.

$$x_0 \to x(t, x_0)$$

$$x_0 + dx_0 \to x(t, x_0 + dx_0) = x(t, x_0) + \overbrace{\frac{\partial x}{\partial x_0} dx_0}^{dx}$$

This implies

$$\rho_0(x_0) dx_0 = \rho(t, x(t, x_0)) \frac{\partial x}{\partial x_0} dx_0$$

Thus,

$$\frac{\partial x}{\partial x_0} = \frac{\rho_0(x_0)}{\rho(t, x(t, x_0))} \tag{6.95}$$

For further use of Equation 6.95, we rewrite Equation 6.94 in the following way:

$$U = \frac{1}{2t}\left(x - x_0 - x_0\left(\frac{\partial x}{\partial x_0} - 1\right)\right) = \frac{1}{2t}\left(x - x_0 \frac{\partial x}{\partial x_0}\right) \tag{6.96}$$

Substituting Equation 6.95 in Equation 6.96, we obtain

$$U = \frac{1}{2t}\left(x - x_0 \frac{\rho_0(x_0)}{\rho(t, x(t, x_0))}\right) \tag{6.97}$$

Equation 6.97 is quite inconvenient since it includes both x and x_0. To eliminate this inconvenience, we treat only the case when the initial markers' distribution was homogeneous:

$$\rho_0(x_0) = \rho_0 = \text{const} \tag{6.98}$$

In addition, we may take the inverse dependence $X_0(t, x)$. At that, one should be very careful as the correspondence $X_0 \leftrightarrow X$ is not completely "reverse." Obviously, for each X_0 one can find the corresponding $X = X_0 + y(t, x_0)$, since the markers do not disappear. But it is not possible to find the corresponding value X_0 for every X.

Indeed, in the case of two K-planes moving in opposite directions, the substance between them is "newly born," therefore it is impossible to find the corresponding X_0 for some given X ($X_{K1} < X < X_{K2}$). Formally, this interval ($X_{K1} < X < X_{K2}$) corresponds to one point $X_0 = 0$. Further on, we exclude such intervals, if any, grounding upon the following consideration. Anyway, there are no markers in such an interval ($\rho(x) = 0$), so we know nothing about the velocity curve inside it.

For the regions, where each time t corresponds to some X_0, we have

$$\frac{\partial x_0}{\partial x} = 1 \Big/ \frac{\partial x}{\partial x_0} = \frac{\rho(t, x)}{\rho_0} \tag{6.99}$$

resulting in

$$\left. \begin{aligned} X_0(t, x) &= X_L + \int_{X_L}^{X} \frac{\rho(t, x')}{\rho_0} dx' \\ X_0(t, x) &= X_R - \int_{X}^{X_R} \frac{\rho(t, x')}{\rho_0} dx' \end{aligned} \right\} \tag{6.100}$$

X_L and X_R may fall beyond the diffusion zone, so that

$$X_0(t, X_L) = X_L (X_0(t, X_R) = X_R)$$

Substituting Equation 6.99 in Equation 6.96, we get

$$U(t, x) = \frac{1}{2t} \left(x - \frac{X_L \rho_0 + \int_{X_L}^{X} \rho(t, x') dx'}{\rho(t, x)} \right) \tag{6.101}$$

It is clear that the results do not depend on the choice of X_L outside the diffusion zone.

Since the stable K-plane acts as an attractor for surrounding markers, $\rho(t, X_K) \to \infty$ and $U \to \frac{X_K}{2t}$. This means that the latter equation can be used only for the left part of the diffusion couple. For the right part (and it can be easily checked), the equation is as follows:

$$U(t, x) = \frac{1}{2t} \left(x - \frac{X_R \rho_0 - \int_{x}^{X_R} \rho(t, x') dx'}{\rho(t, x)} \right) \tag{6.102}$$

The markers' density can be small but the size of the layer (cross section) has to be large in order to obtain sufficient statistics. The results presented here were published first in [48–50].

6.7
Electromigration-Induced Grain Rotation in Anisotropic Conducting Beta Tin

White tin (or β-Sn) has a body-centered tetragonal crystal lattice (with lattice parameters $a = b = 0.583$ nm and $c = 0.318$ nm), for which essentially anisotropic properties are typical (experiments described in this section were performed by Albert Wu and King-Ning Tu [51]). In particular, specific conductivities along the a (b) and c directions are 13.25×10^{-8} (Ωm) and 20.27×10^{-8} (Ωm), respectively. Lloyd [52] reported a voltage drop in tin line under electromigration (atoms transfer due to electron wind) during one day (at constant current). This implies a reduction in resistivity in tin at electromigration. Apparently, some structural changes take place in tin. On the one hand, the grains with lower resistivity may consume (relative to the fixed current direction) those with higher one. On the other hand, a change in grain orientation (grain rotation) may take place resulting in a change in resistivity. To study the evolution of the grain structure in thinfilm tin lines under electromigration, the local microdiffraction of synchrotron radiation was used [53]. Surface morphological changes also indicate that realignment may be caused by grain rotation.

Driving forces and kinetics of grain rotation were treated in [54–58]. According to these studies, the driving force of rotation firstly presents the dependence of GB tension on the angle between the neighboring grains. In this section, we propose a mechanism in which the grain rotation is induced by electromigration. Resistivity of the line consists of bulk resistivity of grains plus resistivity of GBs. Grain rotation simultaneously leads to the drop in bulk conductance and minimizes the misorientation relative to the current direction.

It requires a torque consisting of a force and a moment arm to rotate an object. Owing to anisotropic properties of beta tin, we suggest a plausible mechanism that can generate a torque under an applied electric current. Electromigration-induced vacancy fluxes directed from cathode to anode have different values in neighboring grains with different orientations. This difference (singular flux divergence) leads to supersaturation with vacancies of the boundary at one side of the grain and undersaturation at the other. Since the top surface of the tin line is stress free and can be considered to be a good sink/source of vacancies, it must have the equilibrium vacancy concentration. That is why the vacancy gradients and fluxes should be built up in vertical directions. Vertical vacancy gradients imply vertical stress gradients along the GBs. Thus, they indicate transversal forces (up and down) acting on the atom at the two GBs. This pair of forces is the origin of the torque. Further, we show that the atomic flow or diffusion along the GBs results in grain rotation.

Many studies have been focused on electromigration in metal interconnects [59–65]. However, these studies were based on electrically isotropic materials, such as aluminum or copper. The vacancy flux in such materials causes the inverse gradient of mechanical stress (a so-called back stress) mainly in the direction parallel to the current flow but not in the direction normal to it, hence there is no grain rotation.

In Figure 6.22, we consider a simple and geometrically ideal situation of three grains: a "bad" grain, grain 2, in the middle, with its c axis directed along the current (electron flow) direction, and situated between two "good" grains, grain 1 and 3 at the left and the right, respectively, with their a axis also orientated along the current direction. The electrons flow from left to right. Since both resistivity and diffusivity of grains along a and c axes are different, the electron wind effect

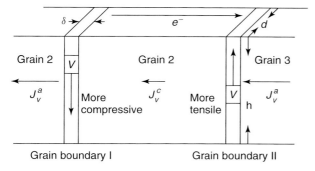

Figure 6.22 Schematic diagram of grain 2 with "bad" orientation between grains 1 and 3, having a "good" orientation. The height of the grain is "h", the width "d," and the width of the grain boundaries "δ." Inside the grain boundaries, "v" marks the vacancies and the arrows indicate the directions of vacancy flow.

and the corresponding vacancy fluxes in the bad and good grains are different:

$$J_v^c = \frac{C_v^{bulk} D_v^{c,bulk}}{kT} Z^* e \rho^c j$$

$$J_v^a = \frac{C_v^{bulk} D_v^{a,bulk}}{kT} Z^* e \rho^a j \qquad (6.103)$$

Here J_v^c and J_v^a are the vacancy fluxes along a and c axes. The reference data for the diffusivity and resistivity of tin atoms along these two directions are

$$D_c = 5.0 \times 10^{-17} \text{ m}^2 \text{ s}^{-1}, \quad \rho_c = 20.33 \times 10^{-8} \, \Omega\text{m}$$

$$D_a = 1.33 \times 10^{-16} \text{ m}^2 \text{ s}^{-1}, \quad \rho_a = 13.33 \times 10^{-8} \, \Omega\text{m}$$

The atomic flux under electromigration should be in the same direction as electron flow. Therefore, a counterflux of vacancies flows from right to left. The effective charge, Z^*, is considered to be the same in both directions (actually, this assumption is not proven). Since $D_c \rho_c < D_a \rho_a$, from Equation 6.103, we find that a larger vacancy flux reaches the GB II from grain 3 to grain 2, yet a smaller vacancy flux leaves GB II going into grain 2. This smaller vacancy flux then goes through grain 2 and reaches GB I. Correspondingly, a larger vacancy flux leaves GB I and goes into grain 1. In grain 2, depletion (undersaturation) of vacancies occurs at GB I and corresponds to compressive stresses near the boundary. On the other hand, supersaturation of vacancies and corresponding tensile stresses occur at GB II. In this model, the divergence of vacancies at GBs is such that the vertical gradient of stresses acts as vertical forces at each atom in the GBs. It acts down (atomic flux moves down) at GB II, but up at GB I. Therefore, this pair of forces generates a clockwise torque and leads to the rotation of the bad grain.

In a slow (quasi-steady-state) process, the vertical gradient of vacancy concentration can be found from the flux balance condition: the number of spare vacancies entering the boundary due to the difference in horizontal fluxes equals the number of vacancies arriving at free surface,

$$\delta D_v^{GB} \frac{\Delta C_v}{h} \cong \frac{C_v^{bulk} Z^* ej}{kT} \left(D_v^{a,bulk} \rho^a - D_v^{c,bulk} \rho^c \right) h \tag{6.104}$$

Since the vertical gradient is what makes the torque rotate the grain, it is important to correlate the vacancy concentration gradient to the stress gradient. Since the stress is in equilibrium with the vacancy concentration,

$$C_V = C_{V0} \cdot \exp\left(\frac{\sigma \Omega}{kT}\right)$$

the stress gradient along the GB is

$$\frac{\partial \sigma}{\partial h} \cong \frac{kT}{\Omega C_v} \frac{\partial C_v}{\partial h} \cong \pm \frac{Z^* ej}{\Omega \delta D_v^{GB}} \left(D_v^{a,bulk} \rho^a - D_v^{c,bulk} \rho^c \right) h \tag{6.105}$$

By multiplying the stress gradient with the atomic volume, the force which acts on an atom along the GB is

$$F_y' = \Omega \times \frac{\partial \sigma}{\partial h} \cong \pm \frac{Z^* ej}{\delta D_v^{GB}} \left(D_v^{a,bulk} \rho^a - D_v^{c,bulk} \rho^c \right) h \tag{6.106}$$

To sum up the effect from every atom, the total force that acts upon all atoms in the GB becomes

$$F_y = \frac{d \times h \times \delta}{\Omega} F_y' \cong \pm \frac{Z^* ej}{\Omega D_v^{GB}} \left(D_v^{a,bulk} \rho^a - D_v^{c,bulk} \rho^c \right) h^2 d \tag{6.107}$$

For the full torque (from both GBs), which acts upon unit width of the grain and causes the grain to rotate is

$$M = 2F_y \frac{h}{2} \frac{1}{d} \cong \pm \frac{Z^* ej}{\Omega D_v^{GB}} \left(D_v^{a,bulk} \rho^a - D_v^{c,bulk} \rho^c \right) h^3 \tag{6.108}$$

For the standard equation for angular velocity of grain rotation,

$$\omega = \frac{d\theta}{dt} = LM \tag{6.109}$$

with a mobility term [45]

$$L = \frac{4096 \Omega \delta D^{GB}}{kT} \frac{1}{d^5} \tag{6.110}$$

where δ is the width of the GB, D_V^{GB} is the vacancy diffusivity at GB, and Ω is the atomic volume. Inserting Equations 6.108 and 6.110 in Equation 6.109, the angular velocity becomes

$$\omega = \frac{d\theta}{dt} = \frac{4096 \delta D^{GB}}{kT} \frac{1}{h^2} \frac{Z^* ej}{\Omega D_v^{GB}} \left(D_v^{a,bulk} \rho^a - D_v^{c,bulk} \rho^c \right) \tag{6.111}$$

Since the diffusivity of vacancy and atomic coefficient of self-diffusion can be correlated both in the bulk and at the surface,

$$D^{bulk} = C_V^{bulk} D_V^{bulk}, \quad D^{GB} = C_V^{GB} D_V^{GB}$$

6.7 Electromigration-Induced Grain Rotation in Anisotropic Conducting Beta Tin

Equation 6.111 becomes

$$\omega = \frac{d\theta}{dt} = \frac{C_V^{GB}}{C_V^{bulk}} \frac{4096\delta}{kTh^2} Z^*ej \left(D_v^{a,bulk} \rho^a - D_v^{c,bulk} \rho^c \right) \quad (6.112)$$

In the general case, where the c axis of the "bad" grain is orientated at an arbitrary angle u to the current direction, the magnitudes of vacancy flux density along the c and a axes are equal to

$$J_v^c = \frac{C_v^{bulk} D_v^{c,bulk}}{kT} Z^*e\rho^c j \cos\theta, \quad J_v^a = \frac{C_v^{bulk} D_v^{a,bulk}}{kT} Z^*e\rho^a j \sin\theta \quad (6.113)$$

Considering that the electron flow direction is along the X axis, only the x direction of the overall vacancy flux contributes to the torque,

$$J_{vx}^{bulk} = -J_v^c \cos\theta - J_v^a \sin\theta$$

By following the same logic, the angular velocity would be

$$\omega = \frac{d\theta}{dt} = K \cos^2\theta$$

where

$$K = \frac{4096\delta D^{GB}}{kT} \frac{1}{h^2} \frac{Z^*ej}{D_v^{GB}} \left(D_v^{a,bulk} \rho^a - D_v^{c,bulk} \rho^c \right)$$

Thus,

$$\omega = \frac{d\theta}{dt} = \frac{C_V^{GB}}{C_V^{bulk}} \frac{4096\delta}{kTh^2} Z^*ej \left(D_v^{a,bulk} \rho^a - D_v^{c,bulk} \rho^c \right) \cos^2\theta \quad (6.114)$$

A real tin strip sample, as can be seen in Figure 6.23, was tested at a current density of 2×10^4 A m^{-2} at 100 °C. The grain size was around 50 μm, and the GB width was about 0.5 nm. The effective charge number of tin is 17. The D^{GB} at 100 °C is 2×10^{-7} cm^2 s^{-1}. The only unknown parameter was the ratio of vacancy concentration in the GB and in the bulk. At 100 °C, it is reasonable to take this ratio equal to 100. The calculated time for the grain to rotate was 447 h.

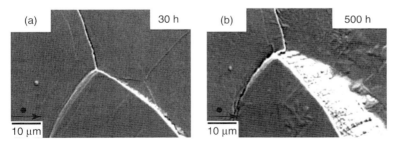

Figure 6.23 (a) Sn strip under electromigration at a current density of 2×10^4 A cm^{-2} at 100 °C for 30 h.
(b) Morphology of the strip shows grain rotation after electromigration for 500 h.

The experiment shown in Figure 6.23 was carried out for 500 h. In [39], it is stated that fine tin grains were also rotated under electromigration. In this sample, the grain size is more equiaxial and should be 1 μm in diameter. If we plug-in all the known values, the calculated time to rotate one grain for a size of 1 μm is about 3 h. The experiment was performed for 7 h. From both results, we can verify that the model is in good agreement with the experimental data.

In summary, a mechanism of grain rotation has been proposed to depict the microstructure evolution in anisotropic conducting materials in electromigration. We propose that the divergence of the vacancy fluxes from two neighboring but differently orientated grains induces a vacancy flux for an atomic flux in the reverse direction along the GB between them. The flux divergence exists because of the anisotropic conduction. The GB diffusion or creep results in a rotation of the grain. The rate of rotation estimated on the basis of the model seems to agree well with the observed experimental data. The results of this section were published first in [51].

6.8
Thermomigration in Eutectic Two-Phase Structures

When an inhomogeneous binary solid solution (or alloy) is annealed at constant temperature, it tends to become homogenous. On the contrary, if a homogeneous binary solid solution is annealed in a temperature gradient, the opposite happens and the homogeneous alloy tends to become inhomogeneous. This tendency of dealloying is called the *Soret effect*. This is due to thermomigration or mass migration driven by a temperature gradient.

Since the transformation of a homogenous into an inhomogeneous solid solution requires uphill diffusion of the components against their concentration gradient, thermomigration is opposed by the concentration gradient. Thus, in the flux equation of thermomigration, the opposing flux is added as

$$J = -C\frac{D}{kT}\frac{Q^*}{T}\frac{\partial T}{\partial x} + D\frac{\partial C}{\partial x} \qquad (6.115)$$

where C is the concentration, D is the diffusivity, and Q^* is the heat of transport. By changing the concentration gradient at a fixed temperature gradient (thus, segregating), the alloy eventually reaches the inhomogeneous steady state at which the diffusant flow is not only constant but equal to zero, that is, $J = 0$. A steady state can be reached, in which a constant concentration gradient is maintained in the solid solution. This was achieved in the experiment of thermomigration in the Fe–C system by Shewmon, for example. The concentration of C in Fe has reached a linear gradient. Thermomigration in a solid solution of PbIn solder has been reported. Thermomigration of small impurities and the problems of heat transfer are featured in a well-known monograph by Kuzmenko [66].

We discuss here thermomigration in eutectic two-phase structures, such as the SnPb solder. It is very different from the Soret effect. The change in concentration in the two-phase structure under a fixed temperature does not mean any change of chemical potential since the two phases coexist in equilibrium with each other. The concentration change or segregation in the mixture means only a change of volume fraction of the two phases; in other words, it leads to a gradient of volume fraction, but not a gradient of chemical potential. Gradient of volume fraction is not a driving force for atomic diffusion; therefore, uphill redistribution of the volume fraction is not counteracted by chemical force. Thus, we may ignore the last term in Equation 6.115. Consequently, segregation in eutectic two-phase mixtures can be enormous, so no steady state of a constant concentration gradient exists, instead a complete segregation occurs at the end of thermomigration. What is also unique is that, due to the lack of counteracting force, the segregation is not smoothened by diffusion processes, so a tendency of stochastic behavior occurs during thermomigration of the eutectic mixture. We do not find a smooth concentration gradient as the Soret effect in a solid solution. Rather, we find a random state of segregation.

Experimentally [67], it was demonstrated first that, in a bulk diffusion couple of two SnPb alloys, there is no interdiffusion at a constant temperature below the eutectic temperature (183 °C). A bulk diffusion couple of 70Pb30Sn and 30Pb70Sn alloys (in wt%) were annealed at 150 °C for one week. Assuming an interdiffusion coefficient of 1×10^{-8} cm^2 s^{-1}, we expect to detect an interdiffusion zone of 550 μm width. Composition distribution curves across the interface of the couple measured by electron microprobe analysis before and after the annealing are shown in Figure 6.24. There is hardly any difference between the two, indicating that no interdiffusion has occurred.

One also finds no interdiffusion or mixing of composition within a flip-chip composite solder joint consisting of 5Sn95Pb and 63SnPb34 (in wt%) after aging at 150 °C for one week. Scanning electron microscopic images of the cross sections of the composite solder joint before and after the isothermal aging are shown in Figure 6.25. In the figures, the light region in the upper and central part of the joint was the high-Pb and the darker phase in the lower part was the eutectic; no mixing was found except for some ripening that had occurred in the eutectic.

However, when the composite solder joints were subjected to thermomigration, a substantial segregation of Sn and Pb occurred in the joint. A temperature gradient of about 1000 K cm^{-1}, or a 10 °C difference across a flip-chip solder

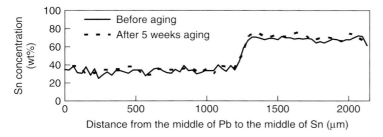

Figure 6.24 Redistribution of the tin fraction along the diffusion couple before and after five weeks annealed at 500 °C.

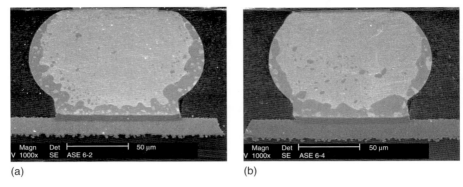

Figure 6.25 Composite solder (high plumbum solder joined with the eutectic one) (a) before and (b) after one week of aging at 150 °C.

joint of 100 μm in diameter, introduced by the Joule heating from applying a current density of around 1×10^4 A cm^{-2} for a few hours at 150 °C, is sufficient to generate electromigration and thermomigration in the powered joints. In addition, we found that those neighboring solder joints, which were unpowered or conducted no current, had also exhibited thermomigration.

Figure 6.26 shows scanning electron microscopic images of a row of six flip-chip solder joints; among them, we have powered the pair on the right-hand side and the other four are unpowered. The latter, however, have shown the segregation of the eutectic phase to the top due to thermomigration. This is because the Al interconnect between the two powered joints was the heat source due to joule heating and the Si chip itself was a very good heat conductor and was able to transfer heat from the powered to the unpowered joints.

Figure 6.27 show a set of the cross-sectional images of unpowered solder joints after thermomigration for 30 min, 2 h, and 12 h, respectively. Before thermomigration, the image is similar to that shown in Figure 6.25. In Figure 6.27a, a random state of phase separation is observed. In Figure 6.27b, the eutectic is segregated toward the hot end. In Figure 6.27c, a near complete phase separation is achieved. Figure 6.28 illustrates the random state of phase separation in the microstructure.

When an electron microprobe was used to measure composition distribution across a polished cross section of the flip-chip sample after thermomigration, a

Figure 6.26 Cross sections of solder joint series. The two right-hand side joints were powered at a current density of 1×10^4 A m^{-2} for 5 h. The four left-hand side joints were unpowered.

6.8 Thermomigration in Eutectic Two-Phase Structures | 181

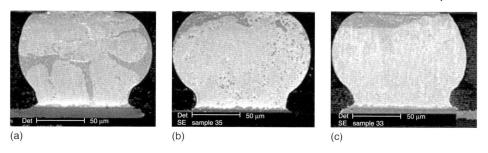

Figure 6.27 Image of cross sections of unpowered solder joints after thermomigration for (a) 30 min, (b) 2 h, (c) 12 h.

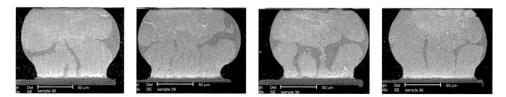

Figure 6.28 For unpowered solder joints after thermomigration for 30 min.

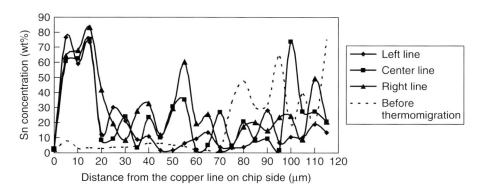

Figure 6.29 Sn fraction distribution from the Si chip side to the basis side on the flip-chip solder joint before and after thermomigration.

highly irregular or stochastic composition distribution was observed as shown in Figure 6.29; in other words, no smooth concentration profile was observed.

There is no other driving force except the temperature gradient that can lead to such phase separation. Creep does not since there is no stress potential gradient along the loading direction between the chip and the substrate. The effect of back stress (which has a gradient) induced by thermomigration (or electromigration) on phase separation will be analyzed later. The polarity effect of IMC formation at the cathode and the anode may have an influence, but the polarity effect does not exist in unpowered joints. In the isothermal annealing of composite joints shown in Figure 6.25, IMC formation did occur, yet no mixing or unmixing was observed.

Here we introduce a formal description of thermomigration in a two-phase structure assuming a coarsened space scale. We do not restrict the alloy composition to that at the eutectic point. Rather, it can be a two-phase mixture below the eutectic temperature having any composition between the two primary phases. The main assumptions are the following: (i) Conservation of volume and shape of samples (equalizing of volume fluxes) at undercritical regimes, meaning no void or hillock formation. Two ways of equalizing volume fluxes are discussed, either by back stress or by Kirkendall lattice shift. (ii) The fluxes do not contain concentration gradient terms as shown in Equation 6.115. Instability issues for the concentration profiles are analyzed, and we show that the concentration profile in eutectic structures under thermomigration should demonstrate the stochastic tendency.

Consider a two-phase mixture of almost pure components; therefore, hereafter the indexes 1 and 2 correspond to phases as well as to species. Since we limit ourselves to the "undercritical" regime, where the shape of the sample remains unchanged, we have the constraint of constant volume in every part of the sample. This means that, in the laboratory reference frame, the sum of volume fluxes of two species should be zero everywhere:

$$\Omega_1 J_1 + \Omega_2 J_2 = 0 \qquad (6.116)$$

where J_1 and J_2 are the fluxes of atoms per unit area per unit time, and Ω_1 and Ω_2 are the atomic volumes. For convenience, we introduce the local volume fractions of phases, p_1, p_2:

$$\Omega_i n_i = \Omega_i \frac{\Delta V_i/\Omega_i}{\Delta V} = \frac{\Delta V_i}{\Delta V} = p_i \qquad (6.117)$$

$$p_1 + p_2 = 1$$

where n_i is the number of atoms of either 1 or 2 species per unit volume. In the coarsened space scale, the unit volume ΔV includes at least a few grains.

The fluxes generated by the temperature gradient are given as

$$\Omega_1 J_1^{TM} = \Omega_1 \frac{n_1 D_1}{kT}\left(-Q_1 \frac{\partial \ln T}{\partial x}\right) = -\frac{p_1 D_1}{kT} Q_1 \frac{\partial \ln T}{\partial x} = p_1 D_1 Q_1 \frac{\partial (1/kT)}{\partial x}$$

$$\Omega_2 J_2^{TM} = \Omega_2 \frac{n_2 D_2}{kT}\left(-Q_2 \frac{\partial \ln T}{\partial x}\right) = -\frac{p_2 D_2}{kT} Q_2 \frac{\partial \ln T}{\partial x} = p_2 D_2 Q_2 \frac{\partial (1/kT)}{\partial x}$$

(6.118)

If they have the same direction (if Q_i have the same sign or the heat of transport of 1 and 2 are both positive), they do not satisfy the constraint in Equation 6.116. In Equation 6.118, we neglect the distribution of thermal field between particles of different phases with different thermal conductivities. And D_i are the tracer diffusivities of species 1 and 2 in the two-phase mixture. The diffusivities D_i and heats of transport Q_i of both species are not just the characteristics of components in single phases, since the migration trajectories of all atoms in a random two-phase mixture may pass through the grains of both phases and interfaces between them.

Therefore, in general, D_i and Q_i are some averaged characteristics that depend on the volume fraction and geometry of the mixture.

To satisfy the constraint of constant volume in Equation 6.116, we assume that there are two alternative ways: back stress and lattice shift, or may be a combination of these two. We now consider different possible cases in detail.

6.8.1
Thermomigration Induced Back Stress in Two-Phase Mixtures

We apply the back stress concept to electromigration in Al interconnects to thermomigration in a mixture of two coexisting phases. Accumulation of excess atoms at the anode and of vacancies at the cathode leads to a stress gradient, changing the fluxes, so that the total flux of volume becomes zero. Namely,

$$\Omega_1 J_1 = \frac{p_1 D_1}{kT} \left(-Q_1 \frac{\partial \ln T}{\partial x} + \Omega_1 \frac{\partial \sigma}{\partial x} \right)$$

$$\Omega_2 J_2 = \frac{p_2 D_2}{kT} \left(-Q_1 \frac{\partial \ln T}{\partial x} + \Omega_2 \frac{\partial \sigma}{\partial x} \right) \tag{6.119}$$

Substituting Equation 6.119 in the constraint of Equation 6.116, we obtain an expression for the arising stress gradient:

$$\frac{\partial \sigma}{\partial x} = \frac{\partial \ln T}{\partial x} \frac{p_1 D_1 Q_1 + p_2 D_2 Q_2}{p_1 D_1 \Omega_1 + p_2 D_2 \Omega_2} \tag{6.120}$$

Owing to this back stress influence, the larger flux becomes less, and the smaller flux changes its sign to the opposite, so that now they compensate each other. Indeed, substituting Equation 6.120 in Equation 6.119, we obtain,

$$\Omega_1 J_1 = \frac{\partial (1/kT)}{\partial x} \Omega_1 \Omega_2 \frac{p_1 p_2 D_1 D_2}{p_1 D_1 \Omega_1 + p_2 D_2 \Omega_2} \left(\frac{Q_1}{\Omega_1} - \frac{Q_2}{\Omega_2} \right) = -\Omega_2 J_2 \tag{6.121}$$

The sign of segregation is determined not by the difference of diffusivities, but instead by the difference in the ratios

$$\left(\frac{Q_1}{\Omega_1} - \frac{Q_2}{\Omega_2} \right)$$

as shown in Equation 6.121. Both Q_1 and Q_2 can be positive, yet it is the difference in the ratio that determines which one will diffuse with or against the temperature gradient. For example, the depletion by Pb (and corresponding enrichment by Sn) at the hot end does not necessarily mean that Pb is a faster diffusant than Sn or the heat of transport of Sn is negative. It only means that

$$\left(\frac{Q_{Pb}}{\Omega_{Pb}} > \frac{Q_{Sn}}{\Omega_{Sn}} \right)$$

The redistribution of volume fractions is determined, of course, by the continuity equation,

$$\frac{\partial p_1}{\partial t} = \Omega_1 \frac{\partial n_1}{\partial t} = -\frac{\partial (\Omega_1 J_1)}{\partial x}$$

so that

$$\frac{\partial p_1}{\partial t} = -\frac{eE}{kT}\Omega_1\Omega_2\left(\frac{Z_1}{\Omega_1} - \frac{Z_2}{\Omega_2}\right)\frac{\partial}{\partial x}\left[\frac{D_1 D_2}{p_1 D_1 \Omega_1 + (1-p_1)D_2\Omega_2} \cdot p_1(1-p_1)\right] \quad (6.122)$$

We discuss the stochastic behavior of the numerical solution of Equation 6.122 later.

6.8.2
Thermomigration-Driven Kirkendall Effect in Binary Mixtures

In the alternative case, there is no back stress at all, implying that all possible stresses are relaxed immediately by the lattice shift. In a usual polycrystalline body with large grains, it is described by a dislocation climb as the source and sink of vacancies, and corresponding construction of extra planes in the region of atoms accumulation, and reconstruction of planes in the region of vacancy accumulation. Vacancy is at equilibrium everywhere in the sample as assumed in Darken's analysis of interdiffusion. Then in the laboratory reference frame

$$\Omega_1 J_1^{TM} = p_1 D_1 Q_1 \frac{\partial(1/kT)}{\partial x} + p_1 U$$

$$\Omega_2 J_2^{TM} = p_2 D_2 Q_2 \frac{\partial(1/kT)}{\partial x} + p_2 U \quad (6.123)$$

Substituting Equation 6.123 in the constraint of Equation 6.116, we obtain the velocity of Kirkendall shift,

$$U = -(p_1 D_1 Q_1 + p_2 D_2 Q_2)\frac{\partial(1/kT)}{\partial x} \quad (6.124)$$

Substituting Equation 6.124 in the flux Equation 6.116, we obtain the final equation for fluxes of both species in the laboratory reference frame; they are equal but opposite, that is,

$$\Omega_1 J_1 = (Q_1 D_1 - Q_2 D_2)p_1 p_2 \frac{\partial}{\partial x}\left(\frac{1}{kT}\right) = -\Omega_2 J_2 \quad (6.125)$$

The segregation rate can be found from the difference of QD products. The redistribution of volume fractions is then determined by the continuity equation

$$\frac{\partial p_1}{\partial t} = -\frac{\partial}{\partial x}\left[(Q_1 D_1 - Q_2 D_2)p_1(1-p_1)\frac{\partial}{\partial x}\left(\frac{1}{kT}\right)\right] \quad (6.126)$$

An interesting point with the second case (thermomigration-driven Kirkendall shift) is that there is no void formation or hillock extrusion at all.

6.8.3
Stochastic Tendencies in Thermomigration

In both the cases, back stress and Kirkendall shift, the equation for redistribution of volume fractions is of the type

$$\frac{\partial p_1}{\partial t} = -\frac{\partial}{\partial x}\left[V(p_1)p_1(1-p_1)\right], \quad (6.127)$$

but with different explicit expressions for the kinetic coefficient V (which has the unit of velocity). It is a first-order nonlinear equation, very different from Fick's second law or from Fick's second law with a drift term. The main difference is that it does not contain the second-order spatial derivatives, which tend to smooth out any local fluctuation of the composition profile by diffusion. If V is a constant, Equation 6.127 is reduced to the well-known Burger's equation in hydrodynamics with a zero viscosity term, having peculiarities of solutions like shocks [14]. This means that even a smooth waviness of composition profile should evolve into sharp breaks of the concentration profile.

Therefore, we might expect, at least, low stability or even instability of solutions against small fluctuations.

To investigate the instability, we first analyze the case V = const in Equation 6.127. Examples of numerical solutions for V = const and for simplest dependencies V(p) are demonstrated in Figure 6.30. We observe the developing waviness of profiles with time. To find out the conditions for this instability, we let $p_0(t, x)$ be the unperturbed solution of Equation 6.127. At some moment t_0, a small perturbation $\delta p(t_0, x)$ is introduced. How this perturbation evolves in the first period after initiation (when the perturbation is still small) has been studied by substituting $p(t, x) = p_0(t, x) + \delta p(t, x)$ in Equation 6.127 with constant V, and by neglecting the quadratic terms δp^2, we obtain with such linear analysis,

$$\frac{\partial \delta p}{\partial t} = -V(1 - 2p_0(t, x))\frac{\partial \delta p}{\partial x} + \left(2V\frac{\partial p_0}{\partial x}\right)\delta p \quad (6.128)$$

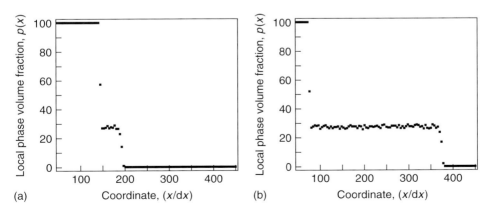

Figure 6.30 Evolution of the eutectic solder concentration profile at thermomigration.

The first term in the right-hand side of Equation 6.128 is responsible for the propagation of the perturbation with the velocity $V(1 - 2p_0(t, x))$ along the axis x (if $p_0(t, x) < 1/2$) or in the opposite direction (if $p_0(t, x) > 1/2$). The second term is important for understanding the thermomigration-driven roughening. This means that the time dependence of perturbation amplitude should be exponential,

$$\delta p(t, x) = \delta p(t_0, x) \cdot \exp\left(2V \frac{\partial p_0}{\partial x}(t - t_0)\right)$$

This exponential change means increasing perturbation in the case of

$$2V \frac{\partial p_0}{\partial x} > 0$$

but decreasing it in the opposite case. The initially smooth and wavy perturbation should transform into sharp shock wave–like disturbances (see Figure 6.30). Numerical simulations demonstrate (i) the segregation of pure components on hotter and cooler ends and (ii) the stochastic character of structure profiles (Figure 6.30a and b). Thus, thermomigration should lead to randomness in the microstructure (Figure 6.28) and to stochastic variation of composition profiles (Figure 6.29). Besides thermomigration, we can use electromigration and stress migration to induce a similar stochastic phase separation in eutectic two-phase structures at a constant temperature but we do not address these cases here. Actually, electromigration and/or thermomigration in solders induce both back stress and Kirkendall effect (lattice movement) simultaneously. The interrelation between them has been analyzed elsewhere.

References

1. Gusak, A.M. and Tu, K.N. (**2002**) *Physical Review B*, **66**, 115403.
2. Gusak, A.M. and Tu, K.N. (**2002**) Theory of flux-driven ripening, in *Diffusion and Thermodynamics of Materials* (Proceedings of the VIII Seminar) (eds J. Cermak and J. Vrestal), Masaryk University, Czech Republic, Brno, p. 139.
3. Tu, K.N., Gusak, A.M., and Li, M. (**2003**) *Journal of Applied Physics*, **93**, 1335.
4. Tu, K.N., Gusak, A.M., and Sobchenko, I. (**2003**) *Physical Review B*, **67**, 245408.
5. Gusak, A.M. and Tu, K.N. (**2003**) *Acta Materalia*, **51**, 3895.
6. Gusak, A.M. and Lutsenko, G.V. (**2003**) *Metallofizika i Noveishie Tekhnologii (Journal of Metal Physics and Advanced Technologies)*, **25**, 381 (in Russian).
7. Lifshitz, I.M. and Slezov, V.V. (**1958**) *Journal of Experimental and Theoretical Physics*, **35**, 479 (in Russian).
8. Lifshitz, I.M. and Slezov, V.V. (**1961**) *Journal of Physics and Chemistry of Solids*, **19**, 35.
9. Wagner, C. (**1961**) *Zeitschrift für Elektrochemie*, **65**, 581.
10. Slezov, V.V. (**1995**) *Theory of Diffusive Decomposition of Solid Solutions*, Harwood Academic, Dordrecht.
11. Ardell, A.J. (**1972**) *Acta Metallurgica*, **20**, 61.
12. Ardell, A.J. (**1988**) Precipitate coarsening in solids: modern theories, chronic disagreement with experiment, in *Phase Transformations'87* (ed. G.W. Lorimer), Institute of Metals, London, pp. 485–494.
13. Voorhees, P.W. (**1992**) *Annual Review of Materials Science*, **22**, 197.

14. Bitti, R.R. and Di Nunzio, P.E. (1998) *Scripta Materialia*, **39**, 335.
15. Marder, M. (1997) *Physical Review A*, **36**, 858.
16. Burke, J.E. and Turnbull, D.R. (1952) *Progress in Metal Physics*, **3**, 220.
17. Hillert, M. (1965) *Acta Metallurgica*, **13**, 227.
18. Fradkov, V.E., Udler, D.G., and Shvindlerman, L.S. (1985) *Scripta Metallurgica*, **19**, 1286.
19. Fradkov, V., Shvindlerman, L., and Udler, D. (1987) *Philosophical Magazine Letters*, **55**, 289.
20. Marder, M. (1987) *Physical Review A*, **36**, 438.
21. Di Nunzio, P.E. (2001) *Acta Materialia*, **49**, 3635.
22. Louat, N.P. (1974) *Acta Metallurgica*, **22**, 721.
23. Pande, C.S. (1987) *Acta Metallurgica*, **35**, 2671.
24. Thompson, C.V. (2000) *Solid State Physics*, **55**, 269.
25. Frear, D., Gravas, D., and Morris, J.W. Jr. (1986) *Journal of Electronic Materials*, **16**, 181.
26. Vianco, P.A., Hlava, P.E., and Kilgo, A.C. (1993) *Journal of Electronic Materials*, **23**, 583.
27. Prakash, K.H. and Sritharan, T. (2001) *Acta Materialia*, **49**, 2481.
28. Chad, S., Laub, W., Sabee, J.M., and Fournelle, R.A. (1996) *Journal of Electronic Materials*, **28**, 1194.
29. Frear, D.R., Kang, J.W., Lin, J.K., and Zhang, C. (2001) *Journal of Metals*, **53**, 28.
30. Gagliano, R.A. and Fine, M.E. (2001) *Journal of Metals*, **53**, 33.
31. Bartels, F., Morris, J.W. Jr, Dalke, G., and Gust, W. (1994) *Journal of Electronic Materials*, **23**, 787.
32. Kim, H. and Tu, K.N. (1996) *Physical Review B*, **53**, 16027.
33. Tu, K.N. and Zeng, K. (2001) *Material Science and Engineering R*, **34**, 1.
34. Tu, K.N., Ku, F., and Lee, T.Y. (2001) *Journal of Electronic Materials*, **30**, 1129.
35. Lee, J.H., Park, J.H., Lee, Y.H. et al. (2001) *Journal of Materials Research*, **16**, 1227.
36. Gorlich, J., Schmitz, G., and Tu, K.N. (2005) *Applied Physics Letters*, **86**, 053106.
37. Suh, J.O., Tu, K.N., Lutsenko, G.V., and Gusak, A.M. (2008) *Acta Materialia*, **56**, 1075.
38. Sobchenko, I., Gusak, A., and Tu, K.N. (2005) *Defect and Diffusion Forum*, **237–240**, 1281.
39. Sobchenko, I., Schmitz, G., Gusak, A., and Baither, D. (2007) *Bulletin of Cherkasy University*, **117**, 51.
40. von Neumann, J. (1952) Written discussion of grain shapes and other metallurgical applications of topology, in *Metal Interfaces* (ed. C. Herring), American Society for Metals, Cleveland, p. 108.
41. Mullins, W.W. (1956) *Journal of Applied Physics*, **27**, 900.
42. Di Nunzio, P. (2003) *Physical Review B*, **68**, 115432.
43. Darken, L.S. (1948) *Transactions of AIME*, **175**, 184.
44. Geguzin, Ya.E. (1979) *Diffusion Zone*, Nauka, Moscow (in Russian).
45. Cornet, J.-F. and Calais, D. (1972) *Journal of Physics and Chemistry of Solids*, **33**, 1675.
46. van Dal, M.J.H., Pleuneekers, M.C.L., Kodentsov, A.A., and van Loo, F.J.J. (2000) *Acta Materialia*, **48**, 385.
47. van Dal, M.J.H., Kodentsov, A.A., and van Loo, F.J.J. (2000) *Solid State Phenomena*, **72**, 111.
48. van Dal, M.J.H., Gusak, A.M., Cserhati, C., Kodentsov, A., and van Loo, F.J.J. (2001) *Physical Review Letters*, **86**, 3352.
49. van Dal, M.J.H., Gusak, A.M., Cserhati, C. et al. (2002) *Philosophical Magazine A* **82**, 943.
50. van Dal, M.J.H. and Gusak, A.M. (2000) *Bulletin of Cherkasy University*, **19**, 5.
51. Wu, A.T., Gusak, A.M., Tu, K.N., and Kao, C.R. (2005) *Applied Physics Letters*, **86**, 241902.
52. Lloyd, J.R. (2003) *Journal of Applied Physics*, **94**, 6483.
53. Wu, A.T., Tu, K.N., Lloyd, J.R. et al. (2004) *Applied Physics Letters*, **85**, 2490.
54. Li, J.C.M. (1962) *Journal of Applied Physics*, **33**, 2958.

55. Hermann, G., Gleiter, H., and Baro, G. (**1976**) *Acta Metallurgica*, **24**, 353.
56. Moldovan, D., Yamakov, V., Wolf, D., and Phillpot, S.R. (**2002**) *Physical Review Letters*, **89**, 206101.
57. Harris, K.E., Singh, V.V., and King, A.H. (**1998**) *Acta Materialia*, **46**, 2623.
58. Haslam, A.J., Phillpot, S.R., Wolf et al. (**2001**) *Materials Sciences and Engineering A*, **318**, 293.
59. Shatzkes, M. and Lloyd, J.R. (**1986**) *Journal of Applied Physics*, **59**, 3890.
60. Clement, J.J. and Lloyd, J.R. (**1992**) *Journal of Applied Physics*, **71**, 1729.
61. Kirchheim, R. (**1992**) *Acta Metallurgica Materialia*, **40**, 309.
62. Korhonen, M.A., Borgesen, P., Tu, K.N., and Li, C.Y. (**1993**) *Journal of Applied Physics*, **73**, 3790.
63. Kao, H.K., Cargill, G.S., and Hu, C.K. (**2001**) *Journal of Applied Physics*, **89**, 2588.
64. Sukharev, V. and Zschech, E. (**2004**) *Journal of Applied Physics*, **96**, 6337.
65. Blech, I.A. (**2004**) *Journal of Applied Physics*, **96**, 6337.
66. Kuzmenko, P.P. (**1983**) *Electromigration Thermomigration and Diffusion in Metals*, High School, Kyiv (in Russian).
67. Huang, A.T., Gusak, A.M., Tu, K.N., and Lai, Y.S. (**2006**) *Applied Physics Letters*, **88**, 141911.

7
Nanovoid Evolution
Tatyana V. Zaporozhets and Andriy M. Gusak

7.1
Introduction

The present chapter also discusses the analysis of flux-induced morphology evolution. However, it has its own independent significance since it is directed toward voids, and voids have always represented a special object to be studied in material science. Void coarsening, described by Lifshitz and Slezov, determined the kinetics of the final stage during sintering of powder systems. Voiding in materials under radiation alters their mechanical properties and determines the reliability of nuclear reactor walls. Voiding accompanies almost all interdiffusion processes. A porous metal is the simplest model of a two-phase alloy where the second phase is "emptiness". In the present chapter, we concentrate on the behavior of nanovoids, the bubble pressure for them reaching giga pascal values. In Sections 7.2 and 7.4, we treat nanovoids in hollow nanoparticles and establish the conditions of their formation, existence, and collapse.

Void formation upon reactive diffusion in the samples with closed geometry (cylindrical or spherical) has been known for a long time. It is a common feature of interdiffusion in binary systems and is most frequently related to the difference in partial diffusivities of the components. Such a difference leads to a difference in intrinsic atomic fluxes (in the lattice reference frame), creating a vacancy flux directed toward the faster component. Since the vacancy flux has some spatial distribution (which is maximum in the vicinity of the initial contact zone), it has a nonzero divergence, leading to a local accumulation of vacancies at the fast component side and depletion with vacancies at the slow component side. If vacancy sinks at the fast component side are not effective enough, then extra vacancies should gather into voids. Such voiding should be especially pronounced in spherical and cylindrical samples when the faster component is located inside.

This possibility was realized by Aldinger [1], who obtained hollow shells of a BeNi alloy after annealing of Be microparticles coated with Ni. Moreover, Ya. Geguzin, in his monograph "Diffusion zone" (1979), reported the formation of a hollow intermetallic wire by the reaction $Cd + Ni \rightarrow Cd_{23}Ni_5$ in the Cd wire covered with

Diffusion-Controlled Solid State Reactions. Andriy M. Gusak
Copyright © 2010 WILEY-VCH Verlag GmbH & Co. KGaA, Weinheim
ISBN: 978-3-527-40884-9

Ni [2, 3]. The authors of [2, 3] connected this phenomenon with the large change of molar volume in the intermetallic formation (about 8%).

In 2004, void formation in spherical samples was rediscovered at the nanolevel and discussed in [4–6]. Hollow nanoshells of cobalt and iron oxides and sulfides have been obtained by means of reaction of metallic nanopowders with oxygen or sulfur. Contrary to [2, 3], these results have been explained by the Frenkel effect – out-diffusion of metal through the formation of a spherical layer of the compound is faster than in-diffusion of oxygen or sulfur through the same phase. This inequality of fluxes generates the inward flux of vacancies, meeting inside and forming the void in the internal part of the system.

The Frenkel effect is often called *Kirkendall voiding*, and this notation sometimes leads to confusion, since Kirkendall voiding and Kirkendall shift are in fact two different effects and, moreover, two competing effects, competing consequences of the same reason (difference of partial diffusivities). To emphasize the competing character of Frenkel voiding and Kirkendall shift, Geguzin even introduced special notations for vacancy sinks/sources – K-sinks and F-sinks. K-sinks are, for example, the kinks at dislocations. Vacancies can be annihilated or generated at these K-sinks, changing the number of lattice sites locally and generating a dislocation climb and corresponding lattice shift (Kirkendall effect). F-sinks are voids. Vacancies can join these F-sinks by not changing the number of the lattice sites but by increasing the total volume. Functioning of F-sinks leads to voiding instead of Kirkendall shift. If a comparatively low pressure of about 5–10 MPa is applied, which does not change the jump frequency considerably but is sufficient to suppress the void nucleation and/or growth, then the Frenkel effect is suppressed and the "pure" Kirkendall effect is observed. Indeed, such experiments were made by the Geguzin group; these demonstrated an approximately twofold increase in the Kirkendall shift with almost full suppression of voiding [2].

In the hollow nanoshell formation, we have the opposite phenomenon – an almost pure Frenkel effect with a fully or partially suppressed Kirkendall shift. Most probably, this happens due to peculiarities of interdiffusion in nanosystems: (1) there is little place for dislocations in a nanoparticle so that K-sinks are just lacking; (2) a common Kirkendall shift in radial direction (simultaneous for all atoms) leads to tangential deformation and corresponding stresses in spherical or cylindrical samples, which are large in the case of nanosystems, and suppress the shift.

Just after publication of the first experimental paper on hollow nanoshell formation [4], Tu and Gösele argued that such a hollow structure should be unstable since shrinkage of a hollow shell is energetically favorable (total surface and surface energy decreases) [7]. The mechanism of shrinkage is related to the Gibbs–Thomson effect (effect of Laplace pressure on the boundary vacancy concentrations): the vacancy concentration on the inner boundary of a nanoshell should be higher than that on the external boundary. This difference of boundary concentrations leads to an outward vacancy flux meaning shrinkage.

In Section 7.2, we argue that, contrary to shrinkage of a single-component shell, shrinkage of a binary compound shell is also affected by the inverse Kirkendall effect [8]. Namely, outward vacancy flux generates opposite fluxes of two components, which are different due to different mobilities. In turn, this leads to a segregation of the faster component near the inner boundary. The corresponding concentration gradient substantially suppresses the vacancy flux. Therefore, the shrinkage rate of the compound shell is controlled by the slow species. Shrinkage of compound shells was experimentally verified by Nakamura *et al.* [9]. In Section 7.3, we analyze the formation of a hollow compound nanoshell. We demonstrate that the Gibbs–Thomson effect, leading to shrinkage of "ready" compound shells, should be important at the formation stage as well – it sometimes may even suppress the nanoshell formation [10].

Both papers [9] and [10] treated compound shells with a very narrow homogeneity range. For this case, the steady state approximation for both vacancies and main components worked well. In Section 7.4, we analyze the formation and collapse of hollow shells in the system with full solubility or, at least, with a broad homogeneity range. In this case, one can expect that the formation and collapse will represent two stages of one whole process – at the first stage (when the chemical driving forces are still large), Kirkendall voiding wins over the curvature effect, then, after arriving at a sufficiently small concentration gradient, the curvature-driven shrinkage will take the revenge. We use phenomenological and MC modeling, giving the possibility to observe both stages (formation and collapse) in one run [11]. In all mentioned models, the peculiarities of diffusion in oxides or sulfides are neglected. These peculiarities are treated elsewhere.

In Section 7.4, we consider the behavior of nanovoids at a metal–dielectric interface under a strong electric current in the integrated circuits. Recent direct experimental observations (in situ) of the processes occurring in the microcircuits when Cu is used instead of Al [12] have shown that void formation occurs on the surface of the copper/dielectric interconnect. Moreover, void formation generally takes place far from the cathode, although eventually the contact breaks at the cathode end. In [12], it was first demonstrated that the voids migrate along the copper/dielectric interface, stop for some time (most likely at grain boundaries), coagulate with the other voids, and reach the cathode where they cause failure by contact break. We will make sure that the "accumulation of emptiness" at electromigration is not necessarily caused by the migration of individual vacancies from cathode to anode; in modern metallic "interconnects" (copper thin films of 100 nm thickness), the emptiness has an inclination to be transported by larger portions in the form of nanovoids [13].

7.2
Kinetic Analysis of the Instability of Hollow Nanoparticles

7.2.1
Introduction

The mechanism of shrinkage is related to the Gibbs–Thomson effect (effect of Laplace pressure on the vacancy concentration): the vacancy concentration on the inner boundary of a nanoshell should be higher than that on the external boundary. This difference of boundary concentrations leads to an outward vacancy flux resulting in a shrinkage. In [8], we argued that, contrary to shrinkage of a single-component shell, shrinkage of a binary compound shell is also affected by the inverse Kirkendall effect. Namely, an outward vacancy flux generates opposite fluxes of two components, which are different due to different mobilities. In its turn, this leads to segregation of the faster component near to the inner boundary. The corresponding concentration gradient substantially suppresses the vacancy flux. Therefore, the shrinkage rate of the compound shell is controlled by the slow species. Shrinkage of compound shells was experimentally verified by Nakamura et al. [9].

We propose some models (both phenomenological and computer ones), which allow one to analyze the influence of a nanoshell's size and structure on the process of its shrinkage into a solid nanoparticle and to estimate the stability of nanoshells. Shrinkage kinetics and collapse time are generally determined by a slowly diffusing component, which makes nanoshells more stable than expected.

7.2.2
Mechanism of Nanoshell Shrinkage

It is energetically favorable for a nanoshell to shrink into a compact nanosphere as the interface area is reduced, i.e., if the inequality

$$\gamma \left(4\pi r_{i0}^2 + 4\pi r_{e0}^2\right) > \gamma 4\pi r_f^2 \tag{7.1}$$

holds while conservation of volume is fulfilled, implying

$$\left(\frac{4\pi}{3}\right)\left(r_{e0}^3 - r_{i0}^3\right) = \left(\frac{4\pi}{3}\right) r_f^3$$

Here γ is the surface tension, r_{i0} and r_{e0} are the initial internal and external radii of a nanoshell, and r_f is the final radius of the collapsed particle.

The proposed atomistic mechanism of shrinkage is the vacancy flux from the internal surface (with radius r_i) to the external surface (with radius r_e). The driving force of the vacancy flux is the difference in vacancy/atoms chemical potentials and the corresponding difference in equilibrium vacancy concentrations at the curved interfaces $c_V(r_i) > c_V(r_e)$. This Gibbs–Thomson effect can be expressed, in a linear approximation so far, as

$$c_V(r_i) = c_V^{eq}\left(1 + \frac{\beta}{r_i}\right), \qquad c_V(r_e) = c_V^{eq}\left(1 - \frac{\beta}{r_e}\right), \qquad \beta = \frac{2\gamma\Omega}{kT} \tag{7.2}$$

where Ω is the atomic volume and c is the concentration, meaning the fraction of vacant lattice sites; c_V^{eq} refers to the equilibrium concentration of vacancies with respect to a planar surface. The linear approximation holds if $r_i \gg \beta$. If the term β/r becomes comparable with unity, we need to reformulate Equation 7.2 using the exponential size dependencies.

Thus, for the case of nanoshells of a single element, the physical picture of shrinkage seems to be very clear. It is analogous to the LSW theory of caking [14],

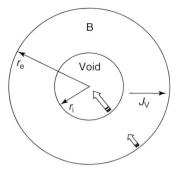

Figure 7.1 Scheme of pure element shell shrinkage.

when voids are dissolved by vacancy fluxes to the external surface. Yet, in the nanoshell case, we are confronted with a much more symmetrical picture: vacancies are generated at the internal surface (void dissolution), migrate to the external surface, causing the influx of atoms and movement of the internal surface, and annihilate at the external surface, causing shrinkage of the latter (Figure 7.1).

In case of a binary alloy or IMC nanoshell, the process is more complicated due to the "inverse Kirkendall effect" [15], which was discussed in the problem of irradiation of an alloy with energetic particles. As is evident, the radiation leads to a generation of a number of Frenkel pairs, "vacancy-interstitial". Interstitials quickly disappear at sinks since their mobility is much higher than that of the vacancies, and they quickly disappear at sinks. Vacancies move to sinks much faster, generating directed vacancy fluxes. Since in an alloy or intermetallic, one of the components may diffuse much faster than the other ($D_B \gg D_A$), the vacancy flux should lead to a redistribution of the components – segregation of B atoms at the internal boundary, and of A atoms near the external surface (Figure 7.2). The arising concentration gradient $\partial c_B / \partial r$ influences and reduces the vacancy flux and hence generates a tendency to suppress the shrinkage process. We deduce below that the slower diffusing species controls the shrinkage rate.

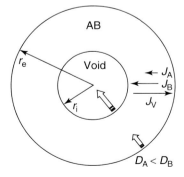

Figure 7.2 Scheme of a binary alloy shell shrinkage, accompanied by B-segregation at the internal surface.

Moreover, segregation is approximately inversely proportional to the square of the internal radius. This means that sooner or later the concentration of B at the internal radius will exceed the homogeneity limit of the IMC, and the pure B component can precipitate at the void–IMC interface. After this, the shrinkage should proceed further, ending with a tiny pure B nanoparticle inside an inhomogeneous (with concentration gradient inside) IMC nanoshell. This kinetic process is discussed in detail later. To develop the kinetic equations, we use the formalism of the classic Darken analysis of interdiffusion and take into account the nonequilibrium vacancies in the interdiffusion. This formalism is in line with the approach developed following the direction developed first by Gurov [16] and later modified by Gurov and Gusak [17–19].

Here we present four models of the shrinkage of hollow nanospheres with increasing degree of complexity. Model 1 describes the shrinkage kinetics of a single element nanoshell in the steady state approximation for vacancies. Model 2 describes the shrinkage of a binary compound nanoshell in the steady state approximation both for vacancies and for B species, taking into account the inverse Kirkendall effect. We demonstrate that the steady state approximation is good for vacancies, but not so well justified for the slow-diffusing B component. Therefore, in model 3, we present a Stephan-like problem for B diffusion between moving boundaries, solving (numerically) Fick's 2nd law for B particles but still using the steady state approximation for the vacancy flux. Model 4 is the most rigorous one (but most complicated and time consuming). Here we solve the set of non–steady state diffusion equations numerically for both vacancies and for species B. We compare the results of the Models 1 to 4 and conclude that in most cases Model 2 is sufficient for a rough estimate, but Model 1 may lead to wrong predictions for the case of compounds. In Section 7.2.3, we consider the case of pure B-segregation at the internal boundary. In Section 7.2.4, we present kinetic Monte Carlo (kMC) simulations of nanoshell shrinkage, comparing the cases of single element and IMC nanoshell with the same average bonding energy, and also comparing the kMC results with the phenomenological analysis. Within Section 7.2.5, we use only the linear approximation of the Gibbs–Thomson relation.

7.2.3
Models of Nanovoid Shrinkage

As already mentioned in the previous section, any hollow nanoparticle should shrink. The general driving force of shrinkage is the same for a pure component shell (Model 1) and for an IMC shell (Models 2–4) – a decrease in the total surface energy (in other formulation – Gibbs–Thomson effect). Yet, the kinetics are different.

The main differences are as follows: (1) segregation of fast-diffusing particles at the internal surface due to the inverse Kirkendall effect, and (2) the resulting decrease in vacancy flux and in the corresponding shrinkage rate. In general, the problem of interdiffusion within a region with moving boundaries is rather complicated to give the possibility for a rigorous solution. So, we try to find a simplified model, which should be effective but, on the other hand, should not lose the main feature of the full process. The main simplifications are the steady state approximations for the fluxes of vacancies and atoms. So, we present here various models with increasing degree of complexity, and then find out which model can be regarded as optimal for practical use for a given specific case.

7.2.3.1 Model 1: Shrinkage of Pure Element Nanoshells

The vacancy outflux is generated by the difference in equilibrium vacancy concentrations at two curved boundaries (Figure 7.1). The equation governing the vacancy concentration c_V (fraction of vacant sites) inside the shell is

$$\frac{\partial c_V}{\partial t} = -\frac{1}{r^2}\frac{\partial}{\partial r}(r^2 \Omega j_V) + 0, \qquad r_i(t) < r < r_e(t) \tag{7.3}$$

j_V is the vacancy flux density and Ω is the atomic volume. The term "0" in Equation 7.3 is just to indicate that vacancies in our model are generated and annihilated only at the internal and external surfaces of the shell. We do not consider any vacancy sinks/sources (dislocation kinks) inside the nanoshell.

The vacancy flux is governed by Fick's first law

$$\Omega j_V = -D_V \frac{\partial c_V}{\partial r} \tag{7.4}$$

The boundary conditions for vacancies were formulated by Equation 7.2.

Internal shell/vacuum (or inert medium) and external surfaces move due to vacancy fluxes as

$$\frac{dr_i}{dt} = -\Omega j_V(r_i) = D_V \frac{\partial c_V}{\partial r} \qquad (r = r_i) \tag{7.5}$$

$$\frac{dr_e}{dt} = -\Omega j_V(r_e) = D_V \frac{\partial c_V}{\partial r} \qquad (r = r_e) \tag{7.6}$$

In the steady state approximation,

$$\frac{\partial c_V}{\partial t} \approx 0 = -\frac{1}{r^2}\frac{\partial}{\partial r}(r^2 \Omega j_V) \tag{7.7}$$

which leads to

$$\Omega j_V(r_i) = -D_V \frac{\partial c_V}{\partial r} = \frac{K_V}{r^2} \tag{7.8}$$

so that

$$c_V(r) = \frac{K_V}{D_V r} + M_V \tag{7.9}$$

where K_V and M_V are independent of r. Substituting the solution of Equation 7.9 into the Gibbs–Thomson boundary conditions of Equation 7.2, we obtain

$$K_V = c_V^{eq} D_V \beta \frac{r_e + r_i}{r_e - r_i} = D^* \beta \frac{r_e + r_i}{r_e - r_i} \tag{7.10}$$

with D^* being the self-diffusivity of the pure element B.

Substituting the expression Equation 7.10 into Equations 7.5 and 7.6 for the boundary movement, we obtain the following simple equations for the shrinkage kinetics

$$\frac{dr_i}{dt} = -\frac{D^*\beta}{r_i^2}\frac{r_e + r_i}{r_e - r_i} \qquad (7.11)$$

$$\frac{dr_e}{dt} = -\frac{D^*\beta}{r_e^2}\frac{r_e + r_i}{r_e - r_i} \qquad (7.12)$$

Equations 7.11 and 7.12 are not independent and already contain the conservation of matter (and volume)

$$r_e^3 = r_i^3 + r_f^3 \qquad (7.13)$$

To obtain an analytical estimate, we use further assumptions. Let

$$\Delta r \equiv r_e - r_i \ll r_i \qquad (7.14)$$

Then conservation of volume gives

$$\Delta r \approx \frac{r_{i0}^2}{r_i^2}\Delta r_0 \qquad (7.15)$$

Substitution of Equation 7.15 in Equation 7.11 and taking account of Equation 7.13 gives

$$\frac{dr_i}{dt} \approx -\frac{2D^*\beta}{r_{i0}^2 \Delta r_0} r_i \qquad (7.16)$$

which provides a shrinkage behavior of the internal radius for the initial stage of shrinkage

$$r_i(t) \approx r_i^0 \exp\left[-\frac{2D^*\beta}{r_{i0}^2 \Delta r_0} t\right] \qquad (7.17)$$

An approximate shrinkage time follows from Equation 7.17 as

$$t_{collapse} \approx \frac{r_{i0}^2 \Delta r_0}{2D^*\beta} = \frac{kT}{4\gamma\Omega}\frac{r_{i0}^2 \Delta r_0}{D^*} \qquad (7.18)$$

Conservation of matter relates the initial parameters of the nanoshell with the radius r_f of the evolving nanoparticle after the collapse of the nanoshell resulting in

$$r_{i0}^2 \Delta r_0 \approx \frac{r_f^3}{3} \qquad (7.19)$$

So, finally

$$t_{collapse} \approx \frac{kT}{12\gamma\Omega}\frac{r_f^3}{D^*} \qquad (7.20)$$

To estimate the time of collapse of a nanoshell, we take the following set of typical parameters

$$r_f = 4\,\text{nm}, \quad \gamma = 0.5\,\text{Jm}^{-2}, \quad \Omega = 4\times 10^{-29}\,\text{m}^3$$

$$D_A^* = 3.25\times 10^{-8} \exp\left(-\frac{1.24\,\text{eV}}{kT}\right)\,\text{m}^2\,\text{s}^{-1}$$

and we find

$$T = 300 \text{ K} \Rightarrow t_{\text{collapse}} \approx 300 \text{ years}, \quad T = 600 \text{ K} \Rightarrow t_{\text{collapse}} \approx 1 \text{ s}$$

7.2.3.2 Model 2: Shrinkage of a Binary Compound Nanoshell with Steady State Approximation for Both Vacancies and B Species

The vacancy outflux should lead to a redistribution of components, i.e., segregation of the fast-diffusing atoms B at the internal boundary, and segregation of the slowly diffusing A atoms near the external surface (the segregation of A is relatively less intensive because of the larger total area of the external surface). At first, the vacancy flux, caused by the difference in curvatures and corresponding vacancy concentrations, generates the inward migration fluxes j_A and j_B of both A and B species (shrinkage of nanoshell). Migration of B proceeds faster, leading to segregation of B near the internal surface (Figure 7.2). The arising concentration gradient of B influences all fluxes and it decreases the vacancy flux, suppressing the shrinkage process; it also changes the fluxes of the components.

Here we use a phenomenological model treating a total local vacancy concentration and total local fluxes of vacancies and atoms. We do not write down the fluxes and concentrations at each sublattice separately. Different mobilities at different sublattices are taken into account implicitly via different intrinsic diffusivities of species A and B.

Basic equations In the equation for vacancies, we consider sources/sinks inside a shell to be absent, therefore

$$\frac{\partial c_V}{\partial t} = -\frac{1}{r^2}\frac{\partial}{\partial r}(r^2 \Omega j_V) + 0, \quad r_i(t) < r < r_e(t) \tag{7.21}$$

The equation for species B with a concentration c_B (local atomic fraction of B) is

$$\frac{\partial c_B}{\partial t} = -\frac{1}{r^2}\frac{\partial}{\partial r}(r^2 \Omega j_B), \quad r_i(t) < r < r_e(t) \tag{7.22}$$

Here the expressions for both fluxes should include the contributions from the vacancy gradient, in contrast to the classical Darken analysis [15–18]

$$\Omega j_V = (D_B - D_A)\frac{\partial c_B}{\partial r} - D_V \frac{\partial c_V}{\partial r} \tag{7.23}$$

$$\Omega j_B = -D_B \frac{\partial c_B}{\partial r} + \frac{c_B D_B^*}{c_V}\frac{\partial c_V}{\partial r} \tag{7.24}$$

All fluxes are written down in the lattice reference frame, D_B, D_B^*, D_V are, respectively, the partial diffusivity of B, the tracer diffusivity of B, and the diffusivity of vacancies inside the nanoshell. In Equations 7.23 and 7.24, we have neglected Manning's corrections (vacancy wind effect). It could be done but it would make the mathematical equations more cumbersome without a substantial change in the main results. Under the same assumption, the mentioned diffusivities are interrelated in the following way

$$D_B = D_B^* \varphi, \quad D_A = D_A^* \varphi, \quad \varphi = \frac{c_A c_B}{kT}\frac{\partial^2 g}{\partial c_B^2}, \quad D_V = \frac{c_A D_A^* + c_B D_B^*}{c_V} \tag{7.25}$$

Here ϕ is a thermodynamic factor and g is the Gibbs free energy per atom.

As in Model 1, the internal and external surfaces move due to vacancy fluxes

$$\frac{dr_i}{dt} = -\Omega j_V(r_i) = -(D_B - D_A)\frac{\partial c_B}{\partial r}(r = r_i) + D_V\frac{\partial c_V}{\partial r}(r = r_i) \tag{7.26}$$

$$\frac{dr_e}{dt} = -\Omega j_V(r_e) = -(D_B - D_A)\frac{\partial c_B}{\partial r}(r = r_e) + D_V\frac{\partial c_V}{\partial r}(r = r_e) \tag{7.27}$$

Boundary conditions for vacancies were formulated in Equation 7.2. Conservation of atoms of B implies that the flux of B across the moving boundaries should be absent

$$\Omega j_B(r_i) - c_B(r_i)\frac{dr_i}{dt} = 0 \tag{7.28}$$

$$\Omega j_B(r_e) - c_B(r_e)\frac{dr_e}{dt} = 0 \tag{7.29}$$

Main assumptions Let the degree of segregation be small, so that we can use c_1 instead of c_B and $1 - c_1$ instead of c_A everywhere except the derivatives (c_1 are the stoichiometric concentrations of B in the IMC). Let $\beta/r \ll 1$, so that we can use c_V^{eq} instead of c_V everywhere except the derivatives. This means that all diffusivities are treated as constant.

We use the steady state approximation both for components and for vacancies

$$\frac{\partial}{\partial r}(r^2 j_V) \approx 0, \quad \frac{\partial}{\partial r}(r^2 j_B) \approx 0$$

so that

$$\Omega j_V \approx \frac{K_V}{r^2}(>0), \quad \Omega j_B \approx \frac{K_B}{r^2}(<0) \tag{7.30}$$

which immediately leads (with almost constant diffusivities) to the same type of dependence for both vacancies and component B

$$\frac{\partial c_V}{\partial r} \approx -\frac{L_V}{r^2}, \quad \frac{\partial c_B}{\partial r} \approx -\frac{L_B}{r^2} \tag{7.31}$$

From Equation 7.31 and the boundary conditions in Equation 7.2, we obtain

$$c_V = M_V + \frac{L_V}{r} \tag{7.32}$$

$$L_V = c_V^{eq}\beta\frac{1/r_i + 1/r_e}{1/r_i - 1/r_e} \quad M_V = c_V^{eq}\left(1 - \frac{2\beta}{r_e - r_i}\right) \tag{7.33}$$

On the other hand, combining Equations 7.26–7.31, we obtain the following set of relations between the parameters K_V, K_B, L_V, and L_B

$$K_V = -(D_B - D_A)L_B + D_V L_V$$

$$K_B = D_B L_B - \frac{c_1 D_B^*}{c_V}L_V, \quad K_B = -K_V c_1 \tag{7.34}$$

This set of algebraic equations, together with Equation 7.33, has the following solution

$$K_V = L_V \frac{D_A^* D_B^*}{c_V^{eq}(c_1 D_A^* + (1-c_1) D_B^*)} = \frac{D_A^* D_B^*}{c_1 D_A^* + (1-c_1) D_B^*} \beta \frac{1/r_i + 1/r_e}{1/r_i - 1/r_e} \quad (7.35)$$

$$K_B = -\frac{c_1 D_A^* D_B^*}{c_1 D_A^* + (1-c_1) D_B^*} \beta \frac{1/r_i + 1/r_e}{1/r_i - 1/r_e} \quad (7.36)$$

$$L_B = -\frac{1}{\varphi} \frac{c_1 (1-c_1)(D_B^* - D_A^*)}{c_1 D_A^* + (1-c_1) D_B^*} \beta \frac{1/r_i + 1/r_e}{1/r_i - 1/r_e} \quad (7.37)$$

To obtain the parameters of the B component redistribution,

$$c_B(r) - c_1 = M_B + \frac{L_B}{r} \quad (7.38)$$

we should take into account conservation of matter

$$\int_{r_i}^{r_e} (c_B(r) - c_1) r^2 dr = 0 \quad (7.39)$$

Substituting Equation 7.38 in Equation 7.39, we obtain

$$M_B = -\frac{3}{2} L_B \frac{r_e^2 - r_i^2}{r_e^3 - r_i^3} = \frac{1}{\varphi} \frac{c_1(1-c_1)(D_B^* - D_A^*)}{c_1 D_A^* + (1-c_1) D_B^*} \beta \frac{(r_e + r_i)^2}{(r_e - r_i)(r_e^2 + r_i^2 + r_e r_i)} \quad (7.40)$$

Thus, the main results of the analysis of Model 2 are the following

1) The kinetics of shrinkage is governed by the slow component diffusion ($D_A^* \ll D_B^*$)

$$\frac{dr_i}{dt} = -\frac{D_A^*}{1 - c_1 + c_1 \frac{D_A^*}{D_B^*}} \frac{r_e + r_i}{r_e - r_i} \frac{\beta}{r_i^2} \approx -\frac{D_A^*}{1 - c_1} \frac{r_e + r_i}{r_e - r_i} \frac{\beta}{r_i^2} \quad (7.41)$$

$$t_{collapse} \approx \frac{kT}{12(1-c_1)\gamma\Omega} \frac{r_f^3}{D^*}$$

$$\frac{dr_e}{dt} = \frac{r_i^2}{r_e^2} \frac{dr_i}{dt} \quad (7.42)$$

2) The vacancy distribution across the shell is

$$c_V = c_V^{eq} \left(1 - \frac{2\beta}{r_e - r_i} + \frac{\beta}{r} \frac{r_e + r_i}{r_e - r_i}\right) \quad (7.43)$$

3) The redistribution of component B across the shell is

$$c_B(r) - c_1 = \frac{c_1(1-c_1)}{\varphi} \frac{(D_B^* - D_A^*)}{c_1 D_A^* + (1-c_1) D_B^*} \frac{r_e + r_i}{r_e - r_i} \left(\frac{\beta}{r} - \frac{3\beta}{2} \frac{r_e^2 - r_i^2}{r_e^3 - r_i^3}\right) \quad (7.44)$$

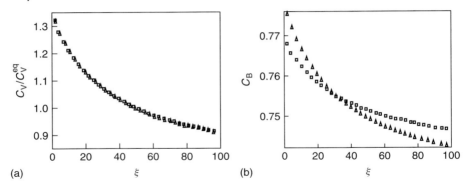

Figure 7.3 Redistribution of vacancies and B species inside the shell, calculated with Model 2 (squares) and Model 3 (triangles). The dimensionless space coordinate is $\xi = 100(r - r_i)/(r_e - r_i)$. (a) Profiles of $(c_V(r)/c_V^{eq})$ practically coincide; (b) profiles of $c_B(r)$ inside the shell for Models 2 and 3 differ substantially but not by the order of magnitude.

In full analogy with Model 1, we can obtain a rough estimate for the shrinkage time

$$t_{collapse} \approx \frac{kT}{12(1-c_1)\gamma\Omega}\frac{r_f^3}{D^*}$$

with

$$D^* \equiv \frac{D_A^* D_B^*}{c_1 D_A^* + (1-c_1) D_B^*} \tag{7.45}$$

Characteristic profiles of $c_V(r)/c_V^{eq}$ and $c_B(r)$ for Model 2 and Model 3 are shown in Figure 7.3: profiles for vacancies $c_V(r)/c_V^{eq}$ practically coincide and differ for the B component, but are of the same order of magnitude.

Time dependencies on the internal and external radii (typical example shown in Figure 7.4) can be found up to a certain minimal radius (3β in this case), since at an internal radius approaching zero the finite-difference methods becomes less accurate. One may stop calculating earlier as the last stage of shrinkage (collapse) is very quick, which is explained by the inverse relation between the velocity and the internal radius. Thus, the total shrinkage time almost equals the time of shrinkage to the above-mentioned minimal radius.

The dependencies of the collapse time on the ratio D_B^*/D_A^* at initial reduced sizes $r_i/\beta = 10, 11, 12, 13, 14, 15$ are shown in Figure 7.5. Dependencies of the collapse time on the initial internal radius r_i/β at $D_B^*/D_A^* = 10, 20, 40, 60, 80, 100$ are shown in Figure 7.6. Figures 7.5 and 7.6 demonstrate that the rough approximation of shrinkage time in Equation 7.45 is quite reasonable.

7.2.3.3 Model 3: Steady State and Non–Steady State Vacancies for Component B

In this model, we still use the steady state approximation for vacancies but solve the non–steady state diffusion equation for the redistribution of component B between

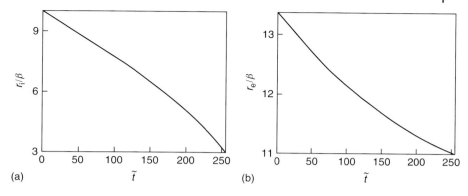

Figure 7.4 Time dependencies of (a) internal and (b) external radii according to Model 2 ($\tilde{t} = D_B^* t/\beta^2$ is the dimensionless time and β is the characteristic size introduced in Equation 7.2).

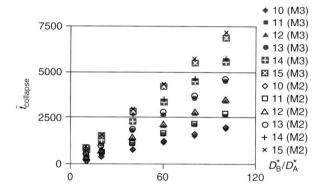

Figure 7.5 Dependence of dimensionless collapse time $\tilde{t}_{collapse} = D_B^* t_{collapse}/\beta^2$ on the ratio of diffusivities D_B^*/D_A^* for Models 2 and 3 at different initial nanoshell sizes r_{i0}/β (for example, 11 (M3) calculated in Model 3 for $r_{i0}/\beta = 11$).

the moving surfaces. As in Model 1 and Model 2, we have $\partial(r^2 j_V)/\partial r \approx 0$, so that $\Omega j_V \approx K_V/r^2 > 0$. Yet, we do not have similar equations for B. Therefore, in this more general case, Equation 7.23 transforms into

$$\frac{\partial c_V}{\partial r} = \frac{D_B - D_A}{D_V} \frac{\partial c_B}{\partial r} - \frac{K_V}{D_V} \frac{1}{r^2} \tag{7.46}$$

Let, as in model 2, D_A, D_B, D_V, and φ be approximately constant inside the nanoshell with a narrow homogeneity range. Integrating Equation 7.46 over all possible radii inside the nanoshell, we obtain the relation between the growth rate parameter K_V and the boundary values of the B-concentration

$$K_V = c_V^{eq} D_V \beta \frac{r_e + r_i}{r_e - r_i} - \frac{(D_B - D_A)(c_B(r_e) - c_B(r_i))}{1/r_i - 1/r_e} \tag{7.47}$$

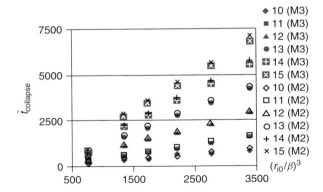

Figure 7.6 Dependence of dimensionless collapse time $\tilde{t}_{collapse} = D_B^* t_{collapse}/\beta^2$ on the initial nanoshell size r_{i0}/β for Models 2 and 3 at different diffusivity ratios D_B^*/D_A^* (for example, 20 (M2) calculated in Model 2 for $D_B^*/D_A^* = 20$).

On the other hand, integrating Equation 7.46 over r from the internal radius to some arbitrary value, we obtain the interrelation between vacancy and B redistributions

$$c_V(r) = c_V(r_i) + \frac{D_B - D_A}{D_V}(c_B(r) - c_B(r_i)) + \frac{K_V}{D_V}\left(\frac{1}{r} - \frac{1}{r_i}\right) \quad (7.48)$$

By combining Equations 7.24 and 7.46, we get

$$\Omega j_B = -\frac{D_A^* D_B^* \varphi}{c_1 D_B^* + (1-c_1) D_A^*} \frac{\partial c_B}{\partial r} - \frac{c_1 D_B^*}{c_1 D_B^* + (1-c_1) D_A^*} \frac{K_V}{r^2} \quad (7.49)$$

Then, Equation 7.23 becomes rather simple for the redistribution of B with steady state vacancies

$$\frac{\partial c_B}{\partial t} = D_{NG} \frac{1}{r^2} \frac{\partial}{\partial r}\left(r^2 \frac{\partial c_B}{\partial r}\right) \quad (7.50)$$

$$D_{NG} = \frac{D_A^* D_B^* \varphi}{c_1 D_B^* + (1-c_1) D_A^*}, \quad r_i(t) < r < r_e(t)$$

For boundary conditions and for boundaries motion, we obtain, using Equations 7.26–7.29

$$\left.\frac{\partial c_B}{\partial r}\right|_{r_i} = -\frac{K_V}{r_i^2 D_{NG}} c_1 \cdot (1-c_1) \frac{D_B^* - D_A^*}{c_1 D_A^* + (1-c_1) D_B^*} \quad (7.51)$$

$$= -\frac{K_V}{r_i^2} c_1 (1-c_1) \frac{D_B^* - D_A^*}{\varphi D_A^* D_B^*}$$

$$\left.\frac{\partial c_B}{\partial r}\right|_{r_e} = -\frac{K_V}{r_e^2} c_1 (1-c_1) \frac{D_B^* - D_A^*}{\varphi D_A^* D_B^*} \quad (7.52)$$

Thus, the smaller the ratio D_A^*/D_B^*, the sharper is the segregation at the boundaries. The shrinkage kinetics is governed by the equations

$$\frac{dr_i}{dt} = -\frac{K_V}{r_i^2}, \quad \frac{dr_e}{dt} = \frac{r_i^2}{r_e^2} \frac{dr_i}{dt} \quad (7.53)$$

7.2 Kinetic Analysis of the Instability of Hollow Nanoparticles

with K_V being determined by Equation 7.47 and the parameter depending on sizes and compositions at the interfaces.

Equations 7.47 and 7.50–7.53 form a full set for a "Stephan-like" problem with diffusion between two moving boundaries in a spherically symmetrical shell. A direct solution (even a numerical one) is complicated, due to the problem of changing grid. So, we propose below to change variables, making equations somewhat cumbersome, but fixing the grid

$$x = \frac{r - r_i(t)}{r_e(t) - r_i(t)}, \quad \tau = c_V^{eq} \frac{Bt}{\beta^2}, \quad \rho_i = \frac{r_i}{\beta}, \quad \rho_e = \frac{r_e}{\beta} \quad (7.54)$$

Here $B = D_B^*/c_V$, which we take as constant and which appears to be a reasonable assumption for an ordered, nearly stoichiometric phase. Let further $A = D_A^*/c_V$, so that

$$D_V = c_A A + c_B B \quad (7.55)$$

The new variable x belongs to the interval $(0, 1)$. Moving boundaries now correspond to $x = 0$ and $x = 1$. Owing to the transformation of derivatives, we get

$$\frac{\partial}{\partial r} = \frac{1}{\beta(\rho_e - \rho_i)} \frac{\partial}{\partial x}, \quad \frac{\partial}{\partial t} = \frac{B}{\beta^2} \left(\frac{\partial}{\partial \tau} - \frac{(1-x)\frac{d\rho_i}{d\tau} + x\frac{d\rho_e}{d\tau}}{(\rho_e - \rho_i)} \frac{\partial}{\partial x} \right) \quad (7.56)$$

The equations have become more extended, but they are now suitable for a numerical solution using the fixed grid. The equation for redistribution of species B within the interval $0 < x < 1$ reads

$$\frac{\partial c_B(\tau, x)}{\partial \tau} = \frac{(1-x)\frac{d\rho_i}{d\tau} + x\frac{d\rho_e}{d\tau}}{(\rho_e - \rho_i)} \frac{\partial c_B}{\partial x} + \frac{1}{(\rho_e - \rho_i)^2 ((\rho_e - \rho_i)x + \rho_i)^2}$$
$$\times \frac{\partial}{\partial x} \left(((\rho_e - \rho_i)x + \rho_i)^2 \, dng \frac{\partial c_B}{\partial x} \right) \quad (7.57)$$

where

$$dng = \varphi \frac{A/B}{c_1 + (1 - c_1)A/B}$$

In dimensionless units, the shrinkage rates are

$$\frac{d\rho_i}{d\tau} = -\frac{\tilde{K}_V}{\rho_i^2}, \quad \frac{d\rho_e}{d\tau} = -\frac{\tilde{K}_V}{\rho_e^2} \quad (7.58)$$

For $\tilde{K}_V$, the expression

$$\tilde{K}_V = dv \frac{\rho_e + \rho_i}{\rho_e - \rho_i} - \varphi \frac{(1 - A/B)(c_B|_{x=1} - c_B|_{x=0})}{1/\rho_i - 1/\rho_e} \quad (7.59)$$

is obtained, The boundary conditions are

$$\left. \frac{\partial c_B}{\partial x} \right|_0 = -\frac{\tilde{K}_V}{(\rho_e - \rho_i)\rho_i^2} c_1 (1 - c_1) \frac{1 - A/B}{\varphi A/B} \quad (7.60)$$

$$\left.\frac{\partial c_B}{\partial x}\right|_1 = -\frac{\tilde{K}_V}{(\rho_e - \rho_i)\rho_e^2} c_1 (1-c_1) \frac{1 - A/B}{\varphi A/B} \tag{7.61}$$

A comparison of the numerical solution for Model 3 and the analytical solution for the less rigorous Model 2 is represented in Figures 7.5 and 7.6. This comparison demonstrates that in most cases Model 2 is quite good for practical estimates.

7.2.3.4 Model 4: Non–Steady State Vacancies and Atoms

This model employs non–steady state equations for vacancies and B species. The basic equations can be formulated as follows

$$\frac{\partial c_V}{\partial t} = \frac{1}{r^2}\frac{\partial}{\partial r}\left(r^2\left[-(D_B - D_A)\frac{\partial c_B}{\partial r} + D_V\frac{\partial c_V}{\partial r}\right]\right) \tag{7.62}$$

$$\frac{\partial c_V}{\partial t} = \frac{1}{r^2}\frac{\partial}{\partial r}\left(r^2\left[D_B\frac{\partial c_B}{\partial r} - \frac{c_B D_B^*}{c_V}\frac{\partial c_V}{\partial r}\right]\right) \tag{7.63}$$

$$\frac{dr_i}{dt} = \left[\frac{D_A D_B}{(c_A D_B + c_B D_A)\varphi}\frac{1}{c_V}\frac{\partial c_V}{\partial r}\right]_{r=r_i} \tag{7.64}$$

$$\frac{dr_e}{dt} = \left[\frac{D_A D_B}{(c_A D_B + c_B D_A)\varphi}\frac{1}{c_V}\frac{\partial c_V}{\partial r}\right]_{r=r_e} \tag{7.65}$$

$$\frac{\partial c_B}{\partial r}(r_i) = \left[c_A c_B \frac{D_B - D_A}{(c_A D_B + c_B D_A)\varphi}\frac{1}{c_V}\frac{\partial c_V}{\partial r}\right]_{r=r_i} \tag{7.66}$$

$$\frac{\partial c_B}{\partial r}(r_e) = \left[c_A c_B \frac{D_B - D_A}{(c_A D_B + c_B D_A)\varphi}\frac{1}{c_V}\frac{\partial c_V}{\partial r}\right]_{r=r_e} \tag{7.67}$$

$$c_V(r_i) = c_V^{eq}\left(1 + \frac{\beta}{r_i}\right) \tag{7.68}$$

$$c_V(r_e) = c_V^{eq}\left(1 - \frac{\beta}{r_e}\right) \tag{7.69}$$

From these relations, one may conclude that the main feature of each solution is that the shrinkage velocity is controlled by a slower component. Indeed, if $D_A \ll D_B$, then

$$\frac{D_A D_B}{c_A D_B + c_B D_A} \approx \frac{D_A D_B}{c_A D_B} = \frac{D_A}{c_A}$$

Thus, the less the ratio D_A/D_B, the slower the shrinkage of a shell proceeds. This means that, although a nanoshell is generally unstable, it may be a long-living object.

Similar to Model 3, let us proceed to new variables, having fixed the boundaries in the following way

$$x = \frac{r - r_i(t)}{r_e(t) - r_i(t)}, \quad \tau = \frac{Bt}{\beta^2}, \quad \rho_i = \frac{r_i}{\beta}, \quad \rho_e = \frac{r_e}{\beta} \tag{7.70}$$

Here $B = D_B^*/c_V$ is treated as constant (for an ordered, almost stoichiometric phase). Also, let $A = D_A^*/c_V$, so that $D_V = c_A A + c_B B$. The new variable x belongs to the interval $(0, 1)$. Moving boundaries correspond to $x = 0$ and $x = 1$. So, Equations 7.62–7.68 can be transformed as follows:

$$\frac{\partial c_V(\tau, x)}{\partial \tau} = \frac{(1-x)\frac{d\rho_i}{d\tau} + x\frac{d\rho_e}{d\tau}}{(\rho_e - \rho_i)} \frac{\partial c_V}{\partial x} + \frac{1}{(\rho_e - \rho_i)^2 ((\rho_e - \rho_i)x + \rho_i)^2}$$
$$\times \frac{\partial}{\partial x}\left(((\rho_e - \rho_i)x + \rho_i)^2 \left(-\varphi\left(1 - \frac{A}{B}\right) c_V \frac{\partial c_B}{\partial x} + \left(c_B + c_A \frac{A}{B}\right) \frac{\partial c_V}{\partial x}\right)\right)$$

$$\frac{\partial c_B(\tau, x)}{\partial \tau} = \frac{(1-x)\frac{d\rho_i}{d\tau} + x\frac{d\rho_e}{d\tau}}{(\rho_e - \rho_i)} \frac{\partial c_B}{\partial x} + \frac{1}{(\rho_e - \rho_i)^2 ((\rho_e - \rho_i)x + \rho_i)^2}$$
$$\times \frac{\partial}{\partial x}\left(((\rho_e - \rho_i)x + \rho_i)^2 \left(\varphi c_V \frac{\partial c_B}{\partial x} - c_B \frac{\partial c_V}{\partial x}\right)\right)$$

$$\frac{d\rho_i}{d\tau} = \frac{A/B}{c_A(0) + c_B(0) A/B} \frac{\partial c_V}{\partial x}, \quad (x = 0)$$

$$\frac{d\rho_e}{d\tau} = \frac{A/B}{\tilde{n}_A(1) + \tilde{n}_B(1) A/B} \frac{\partial c_V}{\partial x}, \quad (x = 1)$$

$$\left.\frac{\partial c_B}{\partial x}\right|_{x=0} = c_A(0) c_B(0) \frac{1 - A/B}{c_A(0) + c_B(0) A/B} \frac{1}{\varphi c_V(0)} \frac{\partial c_V}{\partial x}, \quad (x = 0)$$

$$\left.\frac{\partial c_B}{\partial x}\right|_{x=1} = c_A(1) c_B(1) \frac{1 - A/B}{c_A(1) + c_B(1) A/B} \frac{1}{\varphi c_V(1)} \frac{\partial c_V}{\partial x}, \quad (x = 1)$$

$$c_V(X = 0) = c_V^{eq}(1 + 1/\rho_i), \quad c_V(X = 1) = c_V^{eq}(1 + 1/\rho_e)$$

$$\rho_i(t = 0) = \rho_{i0}, \quad \rho_e(t = 0) = \rho_{e0}$$

The numerical solution of the above-stated problem confirms the main result, according to which the time of a shell shrinkage is approximately inversely proportional to the ratio $A/B = D_A/D_B$.

7.2.4
Segregation of Pure B at the Internal Surface

We can see from the results of Model 2 and Model 3 that, with decreasing internal radius, the segregation of the fast-diffusing atom B increases. Sooner or later, it can exceed the homogeneity range of the intermediate phase, which could lead to precipitation of the almost pure B-phase at the internal surface. Formally, to conserve the spherical symmetry, after the precipitation, the nanoshell should consist of two subshells – pure B inside and inhomogeneous IMC outside. Thus,

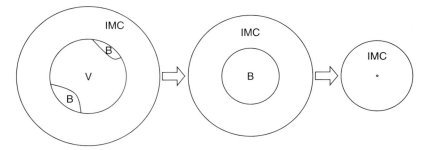

Figure 7.7 Possible scheme of precipitation of B at the internal surface.

we would have three interfaces instead of two – void–pure B, pure B–IMC, and IMC–media.

Yet, this scenario seems to be highly improbable. More probable is that the pure B will precipitate as one or few tiny pieces in the remaining void, adjacent to the IMC shell, as depicted in Figure 7.7. With time, the residual void will disappear, and we will obtain pure B inside the inhomogeneous IMC shell. After some time, the partial homogenization of IMC will proceed, leaving the nonzero difference of composition between the internal and the external interface due to the curvature effect.

The decrease in the internal radius, simultaneously with increasing segregation, causes the shift of the equilibrium boundary concentration of B inside IMC at the curved interface between B and IMC. Let us compare these two equilibrium concentrations. According to Gibbs theory, at curved interfaces, the mentioned shift is equal to

$$\Delta c_B(r_i) = \frac{2\gamma_{B/IMC}\Omega}{g''_{IMC}(1-c_1)}\frac{1}{r_i} \qquad (7.71)$$

On the other hand, according to Equation 7.44, even for a diffusivity of A tending to zero, the segregation at the internal boundary is less than

$$c_B(r_i) - c_1 = \frac{1}{g''_{IMC}(1-c_1)}\left(\frac{r_e+r_i}{r_e-r_i}\right)\frac{2\gamma_{void/IMC}\Omega}{r_i} \qquad (7.72)$$

The difference between Equations 7.62 and 7.62 may be positive, meaning precipitation, but the absolute value of supersaturation is very small, taking into account the usually large second-order derivatives of Gibbs potential over concentration (g'') for a highly ordered intermetallic phase. Thus, we expect that the segregation cannot exceed the size-dependent homogeneity range too much, and the precipitation of B, if attainable at all, will be small.

7.2.5
Kinetic Monte Carlo Simulation of Shrinkage of a Nanoshell

The analyses presented in the previous paragraphs are purely phenomenological and do not take into account the possible peculiarities of diffusion and structure changes in nanoparticles. It is important to check the main ideas of shrinkage of a nanoshell by direct kMC simulation. We have simulated the nanoshell behavior for two cases

1) pure component B with fcc lattice in vacuum in (Model 1MC);
2) an ordered fcc phase of initially homogeneous composition A_1B_3 in vacuum (Model 2MC).

The basic algorithm is the following:

We consider pair interactions only with nearest neighbors. To compare the rates of shrinkage of pure B (case (1)) and of IMC (cases (2), (3)), we used a constraint of equal average pair energy in the random alloy approximation

$$c_A^2 \Phi_{AA}^{IMC} + 2c_A c_B \Phi_{AB}^{IMC} + c_B^2 \Phi_{BB}^{IMC} = \Phi_{BB}^{pure} \tag{7.73}$$

Namely, we have chosen the following pair energies $\Phi_{BB}^{pure}/kT = -0.414$ for case (1) and

$$\Phi_{AA}^{IMC}/kT = -0.18045, \quad \Phi_{AB}^{IMC}/kT = -0.8, \quad \Phi_{BB}^{IMC}/kT = -0.2$$

for case (2).

Different preexponential frequencies v_{0A} and v_{0B} are introduced for A and B species. The candidate for the jumping atom is chosen randomly from the array consisting of N_A A atoms and $v_B/v_A \cdot N_B$ B atoms. Since we have a large number of surface atoms, often an atom has several ($0 \leq Z \leq 12$) possible jump directions. For each of them, the difference in energies after jump and before jump is calculated. Besides, there is one more possibility – no jump (staying in site).

Probabilities of these events are calculated as

$$p(i) = \frac{\exp\left(-\frac{E^{after}(i) - E^{before}(i)}{kT}\right)}{\sum_{j=1}^{Z} \exp\left(-\frac{E^{after}(j) - E^{before}(j)}{kT}\right) + 1}, \quad i = 1, \ldots, Z \tag{7.74}$$

$$p(stay) = \frac{1}{\sum_{j=1}^{Z} \exp\left(-\frac{E^{after}(j) - E^{before}(j)}{kT}\right) + 1} \tag{7.75}$$

7.2.5.1 Model 1MC: Pure B-Shell in Vacuum

There are 76662 atoms of B initially to form a nanoshell with fcc lattice. Initially, the internal radius of a shell is $r_i = 7a$ (a is the lattice parameter) and the external radius is $r_e = 17a$, point defects being absent. Thereafter, the faceting of both internal and outer surfaces of the shell becomes observable, while the bulk becomes saturated with vacancies that migrate from the internal pore outside. As a result of the applied RTA-algorithm [13], the whole shell collapses after 3650 Monte Carlo steps (MCS).

Figures 7.8a–c presents the dynamics of the shrinkage process. The change in the external radius is hard to detect since at given parameters (even neglecting

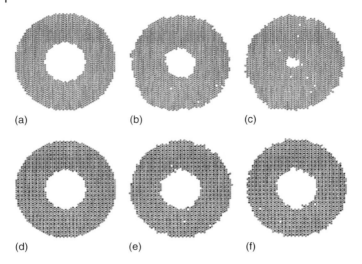

Figure 7.8 Nanoshell shrinkage kinetics (only "equatorial" cross sections are given) in vacuum for the pure element (a, b, c) and for the alloy (d, e, f): a, d – initial configuration, b, e – after 1800 MCS, c, f – after 3600 MCS.

newly formed vacancies) it is approximately equal to $\Delta r_e/r_e \cong (r_i/r_e)^3/3$ and is less than 2.4 %.

7.2.5.2 Model 2MC: Ordered IMC Nanoshell in Vacuum

An ideally ordered defect-free AB_3 compound nanoshell of the same size as in the previous case was used. kMC confirm that

- shrinkage is much slower for a compound shell compared with a pure element shell (Figures 7.8e, f);
- the fast-diffusing component can indeed segregate at the internal surface, but the degree of segregation is small;
- the order is basically conserved during shrinkage.

7.2.6
Influence of Vacancy Segregation on Nanoshell Shrinkage

The above outlined theory has been mathematically developed in [20] and was subjected to some criticism in Refs [21, 22]. First, the authors of [20–22] doubt the possibility of applying the quasi–steady state assumption to the nanoshell shrinkage and, thus, the possible role of the Kirkendall effect in the shrinkage rate. Indeed, if the nanoshell is a solid solution with a wide homogeneity range, the quasi–steady state approximation is not appropriate. But, for the nanoshell with a narrow homogeneity interval, and, therefore, a large thermodynamic factor, the quasi–steady state approximation works well (see Section 7.4). Second, instead of the results on nanoshell time instability, the authors of [21] assume that hollow nanoshells can be metastable due to a specific radial vacancy distribution. On the basis of the results of molecular dynamics simulations, they conclude that the energy of the nanoshell nonmonotonically depends on the location of the vacancies in it. It is energetically most favorable for the vacancy to be located in the bulk between internal and external surfaces. According to [21], this creates the energy barrier for vacancy migration from the internal surface to the external one and provides the metastability of the system (stability of the nanoshell cavity).

It is known that the energy of any spatially constrained system with a vacancy depends on the distance from this vacancy to the surface. For example, in Ref. [23], the exponential spatial dependence was established for both, the energy of vacancy formation and for vacancy migration near the planar free surface. In [24], a nonmonotonic dependence, with the maximum in the vicinity of the second atomic plane, was obtained. Obviously, such depths of nonmonotony appear to be significant only for nanoobjects, since the change of the particle's radius by one atomic plane results in the curvature change of several percent; segregation in several atomic planes causes noticeable gradients of chemical potential.

On the other hand, nanosizes allow one to neglect the work of vacancy sources and sinks, and, therefore, the movement of the lattice (Kirkendall shift) is also negligible. First, the linear size of the nanoparticle is less than the characteristic distance between vacancy sinks in real metal (~ 1 μm). Second, even having assumed that a nanoshell has dislocations or grain boundaries, serving as sources/sinks of vacancies, the disappearance/formation of vacancies on such heterogeneities must cause high stresses in the bulk, which, in their turn, prevent further formation/disappearance of vacancies. Besides that, the nanoshell shrinkage leads to the change in radii and, correspondingly, to the change in the curvature of the surface determining vacancy concentrations on the internal ($c_V(r_i)$) and external ($c_V(r_i)$) surfaces. Therefore, the concentration profile can be quasi–steady state but not equilibrium. So, even knowing the dependence of the system's energy on the location of the vacancy $E_V^f(r)$, one cannot apply the standard relation

$$c_V(r) = \exp\left(-E_V^f(r)/kT\right)$$

For the sake of simplicity, let us consider a one-component hollow nanoshell where vacancy sources/sinks are provided only by internal and external surfaces. At such an assumption, inside the nanoshell the law of conservation of vacancies is fulfilled

$$\frac{\partial c_V}{\partial t} = -\frac{1}{r^2}\frac{\partial (r^2 \Omega j_V)}{\partial r} + 0, \quad r_i < r < r_e \quad (7.76)$$

where j_V is the vacancy flux density. Besides the entropy term (vacancy gradient), the vacancy flux density contains the drift term, connected with energy spatial dependence

$$\Omega j_V = -D_V \frac{\partial c_V}{\partial r} + \frac{c_V D_V}{kT}\left(-\frac{\partial E_V}{\partial r}\right) \quad (7.77)$$

where $E_V(r)$ is the total energy of the system with a vacancy, located at the distance r from the nanoshell's center (different from the formation energy E_V^f). Since vacancies are much more mobile than atoms, after a short transition period, the spatial vacancy distribution must attain the quasi–steady state regime, $\partial c_V/\partial t = 0$. At that, the total flux through an arbitrary sphere is constant, so let us designate $r^2 \Omega j_V = \text{const} = \varpi$.

Hereinafter, we consider the vacancy diffusivity as constant everywhere. Then, the change in migration energy is negligible and we can treat the dependence of the vacancy concentration profile only on the energy profile $E_V(r)$ via

$$\frac{\partial c_V}{\partial r} + c_V \frac{\partial (E_V/kT)}{\partial r} = -\frac{\omega}{r^2}, \quad \omega = \frac{\varpi}{D_V} \tag{7.78}$$

with the boundary conditions

$$c_V(r_i) = \exp\left(-\frac{E_V(r_i) - E_i}{kT}\right), \quad c_V(r_e) = \exp\left(-\frac{E_V(r_e) - E_e}{kT}\right) \tag{7.79}$$

In Equation 7.79, two different system's energies for each surface were introduced:

- $E_V(r_i)$, $E_V(r_e)$, when the vacancy is still in the nanoshell at the subsurface atomic layer; such a site in the plane 111 has three vacancies and in the plane 100 it has four (Figure 7.9a);
- E_i, E_e, when the vacancy disappears at the surface step (Figure 7.9b).

Taking into account the steady state of the vacancy flux for the system of Equations 7.78 and 7.79

$$4\pi r^2 \Omega j_V = 4\pi \varpi = 4\pi D_V \frac{\exp\left(\frac{E_i - E_e}{kT}\right) - 1}{\int_{r_i}^{r_e} \exp\left(\frac{E_V(r') - E_e}{kT}\right) \frac{dr'}{r'^2}} \tag{7.80}$$

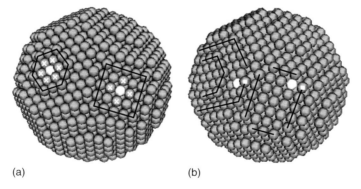

(a) (b)

Figure 7.9 Vacancies (light) at the external surface of the nanoshell inside the surface atomic plane (the empty site is considered as a vacancy and has less than a half of empty neighboring sites): (a) if the surface vacancy exchanges places with the atom marked by crosses, it remains inside the facet with the same number of neighbors and is still considered a vacancy. (b) If the surface vacancy exchanges places with the atom marked by a square and finds itself at the edge of the the surface plane, it is not considered a vacancy anymore – so the aforementioned exchange describes annihilation of vacancies at the surface step.

we get the radial vacancy concentration profile at

$$E_i - E_e = 2\gamma\Omega\,(1/r_i + 1/r_e)$$

as

$$c_V(r) = \exp\left(-\frac{E_V(r) - E_e}{kT}\right) + \frac{\exp\left(\frac{E_i - E_e}{kT}\right) - 1}{\int_{r_i}^{r_e} \exp\left(\frac{E_V(r') - E_e}{kT}\right)\frac{dr'}{r'^2}}$$

$$\times \int_r^{r_e} \exp\left(\frac{E_V(r') - E_V(r)}{kT}\right)\frac{dr'}{r'^2} \qquad (7.81)$$

From Equation 7.80 it is seen that the vacancy flux always exists at any dependence $E_V(r)$. So, vacancy segregation cannot represent the reason for the stability of a hollow nanoshell. The delay can be connected only with the time of attainment of the quasi–steady state regime. But, evidently, the vacancy segregation in the nanoshell cannot present anything special, since similar segregation must take place in macroscopic samples (for example, at sintering of powder mixtures) and, as far as it is known, it in no way prevents, say, void coalescence.

For a qualitative estimation of the radial vacancy concentration profile, we need the corresponding dependence on the system's energy. To find $E_V(r)$, the model in atomic scale was used, which allows us to place the vacancy into the given site and to determine the system's energy directly through atomic interactions. Since the system has free surfaces, it is correct to determine the energy after the system's relaxation. For comparison, we treated a nanoshell both without (all atoms in the sites of an ideal lattice) and with relaxation (after the system has reached equilibrium by the method of molecular statics (MS)).

To obtain the dependence of the system's energy on the position of the vacancy, r, determined by the radius vector from the nanoshell's center, one atom was replaced by a vacancy in turns and the energy of the system was calculated. Owing to the absence of spherical symmetry of the crystal, discreteness of the sites and faceting of the nanoshell's surfaces, the systems with equal distance r from the vacancy to the void's center can have different energies (Figure 7.10). And, conversely, the systems at different r may have equal energies. In the latter case, one more reason arises: the vacancy "feels" the surface up to a certain depth (along with physical reasons the model may reveal artifacts, connected with the cutoff of the many-body Sutton–Chen potential for copper [25]).

As seen from Figure 7.10, the energy dependence near the surfaces has a localized range of values. We have found that the number of empty sites in the vacancy's nearest neighborhood is common for such regions. In what follows, we mark any site with one empty nearest neighbor site as b1, with two nearest neighbor sites as b2, etc. The atom in the bulk is marked as b0. In particular, the vacancy at the internal surface has no more than six empty neighbors, at the external one – up to nine. Since the energy is the least for the site with the largest number of broken bonds (for complete detachment there is the least number of existing bonds left),

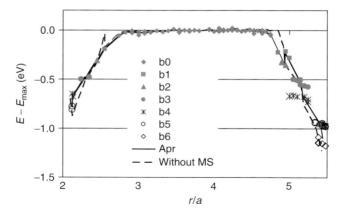

Figure 7.10 Dependence of $E_V(r)$ without relaxation (rigid lattice) and after relaxation (by molecular statics) of the system "2-5.5". The dashed line corresponds to the rigid lattice case. The solid line is an approximation of data for a relaxed system. The data for the energies at various vacancy positions are given with indication of the vacancy type ("bn" corresponds to the vacancy position with n empty neighbor sites).

then the vacancies with the largest number of empty sites determine minimal energies at internal and external surfaces, E_i and E_e. One may conclude that the system's energy with a subsurface vacancy is strongly influenced by the roughness (imperfection) of the surface which, in turn, is determined by local curvature, faceting, etc.

For the small system with $r_i = 2a$ and $r_e = 5.5a$, containing 2639 atoms, an octant with boundary planes (518 atoms in all) was used for the determination of the dependence. Considering the superposition of the dependencies of nanoshell's energies $E_V(r)$ on the position of the vacancy for the system before and after relaxation, we notice that relaxation does not qualitatively change the radial energy dependence for the system with a vacancy (Figure 7.10). The slopes become 11% lower and acquire some smoothness. The localization of atoms by the number of broken bonds is preserved. We qualitatively compare the obtained dependencies below.

Additionally, we analyzed five more systems "$(r_i/a - r_e/a)$" with different internal (r_i) and external (r_e) radii without relaxation (Figure 7.11). As we should have expected, the influence of the curvature is essential near the surface. In the bulk at the depth of the vacancy location (6–7 Å), the energy of the system reaches the plateau with a constant magnitude of E_{bulk}, which can be treated as the energy of the system with a vacancy in the bulk. Then, the differences $E_{bulk} - E_i$ and $E_{bulk} - E_e$ can be interpreted as the energies of vacancy formation on internal and external surfaces, respectively. The plateau is not observed for the system "2-5.5" (Figure 7.10), since the thickness of its nanoshell is 12.6 Å, and the subsurface layers are just overlapping. Such depths can be connected with the radius of potential cutoff $a\sqrt{3} = 6.26$ Å, though accounting for the farther neighborhood when finding that the potential energy does not make any essential contribution.

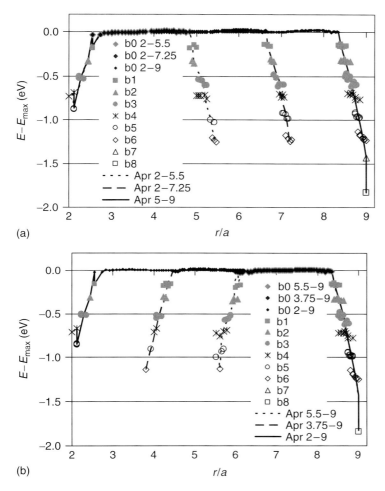

Figure 7.11 The approximation $E_V(r)$ for the system "$(r_i/a - r_e/a)$" with different internal (r_i) and external (r_e) radii without relaxation: (a) for the systems "2-5", "2-7.25" "2-9" and (b) for the systems "3.75-9", "5.5-9", "2-9".

Knowing the dependence $E_V(r)$, from Equation 7.81 one can find the steady state radial profile of vacancy concentration (Figure 7.12).

If we neglect the above-mentioned peculiarities of the vacancy subsystem near the curved surface and consider the energy of the system with a vacancy to be equal everywhere, irrespective of the vacancy position (E_{bulk}), we can simplify Equation 7.81

$$c_V^0(r) = \exp\left(-\frac{(E_{bulk} - E_e)}{kT}\right)\left(1 + \frac{1/r - 1/r_e}{1/r_i - 1/r_e}\left(\exp\left(\frac{(E_i - E_e)}{kT}\right) - 1\right)\right) \tag{7.82}$$

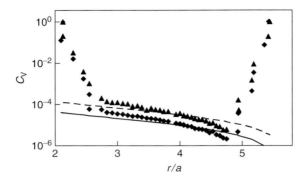

Figure 7.12 Steady state radial dependence of vacancy concentration $c_V(r)$, calculated according to Equation 7.81, for the system "2-5.5" without relaxation (rhombuses and solid line) and after relaxation (triangles and dashed lines). The lines determine the concentration for the profile without segregation, $E_V(r) = E_{bulk}$.

The profile calculated according to Equation 7.82 is depicted as a solid line in Figure 7.12. Apparently, in this case, the concentration dependence is monotonic and there is no surface segregation of vacancies.

To find out how vacancy segregation influences the kinetics of the nanoshell Shrinkage, it is sufficient to compare the vacancy fluxes that correspond to concentration profiles with and without segregation

$$\frac{\frac{dr_i}{dt}^{\text{with segr}}}{\frac{dr_i}{dt}^{\text{without segr}}} = \frac{\left(\frac{1}{r_i} - \frac{1}{r_e}\right) \exp\left(\frac{E_{bulk} - E_e}{kT}\right)}{\int_{r_i}^{r_e} \exp\left(E_V(r') - E_e\right) \frac{dr'}{r'^2}} = \eta \quad (7.83)$$

The ratio Equation 7.83 for nanoshells of different sizes is indicated in Table 7.1. The shrinkage rate increases 1.5 times on average, and twice as much for the relaxed system. Thus, we may conclude that vacancy segregation near the surface accelerates the shrinkage. In other words, the "barrier" understood by the authors of [21] always exists for the nanoshells of any thickness (rise of the energy of the system after embedding the vacancy into it). At transition to nanothickness, this barrier becomes narrower and smoother, which leads to the increase in vacancy flux.

To estimate the correctness of the proposed calculations, we used the phenomenological equations for the energy of vacancy formation for a nonplanar surface

$$E_{bulk} - E_e = E_V^{f,0} + 2\gamma\Omega/r_e, \quad E_{bulk} - E_i = E_V^{f,0} - 2\gamma\Omega/r_i$$

where $E_V^{f,0}$ is the energy of vacancy formation near the planar surface and γ is the surface tension. So, we arrive at

$$E_V^{f,0} = \frac{r_i(E_{bulk} - E_i) + r_e(E_{bulk} - E_e)}{r_i + r_e}, \quad \gamma = \frac{E_i - E_e}{2a^3/4(1/r_i + 1/r_e)} \quad (7.84)$$

Table 7.1 Calculated parameters for the system "$(r_i/a - r_e/a)$" with internal (r_i) and external (r_e) radii. $E_i - E_{bulk}$ is the formation energy at the internal surface, $E_e - E_{bulk}$ is the formation energy at the external surface, and $E_V^{f,0}$, γ, and η are determined according to Equations 7.84 and 7.83, respectively.

$r_i/a - r_e/a$	2–5.5 (MS)	2–5.5	2–7.25	2–9	3.75–9	5.5–9
Thickness $(r_e - r_i)/a$	3.5	3.5	5.25	7	5.25	3.5
Number of atoms	2642	2642	6319	12035	11240	9396
$E_i - E_{bulk}$, (eV)	−0.82	−0.82	−0.82	−0.82	−0.98	−1.04
$E_e - E_{bulk}$, (eV)	−1.04	−1.04	−1.07	−1.09	−1.09	−1.09
$E_V^{f,0}$, (eV)	1.01	1.13	1.13	1.14	1.18	1.19
γ (J m^{-2})	1.18	1.32	1.44	1.47	0.45	0.73
η	2.03	1.75	1.33	1.35	1.33	1.51

Substituting the obtained values of E_i, E_e, r_i, and r_e into Equation 7.84, we can compare $E_V^{f,0}$ and γ, characterizing the proposed model, with experimental data for copper [2, 25] $E_V^{f,0} \approx (1.17 \pm 0.11)$ eV, $\gamma \approx 1.7$ J/m^2 (Table 7.1). We think that the reason for poor coincidence of the surface tension with the experimental value can be explained by a very rough criterion for the choice of surface steps (which determine E_i and E_e) and neglect of "unrecoverable" sources/sinks of vacancies (those that disappear after creation or annihilation of the vacancy).

7.2.7
Summary

Nanoshells, produced in the diffusive reactions of nanoparticles within the ambient phase with a simultaneous formation of Kirkendall voids inside, are unstable in principle, but the shrinkage time can be very large due to their cubic dependence on the radius of the nanoparticle. The mechanism of shrinkage is the out-diffusion of vacancies from the void due to the curvature effect.

Shrinkage of a nanoshell of an IMC with a large difference between diffusivities of the components is accompanied by the segregation of the fast-diffusing particles at the internal boundary due to the inverse Kirkendall effect. The rate of shrinkage is controlled by the slow species, contrary to usual interdiffusion with quasi-equilibrium vacancies.

The steady state approximation for both vacancies and atoms provides a reasonable and simple scheme for predicting the shrinkage kinetics. In order to properly describe the details of segregation, one must take into account the non–steady state redistribution of atoms; however, the steady state for vacancies can also be used. kMC simulations confirm the main predictions of the phenomenological models.

Near planar as well as curved interfaces, there always exist activation barriers for vacancies, connected with the increase in energy at embedding the vacancy into it. At transition to nanothicknesses, this barrier becomes narrower and smoother, which leads to an increase in vacancy flux. Vacancy segregation near the surface accelerates the nanoshell's shrinkage.

7.3
Formation of Compound Hollow Nanoshells

7.3.1
Introduction

As mentioned above, in a hollow nanoshell formation, we have an almost pure Frenkel effect with fully or partially suppressed Kirkendall shift. In [10], we analyzed the formation of a hollow compound nanoshell. We demonstrated that the Gibbs–Thomson effect, leading to the shrinkage of "ready" compound shells, should influence the formation stage as well. Sometimes it may even suppress the nanoshell formation.

7.3.2
Model of Nanoshell Formation

We consider the interdiffusion of A and B in a two-layer nanoshell structure, as shown in Figure 7.13 (actually, the enveloping layer A may just represent the medium in which the B particles are immersed). In this case, both the Kirkendall effect and the inverse Kirkendall effect coexist, and how they interact with each other is not completely clear. If the flux of B, j_B, is bigger than the flux of A, j_A, the balancing vacancy flux j_V diffuses inward, which counters the vacancy flux due to the Gibbs–Thomson effect. On the other hand, if $j_B < j_A$, the two vacancy fluxes move in the same direction, i.e. outwards.

We use the term "Kirkendall effect" to include both Kirkendall shift and Kirkendall voiding (or Frenkel voiding). We present a detailed analysis of the interaction of the Kirkendall effect and the inverse Kirkendall effect in nanoscale

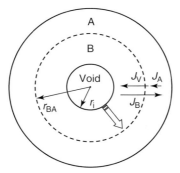

Figure 7.13 Hollow nanoshell consisting of two shells: outflux of B is larger than influx of A, leading to an influx of vacancies.

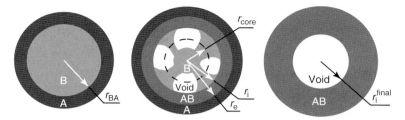

Figure 7.14 Simplified scheme of compound nanoshell formation by reaction between pure B particles with radius r_{BA} and surrounding material A. At the intermediate stage, the remaining core is connected with the growing compound shell by "bridges". This geometry appears due to the formation of voids at the interface at the initial stage of the reaction. The effective core radius is determined according to $(4/3)\pi r_{core}^3 = N_B \Omega_B$. At the end of the reaction, we have $r_{core} = 0$, $r_i = r_i^{final}$.

particles. The interaction is expressed in terms of the interaction of vacancy fluxes. Namely, the arising difference between vacancy concentration at internal and external boundaries (higher at internal boundary and lower at the external one) should oppose the vacancy flux generated by the difference of mobilities of the main components at interdiffusion. Therefore, we expect that under critical conditions void formation in the hollow nanoshell can be suppressed.

We shall consider, for example, the formation of cobalt sulfide (Co_9S_8 transforming to Co_3S_4) hollow nanospheres [4, 5]. Since Co is the dominant diffusing species in cobalt sulfide, the voids form in the internal part of the sphere, between the remaining Co core and growing sulfide shell. The following simplified geometry of the intermediate stages of reaction has been observed experimentally [5]: the remaining metallic core inside, surrounded by an "almost hollow" layer, contains "bridges" connecting the core B and the compound layer AB (Figure 7.14). The bridges provide paths for B atoms to diffuse and react with A atoms in the growing compound nanoshell with an internal radius r_i and an outer radius r_e. This picture is very close to the model of a so-called divided (separated) diffusion couple, which had been proposed by one of the authors (AG) [26, 27]. In this model, the diffusion contact between powders of different materials is kept by surface diffusion and sometimes by vapor phase transfer. In this section, we restrict ourselves to the formation of a single compound. Competition and sequential growth of the phases will be analyzed elsewhere.

In the following, we do not restrict ourselves to compound formation. We discuss the formation of an arbitrary intermediate binary phase with a narrow homogeneity range as the result of a reaction between the initial species B and the surrounding material A. On the other hand, we do not take into account the peculiarities of the diffusion mechanisms via the different sublattices. We treat only general vacancy concentration, assuming that the local redistribution of vacancies between the sublattices proceeds faster than the phase growth.

The following assumptions are made in our analysis of interdiffusion in forming the nanoshells:

1) There is no Kirkendall shift during phase formation, i.e. all vacancy fluxes go to the formation of Kirkendall (or Frenkel) voids instead of being annihilated by internal sinks; hence, they do not causing the lattice shift.
2) At the intermediate stage (after nucleation of several initial voids), the voids surround the remaining B-core, which is connected to the growing compound spherical layer by a few thin bridges, as shown in Figure 7.14. This means that the internal surface of the growing compound shell is mostly free but is provided with a very good diffusion contact with remaining core by surface diffusion along bridges.
3) The effective core radius is determined according to $(4/3)\pi r_{core}^3 = N_B \Omega_B$, where N_B is the number of remaining (nonreacted) atoms of pure B and Ω_B is the atomic volume of pure B; at the end of the reaction, we have $r_{core} = 0, r_i = r_i^{final}$;
4) no internal stress is involved;
5) spherically symmetric geometry;
6) surface tensions of the external and internal boundaries of the growing phase are assumed to be identical (for the sake of simplicity);
7) the peculiarities of chemical bounding and sublattices in the sulfides or oxides are not taken into account, and the compound is treated as a common metallic phase but with a narrow homogeneity range.

The Laplace tension and compression of the compound layer due to curvatures of internal and outer interfaces have been taken into account in the analysis. However, vacancy supersaturation is assumed to be outside the compound layer (in the B-core) at the initial stage of reaction, so voids can be nucleated. Curvature effects on vacancy concentration exist, so the difference of vacancy concentrations at the internal and external boundaries of the compound cannot be neglected. On the basis of the Gibbs–Thomson effect, the equilibrium vacancy concentration at the internal boundary of the compound layer is higher than that at the planar free surface

$$c_V(r_i) = c_{V_0} \exp\left\{\frac{2\gamma\Omega}{kTr_i}\right\}$$

and at the outer boundary it is lower than that at the planar surface

$$c_V(r_e) = c_{V_0} \exp\left\{-\frac{2\gamma\Omega}{kTr_e}\right\}$$

Thus, a vacancy gradient exists in the compound layer. Curvature effects on the driving force of compound formation (and, respectively, on the homogeneity range of the compound) is included as well.

7.3.3
Simplified Analysis of the Competition Between "Kirkendall-Driven" and "Curvature-Driven" Effects

We start with a simplified situation and consider the initial stage of compound nanoshell formation, when the shell thickness $r_e - r_i$ is much less than the initial radius, r_{BA}. In this case, we can use the flux equation for planar interfaces. Moreover, let the curvature effect be linear

$$\frac{2\gamma\Omega}{kTr_{BA}} \ll 1$$

so that

$$\begin{cases} c_V(r_e) = c_{V_0} \exp\left\{-\frac{2\gamma\Omega}{kTr_e}\right\} \simeq c_{V_0}\left(1 - \frac{2\gamma\Omega}{kTr_e}\right) \\ c_V(r_i) = c_{V_0} \exp\left\{\frac{2\gamma\Omega}{kTr_i}\right\} \simeq c_{V_0}\left(1 + \frac{2\gamma\Omega}{kTr_i}\right) \end{cases} \quad (7.85)$$

Then, the product of vacancy flux density and atomic volume is determined by a simple equation, taking into account the vacancy concentration gradient caused by the Gibbs–Thomson effect

$$\Omega j_V = (D_B - D_A)\frac{c_B(r_e) - c_B(r_i)}{\Delta r} - D_V\frac{c_V(r_e) - c_V(r_i)}{\Delta r}$$

$$\approx -(D_B - D_A)\frac{\Delta c_B}{\Delta r} + D_V c_{V_0}\frac{2\gamma\Omega}{kT}\left(\frac{1}{r_i} + \frac{1}{r_e}\right)\frac{1}{\Delta r} \quad (7.86)$$

Here $D_B = D_B^*\varphi$, $D_A = D_A^*\varphi$, where the thermodynamic factor is given by

$$\varphi = \frac{c_B c_A}{kT}\frac{\partial^2 g}{\partial c_B^2}$$

and D_B^* and D_A^* are the tracer diffusivities of both components in the compound. In Equation 7.86, we have taken $c_B D_B^* + c_A D_A^* = c_V D_V$ (thus, we neglect a correlation factor here). Furthermore, $\Delta c_B = c_B(r_i) - c_B(r_e)$ is the homogeneity range of the compound (typically very narrow). Note that the concentrations c_B and c_A are given in mole fractions, so $c_B + c_A = 1$ and $1/\Omega$ is the number of atoms per unit volume.

The first term on the right-hand side of Equation 7.86 is an inward vacancy flux caused by the difference in diffusivities and leading to void formation. The second term in the same equation is an outward vacancy flux caused by the different sign of curvatures at the internal and outer boundaries and counteracting the first term. After all the metal is consumed, this vacancy concentration difference leads to a collapse of the nanoshell. Evidently, if the curvature effect wins, implying

$$(D_B - D_A)\frac{\Delta c_B}{\Delta r} < D_V c_{V_0}\frac{2\gamma\Omega}{kT}\left(\frac{1}{r_i} + \frac{1}{r_e}\right)\frac{1}{\Delta r} \quad (7.87)$$

void formation becomes impossible because the vacancy flux should be directed outward. At the initial stage, when $r_e \approx r_i \approx r_{BA}$, this criterion can be rewritten as

$$r_{BA} < \frac{2\gamma\Omega}{kT} \cdot \frac{2(c_B D_B^* + c_A D_A^*)}{(D_B - D_A)\Delta c_B} \quad (7.88)$$

The criterion in Equation 7.88 is hard to use since the homogeneity range of the compound is very narrow and usually unknown. Yet, this difficulty can be easily circumvented. Indeed, we take

$$D_B \Delta c_B = \int_{c_B(r_e)}^{c_B(r_i)} D_B^* \varphi \, dc_B = \int_{c_B(r_e)}^{c_B(r_i)} D_B^* \frac{c_B c_A}{kT} \frac{\partial^2 g}{\partial c_B^2} dc_B \quad (7.89)$$

$$\simeq \overline{D}_B^* \frac{\overline{c}_B \overline{c}_A}{kT} \left(\frac{\partial g}{\partial c_B} \bigg|_{r_i} - \frac{\partial g}{\partial c_B} \bigg|_{r_e} \right) = \overline{D}_B^* \frac{\Delta g}{kT}$$

Here Δg is the compound formation Gibbs free energy per atom (it has been measured for many compounds). As a result, Equation 7.89 can be reduced to the following form:

$$r_{BA} < \frac{2\gamma \Omega}{\Delta g} 2 c_B \frac{(D_B^*/D_A^* + c_A/c_B)}{(D_B^*/D_A^* - 1)} \quad (7.90)$$

We use the dimensionless parameter

$$G = \frac{2\gamma \Omega}{\Delta g r_{BA}}$$

as a measure of the ratio of the curvature effect and the chemical driving force of reaction. Thus, in our simplified approach, the condition of impossibility for hollow shell to form by a reaction can be represented as

$$G > \frac{1}{2c_B} \frac{D_B^*/D_A^* - 1}{D_B^*/D_A^* + c_A/c_B} \quad (7.91)$$

In this simplified analysis, we neglected the influence of curvature on the chemical driving force (in other words, on the homogeneity range for the main components), nonlinearity of the Gibbs–Thomson effect for vacancy concentration at nanoscale, and spherical geometry. Moreover, the knowledge of the vacancy flux is not sufficient for predicting the reaction rate and the resulting void size. The above has been modified below. Nevertheless, the simplified Equation 7.91 is rather close to the numerical criterion obtained by a more rigorous analytic and numerical analysis (to be shown in Figure 7.17).

7.3.4
Rigorous Kinetic Analysis

We start from the flux balance for the B component (core material) at the outer boundary and the flux balance for the A-component (enveloping material) at the internal boundary. Flux balance means that the product of concentration step across the boundary and the boundary velocity is equal to the difference of fluxes on both sides of this moving boundary. For the external boundary, the flux balance for B is more convenient since we know for sure that both the concentration of B and the outward flux of B are equal to zero. On the other hand, for the internal boundary, the flux balance of A is more convenient since we consider the case when the solubility of A in B can be neglected and no A goes into the void so that the corresponding concentration and flux of A at $r < r_i$ are equal to zero. Thus, in each flux balance equation, we use only one flux – B for the external radius r_e (Equation 7.92) and A for the internal radius r_i (Equation 7.93)

$$(0 - c_B)\frac{dr_e}{dt} = 0 - \left(-D_B \left.\frac{\partial c_B}{\partial r}\right|_{r_e} + \frac{c_B D_B^*}{c_V} \left.\frac{\partial c_V}{\partial r}\right|_{r_e}\right) \quad (7.92)$$

$$(c_A - 0)\frac{dr_i}{dt} = \left(-D_A \left.\frac{\partial c_A}{\partial r}\right|_{r_i} + \frac{c_A D_A^*}{c_V} \left.\frac{\partial c_V}{\partial r}\right|_{r_i}\right) - 0 \quad (7.93)$$

In the expression for both fluxes (the last term), we take into account the input of vacancy gradients due to the Gibbs–Thomson effect [8]. We note that the diffusion of the B-flux is directed out of the shell and the diffusion of the A-flux is directed into it. Concentrations of B and A inside the compound can slightly change in the narrow homogeneity ranges for atomic fractions of B and A inside the compound, respectively ($\Delta c_B = -\Delta c_A$). We assume that

$$c_B + c_A \cong 1, \quad \frac{\partial c_A}{\partial r} \cong -\frac{\partial c_B}{\partial r}$$

Note that we use the fluxes in the lattice reference frame and intrinsic diffusivities (instead of interdiffusivity) according to our basic approximation of no Kirkendall shift. We neglect the correlation factors of Manning's vacancy wind terms [28] since they can change results only quantitatively. Both tracer diffusivities are proportional to vacancy concentration. Here we use one more approximation that we treat only one effective vacancy concentration, without distinguishing between sublattices in the compound. We take

$$D_B^* = c_V K_B, \quad D_A^* = c_V K_A \quad (7.94)$$

where K_B and K_A can be treated as constants. Thus, we rearrange Equations 7.92–7.93 as

$$\frac{dr_e}{dt} = K_B \left(-\frac{c_V \varphi}{c_B} \left.\frac{\partial c_B}{\partial r}\right|_{r_e} + \left.\frac{\partial c_V}{\partial r}\right|_{r_e}\right) \quad (7.95)$$

$$\frac{dr_i}{dt} = K_A \left(\frac{c_V \varphi}{c_A} \left.\frac{\partial c_B}{\partial r}\right|_{r_e} + \left.\frac{\partial c_V}{\partial r}\right|_{r_e}\right) \quad (7.96)$$

Note that in the above equations the values of vacancy concentration at the external and internal boundaries can be very different, due to the Gibbs–Thomson effect, and the vacancy gradient (and the cross terms in Equations 7.92–7.93 and Equations 7.95–7.96) do not tend to zero. At the same time, the concentrations of the main components at the external and internal boundaries remain close to constant because of the narrow homogeneity range of the compound.

Now we use the steady state approximation both for components and for vacancies inside the compound, which has proven to be valid inside the intermediate phases with a narrow homogeneity range [29]. From the continuity equation, we have

$$\frac{\partial c_B}{\partial t} \approx 0 \quad \Rightarrow \quad \text{div}\vec{J}_B \approx 0$$

$$\frac{\partial c_V}{\partial t} = \text{div}\vec{J}_V + \sigma_V \approx \text{div}\vec{J}_V \approx 0 \quad \Rightarrow \quad \text{div}\vec{J}_V \approx 0 \quad (7.97)$$

7 Nanovoid Evolution

In the latter equation, we used our assumption about the absence of effective vacancy sinks/sources inside the growing layer of the compound.

Writing down the fluxes and divergences in spherical coordinates, we obtain

$$\mathrm{div}\vec{j} \approx 0 \quad \Rightarrow \quad \frac{1}{r^2}\frac{\partial}{\partial r}(r^2 j_r) \approx 0 \quad \Rightarrow \quad r^2 j_r \approx \text{constant}$$

$$r^2 \Omega j_B(r) = r^2 \left(-c_V(r)\varphi \frac{\partial c_B}{\partial r} + c_B(r) \frac{\partial c_V}{\partial r}\right) K_B \quad (7.98)$$

$$= r_e^2 \Omega j_B(r_e) = r_e^2 \frac{dr_e}{dt} c_B$$

$$r^2 \Omega j_A(r) = r^2 \left(-c_V(r)\varphi \frac{\partial c_A}{\partial r} + c_A(r) \frac{\partial c_V}{\partial r}\right) K_A \quad (7.99)$$

$$= r_i^2 \cdot \Omega j_A(r_i) = r_i^2 \frac{dr_i}{dt} c_A$$

where

$$\Omega j_B = c_B \frac{dr_e}{dt}, \quad \Omega j_A = c_A \frac{dr_i}{dt}$$

are the fluxes of the metal and surrounding element, respectively, and Ω is the volume per atom of the compound.

Equations 7.98 and 7.99 can be treated as the set of two **differential** equations for two unknown functions $c_B(r)$ and $c_V(r)$ (treating r_e and r_i and their time derivatives as known parameters). These equations are nonlinear because the vacancy concentration $c_V(r)$ and the thermodynamic factor $\varphi(r)$ in the coefficients cannot be treated as constants. On the other hand, Equations 7.98 and 7.99 can be treated as a linear set of **algebraic** equations for determining the quantities $\frac{dc_V}{dr}$ and $c_V(r)\varphi(r)\frac{dc_B}{dr}$. By rearrangement, we change Equation 7.98 and 7.99 to

$$\begin{cases} \frac{dc_V}{dr} = \frac{B_2}{r^2} \\ c_V(r)\varphi(r)\frac{dc_B}{dr} = -\frac{B_1}{r^2} \end{cases} \quad (7.100)$$

with B_1 and B_2 being expressed in terms of the boundary velocities

$$\begin{cases} B_2 = -\frac{c_A}{K_A} r_i^2 \frac{dr_i}{dt} - \frac{c_B}{K_B} r_e^2 \frac{dr_e}{dt} \\ B_1 = c_B c_A \left(\frac{r_e^2}{K_B}\frac{dr_e}{dt} - \frac{r_i^2}{K_A}\frac{dr_i}{dt}\right) \end{cases} \quad (7.101)$$

The solutions of Equation 7.100 give

$$\begin{cases} c_V(r) = -\frac{B_2}{r} + F \\ \varphi \frac{dc_B}{dr} = \frac{B_1}{r^2\left(-\frac{B_2}{r}+F\right)} \end{cases} \quad (7.102)$$

where F is the integration constant. In the first equation in Equation 7.102, the boundary conditions at r_e and r_i, taking into account the Gibbs–Thomson effect on

7.3 Formation of Compound Hollow Nanoshells

vacancy concentration, yield

$$\begin{cases} c_V(r_e) = -\dfrac{B_2}{r_e} + F = c_{V_0} \exp\left(-\dfrac{2\gamma\Omega}{kTr_e}\right) \\ c_V(r_i) = -\dfrac{B_2}{r_i} + F = c_{V_0} \exp\left(\dfrac{2\gamma\Omega}{kTr_i}\right) \end{cases} \quad (7.103)$$

Then, we integrate the second equation in Equation 7.102 over the radius from the internal boundary to the outer one, and obtain

$$\int_{r_i}^{r_e} \varphi \frac{dc_B}{dr} dr = \int_{r_i}^{r_e} \frac{B_1}{r^2 \left(-\dfrac{B_2}{r} + F\right)} dr \quad (7.104)$$

The integration of the right-hand side of Equation 7.104 by substituting Equation 7.103 gives

$$\frac{B_1}{B_2} \int_{r_i}^{r_e} \frac{d\left(-\dfrac{B_2}{r} + F\right)}{-\dfrac{B_2}{r} + F} = \frac{B_1}{B_2} \int_{c_V(r_i)}^{c_V(r_e)} \frac{dc_V}{c_V} = \frac{B_1}{B_2} \left(\ln \frac{c_V(r_e)}{c_V(r_i)}\right) \quad (7.105)$$

$$= -\frac{B_1}{B_2} \frac{2\gamma\Omega}{kT} \left(\frac{1}{r_e} + \frac{1}{r_i}\right)$$

For the integration of the left-hand side of Equation 7.104, we take into account that the thermodynamic factor is given by

$$\varphi = \frac{c_B c_A}{kT} \frac{\partial^2 g}{\partial c_B^2}$$

and that the homogeneity interval is very narrow (concentrations are almost constant within the compound layer but the derivatives of g are very far from being constant), we obtain,

$$\int_{r_i}^{r_e} \varphi \frac{dc_B}{dr} dr = \int_{r_i}^{r_e} \frac{c_B c_A}{kT} \frac{\partial^2 g}{\partial c_B^2} \frac{dc_B}{dr} dr = \frac{\bar{c}_B \bar{c}_A}{kT} \int_{c_B(r_i)}^{c_B(r_e)} \frac{\partial^2 g}{\partial c_B^2} dc_B$$

$$= \frac{\bar{c}_B \bar{c}_A}{kT} \left(\frac{\partial g}{\partial c_B}\bigg|_{r_e} - \frac{\partial g}{\partial c_B}\bigg|_{r_i}\right) \quad (7.106)$$

It is well known for the case of planar layer growth that, for the growth of a single intermediate phase, the difference in the right and the left derivatives would be proportional to the driving force of the reaction,

$$\frac{\partial g}{\partial c_B}\bigg|_{r_e} - \frac{\partial g}{\partial c_B}\bigg|_{r_i} = \frac{\partial g}{\partial c_B}\bigg|_{\infty}^{\text{compound/nonmetal}} - \frac{\partial g}{\partial c_B}\bigg|_{\infty}^{\text{compound/metal}} = -\frac{\Delta g}{\bar{c}_B \bar{c}_A}$$

(7.107)

Yet, in our case, we should take into account the influence of curvature not only on the vacancy concentration but also on the phase equilibrium at the curved boundaries. At the external boundary, the compound is under additional Laplace compression with additional energy per atom $\frac{2\gamma\Omega}{r_e}$, so that, taking into account the common tangent rule, we have

$$\left.\frac{\partial g}{\partial c_B}\right|_{r_e} = \left.\frac{\partial g}{\partial c_B}\right|_\infty^{\text{compound/nonmetal}} + \frac{\frac{2\gamma\Omega}{r_e}}{c_B(r_e) - 0} \quad (7.108)$$

$$\cong \left.\frac{\partial g}{\partial c_B}\right|_\infty^{\text{compound/nonmetal}} + \frac{2\gamma\Omega}{\bar{c}_B r_e}$$

At the internal boundary, the compound shell is under Laplace tension with negative additional energy per atom $\left(-\frac{2\gamma\Omega}{r_i}\right)$, so that

$$\left.\frac{\partial g}{\partial c_B}\right|_{r_i} = \left.\frac{\partial g}{\partial c_B}\right|_\infty^{\text{compound/metal}} + \frac{-\frac{2\gamma\Omega}{r_i}}{c_B(r_i) - 1} \cong \left.\frac{\partial g}{\partial c_B}\right|_\infty^{\text{compound/metal}} + \frac{2\gamma\Omega}{\bar{c}_A r_e} \quad (7.109)$$

Then, instead of Equation 7.107, we obtain

$$\left.\frac{\partial g}{\partial c_B}\right|_{r_e} - \left.\frac{\partial g}{\partial c_B}\right|_{r_i} = -\frac{\Delta g}{\bar{c}_B \bar{c}_A} + \frac{2\gamma\Omega}{\bar{c}_B r_e} - \frac{2\gamma\Omega}{\bar{c}_A r_i} \quad (7.110)$$

Substituting Equation 7.110 in Equation 7.106, we arrive at

$$\int_{r_i}^{r_e} \varphi \frac{dc_B}{dr} dr = \frac{\bar{c}_B \bar{c}_A}{kT}\left(\left.\frac{\partial g}{\partial c_B}\right|_{r_e} - \left.\frac{\partial g}{\partial c_B}\right|_{r_i}\right) = -\frac{\Delta g}{kT} + \frac{2\gamma\Omega}{kT}\left(\frac{\bar{c}_A}{r_e} - \frac{\bar{c}_B}{r_i}\right) \quad (7.111)$$

It is interesting to note that the change in the driving force in Equation 7.111 due to the curvature effect can be negative as well as positive, depending on the stoichiometry and on the ratio of external and internal radii.

Substituting Equations 7.111 and 7.105 in Equation 7.104, we obtain

$$\frac{\Delta g}{2\gamma\Omega} - \frac{\bar{c}_A}{r_e} + \frac{\bar{c}_B}{r_i} = \frac{B_1}{B_2}\left(\frac{1}{r_e} + \frac{1}{r_i}\right) \quad (7.112)$$

After further substituting B_1 and B_2 from Equation 7.101 in Equation 7.112 and after excluding time, we obtain (after simple but long algebra) the main equation of this section

$$\frac{dr_e}{dr_i} = -\frac{r_i^2}{r_e^2}\frac{K_B c_A}{K_A c_B}\left(\frac{\frac{\Delta g}{2\gamma\Omega} - \frac{1}{r_e}}{\frac{\Delta g}{2\gamma\Omega} + \frac{1}{r_i}}\right) \quad (7.113)$$

We now go over to dimensionless parameters

$$x = \frac{r_i}{r_{BA}}, \quad y = \frac{r_e}{r_{BA}}, \quad G = \frac{2\gamma\Omega}{\Delta g r_{BA}}$$

where r_{BA} is the initial radius of the nanoparticle. This "G-parameter", together with the ratio of mobilities

$$\frac{K_B}{K_A} = \frac{D_B^*}{D_A^*} = \frac{D_B}{D_A}(D_{B(A)} = D_{B(A)}^*\varphi)$$

are the two major parameters of our model. Then, Equation 7.113 can be represented in the following form:

$$\frac{dy}{dx} = -\frac{x^3}{y^3} \frac{D_B^*}{D_A^*} \frac{1 - c_B}{c_B} \frac{y - G}{x + G} \quad (7.114)$$

This equation differs from a similar equation derived by Alivisatos et al. [5] by the last term containing $G \neq 0$ (responsible for Laplace pressure and Gibbs–Thomson effect both for vacancies and for main components). When both radii r_e and r_i are large, i.e.

$$r_e \gg \frac{2\gamma\Omega}{\Delta g}, \quad r_i \gg \frac{2\gamma\Omega}{\Delta g}, \quad G \ll 1$$

the equations become similar.

Equation 7.114 can be easily integrated but gives, as a result, a transcendent relation between y and x. It can be solved numerically by the finite-difference method. The governing parameters are $D \equiv D_B^*/D_A^*$, c_B, and G. The equation has been solved with the "initial" condition $x = 1$, $y = 1 + \varepsilon$.

We made calculations in two cases: (1) when all pure metal (core) was consumed by the reaction $c_B (y^3 - x^3) \geq 1$ and (2) when the internal radius of the compound shell became equal to the radius of the remaining metallic core (void in the hollow shell between the compound shell and the remaining metallic core with connecting bridges shrinks prior to the end of the reaction)

$$x^3 \leq r_{core}^3/r_{BA}^3 = 1 - c_B (y^3 - x^3)$$

It is evident from Equation 7.114 that in the case when $G > 1$, nanoshell formation is impossible (taking into account that x and y start from almost 1). This means that, in very small particles, $r_{BA} < 2\gamma\Omega/\Delta g$, the reaction with void formation is impossible. It is interesting that this critical condition coincides with the critical radius for nucleation of a new phase.

7.3.5
Results and Discussion

We present some results for various parameters in Equation 7.114 below. A characteristic solution of Equation 7.114 is represented in Figure 7.15. Here the reduced internal radius x was decremented uniformly and arbitrarily, the reduced external radius y was calculated by a numerical integration of Equation 7.114, and the reduced radius $z = r_{core}/r_{BA}$ of the remaining core was calculated according to conservation of matter (neglecting differences of atomic volumes in compound and metal)

$$z = (1 - c_B (y^3 - x^3))^{1/3}$$

Calculations were stopped when the core or void disappeared ($z = 0$ or $x = z$).

According to the analysis of Alivisatos et al. [5], the resulting radius of the final void is determined only by the ratio of diffusivities and by the stoichiometry of

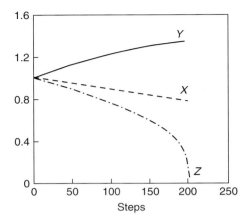

Figure 7.15 Evolution of reduced external radius $y = r_e/r_{BA}$ of the compound nanoshell (solid line) and reduced effective radius $z = r_{core}/r_{BA}$ of the remaining B-core (chain line) with decreasing reduced internal radius of nanoshell $x = r_i/r_{BA}$ (dashed line). The moment of disappearance of the core corresponds to the final radius of the void (end of chain line).

the compound. Ideally, if only B atoms migrate, the final void radius should be just equal to the initial radius of the B particle. But, according to our model, the resulting ratio of the void radius to the initial particle size depends on the initial particle size via the G-factor. Indeed, according to our results, the less the initial radius and the larger the factor G, the less is the ratio of the final void size to the initial particle size, as shown in Figure 7.16. Thus, if the initial radius is less than some critical one, a void cannot be formed at all.

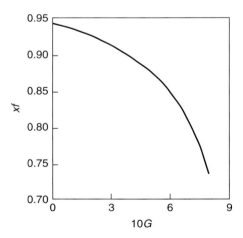

Figure 7.16 Dependence of resulting reduced void radius $xf = r_i^{final}/r_{BA}$ on the G-parameter $\left(G = \frac{2\gamma\Omega}{\Delta g r_{BA}}\right)$ at $D \equiv D_B^*/D_A^* = 10$, $c_B = c_A = 0.5$.

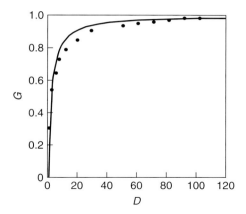

Figure 7.17 Dependence $G(D)$ dividing two regimes: left and up – hollow nanoshell formation impossible; solid line – approximate analytical criterion $G > \frac{D_B^*/D_A^*-1}{D_B^*/D_A^*+1}$ (Equation 7.91 at $c_A = c_B = 1/2$); points – numerical criterion obtained by the solution of Equation 7.114 with the condition $x^3 \leq r_{core}^3/r_{BA}^3 = 1 - c_B(y^3 - x^3)$.

Furthermore, in this case, due to large vacancy gradient, the internal radius converges inside faster than the radius of the remaining B-core, and abolishes the void. In Figure 7.17, we divide the region of parameters G and D

$$G = \frac{2\gamma\Omega}{\Delta g r_{BA}}, \quad D \equiv \frac{D_B^*}{D_A^*}$$

into two subregions (results of our numerical calculations of $y(x)$ with various parameters of G and D). For the left and upper subregion, the formation of a hollow nanoshell is forbidden. This criterion appears to be rather close to the approximate analytic criterion in Equation 7.91, as demonstrated in Figure 7.17.

If necessary, we can also calculate the absolute rate of void formation in real time. Indeed, we can deduce two expressions for the parameter B_2. First, from the set of Equations 7.103, we obtain

$$B_2 = -c_{V_0} \frac{\exp\left(\frac{2\gamma\Omega}{kTr_i}\right) - \exp\left(-\frac{2\gamma\Omega}{kTr_e}\right)}{\frac{1}{r_i} - \frac{1}{r_e}} \qquad (7.115)$$

Second, from Equations 7.101 and 7.113, one gets

$$B_2 = -\frac{c_A}{K_A} r_i^2 \frac{dr_i}{dt} - \frac{c_B}{K_B} r_e^2 \frac{dr_e}{dt} = -\frac{c_A}{K_A} r_i^2 \frac{dr_i}{dt} \cdot \left(1 - \frac{r_i}{r_e} \frac{r_e/r_{BA} - G}{r_i/r_{BA} + G}\right) \qquad (7.116)$$

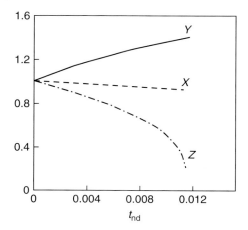

Figure 7.18 Dependence of reduced external ($y = r_e/r_{BA}$) and internal ($x = r_i/r_{BA}$) radii of the nanoshell, and the remaining core radius $\left(z = r_{core}/r_{BA} = (1 - c_B(y^3 - x^3))^{1/3}\right)$ on the dimensionless time $t_{nd} = (D_A^*/r_{BA}^2)t$, calculated according to Equation 7.117 for the case $D \equiv D_B^*/D_A^* = 10$, $c_B = c_A = 1/2$, $\frac{2\gamma\Omega}{kTr_{BA}} = 1$.

Equalizing Equations 7.115 and 7.116 gives

$$\frac{dr_i}{dt} = -\frac{K_A c_{V0}}{c_A} \frac{\exp\left(\frac{2\gamma\Omega}{kTr_i}\right) - \exp\left(-\frac{2\gamma\Omega}{kTr_e}\right)}{r_i^2\left(\frac{1}{r_i} - \frac{1}{r_e}\right)\left(1 - \frac{r_i}{r_e}\frac{r_e/r_{BA} - G}{r_i/r_{BA} + G}\right)} \quad (7.117)$$

Combining Equations 7.113 or 7.114 with Equation 7.117 leads to the full description of the kinetics in real time. A typical result of solving Equation 7.117 together with Equation 7.114 is shown in Figure 7.18.

7.3.6
Summary

The formation of a hollow nanoshell is analyzed by taking into account the competition of three factors:

1) growth of a compound phase due to reaction and tendency for vacancies to go inside (due to difference of diffusivities);
2) shrinkage of the shell due to vacancy gradient between internal and outer boundaries (Gibbs–Thomson effect);
3) change in the driving force by Laplace tension at the internal boundary and Laplace compression at the external boundary. The second factor should win for very small particles.

The critical value for the initial radius is very similar to the result of Gibbs nucleation theory, that is, $r_{BA}^{cr} = 2\gamma\Omega/\Delta g$. Nucleation and phase competition (in the case of a two-phase sequential formation) of the spherical particles have been considered elsewhere along the lines as outlined above.

7.4
Hollow Nanoshell Formation and Collapse in One Run: Model for a Solid Solution

7.4.1
Introduction

In this section, we suggest phenomenological models for both steps of the process – formation of a hollow nanoshell from a core–shell structure with full solubility during interdiffusion and shrinkage of this just-formed nanoshell with a transformation into a compact particle. The description of the shrinkage looks simpler. Therefore, we start with the model of collapse, and then modify this model to describe the formation stage.

7.4.2
Shrinkage

We consider interdiffusion and vacancy fluxes in a spherical hollow nanoparticle, consisting of a binary solid solution with a broad homogeneity range. As explored in Ref. [8], such a shell should shrink due to outward vacancy fluxes generated by the difference in vacancy concentrations at internal and external boundaries. The vacancy flux should lead to a redistribution of the main components with a corresponding feedback of the created concentration gradient on the vacancy flux and on the corresponding shrinkage rate. Contrary to the case of an almost stoichiometric compound [8], in our case of solid solution, we cannot take diffusivities and concentrations as almost constant. Thus, one has to solve the coupled nonlinear set of equations describing vacancy flux, interdiffusion, and boundary motion. The following **approximations** are employed:

A1. Tracer diffusivities of both components are proportional to the local vacancy concentration (fraction of empty sites)

$$D_A^*(c) = c_V K_A(c), \quad D_B^*(c) = c_V K_B(c) \tag{7.118}$$

Here the concentration c is the molar fraction of component B.

A2. Composition dependence of both diffusivities is taken as an exponential one, which is typical. For simplicity (just to limit the number of model parameters), the logarithms of both diffusivities are treated as linear functions of concentration.

$$K_A(c) = K_{A0} \exp(\alpha_A c_B), \quad K_B(c) = K_{B0} \exp(\alpha_B c_B) \tag{7.119}$$

A3. The diffusion fluxes of the main components and of the vacancies take into account the cross terms and are written in the lattice reference frame [2, 31]

$$\Omega j_A(r) = -D_A^* \varphi \frac{\partial c_A}{\partial r} + \frac{c_A D_A^*}{c_V} \frac{\partial c_V}{\partial r} = +K_A \varphi c_V \frac{\partial c_B}{\partial r} + c_A K_A \frac{\partial c_V}{\partial r} \qquad (7.120)$$

$$\Omega j_B(r) = -D_B^* \varphi \frac{\partial c_B}{\partial r} + \frac{c_B D_B^*}{c_V} \frac{\partial c_V}{\partial r} = -K_B \varphi c_V \frac{\partial c_B}{\partial r} + c_B K_B \frac{\partial c_V}{\partial r} \qquad (7.121)$$

$$\Omega j_V(r) = (K_B - K_A) \varphi c_V \frac{\partial c_B}{\partial r} - (c_A K_A + c_B K_B) \frac{\partial c_V}{\partial r} \qquad (7.122)$$

Here φ is a thermodynamic factor,

$$\varphi = \frac{c_A c_B}{kT} \frac{\partial^2 g}{\partial c_B^2}$$

g is the Gibbs free energy per atom of the solution. In the model of regular Solutions, (Z is the coordination number, Φ_{ij} are the pair interaction energies)

$$\varphi = 1 + c_A c_B \underbrace{\frac{2Z}{kT} \left(\frac{\Phi_{AA} + \Phi_{BB}}{2} - \Phi_{AB} \right)}_{-\lambda} \equiv 1 + c_A c_B (-\lambda) \qquad (7.123)$$

Here Manning's corrections due to the effect of vacancy wind are neglected. Our preliminary estimates have shown that these corrections, at least for a disordered solution, do not change the general picture. In this approximation

$$c_A K_A + c_B K_B = D_V$$

(neglecting the correlation factor). The factor λ is proportional to the mixing enthalpy and is negative if the formation of the solution is energetically favorable.

A4. There is no Kirkendall shift inside the nanoshell.

First, as mentioned in the introduction, there is little place for vacancy sinks/sources in a nanovolume. Second, if all atoms of the lattice try to move along the radial direction, this shift would immediately generate a tangential deformation and corresponding stresses. Boundaries of the shell do move but not due to lattice shift – just some atoms leave one boundary and attach to another boundary. Therefore, we can write down the continuity equations (analog of second Fick's law) in the lattice reference frame as

$$\frac{\partial c_V}{\partial t} = -\frac{1}{r^2} \frac{\partial}{\partial r} \left(r^2 \Omega j_V \right) + 0 \qquad (7.124)$$

$$\frac{\partial c_B}{\partial t} = -\frac{1}{r^2} \frac{\partial}{\partial r} \left(r^2 \Omega j_B \right) \qquad (7.125)$$

Zero in Equation 7.124 was put just to remind about the absence or ineffectiveness of vacancy sources and sinks inside the shell.

The boundary conditions can be formulated in the following way. Although $c_V(r, t)$ is a nonequilibrium vacancy distribution, its values at the boundaries are

7.4 Hollow Nanoshell Formation and Collapse in One Run

in equilibrium and fixed by the Gibbs–Thomson relation

$$c_V(r_i) = c_V^{eq} \exp\left(+\frac{2\gamma\Omega}{kT}\frac{1}{r_i}\right) \quad c_V(r_e) = c_V^{eq} \exp\left(-\frac{2\gamma\Omega}{kT}\frac{1}{r_e}\right) \quad (7.126)$$

Flux balance equations at the moving internal and external boundaries depend on the conditions of the experiment. Here we treat the case when evaporation of atoms from the nanoshell is impossible. So, all fluxes outside the shell and inside the central void are absent. Vacancy flux from the internal boundary to the external boundary generates the movement of both these boundaries according to the trivial relations

$$\frac{dr_i}{dt} = -\Omega j_V(r_i), \quad \frac{dr_e}{dt} = -\Omega j_V(r_e) \quad (7.127)$$

The boundary conditions in Equations 7.126 and 7.127 would be sufficient for shrinkage of a single-component shell. In our case of a binary solution, we should have additional conditions. Boundary concentrations of the main components are not fixed (we do not have an analog of Equation 7.126 for A or B), but the conservation laws are valid, of course, implying the conditions on fluxes. The sum of the three fluxes is zero in the lattice reference frame ($j_V + j_A + j_B = 0$). Thus, two fluxes are independent. This means that one should write down the flux balance equations at both moving boundaries for one of the main components, taking into account that fluxes outside the shell are equal to zero

$$(c_B(r_i) - 0)\frac{dr_i}{dt} = \Omega j_B(r_i) - 0, \quad (c_B(r_e) - 0)\frac{dr_e}{dt} = \Omega j_B(r_e) - 0 \quad (7.128)$$

Combining Eqs. (6.3.7, 6.3.8), one gets

$$-c_B(r_i)\Omega j_V = \Omega j_B(r_i), \quad -c_B(r_e)\Omega j_V = \Omega j_B(r_e) \quad (7.129)$$

Below we use the common steady state approximation for fast-diffusing vacancies, which reflects the hierarchy of characteristic times – the vacancy subsystem is fast enough to adapt to the slow redistribution of the main components and to the movement of the boundaries (except the very last stage of collapse, which is very fast)

$$\frac{\partial c_V}{\partial t} \approx 0 \quad \Rightarrow \quad r^2 \Omega j_V(r) = \frac{\Omega j_V}{4\pi} = \text{constant at } r \quad (7.130)$$

Here j_V is the total flux of vacancies, which has to be determined from the boundary conditions.

It is convenient now to make the following transformation of variables

$$t' = t, \quad \xi = \frac{1/r - 1/r_e}{1/r_i - 1/r_e}, \quad 0 < \xi < 1 \quad (7.131)$$

$$\frac{\partial}{\partial t} = \frac{\partial}{\partial t'} + \left(\frac{\xi}{r_i^2}\frac{dr_i}{dt} + \frac{(1-\xi)}{r_e^2}\frac{dr_e}{dt}\right)\frac{r_e r_i}{r_e - r_i}\frac{\partial}{\partial \xi}$$

$$\frac{\partial}{\partial r} = -\frac{[r_i + (r_e - r_i)\xi]^2}{(r_e - r_i) r_e r_i}\frac{\partial}{\partial \xi} \quad (7.132)$$

Using Equations 7.122, 7.131, and 7.132, one obtains

$$\frac{1}{4\pi}\Omega j_V = \frac{r_e r_i}{r_e - r_i}\left(-(K_B - K_A)\varphi c_V \frac{\partial c_B}{\partial \xi} + (c_A K_A + c_B K_B)\frac{\partial c_V}{\partial \xi}\right)$$

or

$$\frac{\partial c_V}{\partial \xi} = \frac{(K_B - K_A)\varphi}{c_A K_A + c_B K_B} c_V \frac{\partial c_B}{\partial \xi} + \frac{(\Omega j_V/4\pi)}{c_A K_A + c_B K_B}\frac{(r_e - r_i)}{r_e r_i} \tag{7.133}$$

If the $c_B(\xi)$ profile is known, then Equation 7.132 has a standard form

$$\frac{dc_V(\xi)}{d\xi} = f(\xi) c_V + \frac{a}{\psi(\xi)} \tag{7.134}$$

with

$$\psi(\xi) = c_A(\xi) K_A(\xi) + c_B(\xi) K_B(\xi), \quad f(\xi) = \frac{(K_B - K_A)\varphi}{c_A K_A + c_B K_B}\frac{\partial c_B}{\partial \xi} \tag{7.135}$$

and can be solved with respect to the unknown function $c_V(\xi)$

$$c_V(\xi) = \exp\left(\int_0^\xi f(\xi') d\xi'\right)\left(c_V^{eq} \exp\left(-\frac{2\gamma\Omega}{kTr_e}\right) + \frac{r_e - r_i}{r_e r_i}\frac{\Omega j_V}{4\pi}\right.$$
$$\left. \times \int_0^\xi \frac{d\xi'}{\psi(\xi')} \exp\left(-\int_0^{\xi'} f(\xi'') d\xi''\right)\right) \tag{7.136}$$

Combining Equation 7.136 with the Gibbs–Thomson relation Equation 7.126 for the boundary vacancy concentrations, we obtain an expression for the total vacancy flux in terms of the unknown B-concentration profile

$$\frac{\Omega j_V}{4\pi} = \frac{r_e r_i}{r_e - r_i} c_V^{eq} \frac{\exp\left(\frac{2\gamma\Omega}{kTr_i}\right)\exp\left(-\int_0^1 f(\xi) d\xi\right) - \exp\left(-\frac{2\gamma\Omega}{kTr_e}\right)}{\int_0^1 \exp\left(-\int_0^\xi f(\xi') d\xi'\right)\frac{d\xi}{\psi(\xi)}} \tag{7.137}$$

Substituting Equation 7.132 in Equation 7.121, one can express the flux of one of the main components in terms of its gradient and of total vacancy flux (which is constant within the layer)

$$r^2 \Omega j_B(r) = -K_B \varphi c_V r^2 \frac{\partial c_B}{\partial r} + c_B K_B r^2 \frac{\partial c_V}{\partial r}$$
$$= \frac{r_e r_i}{(r_e - r_i)}\left(K_B \varphi c_V \frac{\partial c_B}{\partial \xi} - c_B K_B \frac{\partial c_V}{\partial \xi}\right)$$
$$= \frac{r_e r_i}{(r_e - r_i)} \frac{K_A K_B \varphi}{c_A K_A + c_B K_B} c_V \frac{\partial c_B}{\partial \xi} - \frac{c_B K_B}{c_A K_A + c_B K_B}\frac{\Omega j_V}{4\pi} \tag{7.138}$$

Thus, the steady state approximation for vacancies, together with the Gibbs–Thomson boundary conditions, reduces the system of two differential equations, Equations 7.124 and 7.125, to only one, which is an integrodifferential equation

$$\frac{\partial c_B}{\partial t'} = \frac{\xi \frac{d\eta_i}{dt} + (1-\xi) \frac{d\eta_e}{dt}}{\eta_i - \eta_e} \frac{\partial c_B}{\partial \xi} + \frac{(\eta_e + (\eta_i - \eta_e)\xi)^4}{(\eta_i - \eta_e)^2}$$
$$\times \frac{\partial}{\partial \xi} \left(\frac{K_A K_B \varphi(\xi)}{c_A K_A + c_B K_B} c_V(\xi) \frac{\partial c_B}{\partial \xi} - \frac{c_B K_B}{c_A K_A + c_B K_B} (\eta_i - \eta_e) \frac{\Omega j_V}{4\pi} \right)$$
(7.139)

with $\eta_i = 1/r_i$, $\eta_e = 1/r_e$. The total vacancy flux in Equation 7.139 is determined by Equation 7.137. The conditions for B-profile at internal ($\xi = 1$) and external ($\xi = 0$) boundaries can be found from Equation 7.129 with account of Equations 7.137 and 7.138

$$\left. \frac{\partial c_B}{\partial \xi} \right|_{\xi=1} = -\frac{r_e - r_i}{r_e r_i} \frac{\Omega j_V}{4\pi} \left[\frac{c_A c_B (K_B - K_A)}{K_A K_B \varphi} \right]_{\xi=1} \exp\left(-\frac{2\gamma\Omega}{kTr_i}\right) \Big/ c_V^{eq} \quad (7.140)$$

$$\left. \frac{\partial c_B}{\partial \xi} \right|_{\xi=0} = -\frac{r_e - r_i}{r_e r_i} \frac{\Omega j_V}{4\pi} \left[\frac{c_A c_B (K_B - K_A)}{K_A K_B \varphi} \right]_{\xi=0} \exp\left(+\frac{2\gamma\Omega}{kTr_e}\right) \Big/ c_V^{eq} \quad (7.141)$$

The boundary value problem in Equations 7.139–7.141 is solved by the explicit finite difference scheme, with a B-profile at the previous time step used for calculation of the total vacancy flux according to Equation 7.137 and substituted into Equation 7.139 for calculating the profile at the next time step. After this, the just-calculated B-profile is used to calculate the new vacancy profile according to Equation 7.136. The velocities of the boundaries, and new internal and external radii are calculated according to Equation 7.127, also from the just-found value of the total vacancy flux.

7.4.3
Formation of a Hollow Nanoshell from Core–Shell Structure without the Influence of Ambient Atmosphere

To describe the formation of a nanoshell, we do not need to change the model equations – we change only the initial conditions. Namely, let us consider a sphere of pure B of radius r_{BA} surrounded by a shell of pure A. To avoid solving the nucleation problem, let us assume that the initial pure sphere A already contains a small void in the center. Of course, for a very big initial core, this assumption seems unreasonable since the first voids should nucleate in the vicinity of the initial contact between A and B. Yet, for nanoparticles, it is natural that the initial nanovoids coalesce very fast into a single central void. Thus, in our model, the initial B-profile is

$$t' = 0, \quad c_B(\xi) = \begin{cases} 1, & 0 < \xi < \xi_{BA} \\ 0, & \xi_{BA} < \xi < 1 \end{cases}, \quad \xi_{BA} \equiv \frac{1/r_{BA} - 1/r_{e0}}{1/r_{i0} - 1/r_{e0}} \quad (7.142)$$

Evidently, the curvature effect, determining the kinetics of collapse at the shrinkage stage, cannot be neglected at the formation stage as well. The formation of a nanoshell is the result of competition between the Frenkel effect due to chemical forces (and difference of diffusivities) and the Gibbs–Thomson effect, as discussed in Ref. [10]. The smaller the radius, the larger should be the role of curvature-driven effects. Thus, one can expect that, for an initial void radius smaller than some critical value, nanoshell formation will be impossible. This critical radius can be found from Equation 7.137 for the vacancy flux. Namely, the initial pore in the B/A core-shell structure shrinks from the very beginning if the sign of vacancy flux is positive. For this, it is necessary that

$$\exp\left(\frac{2\gamma\Omega}{kTr_i}\right)\exp\left(-\int_0^1 f(\xi)\,d\xi\right) > \exp\left(-\frac{2\gamma\Omega}{kTr_e}\right) \tag{7.143}$$

which means

$$\int_{c_B(r_e)}^{c_B(r_i)} \frac{(K_B - K_A)(1 - \tilde{n}_A c_B \lambda)}{c_A K_A + c_B K_B}\,dc_B < L\left(\frac{1}{r_i} + \frac{1}{r_e}\right), \quad L = \frac{2\gamma\Omega}{kT} \tag{7.144}$$

At the very beginning, $c_B(r_e) = 0$, $c_B(r_i) = 1$. Thus, the formation of a hollow nanoshell from a core–shell structure is forbidden if

$$\frac{1}{r_i} + \frac{1}{r_e} > \frac{1}{L}\int_0^1 \frac{(\kappa e^{\alpha x} - 1)(1 - \lambda x(1 - x))}{1 + x(\kappa e^{\alpha x} - 1)}\,dx \tag{7.145}$$

Here

$$\kappa = \frac{K_{B0}}{K_{A0}}, \quad \alpha = \alpha_B - \alpha_A, \quad x = c_B$$

This criterion is not rigorous since it was obtained under two major assumptions: (a) the steady state approximation for vacancies and (b) the assumption of a single central void preexisting in the center and serving as the sink for all extra vacancies coming into the B-rich core. In reality, at the very beginning the largest supersaturation with vacancies should appear just under the initial contact surface B/A, as seen in Figure 7.22 (vacancy profile at the initial formation stage). It may lead to the formation of several tiny voids under B/A perimeter, as we will see later in the MC simulation and as was observed in experiments at low temperatures [6].

7.4.4
Results of the Phenomenological Model

In Figure 7.19, one can see that the time law of shrinkage correlates with the time dependence of the segregation magnitude measured as the difference between B species concentration at external and internal nanoshell boundaries. Also, one can notice a peculiarity in the time dependence just after the initial stage. The shrinkage rate slows down after rather fast initial segregation, and the $r_i(t)$-dependence has an inflection point. Later, we will see that this peculiarity is not general, and disappears in the crossover regime when nanoshell collapse follows the nanoshell formation at the same temperature. We believe that the reason is related to different initial conditions.

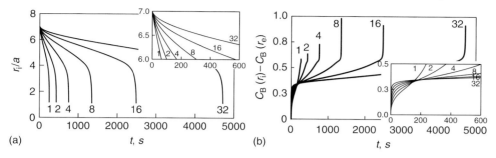

Figure 7.19 Time dependencies of (a) void radius and (b) segregation magnitude during shrinkage at various ratios of diffusivities $\kappa = 1, 2, 4, 8, 16, 32$ (actually, preexponential factors) in the phenomenological model. The parameters are chosen as $c_V^{eq} = 10^{-4}$, $c_B^{init} = 0.5$, $r_i^{init} = a$, $r_e^{init} = 17a$, $L = 10^{-9}$ m, $K = 10$, $\alpha_A = -6$, $\alpha_B = -2$, and $\lambda = 0$.

The results shown in Figure 7.19 are obtained for a rather artificial initial condition, when the initial concentration is taken equal everywhere. So, some short period of adjustment follows. If we take a just-grown nanoshell, it has an inhomogeneous concentration distribution from the very beginning (because a competition occurs between Kirkendall, inverse Kirkendall, and Gibbs–Thomson effects at the formation stage as well). So, one can expect a constant sign of the second-order time derivative in this case (see Figure 7.22).

The time of full collapse depends almost linearly on the ratio k of the pre-exponential factors in Equation 7.119, $t_{collapse} = m\kappa + b$. The magnitudes of the coefficients m and b in this linear approximation change at small mixing enthalpies and reach asymptotic values at large mixing enthalpies (Figure 7.20). Such a linear dependence was predicted by us in the case of compound nanoshells with a very narrow homogeneity range [8] (see also Section 7.2)

$$t_{collapse} \approx \frac{kT}{12\gamma\Omega} r_f^3 \frac{c_A D_B^* + c_B D_A^*}{D_A^* D_B^*} = \frac{kT}{12\gamma\Omega} \frac{r_f^3}{D_B^*} \left((1-c_B) \frac{D_B^*}{D_A^*} + c_B \right) \quad (7.146)$$

Here r_f is the final particle radius after collapse.

In our case

$$\frac{D_B^*}{D_A^*} = \kappa \exp\left((\alpha_B - \alpha_A) c_B \right)$$

Thus, in the case of a narrow compound phase (meaning large mixing enthalpy), one can expect that

$$\frac{m}{b} = \frac{1 - c_B}{c_B} \exp\left((\alpha_B - \alpha_A) c_B \right) \quad (7.147)$$

One can see that the asymptotic values of this ratio in the present case of solid solution are indeed close to the mentioned analytical prediction (Table 7.2).

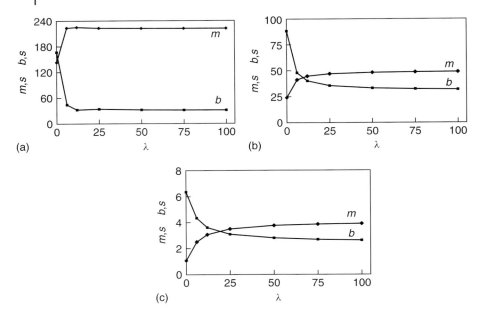

Figure 7.20 Dependencies of the parameters m and b of the linear approximation $t_{collapse} = m\kappa + b$ ($\kappa = 1, 2, 4, 8, 16, 32$) for the collapse time $t_{collapse}$ on the dimensionless mixing enthalpy in the phenomenological model. Parameters: $c_V^{eq} = 10^{-4}$, $c_B^{init} = 0.5$, $r_i^{init} = a$, $r_e^{init} = 17a$, $L = 10^{-9}$ m, $\lambda = 0, 6, 12, 25, 50, 100$, and (a) $\alpha_A = -6$, $\alpha_B = -2$; (b) $\alpha_A = -3$, $\alpha_B = -2$; (c) $\alpha_A = 2, \alpha_B = 3$.

Table 7.2 Asymptotic ratio (at large values of λ) of linear fit coefficients for the dependence of collapse time on the diffusivity ratio. The analytical prediction $((m/b)_{pred})$ via Equation 7.147 is compared with the phenomenological model $((m/b)_{model})$.

α_A	α_B	$(m/b)_{pred}$	$(m/b)_{model}$
−6	−2	7.07	7.39
−3	−2	1.54	1.65
2	3	1.49	1.65

In Figure 7.21, one can see that the validity of the steady state approximation for the main components improves with increasing absolute value of the mixing enthalpy. To demonstrate this, we compare the value of the reduced shrinkage rate

$$RSR = \frac{dr_i}{dt} r_i^2 \left\{ \frac{1/r_i - 1/r_e}{\exp\left(\frac{2\gamma\Omega}{kTr_i}\right) - \exp\left(-\frac{2\gamma\Omega}{kTr_e}\right)} \right\} \quad (7.148)$$

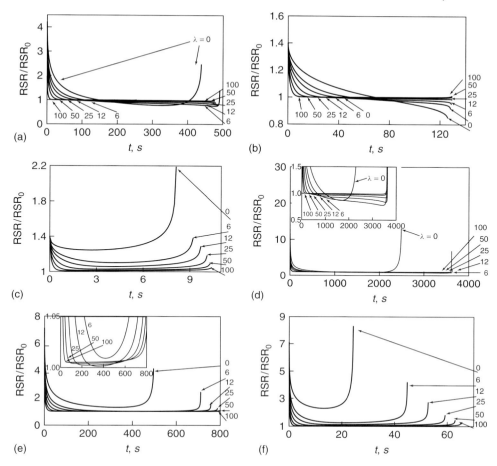

Figure 7.21 Time dependence of reduced shrinkage rate defined by Equation 7.148 at various values of mixing enthalpy $\lambda = 0, 6, 12, 25, 50, 100$ calculated at $c_V^{eq} = 10^{-4}$, $c_B^{init} = 0.5$, $r_i^{init} = a$, $r_e^{init} = 17a$, $L = 10^{-9}$ m, and (a) $\kappa = 2$, $\alpha_A = -6$, $\alpha_B = -2$; (b) $\kappa = 2$, $\alpha_A = -3$, $\alpha_B = -2$; (c) $\kappa = 2$, $\alpha_A = 2$, $\alpha_B = 3$; (d) $\kappa = 16$, $\alpha_A = -6$, $\alpha_B = -2$; (e) $\kappa = 16$, $\alpha_A = -3$, $\alpha_B = -2$; (f) $\kappa = 16$, $\alpha_A = 2$, $\alpha_B = 3$.

obtained in our present model, and the value obtained in the steady state approximation for the compounds in Ref. [8]

$$\mathrm{RSR}_0 = \frac{D_A^* D_B^*}{(c_A D_B^* + c_B D_A^*)} \tag{7.149}$$

with tracer diffusivities calculated for the initial nanoshell composition before shrinkage. One can see that the larger the absolute value of the mixing enthalpy (the larger are the chemical forces), the better the steady state approximation works. This is an answer to the problem stated by the group of Murch [21].

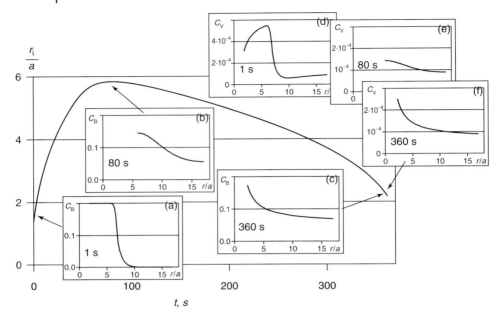

Figure 7.22 Time evolution of an initially tiny void: first, a fast formation of a hollow nanoshell and then its collapse (slow and then faster and faster). The inserts (a), (b), and (c) show the concentration profiles of the faster component at the formation, crossover, and collapse stages respectively. The inserts (d), (e), and (f) show the vacancy concentration profiles in the same stages. Parameters: $c_V^{eq} = 10^{-4}$, $c_B^{init} = 0.075$, $r_i^{init} = 1.4a$, $r_{BA}^{init} = 7a$, $r_e^{init} = 17a$, $L = 0.75 \times 10^{-9}$ m, $\kappa = 10$, $\lambda = 0$, $\alpha_A = -4.5$, $\alpha_B = -2$, $dt = 10^{-4}$ s.

In Figure 7.22, we demonstrate (for the first time, as far as we know) the formation and shrinkage in one run. During the formation stage, the chemical forces in the sharply inhomogeneous system are stronger than the curvature-driven vacancy gradient. After formation of a partially homogeneous hollow nanoshell, the chemical forces become weaker, and the curvature-driven Gibbs–Thomson effect wins, leading to a collapse due to the outward vacancy flux. Profiles of B species in Figure 7.22 demonstrate that the full homogenization is not reached during nanoshell formation – even at maximal void radius, the concentration gradient is not zero, and later it grows back due to the inverse Kirkendall effect. The vacancy profile shows the formation of supersaturation just under the initial B/A interface at the initial stage. Later, this nonmonotonic profile transforms into a monotonic one.

7.4.5
Monte Carlo Simulation of the Vacancy Subsystem Evolution in the Structure "Core–Shell"

Alternatively to the offered phenomenological models, we developed a 3D MC model of nanoshell formation and collapse for the fcc structure (lattice parameter

a) of a binary alloy with vacancies. To be sure that the proposed model is valid and the chosen parameters are reasonable, we should check, at first, if the hollow nanoshell is indeed formed from the initial core–shell structure.

The difference in diffusivities of A and B species (even in the case of an ideal alloy with zero mixing enthalpy) is provided by the probability to choose the atom of the given kind in the Metropolis algorithm, determined by the ratio of frequencies $K = \nu_B/\nu_A$ ($\nu_A \leq \nu_B$). Actually, the difference in mobilities is stipulated not only by different frequencies K but also by thermodynamics and structures of the sublattices: in the ordered phase A_1B_3, the component B has a higher mobility even in the case of equal frequencies. The temperature is, typically for MC simulations, regulated by varying the dimensionless ratio ε of average interaction energy E and kT (larger ε means lower temperature).

$$\varepsilon = c_A^2 \varphi_{AA} + c_B^2 \varphi_{BB} + 2c_A c_B \varphi_{AB} \tag{7.150}$$

In the following, we model the nonideal (with tendency to ordering) as well as ideal solutions, combining the sets of nondimensional pair interaction energies or/and concentrations to satisfy Equation 7.150 at fixed values of ε (fixed temperature).

In Section 7.4.5.1, we simulate the formation of a nanoshell for the case of a nonideal solution. In Section 7.4.5.2, we study the crossover from formation to collapse, as well for nonideal solutions but for another average concentration, to obtain formation and collapse in one run at a reasonable computation time. In Section 7.4.5.3, we investigate mainly the segregation caused (kinetic origin) by the inverse Kirkendall effect; for this reason, we treat an ideal solution in this subsection.

7.4.5.1 Formation of a NanoShell in a MC simulation

To investigate the kinetics of nanoshell formation, a nanoparticle of pure B component (radius $r_{BA} = 15.7a$) was initially enveloped by a pure A component (external radius $r_e = 17a$) (the total number of atoms is 82421), the chosen ratio of internal r_{BA} and external r_e concentric radii providing the necessary volume fractions for the A_1B_3-phase. According to Equation 7.150, pair energies for the A_1B_3-phase are chosen as $\varepsilon_{AA} = \varepsilon_{BB} = 0.7\varepsilon_{AB}$ at low temperatures $\varepsilon_{AA}^L = \varepsilon_{BB}^L = -1.32$, $\varepsilon_{AB}^L = -1.89$ with an average energy $\varepsilon^L = -1.535$ [20]; at high temperatures, $\varepsilon_{AA}^H = \varepsilon_{BB}^H = -0.86$, $\varepsilon_{AB}^H = -1.23$ with an average energy $\varepsilon^H = -0.98$.

As a result of the simulation, the widening of the A/B diffusion zone is observed and small voids start forming at the pure B/alloy interface. In the process of a computer experiment, the voids expand, leaving between themselves the "bridges" of pure B. At high temperature, we observed the subsequent coalescence of voids into a single spherical void in the center of the nanoparticle (Figure 7.23). These simulation results correlate with experiments on the formation of cobalt selenides and sulfides [4, 5, 30].

7.4.5.2 Crossover from Formation to Collapse

In the above-mentioned simulations, the process practically stopped after formation of a hollow nanoshell. The reason for this seems to be quite clear since the formation

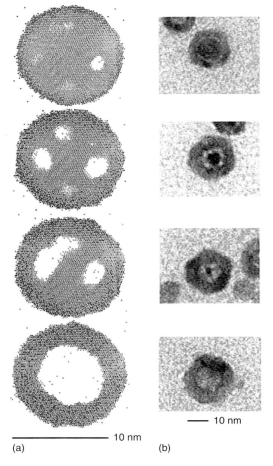

Figure 7.23 Void formation in a nanoparticle: (a) results of MC simulation with $\bar{c}_B = 0.75$, $K = 10$ at high temperature (ε^H) at times 167, 1000, 2000, and 10000 MCS (sample cross section, light dots – B, dark ones – A), the maximal size of the void attained at 8400 MCS; (b) evolution of CoSe hollow nanocrystals with time by injection of a suspension of selenium in dichlorobenzene into a cobalt nanocrystal solution at 455 K: 10 s, 1 min, 2 min, and 30 min [4].

is driven by chemical forces, and collapse by capillary ones. The latter are usually significantly less than chemical forces. Therefore, shrinkage of a nanovoid is typically a much slower process and can be observed at temperatures higher than those for the process of formation.

In order to describe formation and collapse in one simulation, one needs higher temperatures. We took ε_{AA}^H, ε_{BB}^H, ε_{AB}^H and an initial interrelation of the components' radii $r_{BA} = 7a$, $r_e = 17a$, corresponding to an average concentration $\bar{c}_B = 0.075$ (5745 B atoms, 76676 A atoms) and to $\varepsilon = 0.9$. The process of void formation proceeded very quickly (during 1238 MCS). At that, the void formed in the center

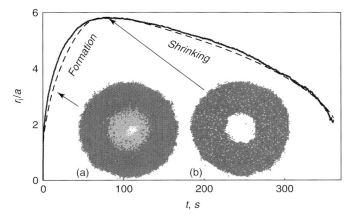

Figure 7.24 Nanoshell formation and collapse in one run: atomistic simulation and phenomenology. MC (solid line) and phenomenological (dashed line) plots are superimposed with respective time rescaling, 1 MCS = 62 ms. MC-modeling at a high temperature (ε^H_{AA}, ε^H_{BB}, ε^H_{AB}), $K = 10$, and a small fraction of a more mobile B component (light dots, $\bar{c}_B = 0.075$): (a) formation stage and (b) crossover. Phenomenological plot calculated at $c_V^{eq} = 10^{-4}$, $c_B^{init} = 0.075$, $r_i^{init} = 1.4a$, $r_{BA}^{init} = 7a$, $r_e^{init} = 17a$, $L = 0.75 \times 10^{-9}$ m, $\kappa = 10$, $\lambda = 0$, $\alpha_A = -4.5$, $\alpha_B = -2$, $dt = 10^{-4}$ s.

of the nanoparticle and appeared to be unstable: just after having reached the maximal size (3314 sites), it collapsed (Figure 7.24). To correlate the simulation results with the phenomenological model, we recalculated the number of empty sites into an effective void radius r_i.

7.4.5.3 Shrinkage and Segregation Kinetics in an MC Simulation

Now we come back to the synergy of curvature-driven effects and inverse Kirkendall effect at the shrinkage stage. The above-described MC-model was used to simulate the collapse of the nanoshell with $r_i = 7a$, $r_e = 17a$ (the total number of atoms is 76676 and the number of empty sites in the void is 5745) for an ideal alloy at $\bar{c}_B = 0.75$ and $K = 10$. At high temperatures (ε^H), after 7000 MCS, the void has shrunk by 9.58 % (Figure 7.25).

To investigate segregation effects, we have used the nanoshells of an ideal alloy with $r_i = 4a$ and $r_e = 8a$ at low temperature (ε^L) with different ratios of frequencies K (10, 50, 100) and concentrations c_B (0.25, 0.50, 0.75, 1.00). We may conclude that the larger the K and/or c_B, the more intensive is the segregation of the more mobile component near the internal surface (Figure 7.26) and the slower is the process of the nanoshell collapse (Figure 7.27).

7.4.6
Summary

The formation of hollow nanoshells and their collapse are presented as two stages of one process with crossover between regimes being determined by the competition

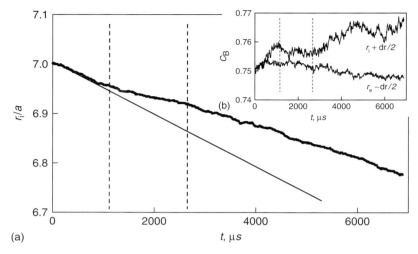

Figure 7.25 MC-time dependence of (a) internal radius and of (b) concentration near moving boundaries at the initial stage of the shrinkage of the A_1B_3 nanoshell ($r_i = 7a$, $r_e = 17a$, $dr \approx 1.7a$, $K = 10$, ε^H): segregation of B species near the internal surface leads to slowing down of the nanoshell shrinkage.

between the Kirkendall-driven Frenkel effect, curvature-driven Gibbs–Thomson effect, and the inverse Kirkendall effect. Such a unified approach appears to be reasonable both in the phenomenological analysis and in MC simulations. The presented analysis is applied (in this chapter) only to the case of solid solutions. One can formulate the following main features and predictions of the mentioned approach:

1) Formation of a hollow shell is driven by the inward vacancy flux, which contains two competing inputs: (a – assisting formation) chemical driving force that is proportional to the main component's gradient and the difference of partial diffusivities (if the faster component is situated inside), (b – suppressing formation) vacancy gradient created by the difference in equilibrium vacancy concentrations at the internal and external boundaries (Gibbs–Thomson effect) after formation of at least the first void inside. This void should appear due to the first driving force (superposition of chemical potential gradient and difference in diffusivities).
2) Conventionally, the critical void size is determined only by vacancy supersaturation and temperature. In our case, the concept of a critical size should be reconsidered. Void behavior, besides local vacancy concentration, is also determined by the gradients of vacancies and the main component, so that a thermodynamically stable void might be kinetically unstable. The approximate criterion of central void growth is given by Equation 7.145.

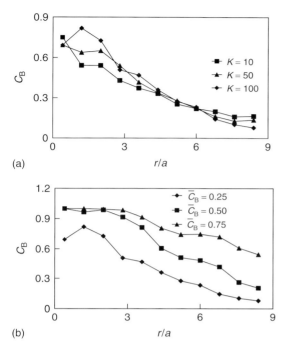

Figure 7.26 Radial concentration profile for a nanoparticle just after collapse (a) at different ratios of the frequencies of components (average concentration $\bar{c}_B = 0.25$); (b) at different average concentrations (at $K = 100$).

3) In the process of internal void growth, the initial reserves of the chemical driving force (concentration and chemical potential gradients) run out, so that at some moment (at some void radius) the chemical driving force becomes less than the capillary forces (and the corresponding vacancy drop between internal and external boundaries). Thus, one has a crossover from formation to shrinkage.

4) In the general case, reaching the maximum void radius (end of formation and crossover to shrinkage) does not mean full homogenization, so that shrinkage starts at once from the inhomogeneous shell. Therefore, a separate treatment of shrinkage starting from the homogeneous initial hollow nanoshell is somewhat artificial and may lead to simulation artifacts, like retardation after some period of segregation. Of course, if formation and shrinkage proceed at different temperatures, some initial "adjustment" of the concentration profile at the beginning of shrinkage might have taken place indeed.

5) The formation stage starts from a vacancy supersaturation just under the interface and formation of multiple tiny nanovoids in this region, with bridges in-between providing mass transfer to the remaining metallic core. This feature was established only by MC simulations, since our phenomenological model is

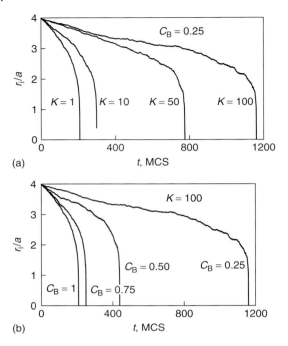

Figure 7.27 Collapse rate influenced by (a) ratio of frequencies and (b) volume fractions of components.

spherically symmetric so far, and cannot describe multiple voids and bridges. This feature fully corresponds to experimental data.

6) The shrinkage stage is characterized by the inverse Kirkendall effect: an outward vacancy flux generates segregation of the faster component near the internal surface, the corresponding concentration gradient decreases the vacancy flux and, respectively, the shrinkage rate. Thus, at the shrinkage stage, one also has a competition of capillary forces, assisting shrinkage, and chemical forces, opposing shrinkage. The peculiarity here is that the chemical force is not "external", it is created by the vacancy flux, generated by capillary forces. This result for solid solutions was obtained both by phenomenological modeling and MC simulation and coincides with results obtained by us earlier [8] for almost stoichiometric compounds with a narrow homogeneity range. The coincidence becomes almost perfect for the case of large absolute values of mixing enthalpy in a solid solution.

7) For the case of large mixing enthalpies, the steady state approximation for the main components (not only for vacancies) appears to be valid. For small and zero mixing enthalpies (ideal solution), the steady state approximation for the main components works worse.

8) In this work, we did not consider the reaction between solid nanoparticles and gases, the respective models be presented elsewhere.

7.5
Void Migration in Metallic Interconnects

Nucleation and growth of voids induced by electromigration in Al lines has been found to be one of the key reliability failure modes in microelectronics [25]. According to the Huntington–Fiks mechanism of electromigration, the electron wind force generates a vacancy flux toward the cathode, where a sufficient supersaturation of vacancies can be reached, resulting in void nucleation, growth, and eventually failure of a microcircuit [31, 32]. This simple picture appears to be invalid at least for void formation in copper interconnects covered by a dielectric as shown in Ref. [33]. Owing to this reason, a more detailed analysis is required.

7.5.1
Hypotheses and Experiments

Migration and coalescence of voids as well as their interactions with grain boundaries (GBs) in the presence of the electric wind force is crucial for understanding the failure mechanism. The first fundamental theory of void migration was developed by Krivoglaz [34] for an isolated spherical void and was later modified by Ho [35] for voids in the vicinity of an external surface. At that, the theory of electron wind force [36, 37] was used to demonstrate a $(1/R)$-size dependence of void velocity. However, the interaction of a void with GBs during electromigration (EM) was not considered.

In [12], in situ scanning electron microscopy (SEM) observations of high current density induced nucleation, migration, and coalescence of voids on the surface of Cu interconnects or at the Cu/dielectric overlayer interface (Figure 7.28) have been modeled by means of analytical treatments and computer simulations. Let us briefly describe the mentioned experiments: A conductive thin parallelepiped film up to 1 µm wide and 0.5 µm thick is treated in a nonconducting medium imposed to a direct current with the density of about 10^{10}–10^{11} A/m^2. The so-called M1 and M2 structures (Figure 7.29) were singled out, indicating transitions from the lower level to the upper one and, conversely, through a "via". The void behavior in these two structures is different. In the M-2 line, the voids first reach the Cu/dielectric interface at some distance from the corner of the via and continue growing and

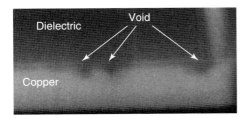

Figure 7.28 Void coarsening at the thin-film copper/dielectric interface (*in situ*).

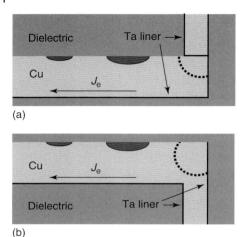

Figure 7.29 Scheme for void migration for (a) the lower M-1 and (b) the upper layer M-2 of the structure [12].

moving along the Cu/dielectric interface toward the via and cathode end of the line; eventually, the void extends and via fails. In the M-1 line, the voids also first nucleate at the Cu/dielectric interface at a certain distance from the via and move along the interface; the ultimate location of voids is indicated by a dotted line.

In order to correctly simulate electromigration, one must constantly recalculate the current distribution at the movement of the void and change of its shape since this is the current redistribution that determines the electron wind act, which in turn influences the shape and location of the void. So, at the least for configuration changes, one must recalculate the distribution of potentials that can alter the result of the next step. As far as we are concerned, only two-dimensional models of such evolution have been developed [38]. One can apply them to real three-dimensional geometries of the microcircuits only qualitatively. We present a three-dimensional atomic simulation for an fcc metal with copper parameters, taking into account grain boundaries. The model had the following basic features:

1) Voids are formed at surface defects and are bounded by the metal (from one side) and dielectric overlayer (from the other side).
2) These surface voids migrate under the electron wind force (Figure 7.29).
3) Migration velocity depends on the void size and on the presence of defects (in particular, GBs).
4) Surface voids can migrate from one GB to another, as well as *along* GBs. In the latter case, the velocity depends on the type of boundary and on the angle between the field and the GB/surface intersection:
5) Voids can be trapped by GBs and/or GB junctions.

6) Void velocity depends on size as well as trapping at GBs and leads to collisions and coalescence of voids (typically, the larger the path traversed, the larger the void).
7) Detrapping of voids from GBs or vertices is possible only after reaching a critical size.
8) Voids tend to stop at the corners above (or below) the via, then grow, thus resulting in a critical decrease in the effective interconnect cross section leading to eventual failure.

An initially hemispherical shaped void, under the influence of an electric current, at the surface of a thin-film confined dielectric is considered. Its planar boundary is a dielectric and the spherical surface is copper. The evolution of the void in the absence of bulk diffusion is governed by two forces: (a) the electron wind force that pushes and eventually redistributes the surface atoms, trying simultaneously to shift the void and change its shape and (b) the surface tension of the void that tries to minimize the void surface energy. Thus, the atomic flux density along the moving surface of the void consists of two terms [39]

$$J_{at} = \frac{c_A D_A}{K_B T} Z_{ef} e E_S - 2\Omega\gamma \frac{\partial k}{\partial S} \qquad (7.151)$$

where c_A is the atomic concentration of mobile surface atoms (adatoms), D_A is the corresponding diffusivity, Z_{ef} is the effective charge of jumping ions (in general, differing from the effective charge of ions in the bulk), e is the elementary charge, E_S is the tangential component of the electric field, Ω is the atomic volume, γ is the surface tension, k is the local surface curvature equal to half of the sum of the inverse radii of curvature, ∂k, and ∂S is the directional derivative along the void surface.

Equation 7.151 has a very simple form, but represents a challenging formulation for calculation, since, to find the local current density (and, thus, the electron wind force), one has to solve the Laplace equation for the electric potential in the whole sample at each new time moment (as the void moves and changes the shape). At that, as shown in Figure 7.30a, the ratio of maximum and minimum current densities near the void can reach 2 orders of magnitude. A schematic representation of shape evolution is shown in Figure 7.30b. Since the current

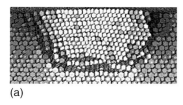

(a) (b)

Figure 7.30 Current density distribution around the void surface: (a) calculated in MC-model, brightness being proportional to the current density; (b) evolution of the initially hemispherical void under the current divergence at the void surface.

density and corresponding atomic surface flux from A to B are less than the current density and flux from B to C, the material is "eroded" and the void extends in location B. On the other hand, since the current and the flux from B to C is larger than that from C to D, a material accumulation is observed at location C. Thus, a change in current density causes the motion of the void and also a change in the shape of the void. This in turn leads to a current redistribution in the new configuration. Thus, we have a self-consistent problem that does not have an analytical solution. Furthermore, numerical modeling is meaningful only for a 3D case, since even for a hemispherical void, a 2D projection (semicircle) does not provide the real picture of current and flux distribution.

7.5.2
The Model

The model employed has the following essential features:

1) The metallic film material is modeled at an atomic level, in an fcc lattice (lattice parameter $a = 3.615$ Å). Bulk diffusion is negligible as the temperature is low ($T = 500$ K).
2) The energy of an atom consists of the effective energy of an ion in the field of the electron wind force $Z_{ef}eU$ ($Z_{ef} = 38$), U is the electric potential; actually, we are interested only in the change in this "energy" (that equals the work of the electron wind force during atomic displacement) plus the sum of pair interactions zE (z is number of nearest neighbors for a surface atom, $0 \leq z \leq 12$, E is the pair interaction energy estimated from the sublimation energy of copper, $E = -0.587$ eV).
3) The interaction energy between copper atoms and dielectric is unknown and is chosen to preserve the void's shape at the boundary (at the absence of currents), corresponding to direct experimental observations.
4) GBs are chosen energetically, i.e. neighboring atoms that belong to different grains are taken to have a different pair interaction energy.
5) Electric potential distribution is found numerically by solving the Laplace equation

$$\nabla \left(\frac{1}{\rho} \nabla U \right) = 0$$

6) Atomic jumps are realized by a modified MC method (for a detailed discussion see Ref. [13]).
7) Void behavior is simulated for different cases (see Figure 7.31): (a) migration along the Cu/dielectric interface far from GBs; (b) interaction with a GB (transverse to current direction and at non-right angle to it), possibility of trapping at and release from the boundary; (c) migration along GBs with varied orientations with respect to the current; (d) possibility of trapping at triple junctions.

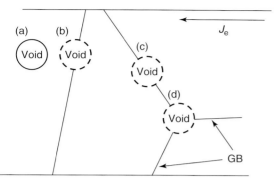

Figure 7.31 Possible locations of the void at the metal–dielectric interface of a thin film with bamboo-like structure (view from the interface): (a) far from GBs; (b) at the GB; (c) migration along GB; (d) at the GB junction (vertices of thin-film bamboolike grains).

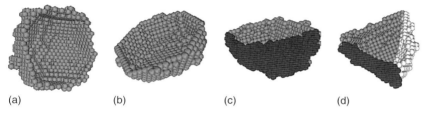

Figure 7.32 Relaxation of an initially spherical void in the bulk grain far from the GBs and the dielectric: (a) equatorial cross section and relaxation of an initially hemispherical void at the copper–dielectric interface; (b) at metal–dielectric interface; (c) at the GB; and (d) at the grain triple junction (the different shades correspond to surface atoms of different grains).

7.5.3
Results

In accordance with the proposed hypothesis of void movement along the metal/dielectric interface and metal's GBs, we investigate the motion of voids along the metal/dielectric interface with and without GBs and GB junctions. To obtain some analytical estimations, we also simulated the motion of initially spherical voids in the bulk. Faceting of the voids along the most dense plane (111) (Figure 7.32)) was observed in all simulations. The shape of the void at different interfaces (metal, dielectric, vacuum) and grains with different orientation was determined by model coefficients of surface energy.

7.5.3.1 Migration of Voids in Bulk Cu and Determination of the Calibration Factor between MCS and Real Time

In our MC scheme, each single MCS does not correspond to elementary atomic jump and may include a combination of different surface diffusion mechanisms.

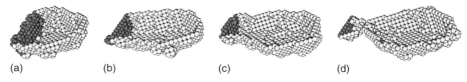

Figure 7.33 Evolution of void delay at the GB transverse to current direction: (a) 0 MCS; (b) 466 MCS; (c) 701 MCS; (d) 944 MCS

To find the relationship between MC steps and real time, we have taken up the motion of spherical voids in bulk metal without GBs. According to [35, 36], the velocity of the void (provided the spherical shape is preserved during migration), in the absence of volume diffusion, is determined by

$$v = \frac{dX}{dt_{real}} = \frac{Z_{ef} eJ}{K_B T} 3 D_A c_A \Omega \frac{1}{R} \tag{7.152}$$

Having used the model for void electromigration at the copper/dielectric interface, we simulated the voids with radii $4a$, $6a$, and $8a$ (a standing for the lattice parameter). Despite the observed faceting, we find that the velocity is inversely proportional to the void radius, which conforms to analytical predictions

$$\tilde{v} = \frac{dX}{dt_{MCS}} = 6.181036 \times 10^{-21} \cdot \frac{1}{R} \cong 9.865871 \times 10^{-14} \text{ m s}^{-1}$$

with an error up to 2.5%.

The constant in this equation is about 3.5% of the linear term for a radius of $6a$. So, we have neglected it for calibration. Using the above-mentioned parameters and neglecting the free term, we can rewrite the latter equation in the form, similar to Equation 7.152

$$\tilde{v} = \frac{dX}{dt_{MCS}} \cong 0.594 \times \frac{Z_{ef} eJ}{K_B T} \Omega \frac{1}{R} \tag{7.153}$$

Comparing Equations 7.152 and 7.153, we obtain the relation between real time and MCS time (in our simulations) in terms of surface diffusivity

$$t_{real} = \frac{0.198}{D_A c_A} \times t_{MCS}$$

7.5.3.2 Void Migration Along the Metal/Dielectric Interface

We simulated an initially hemispherical void containing 2505 empty sites at the interface between Cu and the dielectric layer. After minimization of the surface energy, the void became faceted and more shallow. After having imposed electric current, it started moving and became nonequiaxial and elongated by about 10–30%. In this case, Equation 7.152 for time renormalization is inapplicable. The variation of void velocity with radius (cubic root of volume) was observed to be monotonically increasing, but was far from being linear. Note that the nonlinear dependence was also reported for 2D voids of atomic thicknesses [38]. Owing to the minimization of surface energy, the surface voids prefer to be situated at the GB/capping layer interface or grain interfaces and their contacts with dielectric.

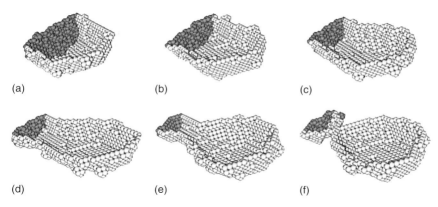

Figure 7.34 Evolution of a void trapped at the GB for $\theta = 60°$ with the current direction: (a) 0 MCS; (b) 463 MCS; (c) 684 MCS; (d) 980 MCS; (e) 1227 MCS; (f) 1277 MCS.

When void migration took place along a GB that was parallel to the current direction, the void velocity was found to be 33% larger than in the case when the void was far from the GB. One of the possible contributors may be the larger tangential component of the electric field E_S for the shallow elongated (boatlike) voids.

The interaction of surface voids with GBs perpendicular to the current direction demonstrates the following mechanism of trapping and detachment. After reaching the GB, the void changes its shape (Figure 7.33a), holding on to the GB and simultaneously elongating toward the cathode and forming a "fishlike" shape (Figure 7.33b) with an elongated "tail-like" structure. The void eventually detaches from the GB, leaving its "tail" at the GB (Figure 7.33c). Since the major part of a void is "pulled" by the field "the fish has to go with the stream," leaving its "tail" at the GB (Figure 7.33d), thus reducing the surface energy of the GBs. In this simulation, the tail (residual void) consists of 43 sites, whereas the detached void consists of 944 sites. According to the simplified estimates (see analytical models below), such a void is overcritical and should indeed detach from the GB. Note that "void tails" can accumulate vacancies or other minor voids, reach the overcritical size, and produce new voids. Thus, such process can be constant, and undercritical voids at the GBs can be considered to be the nuclei of new voids like a Franc–Reed dislocation source.

As the GB makes up an arbitrary angle with the current direction ($0° \leq \theta \leq 90°$), the void moving along the GB experiences two competing factors: on the one hand, the projection of current on the normal to the GB pushes the void along the GB and, on the other hand, the projection of current on the normal to the GB tries to detach the void from the GB. The outcome of this competition depends on the angle between the GB and the current. The critical angle depends on the current density and the surface tension.

The void velocity along the GB decreases with increasing angle. For example, for $\theta = 60°$, the velocity is 40% of the velocity for $\theta = 0°$. Similar to the case when the

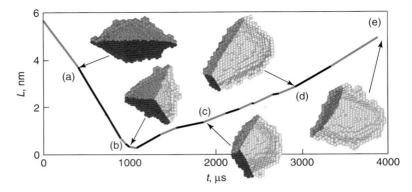

Figure 7.35 Kinetics of void motion in the vicinity of triple junction: (a) moment of approaching the junction; (b) shape relaxation at the junction; (c) transition to a sail-like shape under the current effect; (d) choice of either GB; and (e) motion along the boundary after leaving the junction.

GB is perpendicular to the current, at an arbitrary angle, the detachment of the void is realized by the change of shape (Figure 7.34e) resulting in the leaving of the "tail" at the GB (Figure 7.34f). The void shape appears to be nonsymmetrical before the detachment (Figure 7.34d). In this case, the tail consists of 157 sites, whereas the original void consisted of 1746 sites (such a size is also overcritical according to theoretical estimations).

We also treated the behavior of voids at triple junctions of GBs and the dielectric surface. We considered a marginal case, angles between grain boundaries being equal to 120°. The void, migrating, reaches the triple junction along the interface, which is parallel to the current direction $\theta = 0°$ (Figure 7.35a). Having reached the GB junction, the void undergoes the following stages:

- The void becomes deeper along the junction line and its velocity decreases (Figure 7.35b).
- The current tries to push the void but the void reduces the GB energy at the junction; therefore, it is unfavorable for the system to let the void out of the junction line both toward the dielectric interface without GBs and toward the GB itself; being trapped, the void acquires a more or less symmetrical shape of an isosceles triangular pyramid (Figure 7.35c);
- Despite the favorability of being at the junction, under electric current due to violation of spontaneous symmetry, the void chooses either *one* of the GBs for propagation, and begins to move along the chosen GB (Figure 7.35d); obviously, in the case of nonsymmetrical GB location, the GB that makes a lesser angle with current direction is preferable.
- Subsequently, the void leaves the GB joint and migrates along the GB with an elongation along the current direction (Figure 7.35e).

The time dependence of the distance between the center of the void and the GB junction, L, demonstrates that the velocity of the void increases as the void approaches the GB junction (Figure 7.35a–b). Upon reaching the GB junction, the void virtually stops, deepens, before assuming a pyramidal symmetrical shape (Figure 7.35b, plateau region).

7.5.4
Simplified Analytical Models of Trapping at the GBs and at the GB Junctions

The simulation results described here were found to be in favorable agreement with experimental results [12]. To better understand this phenomenon, we propose herein two simplified models of void trapping at GBs and at GB junctions.

To describe the void trapping at the GB, we consider a hemispherical void, of radius R, situated initially symmetrically around the GB, at the surface making an angle θ with the direction of the current (Figure 7.36). Under the electron wind force, this void tries to simultaneously move along the GB (increasing l) and to detach from the GB (increasing δ). Energetically, this process is determined by the competition between the two factors. Firstly, a motion at some sharp angle to the current direction leads to a decrease in electric energy (work of the electric field). Secondly, increasing δ means the increase in the grain boundary surface area given by

$$(S_{\text{total}} - \pi R^2/2) - (S_{\text{total}} - \pi r^2/2) = \pi \delta^2/2$$

If under the influence of the electron wind force, the void shifts by a distance of

$$x_c = l \cos \theta + \delta \sin \theta$$

along the electric field without change in shape, the energy change of the system can be expressed as

$$W^{\text{el}} = -Z_{\text{ef}} e\rho J \frac{2\pi}{3\Omega} R^3 x_c + \pi \gamma \frac{\delta^2}{2} \tag{7.154}$$

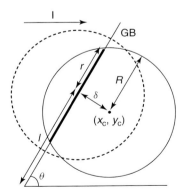

Figure 7.36 Schematic analytical model of a hemispherical void at the GB making the angle θ with the current direction.

A substitution of x_c makes it possible to represent the energy change as the sum of two additive terms, indicating that the shift along the GB and the detachment from the GB are independent events. The minimum of this dependence,

$$\frac{\partial W}{\partial \delta} = 0$$

determines the optimal (equilibrium) shift

$$\delta^* = -\frac{2Z_{ef} e \rho J R^3 \sin \theta}{3 \gamma \Omega} \qquad (7.155)$$

If $\delta^*/R \geq 1$, the void detaches from the GB and migrates further. By using this expression and rearranging the terms in Equation 7.155, one can express the critical radius as

$$R^*_{GB} = \sqrt{\frac{3\gamma\Omega}{2Z_{ef} e \rho J} \times \frac{1}{\sin \theta}} \qquad (7.156)$$

Thus, the void trapped at a GB loses its stability if it becomes larger than R^*. This may happen due to coalescence with other migrating voids or due to the consumption of vacancies. The lesser θ and/or current density, the larger is the critical size for detachment.

It should be noted that this is a simplified model since the inevitable shape changes of the void have not been accounted for.

From Equation 7.156, we have estimated the number of empty sites in the void (through the atomic volume) for the cases of the GB orientations of $\theta = 90°$ and $\theta = 60°$, and obtained respectively 311 and 387 sites. The results of computer simulations confirm such a behavior from below: the voids with 944 ($\theta = 90°$) and 1589 ($\theta = 60°$) sites were overcritical and would detach from the grain boundaries.

To describe the void trapping at the GB junction, we consider a hemispherical void, of radius R, situated initially symmetrically around the GB junction, at the surface making an angle θ with the direction of the current (Figure 7.37). When the void leaves the GB joint along one of the two GBs (GB$_3$), the surface and excess

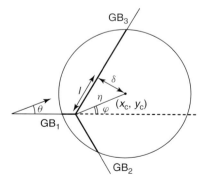

Figure 7.37 Schematic analytical model for a hemispherical void at the triple junction, oriented at an angle θ to the current.

energy of the two remaining GBs (GB$_1$ and GB$_2$) increase (trapping factor), but the surface and surface energy of GB$_3$ decrease. Assuming that the GBs forming the junction have the same surface tension, mechanical equilibrium implies that all angles between the GBs are 120°. Let the current I be directed under some angle θ to GB$_1$ (in the MC simulations, this angle was chosen to be equal to zero). Let δ be the distance between the center of the hemispherical void and GB$_3$, and l be the distance between the projection of this center on GB$_3$ and the joint position. Using polar coordinates η, ε, one can find x_c, y_c, l, and δ.

As discussed in the previous section, the critical void size for detrapping depends upon the work of the electric current and the surface energy changes. In the present case, the surface term includes three terms:

$$S^{GB} = S_0^{GB} - (S_1 + S_2 + S_3)$$

Here S_0^{GB} is the total surface area of all the three grain boundaries in the absence of the void. S_1, S_2, and S_3 are the areas of intersections of the void with the respective GBs. Interaction with the electric field is as described in the previous case if the void center shifts from the junction position (x_0, y_0) to a new position (x_c, y_c) without a change in void shape; the system energy change due to the work of the electric field may be expressed as

$$W^{el} = -Z_{ef} e \rho J \frac{2\pi}{3\Omega} R^3 \times \eta \cos(\varphi - \theta) \tag{7.157}$$

and the critical radius of the void can be estimated as

$$R^*_{joint} = k(\theta) \sqrt{\frac{3\gamma\Omega}{2Z_{ef} e \rho J}} \tag{7.158}$$

where $k(\theta)$ depends on the angle between the direction of the current and GB$_1$.

7.5.5
Summary

The explicit scheme of the MC simulation of void migration gives us a quite realistic description of the process both for the case of migration in the bulk metal far from GBs and the dielectric interface and for migration along the GBs.

While simulating the migration of a void far from the interface, one finds that its velocity is inversely proportional to its radius. The dependence of the velocity on the size for surface voids increases monotonically, but is no longer linear. This may be considered to be the manifestation of the complex shape evolution and the corresponding changes in current distribution around the void. Therefore, we are not able to extrapolate the observed size dependence of velocity to dimensions in the range of 100 nm and larger.

Small (undercritical) surface voids can be trapped at GBs. A rough criterion of trapping is given by Equation 7.155. Detachment of overcritical voids from GBs proceeds with the formation of "fishlike" voids with an elongated "tail-like" structure. This is where the part of the void detaches. The "tail" part of the void remains at the GB, reducing the surface energy of the neighboring grains contact and serving as a nucleus of the new voids' growth and detachment.

Undercritical voids can either move along the GBs or grow consuming the vacancies and other voids. Having reached the overcritical size, they detach from the traps (GBs and GB junctions). After detachment, they leave residual voids that can grow further due to vacancy flux and other voids. In other words, the voids at the GB junctions can be treated as Franc–Reed sources.

The detachment from the GB junction is also possible (Equation 7.157). This mechanism does not lead to the formation of residual voids, and the whole volume of the emptiness moves toward either of the boundaries (with the smaller angle to the current direction).

Voids migrate faster along the GB/interface line and along the metal/dielectric interface than along the bulk/interface regions.

References

1. Aldinger, F. (1974) *Acta Metallurgica*, **22**, 923.
2. Geguzin, Y.E. (1979) *Diffusion zone*, Nauka Publishing, Moscow.
3. Geguzin, Y.E., Klinchuk, Y.I. and Paritskaya, L.N. (1977) *Fizika Metallov I Metallovedenie*, **43**, 602.
4. Yin, Y., Rioux, R.M., Erdonmez, C.K. et al. (2004) *Science*, **30430**, 711–714.
5. Yin, Y., Erdonmez, C.K., Cabot, A. et al. (2006) *Advanced Functional Materials*, **16**, 1389–1399.
6. Wang, C.M., Baer, D.R., Thomas, L.E. et al. (2005) *Journal of Applied Physics*, **98**, 094308.
7. Tu, K.N. and Gösele, U. (2005) *Applied Physics Letters*, **86**, 093111.
8. Gusak, A.M., Zaporozhets, T.V., Tu, K.N. and Gösele, U. (2005) *Philosophical Magazine*, **85**, 4445.
9. Nakamura, R., Tokozakura, D., Lee, J.-G. et al. (2008) *Acta Materialia*, **56**, 5276.
10. Gusak, A.M. and Tu, K.N. (2009) *Acta Materialia*, **57**, 3367.
11. Gusak, A.M. and Zaporozhets, T.V. (2009) *Journal of Physics-Condensed Matter*, **21**, 415303.
12. Vairagar, V., Mhaisalkar, S.G., Krishnamoorthy, A. et al. (2004) *Applied Physics Letters*, **85** (13), 2502.
13. Zaporozhets, T.V., Gusak, A.M., Tu, K.N. and Mhaisalkar, S.G. (2005) *Journal of Applied Physics*, **98**, 103508.
14. Lifshitz, I.M. and Slezov, V.V. (1958) *Sov Phys JETP*, **35**, 479.
15. Marwick, D. (1978) *Journal of Physics F: Metal Physics*, **8**, 1849.
16. Nazarov, A.V. and Gurov, K.P. (1973) *Fizika Metallov I Metallovedenie*, **37**, 496.
17. Gurov, K.P. and Gusak, A.M. (1985) *Fizika Metallov I Metallovedenie*, **59**, 1062.
18. Gusak, A.M. (1994) *Materials Science Forum*, **155-156**, 55.
19. Gusak, A.M. and Gurov, K.P. (1994) *Proceedings of PTM-94*, USA, p. 1133.
20. Evteev, A.V., Levchenko, E.V., Belova, I.V. and Murch, G.E. (2007) *Philosophical Magazine*, **87**, 3787.
21. Evteev, A.V., Levchenko, E.V., Belova, I.V. and Murch, G.E. (2008) *Defect and Diffusion Forum*, **277**, 21.
22. Evteev, A.V., Levchenko, E.V., Belova, I.V. and Murch, G.E. (2008) *Philosophical Magazine*, **88**, 1525.
23. Johnson, W.R.A. and White, P.J. (1978) *Physical Review B*, **18**, 2939.
24. Pabisiak, T. and Kiejna, A. (2007) *Solid State Communications*, **144**, 324.
25. Tu, K.N., Mayer, J.W. and Feldman, L.C. (1992) *Thin Film Science for Electrical Engineers and Materials Scientists*, Wiley, New York.
26. Gusak, A.M. (1989) *Powder metallurgy (Poroshkovaya Metallurgiya, USSR, Kiev)*, **3**, 39 (in Russian).

27. Gusak, A.M. and Lucenko, G.V. (**1998**) *Acta Materialia*, **46**, 3343.
28. Manning, J. (**1967**) *Acta Metallurgica*, **15**, 817.
29. Gusak, A.M. and Yarmolenko, M.V. (**1993**) *Journal of Applied Physics*, **73**, 4881.
30. Fan Hong, J., Knez, M., Scholz, R. *et al.* (**2007**) *Nano Letters*, **7**, 993.
31. Artz, E., Kraft, O., Nix, W.D. and Sanchez, J.E. (**1994**) *Journal of Applied Physics*, **76**, 1563.
32. Wang, W., Suo, Z. and Hao, T.-H. (**1996**) *Journal of Applied Physics*, **79**, 2394.
33. Hodaj, F., Gusak, A.M. and Desre, P.J. (**1998**) *Philosophical Magazine A*, **77**, 147.
34. Krivoglaz, M.A. and Osinowskiy, M.Y. (**1967**) *Fizika Metallov I Metallovedenie*, **24**, 36.
35. Ho, P.S. (**1970**) *Journal of Applied Physics*, **41**, 64.
36. Huntington, H.B. and Crone, A.R. (**1961**) *Journal of Physics and Chemistry of Solids*, **20**, 76.
37. Fiks, V.B. (**1959**) *Solid State Physics*, **3**, 16 (in Russian).
38. Mehl, H., Biham, O., Millo, O. and Karimi, M. (**2000**) *Physical Review B*, **61**, 4975.
39. Gusak, A.M. (**2004**) *Diffusion, Reactions, Coarsening – Some New Ideas*, Cherkasy National University Publ., Cherkasy.

8
Phase Formation via Electromigration
Semen V. Kornienko and Andriy M. Gusak

8.1
Introduction

In recent years the interest in phase formation at high current densities has been renewed. This interest is mainly related to flip-chip technology usage in electronics. In this technology, a spatial matrix of solder small balls for joining the microchips with their substrate is applied [1–5]. This technology meets the requirements of modern microelectronics development: miniaturizing integrated circuits, increasing their power, and strengthening their junction solidity. For good wetting between solder balls and the substrate, the formation of IMCs is of great importance. It can also influence the electric and stress–strain properties of compounds. Therefore, at the present time, an intensive research of Cu-solder and Ni-solder reactions at a current density of 10^8 A m^{-2} and higher is carried out [6, 7]. While carrying out such investigations, many equations obtained by Gurov and Gusak [8] were "discovered anew."

Diffusion processes taking place in such systems have to be regarded as processes that take place in thin films, and not in bulk samples. Reactive diffusion in thin films differs from that in the bulk case: IMCs are formed in sequence ("one by one"), so that the phase spectrum of the diffusion zone differs from the full list of stable intermediate phases. Besides, the growth of new phase layers very often demonstrates a linear time dependence rather than a parabolic one. The best known theory providing an explanation for this fact was offered by Gösele and Tu [9]. However, it was shown that the linear stage of intermetallic layer growth may also be caused by a finite relaxation rate of nonequilibrium vacancies at these interfaces [10].

Quite long ago it was experimentally proved that imposing a direct current through the reaction zone in the processes of diffusion may affect not only the kinetics of the phase growth but the phase composition of a zone as well [11]. The first model (to the best of our knowledge) of phase formation and intermetallic phase competition under current stressing was offered by Gurov and Gusak [8], and in the course of time it was extended to include the initial stages of phase formation [12] (Section 8.2). In these papers, the finite rate of overcoming the interface barrier

Diffusion-Controlled Solid State Reactions. Andriy M. Gusak
Copyright © 2010 WILEY-VCH Verlag GmbH & Co. KGaA, Weinheim
ISBN: 978-3-527-40884-9

was not taken into consideration. Since in thin film systems the initial stages of phase formation are extremely important, and the finite rate of interface reactions are essential, logically it would be appropriate to combine the Gurov–Gusak model and Gösele–Tu theories. This task was recently performed by Orchard and Greer [13] (Section 8.3).

To analyze phase formation at direct currents (DC) imposed on a binary system (Section 8.4), we consider nonequilibrium vacancies and electric current effects simultaneously. Such an approach was first applied to interdiffusion processes in an electric field in solid solutions by Gurov and Gusak [14].

8.2
Theory of Phase Formation and Growth in the Diffusion Zone at interdiffusion in an External Electric Field

Phase formation and growth in the system Au–Al in an external electric field was studied in [15, 16]. The magnitude and direction of the current were proved to have a strong influence on the phase composition of the diffusion zone, and on the growth kinetics of phase layers. The influence of an electric field on phase formation in the diffusion zone was also found in the CaO-Al_2O_3 system [17].

While studying the influence of an external electric field on phase formation, one should distinguish three stages:

1) kinetic stage of nuclei formation;
2) diffusion interaction of critical nuclei between themselves and with already-growing phase layers;
3) growth of the phase layers (purely diffusion stage of the process).

The effect of an external field at the nucleation stage could cause a change in the size of the critical nuclei, as during new phase formation the density of the electric field energy, entering into the expression for the total free energy of a phase, is changed. However, simple estimates prove the contribution of the electric field to the free energy to be quite insignificant, when speaking about metallic systems. Besides, the mentioned effect cannot depend on the field direction and therefore fails to explain the dependence of phase composition and growth kinetics on polarity. We study the influence of external fields at the two last stages. We start from the third one, which is purely planar diffusion and describes the case when thermodynamically allowed phases have already appeared and continue growing in the diffusion zone.

8.2.1
External Field Effects on Intermetallic Compounds Growth at Interdiffusion

The external electric field makes a certain direction preferable for atomic random walk, which is formally expressed by the additional "drift" component in the interdiffusion equation. Let us write down the equations for fluxes in IMC for two

8.2 Phase Formation and Growth in an External Electric Field

components and vacancies in the lattice reference frame taking the effect of the electric field into account. For simplicity, the components are assumed to have (approximately) the same atomic volumes,

$$\Omega j_i = -D_i \frac{\partial c_i}{\partial x} + \frac{c_i D_i^*}{kT} E e_i \quad (i = 1, 2)$$

$$\Omega j_V = (D_1 - D_2) \frac{\partial c_1}{\partial x} - \frac{E}{kT} \left(c_1 D_1^* e_1 + c_2 D_2^* e_2 \right) \quad (8.1)$$

where E is the projection of the electric field intensity over axis OX; e_1 and e_2 are the effective ionic charges of the components, which tends to be negative due to "electron wind" at electromigration; c_1 and c_2 are the concentrations (averaged values) in IMC (homogeneous alloy of a given composition) of atoms; D_1, D_2 are the intrinsic diffusion coefficients of atoms in IMC; D_1^*, D_2^* are the tracer diffusion coefficients of atoms in IMC, respectively.

Let us write down the equation for the fluxes of the components in the laboratory reference frame as

$$\Omega I_i = \Omega j_i + c_i v = \Omega \left(j_i + c_i j_V \right) \quad (i = 1, 2)$$

where v is the velocity of lattice flow (Kirkendall effect under the influence of the electric field). According to conservation laws, the divergence of the sum of the two fluxes should be equal to zero. It means, in the one-dimensional case, that $I_1 + I_2 = \text{const}$ holds along the diffusion couple. If at least one of the margins of this couple is fixed, and is far from the diffusion zone, then this constant should be equal to zero. From this equation and bearing the expressions for j_1 and j_V (Equation 8.1) in mind we arrive at

$$\Omega I_1 = -\Omega I_2 = -\tilde{D} \frac{\partial c}{\partial x} + c(1-c)B$$

$$B = \frac{E}{kT} \left(e_1 D_1^* - e_2 D_2^* \right) = \frac{j}{kT} \rho \left(e_1 D_1^* - e_2 D_2^* \right) \quad (8.2)$$

where I_1 and I_2 are the volume flux densities of the components, $\tilde{D}$ the interdiffusion coefficient, ρ the resistance, and j the current density. Note that the sign of B depends not only on the field direction E but also on the sign of the difference, $e_1 D_1^* - e_2 D_2^*$, and, thus, it can change from one phase to another. Further, we have

$$\frac{\partial c}{\partial t} = \frac{\partial}{\partial x} \left(\tilde{D} \frac{\partial c}{\partial x} \right) - \frac{\partial}{\partial x} \left[c(1-c) B \right] \quad (8.3)$$

(Equation 8.3) is derived from the general continuity equation, into which the densities of components fluxes at interdiffusion (Equation 8.2) are substituted.

First, let us consider the case of one phase growth, the phase having a narrow concentration range $\Delta c = c_R - c_L \ll 1$, at annealing a couple of mutually insoluble metals in the presence of an electric field. We neglect the change of the molar volume during IMC formation, and thereby from the very beginning disregard the effects connected with stresses arising at interfaces. Besides, we consider the condition of steady state diffusion flux over the phase thickness to be satisfied. This assumption will allow us to escape the necessity of solving the nonlinear equation (Equation 8.2) concerning diffusion inside the IMC layer.

Then, flux balance conditions at interfaces yield the following two differential equations:

$$(1 - c_R) \frac{dx_R}{dt} = D \frac{\Delta c}{\Delta x} - B c_R (1 - c_R) \quad (8.4)$$

$$c_L \frac{dx_L}{dt} = -D \frac{\Delta c}{\Delta x} + B c_L (1 - c_L) \quad (8.5)$$

The system of equations (Equations 8.4 and 8.5) turned out to be rather simple, since due to components' insolubility the fluxes are equal to zero outside the growing phase $\partial c/\partial x = 0$, $c(1 - c) = 0$ at $c = 0$ and $c = 1$. From Equations 8.4 and 8.5 one can easily derive the differential equation for the phase thickness $\Delta x(t)$

$$\frac{d\Delta x}{dt} = \frac{a}{\Delta x(t)} - b \quad (8.6)$$

where

$$a = \frac{1 - \Delta c}{(1 - c_R) c_L} D \Delta c \approx \frac{D \Delta c}{c(1 - c)} \quad (8.7)$$

and

$$b = B(1 + \Delta c) \approx B \quad (8.8)$$

The solution of Equation 8.6 is

$$-\frac{b}{a} \Delta x - \ln \left| 1 - \frac{b}{a} \Delta x \right| = \frac{b^2}{a} t \quad (8.9)$$

At small times t and corresponding small thickness ($\Delta x < a/b$), Equation 8.9 is reduced to a parabolic dependence

$$(\Delta x)^2 \approx 2at \quad (8.10)$$

In order to prove this statement, it is sufficient to expand the logarithm into a Taylor series, having preserved the square in Δx terms.

When t becomes higher, the behavior of the function $\Delta x(t)$ strongly depends on the sign of b, that is on the field direction (Figure 8.1). If $b < 0$, later the parabolic growth gradually turns into a linear one at a constant growth rate

$$\frac{d\Delta x}{dt} = -b = |b| \approx |B| = \frac{|E|}{kT} |e_1 D_1^* - e_2 D_2^*| \quad (8.11)$$

If $b > 0$, then the growth of the phase slows down with time and eventually stops, so that at $t \to \infty$ the phase thickness becomes

$$\Delta x = \frac{a}{b} \approx \frac{1}{c(1-c)} \frac{D \Delta c}{B} \quad (8.12)$$

So, whether the phase succeeds in growth or stops growing at all depends on the field direction.

From a physical viewpoint, the mechanism of suppression of phase growth by an electric field is quite clear. As the phase thickness increases, the concentration

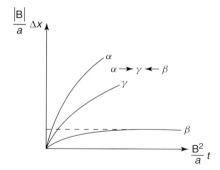

Figure 8.1 Phase thickness dependence on time for different directions of the current: $b < 0$ (α); $b > 0$ (β); $b = 0$ (γ).

gradient in it is reduced, the interdiffusion process slows down, and, at $t \gg a/b^2$, it becomes compensated by electromigration (at a direction of electromigration being opposite to that of *purely* diffusional flux, of course). Compensation occurs due to the drift flux, which, unlike the diffusion flux, is independent of the phase thickness Δx.

Next, let us consider the case of two intermetallic phases 1 and 2 growing between mutually insoluble metals. Equations 8.1 and 8.2 with diffusion coefficients $D^{(1)}$, $D^{(2)}$ and drift components $B^{(1)}$, $B^{(2)}$ remain correct for these intermetallic phases. Again, we employ the assumption of steady state fluxes over the thickness of each phase

$$I_{\text{dif}}^{(1)} = -\frac{D^{(1)} \Delta^{(1)} c}{\Delta^{(1)} x} \qquad I_{\text{dif}}^{(2)} = -\frac{D^{(2)} \Delta^{(2)} c}{\Delta^{(2)} x} \tag{8.13}$$

where $I_{\text{dif}}^{(1)}$ and $I_{\text{dif}}^{(2)}$ are the fluxes of the components per one particle. The balance equations for the number of particles at interfaces have the following form:

$$\begin{aligned}
\left(1 - c_R^{(2)}\right) \frac{dx_R^{(2)}}{dt} &= D^{(2)} \frac{\Delta^{(2)} c}{\Delta^{(2)} x} - B^{(1)} c_R^{(2)} \left(1 - c_R^{(2)}\right) \\
\left(c_L^{(2)} - c_R^{(1)}\right) \frac{dx_L^{(2)}}{dt} &= -D^{(2)} \frac{\Delta^{(2)} c}{\Delta^{(2)} x} + B^{(2)} c_L^{(2)} \left(1 - c_L^{(2)}\right) \\
&\quad + D^{(1)} \frac{\Delta^{(1)} c}{\Delta^{(1)} x} - B^{(1)} c_R^{(1)} \left(1 - c_R^{(1)}\right) \\
c_L^{(1)} \frac{dx_L^{(1)}}{dt} &= -D^{(1)} \frac{\Delta^{(1)} c}{\Delta^{(1)} x} + B^{(1)} c_L^{(1)} \left(1 - c_L^{(1)}\right)
\end{aligned} \tag{8.14}$$

From Equation 8.14, one can easily obtain the system of differential equations for the phase thicknesses:

$$\begin{aligned}
\frac{d\Delta^{(1)} x}{dt} &= a_{11} \frac{D^{(1)} \Delta^{(1)} c}{\Delta^{(1)} x} + a_{12} \frac{D^{(2)} \Delta^{(2)} c}{\Delta^{(2)} x} - b_1 \\
\frac{d\Delta^{(2)} x}{dt} &= a_{21} \frac{D^{(1)} \Delta^{(1)} c}{\Delta^{(1)} x} + a_{22} \frac{D^{(2)} \Delta^{(2)} c}{\Delta^{(2)} x} - b_2
\end{aligned} \tag{8.15}$$

In the accepted approximation (Equation 8.13) and taking into account the smallness of $\Delta^{(1)}c$ and $\Delta^{(2)}c$, the coefficients in Equation 8.15 are defined as

$$a_{ik} = \begin{pmatrix} \frac{c^{(2)}}{c^{(1)}} - 1 & -1 \\ -1 & \frac{1 - c^{(1)}}{1 - c^{(2)}} \end{pmatrix} \frac{1}{c^{(2)} - c^{(1)}}$$

$$b_1 = c^{(2)} \frac{B^{(1)}\left(1 - c^{(1)}\right) - B^{(2)}\left(1 - c^{(2)}\right)}{c^{(2)} - c^{(1)}}$$

$$b_2 = \left(1 - c^{(1)}\right) \frac{B^{(2)} c^{(2)} - B^{(1)} c^{(1)}}{c^{(2)} - c^{(1)}} \tag{8.16}$$

It is impossible to solve the system of nonlinear differential equations (Equation 8.15) analytically. Therefore, we will restrict ourselves to studying the asymptote of the solutions at $t \to \infty$. The behavior of $\Delta^{(1)}c$ and $\Delta^{(2)}c$ strongly depends on the signs of $B^{(1)}$ and $B^{(2)}$, and on their ratio. Let us look at all possible cases:

1) $B^{(1)} > 0$, $B^{(2)} > 0$.
   In such a case the growth of both phases stops with time as they reach their maximal thickness.

$$\Delta^{(1)} x_{max} = \frac{\det a_{ik}}{\det \begin{pmatrix} b_1 & a_{12} \\ b_2 & a_{22} \end{pmatrix}} = \frac{1}{c^{(1)}(1 - c^{(1)})} \frac{D^{(1)} \Delta^{(1)} c}{B^{(1)}} \tag{8.17}$$

$$\Delta^{(2)} x_{max} = \frac{\det a_{ik}}{\det \begin{pmatrix} a_{11} & b_1 \\ a_{21} & b_2 \end{pmatrix}} = \frac{1}{c^{(2)}(1 - c^{(2)})} \frac{D^{(2)} \Delta^{(2)} c}{B^{(2)}}$$

2) $B^{(1)} < 0$, $B^{(2)} < 0$.
   This case is obtained by changing the direction of current. Still, fixed "minus" signs and $B^{(1)}$ and $B^{(2)}$ magnitudes do not provide an unambiguous solution. The ratio between them also appears to be significant. Indeed, the terms b_1 and b_2 that enter into Equation 8.15 may have different signs, depending on the value of the ratio $B^{(1)}/B^{(2)}$.
   (a)

$$b_1 < 0 \qquad b_2 < 0 \qquad \left(\frac{c^{(1)}}{c^{(2)}} < \frac{B^{(2)}}{B^{(1)}} < \frac{1 - c^{(1)}}{1 - c^{(2)}}\right)$$

Here both phases grow infinitely, and the growth becomes linear at $t \to \infty$.

$$\left.\frac{d\Delta^{(1)} x}{dt}\right|_{t \to \infty} = -b_1 = |B^{(1)}| \frac{c^{(2)}\left(1 - c^{(1)}\right)}{c^{(2)} - c^{(1)}} \left[1 - \frac{B^{(2)}}{B^{(1)}} \frac{1 - c^{(2)}}{1 - c^{(1)}}\right] > 0$$

$$\left.\frac{d\Delta^{(2)} x}{dt}\right|_{t \to \infty} = -b_2 = |B^{(2)}| \frac{c^{(2)}\left(1 - c^{(1)}\right)}{c^{(2)} - c^{(1)}} \left[1 - \frac{B^{(1)}}{B^{(2)}} \frac{c^{(1)}}{c^{(2)}}\right] > 0 \tag{8.18}$$

The *velocities* of boundaries movement also approach constant values

$$\frac{dx_L^{(1)}}{dt} = v_L^{(1)} = -|B^{(1)}|\left(1 - c^{(1)}\right)$$

$$\frac{dx_R^{(1)}}{dt} = v_R^{(1)} = v_L^{(2)} = \frac{|B^{(1)}| c^{(1)} \left(1 - c^{(1)}\right) - |B^{(2)}| c^{(2)} \left(1 - c^{(2)}\right)}{c^{(2)} - c^{(1)}}$$

$$\frac{dx_R^{(2)}}{dt} = v_R^{(2)} = |B^{(2)}| c^{(2)}$$

Obviously,

$$\frac{(v_L^{(1)} + v_R^{(1)})}{2} \neq 0 \qquad \frac{(v_L^{(2)} + v_R^{(2)})}{2} \neq 0$$

that is, phases, while extending, simultaneously perform translational motion.

(b) $b_1 > 0$, $b_2 < 0$.

At $t \to \infty$, phase 2 continues growing while the growth of phase 1 stops. The system of equations (Equation 8.15) is then reduced to the following two equations:

$$\left.\frac{d\Delta^{(1)}x}{dt}\right|_{t\to\infty} = 0 = \frac{a_{11}}{\Delta^{(1)}x} - b_1$$

$$\left.\frac{d\Delta^{(2)}x}{dt}\right|_{t\to\infty} = -\frac{|a_{21}|}{\Delta^{(2)}x} + |b_2| \qquad (8.19)$$

From Equation 8.19, we obtain the maximal thickness of the first phase

$$\Delta^{(1)}x_{max} = \frac{a_{11}}{b_1} = \frac{\left(D^{(1)}\Delta^{(1)}c\right)}{|B^{(2)}|(1-c^{(2)}) - |B^{(1)}|(1-c^{(1)})} \qquad (8.20)$$

and the growth rate of the second phase at $t \to \infty$ as

$$\left.\frac{d\Delta^{(2)}x}{dt}\right|_{t\to\infty} = |b_2| - b_1 \frac{|a_{21}|}{a_{11}} = |B^{(2)}| \qquad (8.21)$$

Note that phase 1, having reached the maximal thickness, continues moving as a whole at the following velocity:

$$v^{(1)} = -|B^{(2)}|(1-c^{(2)}) \qquad (8.22)$$

The growing phase 2 in some way "repels" the layer of phase 1, which, moving at constant velocity, just lets through itself the material required for growth of phase layer 2, and its own thickness remains unchanged. The notion of such a phase "repulsion" effect was introduced by Gurov et al. [18, 19]. The total thickness of the phases, $\Delta^{(1)}x + \Delta^{(2)}x$, must increase. Here, the negative sign of both coefficients $B^{(1)}$ and $B^{(2)}$ proves to be important.

In the absence of phase 2, at $B^{(1)} < 0$, phase 1 would constantly grow. So, the presence of another phase, for which

$$B^{(2)} < 0 \qquad \frac{B^{(2)}}{B^{(1)}} > \frac{1-c^{(1)}}{1-c^{(2)}}$$

has the following effect on the first phase: the latter stops growing at $t \to \infty$ despite $B^{(1)}$ being negative, but due to interaction with the other phase it moves as a whole.

(c) $b_1 < 0$, $b_2 > 0$.

At $t \to \infty$, phase 1 continues growing, and phase 2 reaches a maximal thickness

$$\Delta^{(2)} x_{max} = \frac{\left(\frac{D^{(2)} \Delta^{(2)} c}{1-c^{(2)}}\right)}{|B^{(1)}| c^{(1)} - |B^{(2)}| c^{(2)}} \tag{8.23}$$

but continues moving at constant velocity

$$v^{(2)} = v_L^{(2)} = v_R^{(2)} = |B^{(1)}| c^{(1)} \tag{8.24}$$

3) $B^{(1)} > 0$, $B^{(2)} < 0$.

In this case,

$$b_1 = \frac{c^{(2)}}{c^{(2)} - c^{(1)}} \left[B^{(1)} \left(1 - c^{(1)}\right) + |B^{(2)}| \left(1 - c^{(2)}\right) \right] > 0$$

$$b_2 = -\frac{1 - c^{(2)}}{c^{(2)} - c^{(1)}} \left[|B^{(2)}| c^{(2)} + |B^{(1)}| c^{(1)} \right] < 0$$

Thus, the form of the asymptote is the same as in 2b: phase 2 grows at a velocity equal to (at $t \to \infty$)

$$\left.\frac{d\Delta^{(2)} x}{dt}\right|_{t \to \infty} = |b_2| - b_1 \frac{|a_{21}|}{a_{11}} = |B^{(2)}| \tag{8.25}$$

phase 1 reaches maximal thickness

$$\Delta^{(1)} x_{max} = \frac{a_{11}}{b_1} = \frac{\left(\frac{D^{(1)} \Delta^{(1)} c}{c^{(1)}}\right)}{B^{(1)} \left(1 - c^{(1)}\right) + |B^{(2)}| \left(1 - c^{(2)}\right)} \tag{8.26}$$

and then moves as a whole at a velocity

$$v^{(1)} = v_L^{(1)} = v_R^{(1)} = |B^{(2)}| \left(1 - c^{(2)}\right) \tag{8.27}$$

4) $B^{(1)} < 0$, $B^{(2)} > 0$.

This case can be obtained if the direction of current in the system is changed into *the opposite one* compared to case 3. Phase 1 grows, phase 2 reaches maximal thickness, and then moves as a whole.

The obtained results are convenient to present in the form of a diagram (Figure 8.2). The axes correspond to $B^{(1)}$ and $B^{(2)}$, expressed as products of current density j in the sample and the value β, whose sign depends on the values of the parameters of the system, that is,

$$\beta_i = \frac{\rho^{(i)}}{kT} \left(e_1^{(i)} D_1^{(i)} - e_2^{(i)} D_2^{(i)} \right) \tag{8.28}$$

A similar analysis can be performed for a system containing three intermediate phases 1, 2, and 3. In this case, the system of kinetic equations for phase thicknesses is written down in the following way:

$$\frac{d\Delta^{(i)} x}{dt} = \sum_{k=1}^{3} \left[a_{ik} \frac{D^{(k)} \Delta^{(k)} c}{\Delta^{(k)} x} - \gamma_{ik} B^{(k)} \right] \tag{8.29}$$

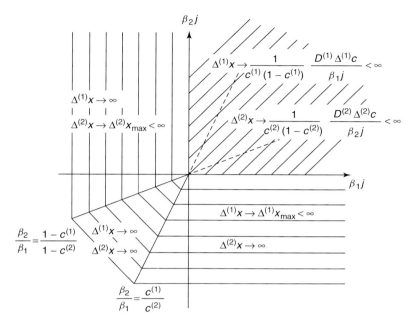

Figure 8.2 The behavior of the intermediate phases in the A-1-2-B system at $t \to \infty$, depending on the current direction and on the difference of components' mobilities in the phases.

$$a_{ik} = \begin{pmatrix} \frac{c^{(2)}}{c^{(1)}} \frac{1}{c^{(2)}-c^{(1)}} & -\frac{1}{c^{(2)}-c^{(1)}} & 0 \\ -\frac{1}{c^{(2)}-c^{(1)}} & \frac{c^{(3)}-c^{(1)}}{\left(c^{(3)}-c^{(2)}\right)\left(c^{(2)}-c^{(1)}\right)} & -\frac{1}{c^{(3)}-c^{(2)}} \\ 0 & -\frac{1}{c^{(3)}-c^{(2)}} & \frac{1-c^{(2)}}{\left(1-c^{(3)}\right)\left(c^{(3)}-c^{(2)}\right)} \end{pmatrix} \quad (8.30)$$

$$\gamma_{ik} = \begin{pmatrix} \frac{\left(1-c^{(1)}\right)c^{(2)}}{c^{(2)}-c^{(1)}} & -\frac{c^{(2)}\left(1-c^{(2)}\right)}{c^{(2)}-c^{(1)}} & 0 \\ -\frac{c^{(1)}\left(1-c^{(1)}\right)}{c^{(2)}-c^{(1)}} & \frac{c^{(2)}\left(1-c^{(2)}\right)\left(c^{(3)}-c^{(1)}\right)}{\left(c^{(3)}-c^{(2)}\right)\left(c^{(2)}-c^{(1)}\right)} & -\frac{c^{(3)}\left(1-c^{(3)}\right)}{c^{(3)}-c^{(2)}} \\ 0 & -\frac{c^{(2)}\left(1-c^{(2)}\right)}{c^{(3)}-c^{(2)}} & \frac{c^{(3)}\left(1-c^{(2)}\right)}{c^{(3)}-c^{(2)}} \end{pmatrix} \quad (8.31)$$

8.2.2
Criteria for Phase Suppression and Growth in an External Field

In the previous sections, we analyzed the growth of phases that have already passed the nucleation stage. Now, let us consider the preceding, second phase of the process when the imposed electric field changes the diffusion interaction both between critical nuclei and between a nucleus and a growing phase. As a consequence, criteria for phase suppression and growth must change.

Let two intermediate phases with narrow homogeneity ranges $\Delta^{(1)}c, \Delta^{(2)}c \ll 1$ exist between insoluble metals A and B on the phase diagram. Taking into account the results obtained in Chapter 3, the rates for phase growth from critical nuclei are equal to

$$\left.\frac{d\Delta^{(1)}x}{dt}\right|_{l_{kp}} = \frac{1}{c^{(2)} - c^{(1)}} \left(\frac{c^{(2)}}{c^{(1)}} q_1 - q_2\right) - b_1 \tag{8.32}$$

$$\left.\frac{d\Delta^{(2)}x}{dt}\right|_{l_{kp}} = \frac{1}{c^{(2)} - c^{(1)}} \left(-q_1 + \frac{1 - c^{(1)}}{1 - c^{(2)}} q_2\right) - b_2 \tag{8.33}$$

respectively, where

$$q_i = \frac{D^{(i)} \Delta^{(1)} c}{l_{kp}^{(i)}} \tag{8.34}$$

$$b_1 = c^{(2)} \frac{B^{(1)}\left(1 - c^{(1)}\right) - B^{(2)}\left(1 - c^{(2)}\right)}{c^{(2)} - c^{(1)}} \tag{8.35}$$

$$b_2 = \left(1 - c^{(1)}\right) \frac{B^{(2)} c^{(2)} - B^{(1)} c^{(1)}}{c^{(2)} - c^{(1)}} \tag{8.36}$$

$$B^{(i)} = \frac{eE^{(i)}}{kT} \left(z_1^{(i)} D_1^* - z_2^{(i)} D_2^*\right) = \frac{ej}{kT} \left[\rho\left(z_1 D_1^* - z_2 D_2^*\right)\right]^{(i)} \tag{8.37}$$

It is convenient to present Equation 8.32 in the following form:

$$\left.\frac{d\Delta^{(1)}x}{dt}\right|_{l_{kp}} = \frac{1}{(c^{(2)} - c^{(1)})} \left\{(q_1 - \alpha_1 j) - \frac{c^{(1)}}{c^{(2)}}(q_2 - \alpha_2 j)\right\}$$

$$\left.\frac{d\Delta^{(2)}x}{dt}\right|_{l_{kp}} = \frac{1}{c^{(2)} - c^{(1)}} \left\{-(q_1 - \alpha_1 j) + \frac{1 - c^{(1)}}{1 - c^{(2)}}(q_2 - \alpha_2 j)\right\} \tag{8.38}$$

where

$$\alpha_i = \frac{c^{(i)}\left(1 - c^{(i)}\right)}{j} B^{(i)} = \frac{c^{(i)}\left(1 - c^{(i)}\right)}{kT} \left[\rho\left(e_1 D_1^* - e_2 D_2^*\right)\right]^{(i)} \tag{8.39}$$

In order to define the criteria of phase suppression and growth, one should, as in Section 3.2, study all possible combinations of signs for the quantities $(d\Delta^{(1)}x/dt)_{lcr}$ and $(d\Delta^{(2)}x/dt)_{lcr}$. A positive sign of $(d\Delta^x/dt)_{lcr}$ means the growth of phase layer starting from the critical nuclei size, and the negative one yields suppression of phase growth.

The system of equations (Equation 8.38) for $(d\Delta^x/dt)_{lcr}$ differs from the corresponding system when an external field is absent, by the substitution of q_1 by $q_1(j) = q_1 - \alpha_1 j$ and q_2 by $q_2(j) = q_2 - \alpha_2 j$. In the absence of the field, diffusion phase competition is defined by the relation

$$\frac{q_1}{q_2} = \frac{D^{(1)} \Delta^{(1)} c \, l_{kp}^{(2)}}{D^{(2)} \Delta^{(2)} c \, l_{kp}^{(1)}}$$

yet, an external field being present, the result of the competition depends not only on diffusion constants but also on the magnitude and the direction of the current

$$\frac{q_1(j)}{q_2(j)} = \frac{q_1 - \alpha_1 j}{q_2 - \alpha_2 j}$$

Thus, changing the magnitude and/or the direction of the current, one can turn from the suppression of one phase to another, or to the growth of both. Note that we still consider the stage of diffusion interaction between two nuclei.

In the next section, we investigate the changes that occur in the later stages, as some phases already grow and suppress the nuclei of other phases. We do not provide all possible cases here, rather we analyze some characteristic situations:

1) In the absence of the field, simultaneous suppression of both phases turned out to be impossible. An imposed field makes the situation different. The condition for simultaneous suppression of both phases is as follows:

$$\left.\frac{d\Delta^{(1)}x}{dt}\right|_{l_{kp}} < 0 \Rightarrow \begin{cases} q_1(j) < \frac{c^{(1)}}{c^{(2)}} q_2(j) \\ q_1(j) > \frac{1-c^{(1)}}{1-c^{(2)}} q_2(j) \end{cases} \quad (8.40)$$

Since

$$\frac{c^{(1)}}{c^{(2)}} < 1 < \frac{(1 - c^{(1)})}{(1 - c^{(2)})}$$

the set of inequalities (Equation 8.40) can be consistent for negative $q_1(j)$ and $q_2(j)$ only, when $|j| > q_1/|\alpha_1|$, $|j| > q_2/|\alpha_2|$. So, the condition of simultaneous suppression can be rewritten in the following form:

$$\frac{c^{(1)}}{c^{(2)}} < \frac{\alpha_1 j - q_1}{\alpha_2 j - q_2} < \frac{1 - c^{(1)}}{1 - c^{(2)}} \quad (8.41)$$

From Equation 8.41, we realize that simultaneous suppression of phases can take place provided the current is rather intensive and the parameters α_1 and α_2 have the same signs (that is, the component which is more mobile in one phase retains its higher mobility in the other phase).

2) Let both phases grow simultaneously when there is no field, that is,

$$\frac{c^{(1)}}{c^{(2)}} < \frac{q_1}{q_2} < \frac{1 - c^{(1)}}{1 - c^{(2)}} \quad (8.42)$$

and suppose $\alpha_1, \alpha_2 > 0$.

$$\text{At } j \to \frac{q_1}{\alpha_1} : \frac{q_1(j)}{q_2(j)} \to 0 < \frac{c^{(1)}}{c^{(2)}}. \text{At } j \to \frac{q_2}{\alpha_2} : \frac{q_1(j)}{q_2(j)} \to \infty > \frac{(1 - c^{(1)})}{(1 - c^{(2)})}$$

In such a way, having imposed currents of different intensity, we are able to attain the total inhibition of phase 2 growth, in one case, and the growth of phase 1, in another. A similar effect was observed when the current had been imposed onto a diffusion couple Au–Al [15, 16]. To visualize this, the diagram

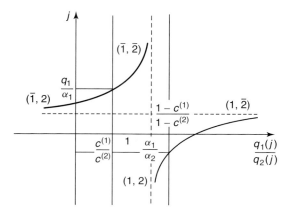

Figure 8.3 Suppression/growth criteria for the phases in the presence of an electric field.

of the $q_1(j)/q_2(j)$ dependence on the current density j is sketched (Figure 8.3) for the cases of

$$\frac{q_1}{q_2} > \frac{1-c^{(1)}}{1-c^{(2)}} \qquad \alpha_1 > 0 \quad \alpha_2 > 0 \qquad \frac{c^{(1)}}{c^{(2)}} < \frac{\alpha_1}{\alpha_2} < \frac{1-c^{(1)}}{1-c^{(2)}}$$

From the figure it is seen that phase composition of the diffusion zone at the initial stage will be different, depending on the magnitude and direction of the field.

8.2.3
Effect of an External Field on the Incubation Time of a Suppressed Phase

Now, let us find out the time dependence of suppression and growth effects in the presence of a field. In the case of two phases growing from the very beginning, the kinetics of their growth, depending on the magnitude and direction of the field, is discussed in Section 8.2.1. In case of simultaneous suppression of both phases they remain suppressed during the whole process of annealing. Thus, there is one case left that describes the situation when at the initial stage one phase (say, phase 1) grows, and the other (phase 2) is suppressed. As it is shown in Section 3.3, when the field is absent, after a certain period $\tau^{(2)}$ (the incubation time), the suppression conditions for phase 2 do not affect the evolution, and the growth of phase 2 begins as a result of diffusion. It is caused by phase 1 having reached a certain thickness, at which the value $(d\Delta^{(2)}x/dt)_{lcr}$ turns from negative into positive. In the presence of an electric field, the conditions for not to suppress the growth of phase 2 will slightly change (Figure 8.4)

$$\frac{d\Delta^{(2)}x}{dt}\bigg|_{l_{cr}} = \frac{1}{c^{(2)} - c^{(1)}} \left\{ -\left[\frac{D^{(1)}\Delta^{(1)}c}{\Delta^{(1)}x} - \alpha_1 j \right] \right.$$
$$\left. + \frac{1-c^{(1)}}{1-c^{(2)}} \left[\frac{D^{(2)}\Delta^{(2)}c}{l_{cr}^{(2)}} - \alpha_2 j \right] \right\} = 0 \qquad (8.43)$$

8.2 Phase Formation and Growth in an External Electric Field

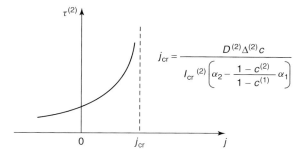

Figure 8.4 Dependence of the incubation time for the suppressed phase on the current density.

that is,

$$\Delta^{(1)} x_{cr} = \frac{D^{(1)} \Delta^{(1)} c}{\frac{1-c^{(1)}}{1-c^{(2)}} \frac{D^{(2)} \Delta^{(2)} c}{l_{cr}^{(2)}} + \left(\alpha_1 - \frac{1-c^{(1)}}{1-c^{(2)}} \alpha_2 \right) j} \tag{8.44}$$

As it is evident from the latter relation, the condition of suppression fails, and therefore, the incubation time depends on the density of the current imposed (both on its magnitude and direction). In particular, if

$$j \to \frac{D^{(2)} \Delta^{(2)} c}{l_{cr}^{(2)} \left(\alpha_2 - \frac{1-c^{(2)}}{1-c^{(1)}} \alpha_1 \right)} \tag{8.45}$$

then $\Delta^{(1)} x_{cr}$ in Equation 8.44 tends to infinity, and, therefore, $\tau^{(2)} \to \infty$. Thus, the presence of an electric field can increase the incubation time for some phases so that they will not be observed in the diffusion zone at any time of annealing.

8.2.4
Conclusions

It has been shown that the kinetics of growth and interaction between the phases strongly depends on the external field Equation 8.1. An imposed external field changes the suppression/growth criteria. The incubation time for the suppressed phases is influenced by the magnitude and direction of the current, and, in some cases, may appear to be infinite. The theory introduced here explains the results of investigations concerning the effects the current regime has upon phase composition of the diffusion zone and the kinetics of phase layers growth in the Au–Al system. However, it disregards heat release at the imposed current in an explicit form, as the current was considered to have an influence only on the temperature of the diffusion zone. Cross-effects between heat transfer and diffusion were not taken into account either. Such effects, basically, can influence the kinetics of phase growth , but fail to explain the dependence of phase composition and growth kinetics on the direction of the current.

8.3
Effects of Electromigration on Compound Growth at the Interfaces

The theory proposed by Orchard and Greer [13] is briefly sketched in this section. In a binary system, containing A and B atoms, let us consider the growth of one phase at a planar interface between the initial phases α (based on A) and γ (based on B) (Figure 8.5a). Suppose the phases α and γ, which can be boundary solid solutions of A and B, are saturated, that is, of constant composition without A/B interdiffusion inside them. The growth of a β layer takes place due to the attachment of A atoms passing through it from α to γ and B atoms moving into the opposite direction. Interdiffusion is considered in the laboratory reference frame where the fluxes of A and B particles are equal in magnitude and directed opposite. For simplicity, A and B are taken to have equal atomic volumes. Interdiffusion is driven by the difference between boundary concentrations $c_{\beta\alpha}$ and $c_{\beta\gamma}$ (Figure 8.5b). In the framework of the Gösele and Tu approach [9], one should take into account the finite reaction rates at the interfaces. It means that boundary concentrations are shifted when compared to their equilibrium values $c_{\beta\alpha}^{eq}$ and $c_{\beta\gamma}^{eq}$.

The flux of A atoms j_β^A for each reaction, which must be equal to the diffusion flux through the phase, is described by the following equation:

$$j_\beta^A = k_{\beta\alpha}\left(c_{\beta\alpha}^{eq} - c_{\beta\alpha}\right) = k_{\beta\gamma}\left(c_{\beta\gamma} - c_{\beta\gamma}^{eq}\right) \tag{8.46}$$

where $k_{\beta\alpha}$ and $k_{\beta\gamma}$ stand for reaction constants (Gösele and Tu terms) [9]. The diffusion flux through the phase is proportional to the concentration gradient

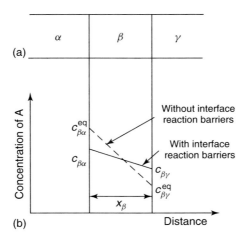

Figure 8.5 a) The β phase appears in the form of a continuous layer at the α and γ interface. b) Concentration profile of A component in the β phase.

(which is taken to be constant)

$$j_\beta^A = \tilde{D}_\beta \frac{(c_{\beta\alpha} - c_{\beta\gamma})}{x_\beta} = \tilde{D}_\beta \frac{\left(\Delta c_\beta^{eq} - \left(\frac{j_\beta^A}{k_\beta^{eff}}\right)\right)}{x_\beta} \quad (8.47)$$

where $\Delta c_\beta^{eq} = c_{\beta\alpha}^{eq} - c_{\beta\gamma}^{eq}$, x_β is the thickness of the β phase, and $\left(k_\beta^{eff}\right)^{-1}$ is the effective interface resistance, which is given as

$$\left(k_\beta^{eff}\right)^{-1} = \left(k_{\beta\alpha}\right)^{-1} + \left(k_{\beta\gamma}\right)^{-1} \quad (8.48)$$

The effective interdiffusion coefficient in the β phase, $\tilde{D}_\beta$, may have contributions from different transport paths including grain and dislocation boundaries in addition to bulk diffusion. The critical phase thickness, below which the reaction barrier at the interface becomes significant, is determined by

$$x_\beta^* = \frac{\tilde{D}_\beta}{k_\beta^{eff}} \quad (8.49)$$

Orchard and Greer combined the above-mentioned approach taking into account the components' drift under electron wind in interdiffusion, which has been considered in [8]. They derived the following expression for the total intermixing flux:

$$j_\beta^A = \frac{\left(\Delta c_\beta^{eq} k_\beta^{eff} + \left(\frac{x_\beta}{x_\beta^*}\right) \frac{c_\beta^A c_\beta^B}{c_0} Z\right)}{\left(\frac{x_\beta}{x_\beta^*} + 1\right)} \quad (8.50)$$

Here,

$$Z = \frac{E|e|}{kT} \left(z_A^* D_A^* - z_B^* D_B^*\right) \quad (8.51)$$

where z_A^* stands for effective charge of an atom and e is the electron charge. The A atoms mobility is D_A^*/kT, where D_A^* is the tracer diffusivity and c_β^A is the concentration (m^{-3}) of A atoms in the β phase (it is the value averaged for $c_{\beta\alpha}$ and $c_{\beta\gamma}$ whose difference will be very small for the phases with a narrow homogeneity range).

Without electromigration ($Z = 0$), we get the expression obtained by Gösele and Tu [9]. The rate of phase growth (extent of thickness) is proportional to the flux

$$\left(\frac{dx_\beta}{dt}\right) = \left[\frac{(1+\beta)^2}{c_0}\right]\left[(\alpha-\beta)^{-1} + (\beta-\gamma)^{-1}\right] j_\beta^A \quad (8.52)$$

where the phase composition is described by the relations $A_\alpha B$, $A_\beta B$, and $A_\gamma B$ [9]. If $\left(x_\beta/x_\beta^*\right) \ll 1$, the value of flux reaches a maximum because of reaction barrier at the interface

$$j_\beta^A \to \Delta c_\beta^{eq} k_\beta^{eff} \quad (8.53)$$

while at $(x_\beta/x_\beta^*) \gg 1$ the maximum value depends on the imposed current as

$$j_\beta^A \to \left(\frac{\Delta c_\beta^{eq} \tilde{D}_\beta}{x_\beta}\right) + \left(\frac{c_\beta^A c_\beta^B}{c_0}\right) Z \tag{8.54}$$

It would be useful to transform Equation 8.50 using dimensionless variables. Having defined the dimensionless flux, J, thickness, X, and electromigration flux, J_{EM}, as

$$J = j_\beta^A/\Delta c_\beta^{eq} k_\beta^{eff} \quad X = \frac{x_\beta}{x_\beta^*} \quad J_{EM} = \left(c_\beta^A c_\beta^B/c_0\right) Z/\Delta c_\beta^{eq} k_\beta^{eff}$$

we obtain Equation 8.50 in the following form:

$$J = \frac{X J_{EM} + 1}{X + 1} \tag{8.55}$$

and its solutions are presented in Figure 8.6 for a certain range of J_{EM} values.

At $J_{EM} = 0$, the curve has the form predicted by Gösele and Tu [9]. At a small phase thickness the flux is limited by the reaction at the interface, and at large thickness, by diffusion. When J_{EM} is positive (that is, when A and B atoms displacements, caused by electromigration, have the same direction as at interdiffusion), there is an increase in the resulting flux J. Deviations of $c_{\beta\alpha}$ and $c_{\beta\gamma}$ from their equilibrium values Equation 8.46 decrease and interdiffusion flux increases, since, as a result of this, the difference $(c_{\beta\alpha} - c_{\beta\gamma})$ is reduced, which is compensated by the electromigration flux. When $J_{EM} = 1$, $(c_{\beta\alpha} - c_{\beta\gamma}) = 0$ and the interdiffusion flux is equal to zero, and in this case the resulting flux J does not

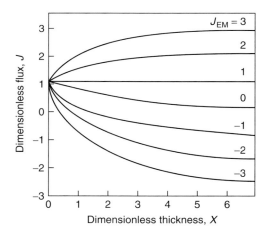

Figure 8.6 Dependence of resulting flux J (in the laboratory reference frame) in the β phase on the phase thickness, X, for selected values of the electromigration flux, J_{EM}. The variables J, X and J_{EM} are dimensionless. The flux J_{EM} is proportional to the electric current density. J_{EM} is positive (negative) when electromigration stimulates (reduces) interdiffusion. (Reproduced with permission from [13].)

depend on the thickness of the layer. When $J_{EM} > 1$, the resulting flux increases as there is an increase in the thickness of the layer. This unusual result occurs, since the expression $(c_{\beta\alpha} - c_{\beta\gamma})$ in this case is negative, causing the flux opposite to the interdiffusion flux, which is less significant at larger phase thickness. When J_{EM} is negative, J, still being positive for small thicknesses, reduces to zero. The condition $J_{EM} = 0$ determines the critical thickness X_{lim} of the β phase from Equation 8.55, $X_{lim} = -1/J_{EM}$. The growth of the phase layer at fixed J_{EM} must stop, having reached the thickness of X_{lim}, and if we increase J_{EM}, this thickness will decrease. Also note that as large as the flux J_{EM} may be, it is unable to remove (take away) the phase, and this has been proved by observations, where although an electromigration effect on the reaction kinetics at the boundary was apparent, there were no changes in the phase composition of the diffusion zone [7, 20].

8.4
Reactive Diffusion in a Binary System at an Imposed Electric Current at Nonequilibrium Vacancies

8.4.1
Equation for the Growth of an Intermediate Phase taking into Account Nonequilibrium Vacancies

We consider now a binary diffusion couple A|B as a model system. The components A and B are mutually insoluble. The system has one intermediate phase, α. In the process of isothermal annealing of such a diffusion couple, the growth of the intermediate phase α starts in the form of a continuous layer at the interface A|B. Let us consider this process in the presence of an electric field in the system. Let the electric field $\vec{E}$ be directed toward X direction. Let us also assume that the A component is more mobile than the B component ($D_A > D_B$), then the total flux of substance flows along the X direction, and the total vacancy flux, in the opposite direction (Figure 8.7). Since A and B are insoluble, substance and vacancy fluxes exist only in the α phase. We will start considering the growth process of intermediate α phase when its thickness equals the critical nucleus width Δx_{cr} (nucleation and nucleus growth stages are not considered here).

Let us write down the equations for the fluxes in the α phase for A and B components in the lattice reference frame taking into account the nonequilibrium vacancy distribution and the effects of electric field [1]. For simplicity, A and B are assumed to have (approximately) the same atomic volumes. If one neglects correlation effects (leading, in the case of random alloys, to Manning's corrections [21]), then

$$\Omega j_A = -D_A \frac{\partial c_A}{\partial x} + \frac{c_A D_A^*}{kT} E_x z_A e + \frac{c_A D_A^*}{c_V} \frac{\partial c_V}{\partial x} \quad (8.56)$$

$$\Omega j_B = -D_B \frac{\partial c_B}{\partial x} + \frac{c_B D_B^*}{kT} E_x z_B e + \frac{c_B D_B^*}{c_V} \frac{\partial c_V}{\partial x} \quad (8.57)$$

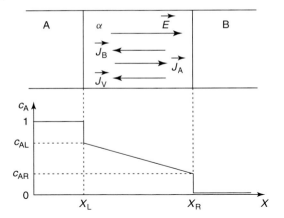

Figure 8.7 At the A|B interface, the phase α appears in the form of a continuous layer. The concentration profile of the component A in the α phase is shown.

The vacancy flux in the lattice reference frame is expressed by the following equation:

$$\Omega j_V = -\Omega \left(j_A + j_B \right) = (D_A - D_B) \frac{\partial c_A}{\partial x}$$
$$- \frac{E_x e}{kT} \left(c_A D_A^* z_A + c_B D_B^* z_B \right) - \frac{\left(c_A D_A^* + c_B D_B^* \right)}{c_V} \frac{\partial c_V}{\partial x} \quad (8.58)$$

where E_x is the OX projection of the electric field intensity; z_A and z_B are the effective charges of A and B atoms, respectively; c_A, c_B, and c_V are the concentrations (averaged values) in the α phase of A and B atoms and vacancies; D_A, D_B are the partial diffusion coefficients of A and B components in the α phase; D_A^* and D_B^* are the isotope diffusion coefficients of A and B components in the α phase.

Let the α phase have a narrow range of concentration homogeneity. In that case, we can assume the following approximation:

$$\frac{\partial c_A}{\partial x} \approx \frac{\Delta c}{\Delta x} \quad \frac{\partial c_B}{\partial x} = -\frac{\partial c_A}{\partial x} \approx -\frac{\Delta c}{\Delta x} \quad \frac{\partial c_V}{\partial x} \approx \frac{c_{VR} - c_{VR}}{\Delta x} \quad (8.59)$$

where Δx is the thickness of the α phase; Δc is the concentration interval of α phase homogeneity in the A component ($\Delta c = c_{AR} - c_{AL}$ and $c_A = (c_{AR} + c_{AL})/2$, c_{AR} and c_{AL} are the A component concentration at the left and right boundary in the α phase, cf Figure 8.7); c_{VR}, c_{VL} are the vacancy concentrations in the α phase at the left and right boundary, respectively.

Taking into consideration Equation 8.59, we rewrite the equations for the fluxes as

$$\Omega j_A = -D_A \frac{\Delta c}{\Delta x} + \frac{c_A D_A^*}{kT} E_x z_A e + \frac{c_A D_A^* (c_{VR} - c_{VL})}{c_V \Delta x} \quad (8.60)$$

$$\Omega j_B = D_B \frac{\Delta c}{\Delta x} + \frac{c_B D_B^*}{kT} E_x z_B e + \frac{c_B D_B^* (c_{VR} - c_{VL})}{c_V \Delta x} \quad (8.61)$$

8.4 Reactive Diffusion in a Binary System

$$\Omega j_V = (D_A - D_B)\frac{\Delta c}{\Delta x} - \frac{E_x e}{kT}(c_A D_A^* z_A + c_B D_B^* z_B)$$
$$- \frac{(c_A D_A^* + c_B D_B^*)}{c_V}\frac{(c_{VR} - c_{VL})}{\Delta x} \tag{8.62}$$

Equations 8.60–8.62 include the difference of concentrations of nonequilibrium vacancies at the left and right boundaries in the α phase, that is, $(c_{VR} - c_{VL})$. To express this value by involving other parameters of the system, we write down the law of vacancy balance for narrow zones of the α phase at its left and right boundaries. At that we consider vacancy sinks/sources to be nonideal. We get

L: $\quad \dfrac{dc_V}{dt} = \dfrac{\Omega j_V}{\delta} - \dfrac{(c_V^{eq} - c_{VL})}{\tau_V}$

R: $\quad \dfrac{dc_V}{dt} = \dfrac{(c_{VR} - c_V^{eq})}{\tau_V} - \dfrac{\Omega j_V}{\delta}$ $\hspace{3em}$ (8.63)

where τ_V is the vacancy relaxation time; δ is the interface thickness; c_V^{eq} is the vacancy concentration at equilibrium (average value in the α phase); j_V is the vacancy flux in the α phase; $(c_V^{eq} - c_{VL})/\tau_V$ describes the number of vacancies, disappearing at the A|α interface in a unit of time; $(c_{VR} - c_V^{eq})/\tau_V$ are the vacancies, appearing at α|B interface in a unit of time.

Using quasi–steady state conditions at the interface, that is, $dc_V/dt = 0$, we get

$$\frac{\Omega j_V}{\delta} = \frac{(c_V^{eq} - c_{VL})}{\tau_V} \qquad \frac{\Omega j_V}{\delta} = \frac{(c_{VR} - c_V^{eq})}{\tau_V} \tag{8.64}$$

The sum of the two equations gives

$$(c_{VR} - c_{VL}) = \frac{2\tau_V \Omega j_V}{\delta} \tag{8.65}$$

If we substitute the expression for Ωj_V in Equation 8.62 in Equation 8.65, we obtain

$$(c_{VR} - c_{VR}) = \frac{\dfrac{2\tau_V}{\delta}\left\{(D_A - D_B)\dfrac{\Delta c}{\Delta x} - \dfrac{E_x e}{kT}(c_A D_A^* z_A + c_B D_B^* z_B)\right\}}{\left(1 + L_V/\Delta x\right)} \tag{8.66}$$

where

$$L_V = \frac{2\tau_V D_V}{\delta}$$

is a characteristic length.

Let us write down now the equation for the flux of the A component in the laboratory reference frame ($J_A = -J_B$)

$$J_A = j_A + c_A j_V \tag{8.67}$$

From Equation 8.67 and bearing the expressions for $(c_{VR} - c_{VR})$ Equation 8.59, j_A (Equation 8.60), and j_V (Equation 8.62) in mind, we arrive at the following result:

$$J_A = -\tilde{D}\Delta c\frac{(1 + L_0/\Delta x)}{(\Delta x + L_V)} - \frac{\tilde{D}}{(\Delta x + L_V)}\frac{(L_V - L_0)}{(D_A - D_B)}\frac{E_x e}{kT}(c_A D_A^* z_A + c_B D_B^* z_B)$$
$$+ \frac{c_A c_B E_x e}{kT}(D_A^* z_A - D_B^* z_B) \tag{8.68}$$

where $L_0 = \left(D_{NG}/\tilde{D}\right) L_V$ is the characteristic length (as $D_{NG} < \tilde{D}$, then $L_0 < L_V$); $\tilde{D}$ is the diffusion coefficient according to Darken ($\tilde{D} = c_A D_B + c_B D_A$); D_{NG} is the diffusion coefficient according to Nernst–Planck (or Nazarov–Gurov, who derived it for the case of nonequilibrium vacancies [22]) given by the relation

$$D_{NG} = \frac{D_A D_B}{(c_A D_A + c_B D_B)} \qquad (8.69)$$

The physical meaning of the characteristic lengths L_0 and L_V becomes clear, if one considers phase growth without currents [10, 11]. From Equation 8.64, we arrive at the following expression:

$$J_A = -\tilde{D}\Delta c(1 + L_0/\Delta x)/(\Delta x + L_V) \qquad (8.70)$$

If $\Delta x \ll L_0$, then

$$J_A = -D_{NG}\frac{\Delta c}{\Delta x} \qquad (8.71)$$

In this case, the nonequilibrium redistribution of vacancies effectively decreases the flux by means of the inverse Kirkendall effect, leading to intermixing controlled by the slow component. If $L_0 < \Delta x \ll L_V$, then

$$J_A = -\tilde{D}\frac{\Delta c}{L_V} \qquad (8.72)$$

making the flux almost equal to zero.

To obtain the equation for the growth of the intermediate phase α at reactive diffusion, we formulate the equation for flux balance of the A component at the right and left boundaries of α (Figure 8.8)

$$(c_A - 1)\frac{dx_L}{dt} = J_A - 0 \Rightarrow -\frac{dx_L}{dt} = \frac{1}{(1 - c_A)}J_A \qquad (8.73)$$

$$(0 - c_A)\frac{dx_R}{dt} = 0 - J_A \Rightarrow \frac{dx_R}{dt} = \frac{1}{c_A}J_A \qquad (8.74)$$

Addition of Equations 8.73 and 8.74 yields

$$\frac{d(x_R - x_L)}{dt} = \frac{1}{c_A(1 - c_A)}J_A \qquad (8.75)$$

Since $x_R - x_L = \Delta x$, and $(1 - c_A) = c_B$, we get

$$\frac{d(\Delta x)}{dt} = \frac{1}{c_A c_B}J_A \qquad (8.76)$$

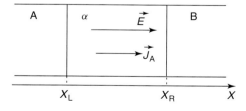

Figure 8.8 Schematic flux of the A component and the vector of electric field in α phase.

8.4 Reactive Diffusion in a Binary System

Using Equations 8.68 and 8.76, we arrive at the following form:

$$\frac{d(\Delta x)}{dt} = -\frac{\tilde{D}\Delta c\,(1+L_0/\Delta x)}{c_A c_B\,(\Delta x + L_V)}$$

$$-\frac{\tilde{D}}{(\Delta x + L_V)}\frac{(L_1 - L_0)}{(D_A - D_B)}\frac{E_x e}{c_A c_B kT}\left(c_A D_A^* z_A + c_B D_B^* z_B\right)$$

$$+\frac{E_x e}{kT}\left(D_A^* z_A - D_B^* z_B\right) \tag{8.77}$$

Equation 8.77 gives the rate of change of thickness of the α phase in the case of nonequilibrium vacancy distribution. If there are no nonequilibrium vacancies in the system, the formula for the thickness rate change of the α phase becomes

$$\frac{d(\Delta x)}{dt} = -\frac{\tilde{D}}{c_A c_B}\frac{\Delta c}{\Delta x} + \frac{E_x e}{kT}\left(D_A^* z_A - D_B^* z_B\right) \tag{8.78}$$

Note that in this and in the following section Δc is negative.

8.4.2
Analysis of the Equation for the Rate of Intermediate Phase Growth in Limiting Cases

Let us consider several different cases:

1) $\Delta x \gg L_V$.

   The thickness Δx of the α-phase is much greater than the characteristic length L_V (for example, at advanced stages of the reaction–diffusion process). In this case, Equation 8.77 is transformed into Equation 8.78. Therefore, we may neglect the nonequilibrium vacancy distribution. Let us consider two extreme cases:

   (a) For

   $$\left|\frac{\tilde{D}}{c_A c_B}\frac{\Delta c}{\Delta x}\right| \gg \left|\frac{E_x e}{kT}\left(D_A^* z_A - D_B^* z_B\right)\right|$$

   (low current density in the system), the expression for the phase growth rate is as follows:

   $$\frac{d(\Delta x)}{dt} = -\frac{\tilde{D}}{c_A c_B}\frac{\Delta c}{\Delta x} \Rightarrow \Delta x = \sqrt{k_1 t + A_1} \tag{8.79}$$

   where

   $$k_1 = -\frac{2\tilde{D}\Delta c}{c_A c_B}$$

   and $\sqrt{A_1}$ is the initial α phase thickness. In this case, the growth of a new α phase obeys a parabolic dependence, since diffusion processes dominate.

   (b) If

   $$\left|\frac{\tilde{D}}{c_A c_B}\frac{\Delta c}{\Delta x}\right| \ll \left|\frac{E_x e}{kT}\left(D_A^* z_A - D_B^* z_B\right)\right|$$

(high current density in the system or rather large phase thickness Δx), the expression for the phase growth rate is as follows:

$$\frac{d(\Delta x)}{dt} = \frac{E_x e}{kT}(D_A^* z_A - D_B^* z_B) \quad \Rightarrow \quad \Delta x = k_2 t + A_2 \tag{8.80}$$

where

$$k_2 = \frac{E_x e}{kT}(D_A^* z_A - D_B^* z_B)$$

and A_2 is the initial α phase thickness. In this case, the growth of a new α phase exhibits a linear dependence, since electromigration prevails.

2) $L_0 \gg \Delta x$ (and then $L_V \gg \Delta x$).

The thickness Δx of the α-phase is much less than the characteristic length L_0. Now Equation 8.81 becomes governing the behavior

$$\frac{d(\Delta x)}{dt} = -\frac{\tilde{D}}{c_A c_B}\frac{\Delta c}{\Delta x}\frac{L_0}{L_V}$$

$$-\frac{\tilde{D}}{(D_A - D_B)}\left(1 - \frac{L_0}{L_V}\right)\frac{E_x e}{c_A c_B kT}(c_A D_A^* z_A + c_B D_B^* z_B)$$

$$+\frac{E_x e}{kT}(D_A^* z_A - D_B^* z_B) \tag{8.81}$$

Let us analyze the obtained expression for limiting cases:

(a) If $L_V \approx L_0$ (so then $D_{NG} \approx \tilde{D}$), then from Equation 8.81 we get the same equations and correspondingly the results as from the case 1. The condition $L_V \approx L_0$ is satisfied when the components' mobilities are equal ($D_A = D_B$), or when the concentration of one of the components is rather low ($c_A \gg c_B$ or $c_A \ll c_B$).

(b) If $L_V \gg L_0$ (so then $D_{NG} \ll \tilde{D}$), then from Equation 8.81 we get the following expression:

$$\frac{d(\Delta x)}{dt} = -\frac{D_A}{c_A}\frac{L_0}{L_V}\frac{\Delta c}{\Delta x} \quad \Rightarrow \quad \Delta x = \sqrt{k_3 t + A_3} \tag{8.82}$$

where

$$k_3 = -\frac{2D_A \Delta c}{c_A}\frac{L_0}{L_V}$$

and $\sqrt{A_3}$ is the initial α phase thickness. In this case, the growth of the new α phase will take place via a parabolic dependence. The condition $L_V \gg L_0$ is satisfied when the difference between the mobilities of the components is significant ($D_A \gg D_B$ or $D_A \ll D_B$).

3) $D_B^* \approx 0$ (the mobility of the B component is low).

As it appears from the above condition, $D_{NG} \approx 0$, $\tilde{D} \approx c_B D_A$, and $L_0 \approx 0$. Let us rewrite the equation for A component flux in this case. It reads

$$J_A = \frac{(\Delta x + L_E)}{(\Delta x + L_V)}\frac{c_A c_B D_A^* E_x e z_A}{kT} \tag{8.83}$$

where

$$L_E = \frac{D_A c_B(-\Delta c)kT}{c_A c_B D_A^* E_x e z_A} = \frac{\varphi(-\Delta c)kT}{c_A E_x e z_A}$$

Let us now analyze the obtained expression.
As in asymptotic approximation of large times, $\Delta x \gg L_V$ and $\Delta x \gg L_E$; hence,

$$J_A \to J_A^{\text{asympt}} = \frac{c_A c_B D_A^* E_x e z_A}{kT}$$

(a) If the electric field intensity $E_x = E_x^{\text{crit}}$, where

$$E_x^{\text{crit}} = \frac{\varphi(-\Delta c)kT}{c_A e z_A L_V}$$

then $L_E = L_V$, and the expression for the flux equation (Equation 8.77) reads

$$J_A = \frac{c_A c_B D_A^* E_x e z_A}{kT} \tag{8.84}$$

This result means that the substance flux must be constant at any phase thickness.

(b) If the electric field intensity $E_x < E_x^{\text{crit}}$ then $L_E > L_V$, and hence at the beginning of the process $J_A > J_A^{\text{asympt}}$, and further as the α phase thickens (Δx) the value of the flux decreases.

(c) If the electric field intensity $E_x > E_x^{\text{crit}}$ then $L_E < L_V$. Therefore, at the beginning of the process $J_A < J_A^{\text{asympt}}$, and further as the α phase thickens (Δx) the value of the flux grows.

This behavior of the substance flux appears to be unusual. In general, as the phase thickens, a decrease in the substance flux is observed. Orchard and Greer managed to obtain the same results for the value of the substance flux (depending on the phase thickness) [13]. However, they used the Gösele and Tu theory for the description of the reaction–diffusion process [9].

4) $L_0 \ll \Delta x \ll L_V$.
The above condition is satisfied when there is a considerable difference of the components' mobilities. Let, for instance, $D_A \gg D_B$. After appropriate simplifications from Equation 8.81, we obtain the following relation:

$$\frac{d(\Delta x)}{dt} = -\frac{D_A}{c_A} \frac{\Delta c}{L_V} = \text{const} \tag{8.85}$$

In this case, the growth of a new α phase proceeds with a linear dependence.

8.4.3
Numerical Solution of the Equation for the Intermediate Phase Rate of Growth

A comparison of the numerical solutions of Equations 8.77 and 8.78 was carried out. For the diffusion parameters of the system, the value and direction of the current density (electric intensity E_x) varied.

Let us first consider the situation that the electric field stimulates the growth of the α phase (it is directed just as the total substance flux). If we take

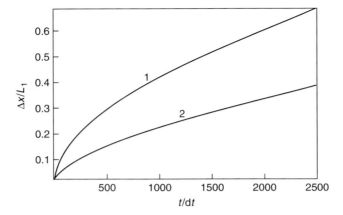

Figure 8.9 (1) Graph for the system with equilibrium vacancies; (2) graph for the system with nonequilibrium vacancies. The presence of an electric field stimulates the growth of the phase. The experiments were carried out under the following conditions: $j = 4 \times 10^8$ A m^{-2}, $D_{NG}/\tilde{D} = 0.209$, $L_1 = 1.6 \times 10^{-4}$ m (Δx, phase width; L_1, characteristic length).

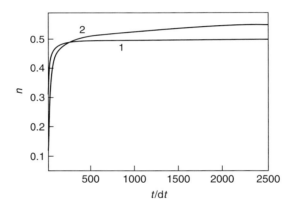

Figure 8.10 (1) Graph for the system with equilibrium vacancies; (2) graph for the system with nonequilibrium vacancies. The presence of an electric field supports the growth of the phase. The experiments were carried out under the following conditions: $j = 4 \times 10^8$ A m^{-2}, $D_{NG}/\tilde{D} = 0.209$, $L_1 = 4 \times 10^{-5}$ m (n, index entering the equation for the thickness of growing phase; $x = kt^n$).

the system with a nonequilibrium vacancy distribution into consideration, the α-phase growth takes place at a lower rate than that in the system with equilibrium vacancies (Figures 8.9–8.11). The larger the vacancy relaxation time τ_V is, the more graphs 1 and 2 (Figure 8.9) differ from each other. If $\tau_V \Rightarrow 0$, then graph 2 coincides with graph 1, as $L_V \to 0$ and Equation 8.77 is transferred into Equation 8.78. When the current density increases, the rate of the phase growth also increases.

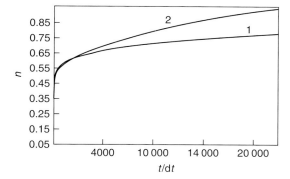

Figure 8.11 (1) Graph for the system with equilibrium vacancies; (2) graph for the system with nonequilibrium vacancies. The presence of an electric field supports the growth of the phase. The experiments were carried out under the following conditions: $j = 9 \times 10^9$ A m^{-2}, $D_{NG}/\tilde{D} = 0.209$, $L_1 = 4 \times 10^{-5}$ m (n, time exponent entering the equation for the thickness of growing phase; $x = kt^n$; $n \equiv (d \ln x/d \ln t)$.

Let us now consider the case that the electric field inhibits the growth of the α phase. When the current density is low, the results are similar to those of case 1 (Figures 8.9–8.11); here, the growth of the α phase proceeds at a higher rate when nonequilibrium vacancies are absent. Still this growth is slower than that in case 1. It means that the terms in Equations 8.77 and 8.78 (that are connected with electromigration) are smaller in value than those connected with interdiffusion.

At higher current density (above the "critical" value), the phase keeps growing only up to a certain thickness, Δx_{max}. Then, the total flux through it becomes equal to zero (Figures 8.12–8.14). The higher the current density, the smaller the Δx_{max}. At a certain current density, no growth of α phase is observed. But, if there are

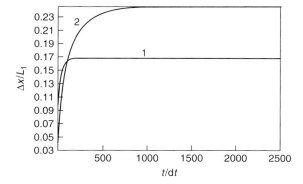

Figure 8.12 (1) Graph for the system with equilibrium vacancies; (2) graph for the system with nonequilibrium vacancies. The presence of an electric field inhibits the growth of the phase. The experiments were carried out under the following conditions: $j = 4 \times 10^8$ A m^{-2}, $D_{NG}/\tilde{D} = 0.209$, $L_1 = 1.6 \times 10^{-4}$ m (Δx, phase width; L_1, characteristic length).

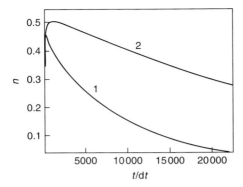

Figure 8.13 (1) Graph for the system with equilibrium vacancies; (2) graph for the system with nonequilibrium vacancies. The presence of an electric field inhibits the growth of the phase. The experiments were carried out under the following conditions: $j = 4 \times 10^8$ A m^{-2}, $D_{NG}/\tilde{D} = 0.209$, $L_1 = 4 \times 10^{-5}$ m (n, index entering the equation for the thickness of the growing phase; $x = kt^n$).

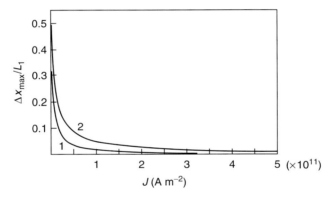

Figure 8.14 (1) Graph for the system with equilibrium vacancies; (2) graph for the system with nonequilibrium vacancies. The presence of an electric field inhibits the growth of the phase. The experiments were carried out under the following conditions: $D_{NG}/\tilde{D} = 0.278$, $L_1 = 4 \times 10^{-5}$ m (Δx_{max}, maximal phase thickness (for a certain current density); L_1, characteristic length; j, current density).

nonequilibrium vacancies, the growth of α phase takes place up to larger values of Δx_{max}.

The dependence of the substance flux on the phase thickness was analyzed for the case of stimulation of α-phase growth and $D_B^* \approx 0$ (Figure 8.10). The result was obtained by the numerical solution of Equation 8.81. For the case of equilibrium vacancies being present in the system, the dependence appears to be monotonically decreasing, as shown by graph 1 in Figure 8.10. If there are nonequilibrium vacancies, the substance flux depends on the phase thickness nonmonotonically; that is, in the form as described by graph 2 in Figure 8.10. This kind of behavior corresponds to results obtained from the analysis of Equation 8.81. From graph

2, it can be seen that the substance flux increase is preceded by its sharp decrease. Moreover, the flux increase takes place up to the value at which there is a decrease in the case of equilibrium vacancies (i.e., when the phase thickness we may neglect the influence of nonequilibrium vacancies).

From the viewpoint of physics, this fact may be explained as follows: At the beginning of the growth process, when the phase thickness is small as compared to the characteristic length L_V, there is a quick increase in the vacancy gradient, which has the same direction as that of the A component (fast supersaturation with vacancies at the left boundary and depletion at the right boundary, see Figure 8.15).

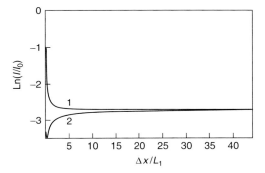

Figure 8.15 (1) Graph for the system with equilibrium vacancies; (2) graph for the system with nonequilibrium vacancies. The presence of an electric field supports the growth of the phase. The experiments were carried out under the following conditions: $j = 9 \times 10^9$ A m^{-2}, $D_{NG}/\tilde{D} = 0.046$, $L_1 = 4 \times 10^{-5}$ m (I, substance flux (A component); I_0, initial value of substance flux (A component) for the system with equilibrium vacancies; L_1, characteristic length, Δx, phase thickness).

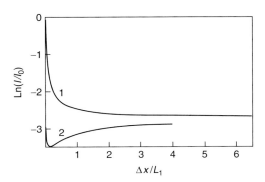

Figure 8.16 (1) Graph for the system with equilibrium vacancies; (2) graph for the system with nonequilibrium vacancies. The presence of an electric field supports the growth of the phase. The experiments were carried under the following conditions: $j = 9 \times 10^9$ A m^{-2}, $D_{NG}/\tilde{D} = 0.046$, $L_1 = 4 \times 10^{-5}$ m (I, substance flux (A component); I_0, initial value of substance flux (A component) for the system with equilibrium vacancies; L_1, characteristic length, Δx, phase thickness).

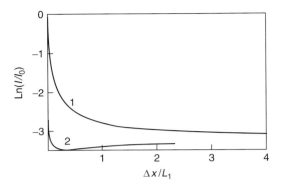

Figure 8.17 (1) Graph for the system with equilibrium vacancies; (2) graph for the system with nonequilibrium vacancies. The presence of an electric field supports the growth of phase. The experiments were carried out under the following conditions: $j = 4.5 \times 10^9$ A m^{-2}, $D_{NG}/\tilde{D} = 0.046$, $L_1 = 4 \times 10^{-5}$ m (I, substance flux (A component); I_0, initial value of substance flux (A component) for the system with equilibrium vacancies; L_1, characteristic length, Δx, phase thickness).

This kind of behavior implies a decrease of the total flux of substance. This stage in conformance with the part of graph 2 (Figures 8.15–8.17), where there is a sharp decrease in the substance flux. Further as the phase grows (the growth, although slowly, continues, since the total flux has diminished considerably, yet has not become equal to zero) the saturation with vacancies decreases, so the vacancy flux (and hence the substance flux) increases.

8.4.4
Conclusion

On the basis of the results obtained by the numerical solution of Equations 8.77 and 8.78, we may conclude that the system may compensate (or, to be more exact, reduce or "damp") the outer action on it (in our case, electromigration) with the help of a nonequilibrium vacancy distribution. The latter may slow down the phase growth if the electric field in the diffusion system supports it. And, conversely, it can stimulate the phase growth if the electric field opposes it.

References

1. Tu, K.N. (2003) *Journal of Applied Physics*, **94**, 5451.
2. Gan, H., Choi, W.J., Xu, G. and Tu, K.N. (2002) *Journal of Metals*, **54**, 34.
3. Liu, C.Y., Chen, C., Mal, A.K., and Tu, K.N. (1999) *Journal of Applied Physics*, **85**, 3882.
4. Lee, T.Y., Tu, K.N. and Frear, D.R. (2001) *Journal of Applied Physics*, **90**, 4502.
5. Choi, W.J., Yeh, E.C.C. and Tu, K.N. (2003) *Journal of Applied Physics*, **94**, 5665.
6. Chen, C.M. and Chen, S.W. (2001) *Journal of Applied Physics*, **90**, 1208.

7. Chen, S.W. and Chen, C.M. (**2003**) *Journal of Metals*, **55**, 62.
8. Gurov, K.P. and Gusak, A.M. (**1981**) *Fizika Metallov i Metallovedeniye (Physics of Metals and Metallography)*, **52**, 7967 (in Russian).
9. Gösele, U. and Tu, K.N. (**1982**) *Journal of Applied Physics*, **53**, 3252.
10. Gusak, A.M. (**1992**) *Metallofizika*, **14**(9), 3 (in Russian).
11. Gurov, K.P., Dol'nikov, S.S., Miller, Yu.G. *et al.* (**1978**) *Fizika i Khimiya Obrabotki Materialov*, **1**, 107 (in Russian).
12. Gurov, K.P. and Gusak, A.M. (**1982**) *Fizika Metallov i Metallovedeniye (Physics of Metals and Metallography)*, **53**(5), 12 (in Russian).
13. Orchard, T. and Greer, A.L. (**2005**) *Applied Physics Letters*, **86**, 231906.
14. Gurov, K.P. and Gusak, A.M. (**1981**) *Fizika Metallov i Metallovedeniye (Physics of Metals and Metallography)*, **52**, 131 (in Russian).
15. Pimenov, V.N., Gurov, K.P., Hudolyakov, K.I. *et al.* (**1978**) *Fizika i Khimiya Obrabotki Materialov*, **1**, 107 (in Russian).
16. Gusev, O.V., Gurov, K.P., Dol'nikov, S.S *et al.* (**1980**) *Fizika i Khimiya Obrabotki Materialov*, **2**, 79 (in Russian).
17. Mackenzie, K.J.D., Banezjee, R.K. and Kasaai, M.R. (**1979**) *Journal of Materials Science*, **14**, 333.
18. Borovskii, I.B., Gurov, K.P., Marchukova, I.D. and Ugaste, Yu.E. (**1971**) *Inter-diffusion Processes in Metals*, Nauka, Moscow (in Russian).
19. Gurov, K.P., Pimenov, V.N., Ugaste, Yu.E. and Shelest, A.E. (**1971**) *Fizika Metallov i Metallovedeniye (Physics of Metals and Metallography)*, **321**, 103 (in Russian).
20. Chen, S.-W., Chen, C.-M. and Liu, W.-C. (**1998**) *Journal of Electronic Materials*, **27**, 1193.
21. Manning, J. (**1967**) *Acta Metallurgica*, **15**, 817.
22. Nazarov, A.V. and Gurov, K.P. (**1974**) *Fizika Metallov i Metallovedeniye (Physics of Metals and Metallography)*, **37**, 496 (in Russian).

9
Diffusion Phase Competition in Ternary Systems
Semen V. Kornienko, Yuriy A. Lyashenko, and Andriy M. Gusak

9.1
Introduction

Phase competition in ternary systems appears to be much more sophisticated than in binary systems if we consider the number of possible regimes. In particular, for binary and ternary systems, it is possible to obtain phase growth or suppression criteria when some phases are either absent in the diffusion zone or their growth can be considerably slowed down by adding the third component to a binary system (Section 9.2). Besides, in the case of binary systems at the asymptotic stage of large annealing times, all intermediate phases (allowed by the phase diagram) must appear and grow parabolically but in the case of ternary systems, the reaction regime turns out to be determined ambiguously even at advanced stages (Section 9.3). The problem of regime choice must be solved at the initial stage during the "natural selection" of phases. The criteria of such selection at the nucleation stage are analyzed in Section 9.4.

9.2
Phase Competition in the Diffusion Zone of a Ternary System

Peculiarities of solid-state reactions at interfaces between a metal film and a substrate (Si, for example) lead to a certain succession of phase formation and growth. Moreover, some phases, which must appear according to the equilibrium phase diagram, can be absent in the diffusion zone. On the other hand, some metastable phases absent at the phase diagram, sometimes are formed instead of the stable ("legal") phases [1–8]. In the majority of cases, the initial stage of the solid-state reaction results in the formation of an intermediate phase in the form of a thin layer. This phase may either grow parabolically ($\Delta y = y_R - y_L = k\sqrt{t}$) at certain rate constants k_R, k_L of the right and left boundaries ($y_{R,L} = k_{R,L}\sqrt{t}$), or deviate from the parabolic law, the deviation being determined by the specific manner in which the components are added to the transformation front and the way they pass through the front, into neighboring phases.

Diffusion-Controlled Solid State Reactions. Andriy M. Gusak
Copyright © 2010 WILEY-VCH Verlag GmbH & Co. KGaA, Weinheim
ISBN: 978-3-527-40884-9

Processes of formation and growth of intermediate phases in the diffusion zone of binary and multicomponent systems are controlled by the following multiscale processes [2, 4, 6, 8]:

1) nucleation of an intermediate phase at the interface of initial components or solid solutions;
2) diffusion through parent and intermediate growth phases;
3) transport of atoms through the transformation front (boundary kinetics).

Processes of the first type include formation of the new phase nuclei in the contact zone as a result of heterophase fluctuations under chemical potential and concentration gradient. The description of solid-state reactions of the second and third types, namely those at the moving boundaries between phases that already exist in the diffusion zone, is reduced to the following consecutive stages:

1) supply of atoms to the transformation boundary;
2) detachment of atoms from the lattice of the phase the transformation front moves to;
3) transition of atoms through the transformation interphase boundary;
4) attachment of atoms to the lattice of the growing phase;
5) removal of atoms from the transformation boundary.

The boundary kinetics (stages 2–4) may result in changing the constant of solid-state reaction rate and deviation from the parabolic law of phase growth controlled by stages 1 and 5 [8].

Introducing the third component into the binary system may also lead to the change of constant of the intermediate phase growth rate. This is possible when the third component alters the diffusion mobility of the atoms in the phases and when there is a solubility limit for this component in the reacting phases [8–10].

9.2.1
Phase Competition in the Diffusion Zone of a Ternary System with Two Intermediate Phases

Let us consider the influence of alloying a binary system with a third component on the formation of the diffusion zone in a ternary system [8, 9]. We analyze a case for an Al–Cd–Mg–like phase diagram of a ternary system (Figure 9.1). In the phase diagram there are two solid solutions: η (diluted solution of B and C in A) and α (unlimited solubility of B and C in each other and the negligible solubility of A in them). There are also two binary intermediate phases: β (the solubility limits are c_{BL}^{β} and c_{BR}^{β}) and γ (with c_{BL}^{γ} and c_{BR}^{γ}, respectively). The intermediate phases are in equilibrium with the γ-phase and the α-phase.

The sandwich-like sample Al–Cd–Mg was treated at 575 K by the method of diffusion zones superposition [11]. Al and Mg interacted through the Cd layer (0.2 μm). After isothermal annealing for t = 10, 100, 200, 500, and 900 h and measuring the concentration distributions, the diffusion paths were built and thus the system's phase diagram was constructed (Figure 9.1b). In that, the first phase,

9.2 Phase Competition in the Diffusion Zone of a Ternary System

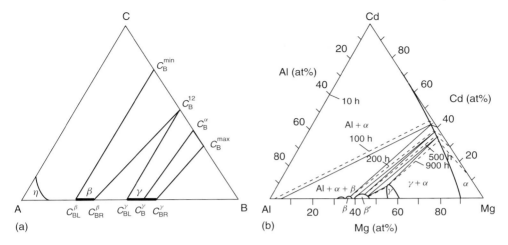

Figure 9.1 (a) Phase diagram of the model ternary system with η- and α- solid solutions and binary β- and γ-intermetallics and positions of boundary conodes. (b) Isothermal cross section of the Al–Mg–Cd phase diagram at 575 K and obtained diffusion paths.

$Al_{12}Mg_{17}$, formed only after 200 h of annealing (the Cd layer dissolved in 100 h). After 500 h, besides the β-phase, the β'-phase appeared in the diffusion zone, and, finally, after 900 h of annealing the layer of the γ-phase appeared in the transition zone (with a solubility of up to 7% Cd). The intermediate phases β and β' are practically binary. So, changing the boundary concentration of the α-solid solution, we can get a diffusion zone with one to three intermetallides.

Let us ensure that by changing the initial concentration of the BC alloy, one can control the phase composition of the diffusion zone. We proceed from the flux balance equations at the interfaces $\eta - \beta$, $\beta - \gamma$, and $\gamma - \alpha$:

$$c_{BL}^\beta \frac{d\gamma_L}{dt} = -\frac{\tilde{D}^\beta \Delta c_1}{\Delta_1 \gamma} \tag{9.1}$$

$$(c_{BL}^\gamma - c_{BL}^\beta)\frac{d\gamma}{dt} = \frac{\tilde{D}^\beta \Delta c_1}{\Delta_1 \gamma} - \frac{\tilde{D}^\gamma \Delta c_2'}{\Delta_2 \gamma} \tag{9.2}$$

$$(c_B^\alpha - c_B^\gamma)\frac{d\gamma_R}{dt} \frac{\tilde{D}^\gamma \Delta c_2'}{\Delta_2 \gamma} - J_B^\alpha\big|_{\gamma_R} \tag{9.3}$$

$$(1 - c_B^\alpha)\frac{d\gamma_R}{dt} = J_B^\alpha\big|_{\gamma_R} \tag{9.4}$$

where $\tilde{D}^\beta \Delta c_1$ and $\tilde{D}^\gamma \Delta c_2'$ are the effective diffusivities of the β- and γ-phases; here, $\Delta c_1 = c_{BR}^\beta - c_{BL}^\beta$ is the concentration width of the β-phase.

Note, at once, that the total concentration width Δc_2 of the γ-phase does not enter the equation, instead we have a part of it, $\Delta c_2' = c_B^\gamma - c_{BL}^\gamma$, determined by the degree of alloying (initial content of the C-component at the BC-side of the diffusion couple; Figure 9.1a). Here, according to the lever rule,

$$\frac{\Delta c_2'}{\Delta c_2} = \frac{c_B^\alpha - c_B^{12}}{c_B^{max} - c_B^{12}} \qquad (9.5)$$

This is the change $\Delta c_2 \to \Delta c_2'$ that influences the composition of the diffusion zone due to alloying.

The flux of the B component in the α-phase becomes

$$J_B^\alpha = (c_B^\alpha(\infty) - c_B^\alpha)\sqrt{\frac{\tilde{D}^\alpha}{\pi}} \frac{\exp\left(-\frac{k_R^2}{4\tilde{D}^\alpha}\right)}{1 - \operatorname{erf}\left(\frac{k_R}{2\sqrt{\tilde{D}^\alpha}}\right)} \qquad (9.6)$$

The concentration distribution of B in the reaction zone between the moving interfaces y_L, y, and y_R is schematically depicted in Figure 9.2.

Combining Equations 9.1 and 9.4, we get the expressions for the growth rates of β- and γ-phases:

$$\frac{d\Delta_1 y}{dt} = \frac{dy}{dt} - \frac{dy_L}{dt} = \left[\frac{1}{c_{BL}^\gamma - c_{BL}^\beta} + \frac{1}{c_{BL}^\beta}\right] \frac{\tilde{D}^\beta \Delta c_1}{\Delta_1 y} - \frac{1}{c_{BL}^\gamma - c_{BL}^\beta} \frac{\tilde{D}^\gamma \Delta c_2'}{\Delta_2 y} \qquad (9.7)$$

$$\frac{d\Delta_2 y}{dt} = \frac{dy_R}{dt} - \frac{dy}{dt} = \frac{-1}{c_{BL}^\gamma - c_{BL}^\beta} \frac{\tilde{D}^\beta \Delta c_1}{\Delta_1 y} + \left[\frac{1}{c_{BL}^\gamma - c_{BL}^\beta} + \frac{1}{1 - c_B^\gamma}\right] \frac{\tilde{D}^\gamma \Delta c_2'}{\Delta_2 y} \qquad (9.8)$$

Analyzing Equations 9.7 and 9.8, let us consider the possibility of growth suppression (for some time) of one, say, the γ-phase [1, 8, 9].

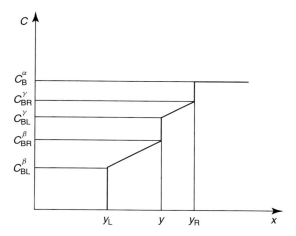

Figure 9.2 Concentration profile of B species corresponding to a diffusion path from composition c_{BL}^β to c_B^α via a pseudobinary β-phase and the intermetallic γ-phase.

9.2 Phase Competition in the Diffusion Zone of a Ternary System

In this case, the γ-phase will not grow until the growth rate of its critical nuclei becomes positive:

$$\frac{d\Delta_2\gamma}{dt} = \frac{-1}{c_{BL}^{\gamma} - c_{BL}^{\beta}} \frac{\tilde{D}^{\beta}\Delta c_1}{\Delta_1\gamma} + \frac{1 - c_{BL}^{\beta}}{(c_{BL}^{\gamma} - c_{BL}^{\beta})(1 - c_{BL}^{\gamma})} \frac{\tilde{D}^{\gamma}\Delta c_2'}{l_{cr}} > 0 \tag{9.9}$$

Here, we use the assumption that the thickness of the suppressed phase equals the size of its critical nuclei l_{cr}.

To analyze the growth of the β-phase, we use the parabolic substitution $\Delta_1\gamma = k_1\sqrt{t}$ in the equation for the growth of the first phase (Equation 9.7):

$$\frac{d\Delta_1\gamma}{d\gamma} = \frac{1}{c_{BL}^{\beta}(1 - c_{BL}^{\beta})} \frac{\tilde{D}^{\beta}\Delta c_1'}{\Delta_1\gamma} \tag{9.10}$$

The equation for the first phase growth includes not the whole concentration range of the β-phase but its change

$$\Delta c_1' = c_B^{\beta} - c_{BL}^{\beta} = \frac{\Delta c_1(c_B^{\alpha} - c_B^{\min})}{c_B^{12} - c_B^{\min}} \tag{9.11}$$

Finally, for the growth constant of β-phase we get

$$k_1 = \sqrt{2\tilde{D}^{\beta} \frac{\Delta c_1(c_B^{\alpha} - c_B^{\min})}{c_{BL}^{\beta}(1 - c_{BL}^{\beta})(c_B^{12} - c_b^{\min})}} \tag{9.12}$$

Due to a variation of c_B^{α}, the influence on the intermediate phase growth is possible. Note that c_B^{α} and the boundary concentration of the α-solid solution are interrelated as

$$c_B^{\alpha} = \frac{\left(c_B^{\alpha}(\infty)\sqrt{\frac{\tilde{D}^{\alpha}}{\pi}} \frac{\exp(-k_R^2/4\tilde{D}^{\alpha})}{1 - \text{erf}(k_R/2\sqrt{\tilde{D}^{\alpha}})} - \frac{k_R}{2}\right)}{\left(\sqrt{\frac{\tilde{D}^{\alpha}}{\pi}} \frac{\exp(-k_R^2/4\tilde{D}^{\alpha})}{1 - \text{erf}(k_R/2\sqrt{\tilde{D}^{\alpha}})} - \frac{k_R}{2}\right)} \tag{9.13}$$

which arises from Equation 9.4 and the explicit form of the flux $J_B^{\alpha}|_{\gamma_R}$ (Equation 9.6). In one of the marginal cases, when $\tilde{D}^{\alpha} \gg k_R^2$ holds, the relation

$$\frac{\exp(-k_R^2/(4\tilde{D}^{\alpha}))}{1 - \text{erf}(k_R/2\sqrt{\tilde{D}^{\alpha}})} \approx 1 \tag{9.14}$$

is fulfilled and we get a quick adjustment of the concentration to its boundary magnitude $c_B^{\alpha} \approx c_B(\infty)$ at the interphase boundary.

In this case, it is possible to effectively influence the kinetics of the β-phase growth, since by changing the boundary composition of the α-phase we alter c_B^{α} to the same extent, and thus,

$$k_1 = \sqrt{2\tilde{D}^{\beta} \frac{\Delta c_1(c_B(\infty) - c_B^{\min})}{c_{BL}^{\beta}(1 - c_{BL}^{\beta})(c_B^{12} - c_B^{\min})}} = f(c_B(\infty)) \tag{9.15}$$

As seen from Equation 9.15, if the boundary concentration c_B^α of the BC alloy falls within the range $c_B^{\min} - c_B^{12}$, only the β-phase grows in the diffusion zone. If $c_B^\alpha - c_B^{12}$ is sufficiently small and c_B^α falls within $c_B^{12} - c_B^{\max}$, we get the case of suppression of the γ-phase growth. So, the γ-phase will be suppressed until the width of the β-phase equals the following (see Equation 9.9):

$$\Delta_1 y = \frac{1 - c_{BL}^\gamma}{1 - c_{BL}^\beta} \frac{\tilde{D}^\beta \Delta c_1}{\tilde{D}^\gamma \Delta c_2'} l_{cr} \qquad (9.16)$$

The time of diffusion suppression of the γ-phase nuclei must equal the time of β-phase growth, up to the thickness determined according to Equation 9.16.

Let us estimate the incubation time for the γ-phase. The equation of β-phase growth during this period is given by

$$\frac{d\Delta_1 y}{dt} = \frac{1}{c_{BL}^\beta (1 - c_{BL}^\beta)} \frac{\tilde{D}^\beta \Delta c_1}{\Delta_1 y} \qquad (9.17)$$

Equation 9.17 differs from Equation 9.10 by the whole homogeneity range of the β-phase entering the calculations.

Assuming the parabolic growth of the β-phase $\Delta_1 y = k_1 \sqrt{t}$, we get for the growth constant

$$k_1 = \sqrt{2\tilde{D}^\beta \frac{\Delta c_1}{c_{BL}^\beta (1 - c_{BL}^\beta)}} \qquad (9.18)$$

From Equation 9.18, we find the incubation time

$$\tau = 2c_{BL}^\gamma (1 - c_{BL}^\gamma) \frac{\tilde{D}^\beta \Delta c_1}{(\tilde{D}^\gamma \Delta c_2')^2} l_{cr}^2 \qquad (9.19)$$

Thus, it becomes clear that the incubation time of the γ-phase depends on the boundary concentration of the α-phase. Further, it turns out that in case of initial suppression of the γ-phase the thickness of the β-phase, at which the suppression stops, is inversely proportional to the difference between the boundary concentration c_B^α and the value of c_B^{12}. The incubation time of the γ-phase is inversely proportional to the square of this difference.

Let us now consider a case when both phases are already growing in the diffusion zone. We use the parabolic substitution for the growth rate $\Delta_1 = k_1\sqrt{t}$ and $\Delta_2 = k_2\sqrt{t}$. Dividing Equation 9.7 by Equation 9.8, we get a quadratic equation for $\frac{k_1}{k_2}$, and the solution is

$$\frac{k_1}{k_2} = \frac{1 - c_{BL}^\gamma}{2(1 - c_{BL}^\beta)} \left(R - 1 \pm \sqrt{1 + R^2 - 2R\left(1 - \frac{2c_{BL}^\gamma(1 - c_{BL}^\beta)}{c_{BL}^\gamma(1 - c_{BL}^\gamma)}\right)} \right)$$

where

$$R = \frac{\tilde{D}^\beta \Delta c_1}{\tilde{D}^\gamma \Delta c_2'} \qquad (9.20)$$

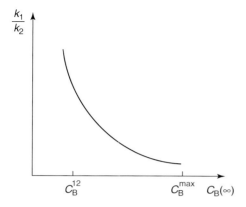

Figure 9.3 Dependence of the ratio of growth constants for β- and γ-phases on the boundary composition of the α-phase.

Here, we employed the reasoning used in the introduction of the relation Equation (9.5) with respect to $\Delta c_2'$. The dependence of $\frac{k_1}{k_2}$ on concentration of the α-phase is given in Figure 9.3.

So, by changing the concentration of the α-solid solution within $c_B^{max} - c_B^{min}$, one can obtain the layers of one or two intermetallics in the resulting diffusion zone. In addition, the rate of phase growth also depends on the initial composition. So, the change of c_B^α within the indicated interval leads to a considerable change in the growth rate relation for β- and γ-phases (Figure 9.3). Therefore, alloying with the third component allows one to control the relative growth rates for β- and γ-phases (up to complete suppression of one of them).

9.2.2
Influence of Pt on Phase Competition in the Diffusion Zone of the Ternary (NiPt)–Si System

9.2.2.1 Basic Considerations

Let us consider a small alloying addition of Pt (about 5 at.%) to Ni and examine its influence on the result of a solid-state reaction with Si, leading to the growth of intermediate phases in the diffusion zone [10, 12–16]. Such silicides are widely used in microelectronics while manufacturing interconnects. Silicides of different stoichiometry are formed as a result of solid state-reactions at the interface of a metallic film and silicon substrate.

Due to the tendency of microelectronic devices to reduce their sizes it is necessary to take into account the peculiarities of solid-state reactions between Ni nanofilms and silicon substrate. Industrial implementation of metallic silicides requires the reduction of contact layers and resistance of source/sink transitions while producing high-speed multielement integrated microcircuits. This is the NiSi that is most frequently used in manufacture of CMOS (Complementary metal–oxide–semiconductor) transistors since it provides high electroconductivity and the required mechanical properties [12–16]. However, the ways of suppressing a NiSi solid-state transformation into, first of all, a low-conductive and fragile $NiSi_2$ phase at relatively low temperatures must be worked out in the techniques of manufacturing microchips.

Interphase equilibrium (Figure 9.4) and solid-state reactions in the binary thin film Ni–Si system have been experimentally studied in detail [12, 17, 18]. It has been established that in the case of solid-state reaction between the Ni film and Si substrate, most often, the layers of silicides are formed in series: at low (200–300 °C) temperatures, Ni_2Si and NiSi grow one after another; at temperatures higher than 750 °C, the $NiSi_2$-phase starts growing at the interface between the NiSi film and the silicon substrate [13–15]. The $NiSi_2$-phase is considered to start growing at rather high homologous temperatures (T/T_{melt}) due to lowering of the nucleation barrier, which results from the relatively small change of free energy at the $NiSi + Si \rightarrow NiSi_2$ solid-state reaction [12–15]. It has been experimentally found that adding the third component, Pt [13–15], can lead to suppression of nucleation and growth of the $NiSi_2$ phase by the stable NiSi phase.

When considering a phase composition of the diffusion zone in the process of heating a binary Ni–Si couple, one can notice that Pt addition leads to the rise of temperature for $NiSi_2$ phase formation. A possible explanation for this is the

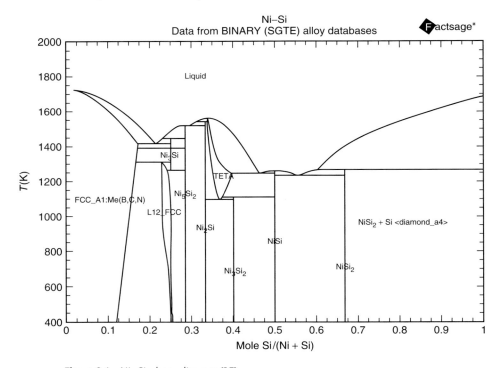

Figure 9.4 Ni–Si phase diagram [17].

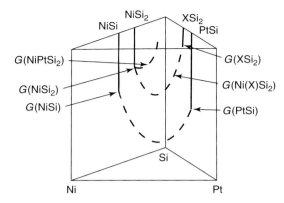

Figure 9.5 Schematic Gibbs potentials for the phases of Ni–Pt–Si system [16].

effect of the increase of critical temperature for NiSi$_2$ nuclei formation [16]. Here, the enhanced stability of the NiSi phase is considered to be connected with the decrease of the Gibbs potential for the NiSi phase due to addition of the third component as compared to a case of Gibbs potential for a binary NiSi$_2$ phase without the third component. The paper [16] demonstrated that the calculated phase diagram for a model ternary system could be used for describing the change in the nucleation driving force and thus, the temperature of NiSi$_2$ phase formation (Figure 9.5). We demonstrate below that the effect of growth suppression of the NiSi$_2$-phase at the NiSi–Si interface can also be described on the basis of diffusion kinetics of phase growth [10]. In particular, in the case of low (compared to that of Ni) diffusivity of Pt in the ternary intermediate phase (NiPt)Si, it is possible to influence the growth rate of the NiSi$_2$ phase by changing the concentration of Pt.

9.2.2.2 Effect of Pt on Phase Competition in the Diffusion Zone of Ni–Si

Let us analyze theoretically how alloying influences the morphology of the diffusion zone in case of a ternary Si–Ni–Pt-like system [10] (Figure 9.6).

The diagram contains two solid solutions: η (dilute solution of B (Ni) and C (Pt) in A (Si)) and α (unlimited solubility of B and C species in each other at low solubilities of component A in them). There are also two intermediate phases: the binary β-phase, corresponding to intermetallic Ni$_2$Si (solubility limits are c_{BL}^β and c_{BR}^β), and the ternary γ-phase with (Ni$_x$–Pt$_{0.5-x}$)–Si composition ($0 \leq x \leq 0.5$, solubility limits are c_{BL}^γ and c_{BR}^γ on the binary side of Si–Ni). The intermediate binary β-phase is in equilibrium both with the η- and γ-phases. The conodes, linking the respective boundary concentrations of β- and γ-phases, are calculated from the lever rule. The concentration interval for β-phase existence, $\Delta c_1 = c_{BR}^\beta - c_{BL}^\beta$, corresponds to that of B species in the γ-phase within c_B^{min} and c_B^{max}, which determined the position of the boundary conodes. Besides, for any other intermediate position of the conode, the following relation is valid

$$\frac{c_B^\beta - c_{BL}^\beta}{c_{BR}^\beta - c_{BL}^\beta} = \frac{c_{BL}^\gamma - c_B^{min}}{c_B^{max} - c_B^{min}} \tag{9.21}$$

where c_B^β and c_{BL}^γ are the respective concentrations in the β- and γ-phases, which are in thermodynamic equilibrium with each other; that is, they belong to the same conode.

Consider a possible system configuration: a homogeneous, fairly thick Ni film with Pt addition (say, 5 at.%) is deposited onto a Si substrate. According to experimental data [13–15], during diffusion annealing a continuous layer of the ternary intermediate γ-phase with $(Ni_x-Pt_{0.5-x})$–Si composition is formed from the Ni–5% Pt solid solution as a result of a solid-state reaction with Si. Next, in the treated system between pure Si and γ-phase, a layer of the intermediate binary β-phase Ni_2Si starts growing, the solubility of Pt in it being low. During the growth of the β-phase from the γ-phase, Pt must either segregate at the growth front or be pushed inside the γ-phase by the moving front. If the diffusivity of Pt in the ternary intermediate γ-phase appears to be too small to accomplish such pushing, this will lead to a slowing-down of the transformation front velocity and thus, the growth rate of the β-phase will be reduced.

Figure 9.6 gives the spatial arrangement of phases and corresponding concentration profiles for the $Si–(Ni_{0.475}Pt_{0.025})$ system. Consider Pt to be capable of redistribution by diffusion only with Ni at one of the sublattices, and Si to be located at the other sublattice and to not participate in the diffusion redistribution. Here, the diffusion redistribution of low-mobility Pt is the dominant process at the competitive phase growth. The point on the concentration triangle, corresponding to a concentration c_{BL}^γ at the $\beta-\gamma$ boundary with the y coordinate, may be shifted along the line of phase equilibrium within the range of c_B^{min} and c_B^{max}. Then, the corresponding concentration in the β-phase will determine the homogeneity

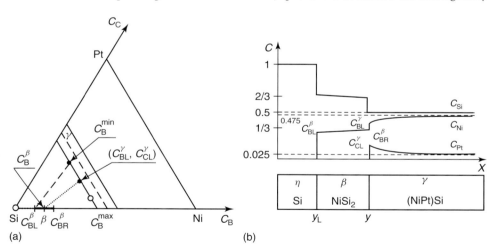

Figure 9.6 (a) Model Si–Ni–Pt phase diagram with binary intermediate β- and ternary γ-phase. One of boundary conodes is given by a dashed line. The diffusion path is given by a thick dotted line between two empty circles, depicting the initial composition of the samples. (b) Spatial configuration and concentration profiles of the growing diffusion zone with demarcated boundaries of the binary intermediate phase.

interval $\Delta c_1^* = c_B^\beta - c_{BL}^\beta$, which will influence the diffusion flux through this phase. The aforementioned influence fixes the flux balance at the interphase boundaries of the β-phase, and thus its growth rate. Let us ensure that by changing the initial concentration of the BC alloy we can control the phase composition of the diffusion zone.

We will depend upon the equations of flux balance at interface boundaries, $\eta - \beta$, $\beta - \gamma$:

$$\left(c_{BL}^\beta - 0\right) \frac{dy_L}{dt} = -\frac{D_{int}^\beta \Delta c_1^*}{\Delta c_1 \Delta_{1\gamma}}$$

$$\left(c_{BL}^\gamma - c_B^\beta\right) \frac{dy}{dt} = \frac{D_{int}^\beta \Delta c_1^*}{\Delta c_1 \Delta_{1\gamma}} + \left. J_B^\gamma \right|_y$$

$$\left(c_{CL}^\gamma - 0\right) \frac{dy}{dt} = \left. J_C^\gamma \right|_y = -\left. J_B^\gamma \right|_y \tag{9.22}$$

where $\Delta_{1\gamma}$ is the thickness of the β-phase; D_{int}^β is the integral diffusion coefficient (Wagner diffusivity) in the binary intermediate β-phase (i.e., the effective diffusivity), determined from the relation [18]:

$$D_{int}^\beta = \int_{c_{BL}^\beta}^{c_{CL}^\beta} \tilde{D}^\beta dc \tag{9.23}$$

In Equations 9.22 it is important that we take into account only some part of the total effective diffusivity of the β-phase, found from the ratio of the concentration intervals

$$\delta = \frac{\Delta c_1^*}{\Delta c_1} = \frac{c_{BL}^\gamma - c_B^{min}}{c_B^{max} - c_B^{min}}$$

and that is connected with the effect of alloying. The flux of the B component in the γ-phase is expressed in the approximation of concentration gradients (difference) only for B and C species, and neglecting the mobility of the A component (in this case in the laboratory reference frame $J_{Si} = 0$).

Having examined the interdiffusion only at the Ni + Pt sublattice, we obtain the expression for fluxes as

$$\left. J_B^\gamma \right|_y = -\left. J_C^\gamma \right|_y = \left(c_{BL}^\gamma(\infty) - c_{BL}^\gamma\right) \sqrt{\frac{\tilde{D}^\gamma}{\pi t}} \frac{\exp\left(-\frac{k^2}{4\tilde{D}^\gamma}\right)}{1 - \text{erf}\left(\frac{k}{2\sqrt{\tilde{D}^\gamma}}\right)} \tag{9.24}$$

where the Boltzmann substitution is used for the parabolic growth of the $\beta - \gamma$ interphase boundary $y = k\sqrt{t}$.

Taking into account the parabolic movement of the $\eta - \beta$ interphase boundary, $y_L = -k_L\sqrt{t}$, and combining Equation 9.22, we can write down the growth rate of the β-phase as

$$\frac{d\Delta_{1\gamma}}{dt} = \frac{dy}{dt} - \frac{dy_L}{dt} = \frac{k + k_L}{2} = \left(\frac{1}{1/2 - c_B^\beta} + \frac{1}{c_{BL}^\beta}\right) \frac{D_{int}^\beta \Delta c_1^*}{\Delta c_1 (k + k_L)} \tag{9.25}$$

The problem is generally reduced to determining two unknown parameters, k and c_{BL}^γ, from two equations obtained from Equations 9.22–9.25:

$$k = \sqrt{2\left(\frac{1}{1/2 - c_B^\beta} + \frac{1}{c_{BL}^\beta}\right) D_{int}^\beta \delta - \frac{2D_{int}^\beta \delta}{c_{BL}^\beta}} \qquad (9.26)$$

$$c_{BL}^\gamma = \frac{\dfrac{k}{4} - c_{BL}^\gamma(\infty)\sqrt{\dfrac{\tilde{D}^\gamma}{\pi}}\,\dfrac{\exp\left(\frac{k^2}{4\tilde{D}^\gamma}\right)}{1 - \mathrm{erf}\left(\frac{k}{2\sqrt{\tilde{D}^\gamma}}\right)}}{\dfrac{k}{2} - \sqrt{\dfrac{\tilde{D}^\gamma}{\pi}}\,\dfrac{\exp\left(-\frac{k^2}{4\tilde{D}^\gamma}\right)}{1 - \mathrm{erf}\left(\frac{k}{2\sqrt{\tilde{D}^\gamma}}\right)}} \qquad (9.27)$$

The growth rate constant of the left η–β interphase boundary is given by the relation:

$$k_L = \sqrt{\frac{2D_{int}^\beta \delta}{c_{BL}^\beta}} \qquad (9.28)$$

The unknown parameters which enter Equations 9.26–9.27 are found as follows:

$$c_{CL}^\gamma = 1/2 - c_{BL}^\gamma \qquad (9.29)$$

$$c_B^\beta = c_{BL}^\beta + \left(c_{BR}^\beta - c_{BL}^\beta\right)\frac{c_{BL}^\gamma - c_B^{min}}{c_B^{max} - c_B^{min}} \qquad (9.30)$$

Analyzing Equations 9.26–9.30, let us examine now, the effect of the third component (whose diffusivity in the ternary intermediate γ-phase is different from that of the main component) on the growth rate of the β phase.

9.2.2.3 Calculations and Discussion

Let us accomplish the calculations for the model Si–Ni–Pt system (for detailed experimental investigations, see [13–15]). For a successful analysis we must fix the diffusivities in the intermediate phases D_{int}^β and $\tilde{D}^\gamma$, the boundary equilibrium concentrations $c_{BL}^\beta, c_{BR}^\beta, c_B^{min}, c_B^{max}$ on the phase diagram, and the initial concentration of Pt in the ternary intermediate phase $c_{BL}^\gamma(\infty)$. The integral diffusivity for the β-phase (see Equation 9.23) will be taken from [18], where diffusion processes in intermediate phases and solid solutions of the Si–Ni system have been experimentally studied. For our calculations we must use the concentration interval where the intermediate binary β-phase exists, and we have taken it as 1 at.%. As far as we know, the Si–Ni–Pt phase diagram has not been studied in detail, so the limits of equilibrium coexistence of β- and γ-phases are unknown. For the sake of clarity, we make the calculations at an arbitrary concentration c_B^{min} and analyze its influence on the obtained results.

Table 9.1 Calculated parameters for the growth of a binary intermediate phase, depending on the interrelation between diffusion characteristics of intermediate phases at $D_{int}^{\beta} = 7 \times 10^{-16}$ m^2 s^{-1} [18].

$\tilde{D}^{\gamma}$ m^2 s^{-1}	$k_L \times 10^8$ ms$^{-1/2}$	$k \times 10^8$ ms$^{-1/2}$	$(k + k_L) \times 10^8$ ms$^{-1/2}$	c_{BL}^{γ}	c_B^{β}	δ
1.0E−12	6.36	5.33	11.69	0.473	0.3549	0.997
1.0E−13	6.34	5.31	11.65	0.471	0.3548	0.989
1.0E−14	6.25	5.23	11.48	0.461	0.3546	0.964
1.0E−15	5.89	4.90	10.79	0.421	0.3535	0.856
1.0E−16	4.38	3.57	7.95	0.277	0.3497	0.474
5.0E−17	3.62	2.93	6.55	0.221	0.3482	0.323
4.0E−17	3.36	2.71	6.08	0.204	0.3477	0.279
3.5E−17	3.21	2.59	5.80	0.195	0.3475	0.254

It should be noted once again that in our analysis, the diffusion characteristics of the intermediate ternary γ-phase are described using the interdiffusion coefficient $\tilde{D}^{\gamma}$ for Ni and Pt species at their own sublattices in the γ phase, while the diffusion characteristics of the β phase are expressed through the effective diffusivity D_{int}^{β} when the real diffusion coefficient is normalized to the concentration interval of β-phase existence.

Thus, solving Equations 9.26–9.30, we get the constants for the moving β-phase boundaries k and k_L, the concentrations c_{BL}^{γ} and c_B^{β} at $\beta - \gamma$ interphase boundary in both phases, the respective concentration c_{CL}^{γ}, and the coefficient δ, which describes the change of diffusivity for the β-phase due to alloying. The results of the calculations depending on the interrelation between diffusivities of intermediate phases are presented in Table 9.1. Here, the following set of main parameters have been used.

From Table 9.1, it is seen that at a reduction of the interdiffusion coefficient in the ternary γ phase, the growth rate constant of the binary β phase decreases, the concentration c_B^{β} in the β phase at the β–γ interphase boundary approaches the boundary concentration c_{BL}^{β}, the concentration c_{BL}^{γ} in the γ phase at the same interphase boundary becomes more and more different from $c_{BL}^{\gamma}(\infty)$, and the coefficient δ, which characterizes the diffusivity of the β phase, decreases. Thus, as the interdiffusion coefficient in the ternary γ phase diminishes (which is achieved due to introduced Pt or other additions), the possibility of growth suppression of the binary intermediate β phase becomes evident. At that, the magnitudes of the boundary concentrations at the β–γ interphase boundary correspond to the conode, which shifts along the phase diagram toward c_B^{min}, and the respective value c_B^{β} determines the concentration interval of the β phase, which allows one to find that the effective diffusivity of the β phase is less than the maximal one D_{int}^{β} by $1/\delta$. The concentration distributions and locations of phase boundaries y_L and y obtained from the calculations for the model Si–Ni–Pt system are given in Figure 9.7.

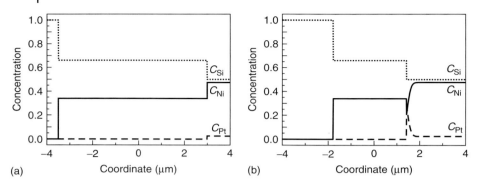

Figure 9.7 Concentration distributions in the diffusion zone and positions of phase boundaries, y_L and y at $D^\beta_{int} = 7 \cdot 10^{-16}$ m² s⁻¹ and (a) $\tilde{D}^\gamma = 3 \cdot 10^{-12}$ m² s⁻¹, (b) $\tilde{D}^\gamma = 3.5 \cdot 10^{-17}$ m² s⁻¹.

So, as the interdiffusion coefficient in the ternary γ phase diminishes, it becomes possible to observe the growth suppression of the binary intermediate β phase. Further, the magnitudes of boundary concentrations at the β–γ interphase boundary correspond to the conode, which shifts along the phase diagram toward c_B^{min}, and the respective value of c_B^β determines the concentration interval of the β phase, which allows one to find that the effective diffusivity of the β phase is less than the maximal one D^β_{int} by $1/\delta$, where

$$\delta = \frac{\Delta c_1^*}{\Delta c_1} = \frac{c_{BL}^\gamma - c_B^{min}}{c_B^{max} - c_B^{min}}$$

Here, the important physical factor is the effect of the slow pushing out of the third component by the moving phase boundary because of its low solubility in the neighboring phase and with respect to the low diffusivity in the parent ternary phase.

9.3
Ambiguity and the Problem of Selection of the Diffusion Path

9.3.1
General Remarks

The phase composition of the diffusion zone for reactive diffusion in multicomponent systems is rather difficult to predict [19–21]. The reason is that some of the phases which are present in the equilibrium phase diagram do not appear even after a long period of annealing [22, 23]. This effect depends on the initial composition of the samples and on the "regime" of diffusion interaction. Note that in an unlimited binary system, all intermediate phases will sooner or later definitely appear. The increase of the number of components participating in diffusion results in an increase in the number of the "possible" regimes for

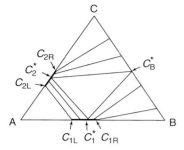

Figure 9.8 Isothermal cross section of the phase diagram for the ternary system chosen as a model of investigation.

diffusion paths for the system. We will try to understand this situation with the example of a ternary system with a simple phase diagram by considering, whenever possible, the most complete spectrum of possible regimes of diffusion (morphologies of the diffusion zone), and we will analyze the possibility of their realization.

The system with an isothermal section of the phase diagram shown in Figure 9.8 is chosen as the model. The components B and C may form a binary solid solution (β) of any composition. The intermediate phase 1 is practically a binary A–B phase, $[c_{1L}, c_{1R}]$ is the homogeneity range for the component B in phase 1. The intermediate phase 2 is practically the binary A–C phase with a homogeneity range $[c_{2L}, c_{2R}]$ of component C. The simplified diagram of a triple system considered here is similar to the diagram Mo–Fe–Co at $t \approx 1200\,°C$ (Figure 9.9).

The following systems can be in thermodynamic equilibrium:

1) (a) a pure component A (strictly speaking, a very dilute solution of B and C in A) and the intermediate phase 1;
   (b) a pure component A and an intermediate phase 2;
2) intermediate phases 1 and 2 (in concentration intervals for the phase 1 on the component B $[c_{1L}, c_1^*]$ and for the phase 2 on the component C $[c_{2L}, c_2^*]$), due to the inevitable (but low) solubility of C in AB and of B in AC;
3) (a) intermediate phase 1 (in an interval of concentrations $[c_1^*, c_{1R}]$) with the solid solution β (in an interval of concentrations on the component B $[c_B^*, 1]$);

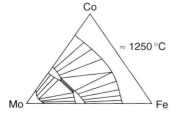

Figure 9.9 Isothermal cross section of the Mo–Co–Fe phase diagram.

(b) an intermediate phase 2 (in an interval of concentrations $[c_2, c_{2R}]$) with a solid solution β (in an interval of concentrations on the component B $[0, c_B^*]$).

The simplest form of the equations for the conode is chosen.
For equilibrium of intermediate phases 1 and 2 (case 2)

$$\frac{c_2' - c_{2L}}{c_2^* - c_{2L}} = \frac{c_1' - c_{1L}}{c_1^* - c_{1L}}$$

where c_2' is the concentration of component C in the phase 2 on the boundary with phase 1; c_1' is the concentration of component B in phase 1 on the boundary with phase 2; and c_2' and c_1' lie on the same conode. As the equilibrium of phases requires equality of chemical potentials of each component in these phases, from the equilibrium between intermediate phases 1 and 2, it follows that phase 1 must contain the component C even if it is a very small amount, and phase 2, the component B.

For phase 1 and β-solution (case 3a), we have

$$\frac{c_1' - c_1^*}{c_{1R} - c_1^*} = \frac{c_B' - c_B^*}{1 - c_B^*}$$

where c_B' is the concentration of B in the solution β on the boundary with phase 1; c_1 is the concentration of B in phase 1 on the boundary with the solution β; and c_B', c_1' lie on the same conode.

For phase 2 and β-solution (case 3b), we have

$$\frac{c_2' - c_2^*}{c_{2R} - c_2^*} = \frac{c_C' - c_C^*}{1 - c_C^*}$$

where c_2' is the concentration of C in phase 2 on the boundary with a solution β. c_C is the concentration of C in the solution β on the boundary with phase 2:

$$c_C^* = 1 - c_B^*$$

where c_2' and c_C' lie on the same conode.

The diffusion couple "pure A–solid solution β (with a concentration of component B equal to c_B^∞ and zero concentration of A)" is considered. At the annealing of such a (diffusion) pair, theoretically, there can be several alternative choices for the diffusion path; for example, A-1-β, A-2-β, A-1-2-β, and A-2-1-β. The formation of the two-phase zone as a result of diffusion is also possible. Our aim is to find out the following:

1) Which diffusion path does the system choose?
2) How does the possibility of realization of that or another diffusion path depend on the parameters of the system?

9.3.2
Analytical Solution of the Simplified Symmetric Model

Let us start the analysis by finding the analytical solution for the simplified symmetric problem $c_1 = c_2 = 1/2$ (c_1 is the mean concentration of a component

9.3 Ambiguity and the Problem of Selection of the Diffusion Path

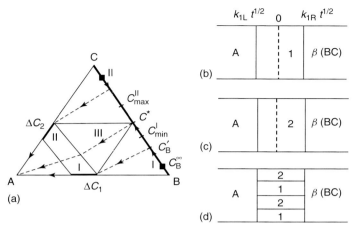

Figure 9.10 Possible modes of phase formation at the reaction A + (BC): (a) possible diffusion paths; (b) formation of layer of phase 1; (c) formation of layer of phase 2; (d) formation of a two-phase zone with parallel connection.

B in the intermediate phase 1 and c_2 is the mean concentration of a component C in the intermediate phase 2; Figure 9.10a). Consider at first two possibilities for reactive diffusion without formation of a two-phase zone for such system: A-1-β (I) and A-2-β (II).

For the regime A-1-β (Figure 9.10b) let us write three flux balance equations (in the laboratory coordinate system): one of them for component B on the interphase A–1 and two equations for B and C on the interface 1–β. Consider the problem for a stage of parabolic growth (i.e., at long periods of annealing). Then, these equations determine three parameters, the constants of growth rate k_{1L}, k_{1R} ($x_{A1} = k_{1L} t^{1/2}$, $x_{1\beta} = k_{1R} t^{1/2}$), and the concentration of component B in the β-solution c'_B on the moving interphase 1–β. Let the coefficients of interdiffusion in the β solution, $\tilde{D}$, and in phase 1, D_1, be constant. Then, the equations of balance and expressions for fluxes will be as follows:

$$(c_1 - 0)\frac{dx_L}{dt} = -\frac{D_1 \Delta c_1}{\Delta x_1} \tag{9.31}$$

$$(c'_B - c_1)\frac{dx_R}{dt} = \frac{D_1 \Delta c_1}{\Delta x_1} + I_B^{(\beta)} \tag{9.32}$$

$$I_B^{(\beta)} = -I_C^{(\beta)} = -\tilde{D}\frac{\partial c_B}{\partial x}\bigg|_{x_R+0}$$

$$(c'_C - 0)\frac{dx_R}{dt} = 0 + I_C^{(\beta)} \tag{9.33}$$

Equations 9.32 and 9.33 yield

$$(1 - c_1)\frac{dx_R}{dt} = \frac{D_1 \Delta c_1}{\Delta x_1} \tag{9.34}$$

Here $I_B^{(\beta)}$ is the flux of component B in β-solution; $I_C^{(\beta)}$ is the flux of component C in β-solution; dx_R/dt is the rate of motion of a boundary between the phase 1 and β solution; dx_L/dt is the rate of motion of the boundary between phase 1 and A.

The solution of the system of Equations 9.31–9.34, together with the diffusion equation in the β solution gives the following results:

$$\Delta x_1 = k_1 \sqrt{t} = \sqrt{\frac{2 D_1 \Delta c_1}{c_1 (1-c_1)} t} \tag{9.35}$$

$$x_R = k_R \sqrt{t}, \quad k_R = \frac{2}{(1-c_1)} \frac{D_1 \Delta c_1}{k_1} = 2 \sqrt{\frac{2 c_1}{(1-c_1)} D_1 \Delta c_1} \tag{9.36}$$

Let us now compute the boundary concentration c'_B. For that, let us determine the flow of B in the β phase near the moving boundary x_R in terms of c'_B:

$$I_B^\beta = -\sqrt{\frac{\tilde{D}}{\pi t}} \left\{ \frac{c'_B - c_B^\infty}{\mathrm{erfc}\left(\frac{k_R}{2\sqrt{\tilde{D}}}\right)} \right\} \exp\left(-\frac{k_R^2}{4\tilde{D}}\right) \tag{9.37}$$

Let us consider the case when $k_R^2 \ll 2\tilde{D}$ ($D_1 \Delta c_1 \ll 2\tilde{D}$). Then

$$I_B^\beta = -\sqrt{\frac{\tilde{D}}{\pi t}} (c'_B - c_B^\infty) \tag{9.38}$$

In order that Equations 9.32 and 9.33 be consistent, it is necessary that

$$\frac{c'_B - c_1}{c'_C - 0} = \frac{D_1 \Delta c_1 / \Delta x_1 + I_B^\beta}{-I_B^\beta}$$

and accounting for Equations 9.37 and 9.35, it can be reduced to the form

$$c'_B = \frac{c_B^\infty - \sqrt{\frac{\pi c_1}{2(1-c_1)} \frac{D_1 \Delta c_1}{\tilde{D}}}}{1 - \sqrt{\frac{\pi c_1}{2(1-c_1)} \frac{D_1 \Delta c_1}{\tilde{D}}}} < c_B^\infty \tag{9.39}$$

Equation 9.39 is obtained for the case

$$\frac{D_1 \Delta c_1}{\tilde{D}} \ll \frac{1 - c_1}{c_1}$$

It means that the atoms C can not penetrate almost into the growing phase 1 and consequently, they are accumulated in the β solution near the interphase. Thus $c'_C > c_C^\infty$ ($c'_B < c_B^\infty$).

It is obvious that the aforementioned solution is possible only if $c'_B > c_B^*$. If this condition is not fulfilled, then c'_B falls in a region of concentrations of the β solution which is in equilibrium with a phase 2, and the phase 1 is not formed. Therefore,

$$c_B^\infty - \sqrt{\frac{\pi c_1}{2(1-c_1)} \frac{D_1 \Delta c_1}{\tilde{D}}} > c_B^* \left(1 - \sqrt{\frac{\pi c_1}{2(1-c_1)} \frac{D_1 \Delta c_1}{\tilde{D}}}\right)$$

that is,

$$c_B^\infty > c_B^* + (1 - c_B^*)\sqrt{\frac{\pi c_1}{2(1-c_1)}\frac{D_1 \Delta c_1}{\tilde{D}}} \tag{9.40}$$

Thus, the choice of the diffusion path I (A-1-β) made by the system is possible only at the realization of condition 9.40. By analogy, the choice of a diffusion path II, A-2-β (Figure 9.10b) is possible only under the condition

$$c_B^\infty = 1 - c_C^\infty < c_B^* - c_B^*\sqrt{\frac{\pi c_2}{2(1-c_2)}\frac{D_2 \Delta c_2}{\tilde{D}}} \tag{9.41}$$

Thus, there exists an interval of initial concentrations for the β solution, in which both considered conditions are prohibited:

$$c_B^* - c_B^*\sqrt{\frac{\pi c_2}{2(1-c_2)}\frac{D_2 \Delta c_2}{\tilde{D}}} < c_B^\infty < c_B^* + (1 - c_B^*)\sqrt{\frac{\pi c_1}{2(1-c_1)}\frac{D_1 \Delta c_1}{\tilde{D}}} \tag{9.42}$$

This interval is rather narrow, as $D_i \Delta c_i \ll \tilde{D}$ (according to our assumption), but also it is not necessarily too narrow (≤ 0.1).

If the initial composition of the p solution falls in this interval, then we assume that the two-phase zone will arise and will grow; this is the diffusion regime III (Figure 9.10d). The diffusion path in this case will pass along the side BC of a concentration triangle up to a point c_B^* and will "jump" on the conode between phases 1 and 2, which is one of the sides of a three-phase triangle (Figure 9.10a). We will assume morphology of parallel connection for the two-phase region. The equations of balance for the average fluxes of components at the left and right boundaries of a two-phase region will be the following:

BL:
$$(p_{1L}c_1 - 0)\frac{dx_L}{dt} = 0 - \frac{p_{1L} + p_{1R}}{2}\frac{D_1 \Delta c_1}{\Delta x} \tag{9.43}$$

CL:
$$(p_{2L}c_2 - 0)\frac{dx_L}{dt} \cong 0 - \frac{p_{2L} + p_{2R}}{2}\frac{D_2 \Delta c_2}{\Delta x} \tag{9.44}$$

BR:
$$(c_B^* - c_1 p_{1R})\frac{dx_R}{dt} = \frac{p_{1L} + p_{1R}}{2}\frac{D_1 \Delta c_1}{\Delta x} + I_B^{(\beta)} \tag{9.45}$$

CR:
$$(1 - c_B^* - c_2 p_{2R})\frac{dx_R}{dt} \cong \frac{p_{2L} + p_{2R}}{2}\frac{D_2 \Delta c_2}{\Delta x} + I_C^{(\beta)} \tag{9.46}$$

where

$$I_B^{(\beta)} = -I_C^{(\beta)} \approx \sqrt{\frac{\tilde{D}}{\pi t}}(c_B^* - c_B^\infty) \tag{9.47}$$

p_{1L} and p_{2L} are the volume fractions of phases 1 and 2, respectively, in the two-phase region on its left boundary; p_{1R} and p_{2R} are the volume fractions of phases 1 and 2, respectively, in the two-phase region on its right boundary.

Equation 9.47 implies that to conserve the quasi-equilibrium on the two-phase zone–β phase boundary, a diffusion path should go through c_B^*, and then the system will not choose a concentration in phases 1, 2, but their fractions on the interphase (p_1, p_2): p_{1R} $(p_{2R} = 1 - p_{1R})$ and p_{1L} are the places where the diffusion path comes into a two-phase region and leaves it.

We will consider a particular symmetrical case

$$D_1 \Delta c_1 = D_2 \Delta c_2 = D\Delta c, \qquad c_1 = c_2 = 1/2, \qquad c_B^* = 1/2$$

Then, from Equations 9.43 and 9.44 it follows that

$$\frac{p_{2L}}{p_{1L}} = \frac{p_{2L} + p_{2R}}{p_{1L} + p_{1R}}$$

so $p_{1L} = p_{2L}$, $p_{1R} = p_{2R}$ (the fraction of phases does not change along the axis X). From this, it is easy to obtain that

$$k = \sqrt{2D\Delta c \cdot \frac{1}{\left(1 - \frac{p_1 + p_2}{2}\right)\frac{p_1 + p_2}{2}}} = \sqrt{8D\Delta c} \qquad (9.48)$$

$$p_2 = \frac{1}{2} + \left(\frac{1}{2} - c_B^\infty\right)\sqrt{\frac{2}{D\Delta c}\frac{\tilde{D}}{\pi}}$$

$$p_1 = \frac{1}{2} - \left(\frac{1}{2} - c_B^\infty\right)\sqrt{\frac{2}{D\Delta c}\frac{\tilde{D}}{\pi}}$$

This result has a physical sense $(0 < p_1, p_2 < 1)$ if

$$\left|\frac{1}{2} - c_B^\infty\right| < \sqrt{\frac{\pi}{8}\frac{D\Delta c}{\tilde{D}}}$$

which is the condition of diffusion regime III.

Note, that according to Equation 9.42 the diffusion paths I, II are not possible if

$$\left|\frac{1}{2} - c_B^\infty\right| < \sqrt{\frac{\pi}{8}\frac{D\Delta c}{\tilde{D}}}$$

$$D_1 \Delta c_1 = D_2 \Delta c_2, \qquad c_1 = c_2 = \frac{1}{2}, \qquad c_B^* = \frac{1}{2}$$

Thus, the side (BC) is divided into three intervals:

Interval I:

$$0 < c_B^\infty < \frac{1}{2} - \sqrt{\frac{\pi}{8}\frac{D\Delta c}{\tilde{D}}}$$

Here, only Regime I is possible (formation and growth of an intermediate phase 1 (Figure 9.10b)).

Interval II:

$$\frac{1}{2} - \sqrt{\frac{\pi}{8}\frac{D\Delta c}{\tilde{D}}} < c_B^\infty < \frac{1}{2} + \sqrt{\frac{\pi}{8}\frac{D\Delta c}{\tilde{D}}}$$

Here only Regime III is possible (formation and growth of a two-phase zone (Figure 9.10d)).

Interval III:

$$c_B^\infty > \frac{1}{2} + \sqrt{\frac{\pi}{8}\frac{D\Delta c}{\tilde{D}}}$$

Here only Regime II is possible (formation and growth of an intermediate phase 2 (Figure 9.10c)).

The analytical solution obtained has allowed us to trace the possibility of various choices for diffusion paths (modes) and to also find the requirements which the system should satisfy for these modes to be realized. But for this purpose, it was necessary to simplify the model and to use a series of approximations. As the interval of concentrations for c_B^∞, within which the formation of a two-phase region is possible is too narrow, it may seem that the existence of diffusion Mode III also grows out of crude approximations (Equation 9.38) for the diffusion fluxes. Therefore, we also offer the more rigorous approach which requires numerical calculations.

9.3.3
Numerical Calculations for a Complex Model

Let us model the isothermal diffusion interaction in a three-component system which corresponds to the state diagram in Figure 9.8. Let the diffusion path at an annealing of the diffusion pair $A\beta$ pass through an intermediate phase 1 as shown in Figure 9.11.

Let us write down the equations of the balance of fluxes of components on the interfaces in the laboratory frame of reference and the equation of the conode for phase 1 and the β solution:

$$(c_{1L} - 0)\frac{dx_{A1}}{dt} = J_B^{(1)} - 0 \qquad (9.49)$$

$$(c_1' - c_B')\frac{dx_{\beta 1}}{dt} = J_B^{(1)} - J_B^{(\beta)}$$

$$(1 - c_B')\frac{dx_{\beta 1}}{dt} = J_C^{(\beta)} - 0$$

$$\frac{c_1' - c_1^*}{c_{1R} - c_1^*} = \frac{c_B' - c_B^*}{1 - c_B^*}$$

where $J_B^{(1)}$ is the flux of the component B in phase 1; $J_B^{(\beta)}$ is the flux of a component B in the β solution; $J_C^{(\beta)}$ is the flux of a component C in the β solution; $dx_{\beta 1}/dt$ is the rate of motion of the interphase 1-β; and dx_{A1}/dt is the rate of motion of the interphase 1-A.

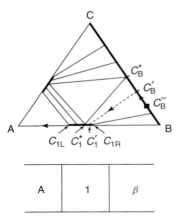

Figure 9.11 Schematic diagram of diffusion, Mode I.

Consider the problem in an approximation of parabolic growth (i.e., at long periods of annealing). Then $x_{A1} = k_{A1} t^{1/2}$, $x_{\beta 1} = k_{\beta 1} t^{1/2}$ and the width of phase 1 $\Delta x_1 = (k_{\beta 1} - k_{A1}) t^{1/2}$. Let the diffusion coefficients $\tilde{D}$ and D_1 be constant. Then the expressions for fluxes take the following form:

$$J_B^{(1)} \approx -D_1 \frac{c_1' - c_{1L}}{\Delta x_1} = -D_1 \frac{c_1' - c_{1L}}{(k_{\beta 1} - k_{A1})\sqrt{t}} \tag{9.50}$$

$$J_B^{(\beta)} = -\tilde{D} \left. \frac{\partial c}{\partial x} \right|_{x_{\beta 1}} = -\tilde{D} \frac{(c_B^\infty - c_B')}{1 - \mathrm{erf}\left(\frac{k_{\beta 1}}{2\sqrt{\tilde{D}}}\right)} \frac{1}{\sqrt{\tilde{D} \pi t}} \exp\left(-\frac{k_{\beta 1}^2}{4\tilde{D}}\right)$$

$$J_C^{(\beta)} = -J_B^{(\beta)}$$

where D_1 is the interdiffusion coefficient in phase 1; $\tilde{D}$ is the interdiffusion coefficient in the β solution.

In view of these expressions, the set of equations becomes

$$c_{1L} \frac{k_{A1}}{2} = -D_1 \frac{c_1' - c_{1L}}{(k_{\beta 1} - k_{A1})} \tag{9.51}$$

$$(c_1' - c_B') \frac{k_{\beta 1}}{2} = \sqrt{\frac{\tilde{D}}{\pi}} \frac{(c_B^\infty - c_B')}{\left[1 - \mathrm{erf}\left(\frac{k_{\beta 1}}{2\sqrt{\tilde{D}}}\right)\right]} \exp\left(-\frac{k_{\beta 1}^2}{4\tilde{D}}\right) - D_1 \frac{c_1' - c_{1L}}{(k_{\beta 1} - k_{A1})}$$

$$(1 - c_B') \frac{k_{\beta 1}}{2} = \sqrt{\frac{\tilde{D}}{\pi}} \frac{(c_B^\infty - c_B')}{\left[1 - \mathrm{erf}\left(\frac{k_{\beta 1}}{2\sqrt{\tilde{D}}}\right)\right]} \exp\left(-\frac{k_{\beta 1}^2}{4\tilde{D}}\right)$$

$$c_B' = c_1' \frac{(1 - c_B^*)}{(c_{1R} - c_1^*)} + \frac{(c_{1R} c_B^* - c_1^*)}{(c_{1R} - c_1^*)}$$

The obtained set of equations allows us to find concentrations lying on conode c_1', c_B' (i.e., the concentrations of component B on the interphase 1–β in phase 1 and

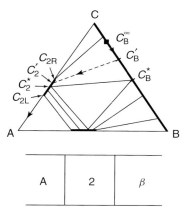

Figure 9.12 Schematic diagram of diffusion, Mode II.

in the β solution respectively), and also the constants of motion of the boundaries $k_{\beta 1}$ and k_{A1}. By substituting the concrete parameters of a diffusion pair into the given equations, we will obtain the values for the magnitudes mentioned above. If at given parameters the system has solutions, and these solutions correspond to the phase diagram and physical sense (the growth rate of a phase is positive, and the concentrations belong to regions of homogeneity of the phases), then we consider the realization of such a diffusion path to be possible. An investigation of the set of equations (by varying the parameters of a diffusion couple) has been carried out by numerical methods. As the choice of diffusion path depends obviously only on the ratio of diffusion coefficients, we set the diffusion coefficients as dimensionless numbers.

One more possibility is the diffusion path which passes through the intermediate phase 2 (Figure 9.12). The flux balance equations on the phase boundaries and equation of conode for phase 2 and β solution have a form similar to the previous case. The parameter which can influence the result of a diffusion process in the simplest way is the initial composition of a diffusion pair (in our model it is c_B^∞). Thus, we have investigated the possibility of Modes I (Figure 9.11) and II (Figure 9.12) depending on c_B^∞. The decrease of the interdiffusion coefficient in the initial solid solution β, as is seen from Figure 9.13, results in the narrowing of the region of single-phase modes. That is, the slow diffusion in the solution (in comparison to diffusion in intermediate phases) does not allow the system to arrange the distribution of concentration with the growth of phases in such a way as to accept the conditions mentioned above.

The calculations have also shown that the width of the region of realization of Modes I and II depends also on other parameters of the system. The interval for c_B^∞, in which the realization of Mode I is possible, depends on the ratio

$$\frac{D_1 \left(c_1^* - c_{1L}\right)}{\tilde{D} \left(c_B^\infty - c_B^*\right)}$$

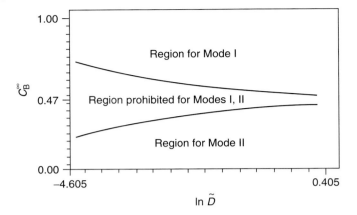

Figure 9.13 Regions of realization of diffusion, Modes I and II, depending on the initial compositions of the β solution and its interdiffusion coefficient (here $c_B^* = 0.47$, $c_{1L} = 0.48$, $c_1^* = 0.52$, $c_{1R} = 0.53$, $c_{2L} = 0.38$, $c_2^* = 0.48$, $c_{2R} = 0.49$, $D_1 = 0.7$, $D_2 = 0.01$).

and of Mode II on the ratio

$$\frac{D_2 \left(c_2^* - c_{2L}\right)}{\tilde{D}\left(c_B^\infty - c_B^*\right)}$$

The larger these values, the narrower is the region of realization of Mode I or II, respectively. Thus, there exists an interval of concentrations for c_B^∞, in which neither Mode I nor Mode II can be realized. And therefore, we assume that the diffusion path in the system will be different if the initial composition of the diffusion pair falls into regions "forbidden" for Modes I and II.

The following possible modes for our model are the "consecutive connection modes" with intermediate phases 1 and 2 (A-1-2-β, A-2-1-β) sequentially distributed in a diffusion zone, when the diffusion path jumps from one side of a concentration triangle to another. Consider the case A-2-1-β (Figure 9.14). The natural question is how does component C penetrate from the β phase through phase 1 into phase 2? C comes into phase 2 due to the flux of this component in phase 1. In the introduction, the reason due to which the phase 1 nevertheless contains the component C, although in a very small amount, has been discussed and this ensures the flux of this component in it. This flux provides the small but finite concentration of component C on the interface 1–2 in phase 1, and due to the condition of quasi-equilibrium the significantly larger concentration of this component on the other side of the interface in phase 2. This concentration is larger than the concentration on the interface A–2, which ensures the necessary direction of a flux of C in phase 2 and consequently, the possibility of growth of this phase. Thus, the chemical quasi-equilibrium on the interface 1–2 serves as such a "pump" which draws out atoms C from the β solution through a phase 1 (despite this phase itself may have small concentration of component C).

9.3 Ambiguity and the Problem of Selection of the Diffusion Path

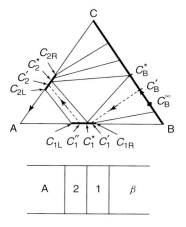

Figure 9.14 Schematic diagram of diffusion, Mode V.

In view of these considerations, the flux balance equations for components B and C on the phase boundaries for diffusion, Mode V, will be as follows:

$$(c_{2L} - 0)\frac{dx_{A2}}{dt} = J_C^{(2)} - 0 \qquad (9.52)$$

$$\left(c_2^{(1)} - c_2'\right)\frac{dx_{21}}{dt} = J_C^{(1)} - J_C^{(2)} \qquad \left(c_1'' - c_1^{(2)}\right)\frac{dx_{21}}{dt} = J_B^{(1)} - J_B^{(2)}$$

$$\left(c_B' - c_1'\right)\frac{dx_{1\beta}}{dt} = J_B^{(\beta)} - J_B^{(1)} \qquad \left(1 - c_B' - c_2^{(1)'}\right)\frac{dx_{1\beta}}{dt} = J_C^{(\beta)} - J_C^{(1)}$$

where $J_C^{(2)}$ is the flux of component C in phase 2; $J_C^{(1)}$ is the flux of component C in phase 1; $J_B^{(1)}$ is the flux of component B in phase 1; $J_B^{(2)}$ is the flux of component B in phase 2; $J_B^{(\beta)}$ is the flux of component B in β solution; $J_C^{(\beta)}$ is the flux of component C in β solution; $c_1^{(2)}$ is the concentration of component B in phase 2 on the interphase 2–1; $c_2^{(1)}$ is the concentration of component C in phase 1 on the interphase 2–1; $c_2^{(1)'}$ is the concentration of component C in phase 1 on the interphase 1–β. As $c_1^{(2)} \ll c_1''$, $c_2^{(1)} \ll c_2'$, $c_2^{(1)'} \ll c_B'$, then $J_C^{(1)} \ll J_C^{(2)}$, $J_B^{(2)} \ll J_B^{(1)}$, $J_C^{(1)} \ll J_C^{(\beta)}$ (but for this purpose, the existence of a particular relationship is also necessary between diffusion parameters of phases 1 and 2, the nature of which will be further determined from the results for Modes V and VI). That is why the fluxes of components C in phase 1 and B in phase 2 can be neglected in the equations of balance at the calculation of the rates of motion of the phase boundaries. This has allowed us to simplify the flux balance equations on the phase boundaries:

$$(c_{2L} - 0)\frac{dx_{A2}}{dt} = J_C^{(2)} - 0, \qquad (0 - c_2')\frac{dx_{21}}{dt} = 0 - J_C^{(2)} \qquad (9.53)$$

$$(c_1'' - 0)\frac{dx_{21}}{dt} = J_B^{(1)} - 0, \qquad (c_B' - c_1')\frac{dx_{1\beta}}{dt} = J_B^{(\beta)} - J_B^{(1)}$$

$$(1 - c_B')\frac{dx_{1\beta}}{dt} = J_C^{(\beta)} - 0, \qquad \frac{c_B' - c_B^*}{1 - c_B^*} = \frac{c_1' - c_1^*}{c_{1R} - c_1^*}$$

$$\frac{c_2^* - c_2'}{c_2^* - c_{2L}} = \frac{c_1^* - c_1''}{c_1^* - c_{1L}}$$

where $J_C^{(2)}$ is the flux of a component C in phase 2; $J_B^{(1)}$ is the flux of a component B in phase 1; $J_B^{(\beta)}$ is the flux of component B in the β solution; $J_C^{(\beta)}$ is the flux of a component C in β solution; $dx_{1\beta}/dt$ is the velocity of the boundary between phase 1 and β solution; dx_{21}/dt is the velocity of the boundary between phases 2–1; dx_{A2}/dt is the velocity of the boundary between phases 2–A.

Let us consider the problem in the approximation of parabolic growth:

$$x_{A2} = k_{A2}t^{1/2}, \qquad x_{21} = k_{21}t^{1/2}, \qquad x_{1\beta} = k_{1\beta}t^{1/2}$$

and the width of phase 1

$$\Delta x_1 = (k_{1\beta} - k_{21})t^{1/2}$$

and of phase 2

$$\Delta x_2 = (k_{21} - k_{A2})t^{1/2}$$

Let the diffusion coefficients $\tilde{D}$, D_1, and D_2 be constant. Then the expressions for the fluxes will be as follows:

$$J_B^{(1)} \approx -D_1 \frac{c_1' - c_1''}{\Delta x_1} = -D_1 \frac{c_1' - c_1''}{(k_{1\beta} - k_{21})\sqrt{t}} \tag{9.54}$$

$$J_C^{(2)} \approx -D_2 \frac{c_2' - c_{2L}''}{\Delta x_2} = -D_1 \frac{c_1' - c_1''}{(k_{21} - k_{A2})\sqrt{t}}$$

$$J_B^{(\beta)} = -\tilde{D}\frac{\partial c}{\partial x}\bigg|_{x_{\beta 1}} = -\tilde{D}\frac{(c_B^\infty - c_B')}{1 - \mathrm{erf}\left(\frac{k_{1\beta}}{2\sqrt{\tilde{D}}}\right)} \frac{1}{\sqrt{\tilde{D}\pi t}} \exp\left(-\frac{k_{1\beta}^2}{4\tilde{D}}\right)$$

$$J_C^{(\beta)} = -J_B^{(\beta)}$$

where D_1 is the inter-diffusion coefficient in phase 1; D_2 is the interdiffusion coefficient in phase 2; $\tilde{D}$ is the interdiffusion coefficient in the β solution.

From the first two equations of the set, the relation between the constants of the rate of motion of the right and left boundaries of a phase 2 is as follows: $k_{A2} = (c_2'/c_{2L})k_{21}$. Both constants have an identical (negative) sign and thus are close in value, because c_2' differs only a little from c_{2L}. This means that phase 2 moves as a whole to the left, and is "repelled" by the growing phase 1. Thus, the rate of growth of the thickness of phase 2 is small (if this condition is generally possible). A similar reasoning can be applied for case A-1-2-β (Figure 9.15) of the Mode VI, and we will obtain a set of equations similar to the set for the Mode V.

The investigations of Modes V and VI have shown that they can be realized only at particular relationships between the parameters of the system. First of all, it concerns the interdiffusion coefficients of intermediate phases D_1 and D_2. So the realization of condition V is possible when D_1 is significantly larger than D_2,

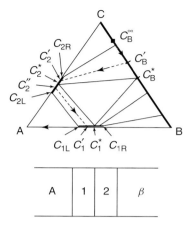

Figure 9.15 Schematic diagram of diffusion, Mode VI.

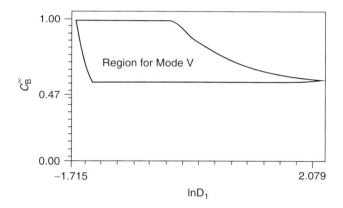

Figure 9.16 The region of realization of diffusion, Mode V (A-2-1-β), depending on the initial composition of the β solution and the interdiffusion coefficient in the intermediate phase.

and the realization of condition VI, when D_2 is significantly larger than D_1. As in the previous cases, the initial composition of the β solution c_B^∞ is considered a parameter, with the help of which it is possible to influence the diffusion path and its choice by the system. Figure 9.16 presents the dependence of the interval for c_B^∞, within which the Mode V is realized on D_1.

The width of intervals for c_B^∞ also depends on $[c_{1L}, c_1^*]$ in the case of a Mode V (Figure 9.17). Figure 9.18 shows how the interval of realization of a mode depends on the interdiffusion coefficient $\tilde{D}$ in β solution. Similar graphs have been obtained for Mode VI.

Calculations for the Modes V and VI have been carried out with the following parameters of the system: $D = 1$, $c_{1L} = 0.48$, $c_{2L} = 0.38$, $c_{1R} = 0.53$, $c_{2R} = 0.49$,

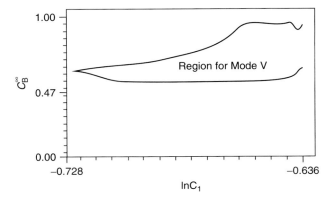

Figure 9.17 Region of realization of diffusion, Mode V (A-2-1-β), depending on the initial composition of the β solution and region of equilibrium between phase 1 and phase 2.

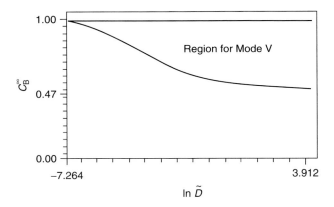

Figure 9.18 Region of realization of diffusion, Mode V (A-2-1-β), depending on the initial composition of the β solution and interdiffusion coefficient in the β solution.

$c_1^* = 0.52$, $c_2^* = 0.48$, and $c_B^* = 0.47$. For condition V $D_1 = 0.5$, $D_2 = 0.01$; for condition VI $D_2 = 0.5$, $D_1 = 0.01$. Consequently, there exist regions of parameters, within which the diffusion conditions mentioned above are impossible with growth of the phase layers. In this case, it is natural to assume that the system can choose a condition with formation of the two-phase region. The model of a two-phase zone which we accept for our study is presented in Figure 9.19.

This zone represents the "parallel" connection of the layers of intermediate phases 1 and 2 of varying thickness. In the description of this model we use a series of approximations for its simplification:

1) The boundaries between the two-phase region and pure A, two-phase region and solid solution β are rectilinear.
2) The diffusion coefficients $\tilde{D}$, D_1, and D_2 are constant.

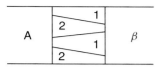

Figure 9.19 Initial model of a two-phase zone.

3) As for Modes I and II, we consider a model with parabolic growth approximation.
4) The condition of quasi-equilibrium is considered to be satisfied in each thin layer. It means that for each x inside the two-phase zone, the concentrations in both phases 1 and 2, $c_1(x)$ and $c_2(x)$, lie on the same conode; that is, they correspond to a condition of equality of chemical potentials for each component. Hence, these concentrations are not independent and can be expressed through a certain general parameter $k(x)$ which we call the parameter of the conode [24]. For example, the concentration of the component B in a phase 1 can be chosen as such a parameter.

Let us write down the equations of flows on the left and right boundaries of a two-phase zone for components B and C:

$$(p_{1L}c_{1L} - 0)\frac{k_L}{2\sqrt{t}} = p_{1L}J_B^{(1)} - 0 = -p_{1L}D_1\frac{dc_1}{dk}\frac{\partial k}{\partial x} \qquad (9.55)$$

$$(p_{2L}c_{2L} - 0)\frac{k_L}{2\sqrt{t}} = p_{2L}J_C^{(2)} - 0 = -p_{2L}D_2\frac{dc_2}{dk}\frac{\partial k}{\partial x}$$

$$(c_B^* - p_{1R}c_1^*)\frac{k_R}{2\sqrt{t}} = J_B^{(\beta)} - p_{1R}J_B^{(1)} = -\tilde{D}\frac{\partial c_1}{\partial x} - p_{1R}D_1\frac{dc_1}{dk}\frac{\partial k}{\partial x}$$

$$(1 - c_B^* - p_{2R}c_2^*)\frac{k_R}{2\sqrt{t}} = J_C^{(\beta)} - p_{1R}J_C^{(2)} = \tilde{D}\frac{\partial c_1}{\partial x} + p_{2R}D_2\frac{dc_2}{dk}\frac{\partial k}{\partial x}$$

where p_{1L} and p_{1R} are the volume fractions of the first phase on the left and right boundaries of the two-phase region, respectively; p_{2L} and p_{2R} are the volume fractions of the second phase on the left and right boundaries of a two-phase region respectively, with

$$p_{1L} + p_{2L} = 1, \qquad p_{1R} + p_{2R} = 1$$

where k_L and k_R are the constants of motion of the left and right boundaries of a two-phase region; k is the parameter of the conode; D_1, D_2, and $\tilde{D}$ are the interdiffusion coefficients of phases 1, 2 and of the β solution. Analyzing the first two equations of the set, it is easy to see that they are simultaneously fulfilled at the following conditions: (i) if the condition $(c_{1R}/c_{2R} = D_1 dc_1/D_2 dc_2)$ is satisfied (if the phase equilibrium 1–2 corresponds to the linear connection between c_1 and c_2, then $c_1/c_2 = \Delta c_1/(\Delta c_2)$), which would be an explicit exception (extremely impossible case); (ii) if $p_{1L} = 0$ or $p_{2L} = 0$.

Let us consider in greater detail the case, when $p_{2L} = 0$ and $p_{1L} = 1$ (Mode III). Then our model of the two-phase region will look a little different (Figure 9.20).

9 Diffusion Phase Competition in Ternary Systems

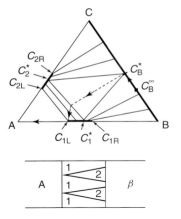

Figure 9.20 Schematic diagram of diffusion, Mode III.

The equations of balance read

$$c_{1L}\frac{k_L}{2} = -D_1\frac{\Delta c_1}{(k_R - k_L)} \tag{9.56}$$

$$(c_B^* - p_{1R}c_1^*)\frac{k_R}{2} = -\sqrt{\frac{\tilde{D}}{\pi}}\frac{c_B^\infty - c_B^*}{1 - \mathrm{erf}\left(\frac{k_R}{2\sqrt{\tilde{D}}}\right)}\exp\left(-\frac{k_R^2}{4\tilde{D}}\right) + p_{1R}D_1\frac{\Delta c_1}{(k_R - k_L)}$$

$$(1 - c_B^* - p_{2R}c_2^*)\frac{k_R}{2} = \sqrt{\frac{\tilde{D}}{\pi}}\frac{c_B^\infty - c_B^*}{1 - \mathrm{erf}\left(\frac{k_R}{2\sqrt{\tilde{D}}}\right)}\exp\left(-\frac{k_R^2}{4\tilde{D}}\right) + p_{2R}D_2\frac{\Delta c_2}{(k_R - k_L)}$$

where $\Delta c_1 = c_1^* - c_{1L}$, $\Delta c_2 = c_2^* - c_{2L}$. The solution of this set of equations will also allow us to find intervals of concentrations for c_B^∞, at which the formation of a two-phase region of such a form is possible. If we take $p_{2L} = 1$ and $p_{1L} = 0$ (Mode IV), then the two-phase region is passed (Figure 9.21) and the diffusion path will be different, while the equations of balance will be analogous.

It has been established that the values of interdiffusion coefficients for the first and second phases influence the width of intervals of initial concentrations within which the conditions III and IV can be realized (see Figure 9.22, the graph of the dependence of regions of realization of a condition III on the initial composition of the diffusion pair and $\tilde{D}$). The value of the interdiffusion coefficient influences to the largest extent, the width of intervals of concentrations c_B^∞ for diffusion by Modes III and IV. The larger $\tilde{D}$, the narrower are these intervals.

So we have considered several conditions for diffusion paths for a ternary system (Figure 9.8), have studied the possibilities of their realization depending on various diffusion parameters, have shown that the initial composition of a diffusion pair can influence not only the diffusion path within the framework of a given condition, but also the choice of such a condition. This permits us to control the result of the diffusion process.

9.3 Ambiguity and the Problem of Selection of the Diffusion Path

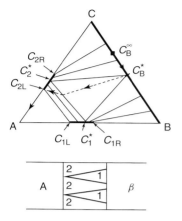

Figure 9.21 Schematic diagram of diffusion, Mode IV.

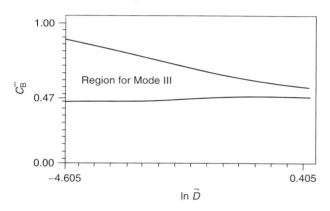

Figure 9.22 The region of realization of diffusion, Mode III, depending on the initial composition of the β solution and its interdiffusion coefficient.

Now we apply the results obtained for the analysis of development of the diffusion process in a system with given diffusion parameters ($0.01 \leq \tilde{D} \leq 5$, $D_1 = 0.7$, $D_2 = 0.01$, $c_{1L} = 0.48$, $c_{2L} = 0.38$, $c_{1R} = 0.53$, $c_{2R} = 0.49$, $c_1^* = 0.52$, $c_B^* = 0.47$, and $c_2^* = 0.48$). We have investigated this system on the possibility of realization of the diffusion conditions considered above. As a result, the following diagram (Figure 9.23) was obtained. It is possible to demarcate some regions in it: (1) the realization of Modes V, IV, and III is possible; (2) the realization of Modes V, IV, I is possible; (3) the realization of Modes V and I is possible; (4) only the Mode IV can be realized; (5) the realization of Modes IV and III is possible; (6) the realization of Modes V and IV is possible; (7) none of the considered modes is realized; and (8) only the Mode II can be realized.

The existence of a range of parameters leading to item (7) indicates that in the given system diffusion modes besides those considered above are possible. For

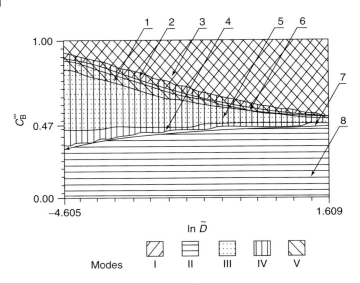

Figure 9.23 Regions of realization of diffusion Modes I, II, III, IV, and V depending on the initial composition of the β solution and its interdiffusion coefficient.

several regions on the graph (1, 2, 3, 5, and 6) ambiguous results are obtained. In these regions (according to our solutions) several conditions can be realized simultaneously. The ambiguity of diffusion conditions for interdiffusion in ternary systems with a two-phase zone has been obtained earlier in [25]. Therefore, there is a "problem" of the choice of a diffusion path out of several possible modes. Hence, in such a form, this approach does not allow us to determine unambiguously the diffusion path of the system. For the solution of the problem, this approach should be supported by some criterion of choice, or it is necessary to search for another approach. As the criterion of choice from several possible diffusion modes, the rate of change of Gibbs potential of a system may be used: the most favorable diffusion mode corresponds to the largest (minimum by absolute value) magnitude of (dG/dt).

For the first time, a similar hypothesis for the description of the criterion of solid-phase amorphization of binary systems has been offered in [26]. Another approach to the problem of searching for the diffusion path in a ternary system can be the consideration of the initial stages of diffusion; in particular, of the competition between the critical nuclei of different phases [1, 27].

9.3.4
Conclusions

1) The choice of the diffusion path at reactive diffusion in a diffusion pair A–(BC) is determined by diffusion parameters of the solution (BC), intermediate phases, and the initial concentration of the solution (BC). Thus, diffusion

modes involving formation of one or both intermediate phases are basically possible (in the form of a two-phase zone as well as a sequential connection).

2) The increase of the diffusion coefficient in the β solution expands the range of initial compositions of a solution at which the modes with growth of only one intermediate phase can be realized.

3) The region of initial compositions of the β solution, at which the growth of sequential phases A-2-1-β is possible, is narrowed down with a decrease of the interdiffusion coefficient in β solution and with increase of the interdiffusion coefficient in the intermediate phase 1.

4) A decrease of the interdiffusion coefficient in the β solution results in an expansion of the region of initial compositions of a solution at which the realization of the modes with formation of a two-phase zone is possible.

5) There are regions of initial compositions of the solution β, at which the simultaneous realization of several diffusion modes is possible. The solution of this problem needs additional research.

9.4 Nucleation in the Diffusion Zone of a Ternary System

At reaction–diffusion processes in ternary systems, the phase spectrum of the diffusion zone, even for small annealing times, depends on the initial composition of the diffusion couple and on the set of diffusion parameters [28–33]. Furthermore, the choice of the diffusion path may appear to be ambiguous [34, 35]. In such a case, the stage of nucleation becomes a decisive one.

Here we shall treat only such initial alloys of the diffusion couple that the transition between them is possible via a region of solid solutions (stable or metastable). In the literature [36–38], the theory of intermediate phase nucleation in a concentration gradient field at interdiffusion in a binary system was developed by a polymorphic transformation. In that, the formation of a nucleus occurs in the concentration-prepared region of a metastable solid solution without change in concentration. This model cannot be applied to the general case of a ternary system, since the diffusion path in a metastable solid solution may not proceed via that region of the concentration triangle where the intermediate phase exists. And formation of a nucleus of such a phase by heterophase fluctuation must be connected not only with the lattice reconstruction but also with a concentration change. This kind of nucleation for a case of binary alloys was analyzed in [39–41]. This approach is peculiar for the local composition of a nucleus being determined by a parallel tangent rule.

9.4.1 Model Description

To investigate the process of nucleation in a ternary system, we have derived the dependence of Gibbs potential change ΔG at formation of a new phase nucleus on its volume, shape, concentration composition, and on the gradients of the components' concentrations in a parent phase. The general expression for the change of the Gibbs potential at formation of a nucleus with some center x_0 having the longitudinal dimension $2R_1$ is the following:

$$\Delta G = n \int_{x_0-R_1}^{x_0+R_1} s(x) \Delta g \left(c_1^o(x), c_2^o(x) \to c_1^n(x), c_2^n(x)\right) dx + \sigma S \tag{9.57}$$

where $s(x)$ is the cross-sectional area of a nucleus with the plane X, perpendicular to the direction of diffusion fluxes; S is the surface area of the nucleus; n is the number of atoms per unit volume; σ is the interface energy per unit area of the nucleus' surface; and Δg is thermodynamic driving force (per one atom) of transformation of the old phase layer with concentrations c_1^o, c_2^o into the new phase with concentrations c_1^n, c_2^n, determined by the rule of parallel tangent planes (Figure 9.24) and obtained due to unlimited transversal diffusion in each narrow layer dx.

To get the expression for ΔG from Equation 9.57, where the concentration composition of the parent phase and nucleus size enter explicitly, some approximations such as those given below, have been made.

1) The surfaces $g(c_1, c_2)$ of the old and new phases are taken to be paraboloids:

$$g^o = g_0^o + \frac{1}{2} \frac{\partial^2 g^o}{\partial^2 c_1^2} \left(c_1^o - c_{01}^o\right)^2 + \frac{1}{2} \frac{\partial^2 g^o}{\partial^2 c_2^2} \left(c_2^o - c_{02}^o\right)^2$$

$$+ \frac{\partial^2 g^o}{\partial c_1 \partial c_2} \left(c_1^o - c_{01}^o\right) \left(c_2^o - c_{02}^o\right) \tag{9.58}$$

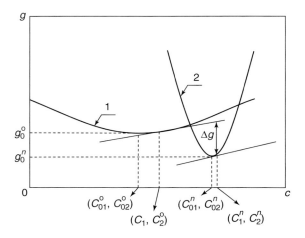

Figure 9.24 Schematic dependence of Gibbs potential per atom, g, of a ternary system on composition for solid solution (parent phase) - line 1 and intermediate (new) phase - line 2.

9.4 Nucleation in the Diffusion Zone of a Ternary System

$$g^n = g_0^n + \frac{1}{2}\frac{\partial^2 g^n}{\partial^2 c_1^2}(c_1^n - c_{01}^n)^2 + \frac{1}{2}\frac{\partial^2 g^n}{\partial^2 c_2^2}(c_2^n - c_{02}^n)^2 + \frac{\partial^2 g^n}{\partial c_1 \partial c_2}(c_1^n - c_{01}^n)(c_2^n - c_{02}^n)$$

where g^o is Gibbs potential per atom of the old phase; g^n is Gibbs potential per atom of the new phase; g_0^o, g_0^n are the minimal Gibbs potentials per atom for the old and the new phases, respectively; c_{01}^o, c_{02}^o are the concentrations of components 1, 2 in the old phase, when $g^o = g_0^o$; c_{01}^n, c_{02}^n are the concentrations of components 1, 2 in the new phase, when $g^n = g_0^n$; c_1^o, c_2^o are the concentrations of components 1, 2 in the old (parent) phase; and c_1^n, c_2^n are the concentrations of components 1, 2 in the new phase (nucleus).

2) A nucleus has a shape of spheroid (ellipsoid having two axes of equal length, making it a surface of revolution).
3) The concentration profiles in the region of formation of the new phase nucleus are approximated by the linear functions

$$c_i(x) = c_i(x_0) + \frac{x}{L_i}, \quad \left(\frac{1}{L_i} = \left.\frac{\partial c_i}{\partial x}\right|_{x=x_0}\right)$$

The concentrations of the components in the new phase nucleus were found using the parallel tangent rule (Figure 9.24):

$$\frac{\partial g^o}{\partial c_1} = \frac{\partial g^n}{\partial c_1}, \quad \frac{\partial g^o}{\partial c_2} = \frac{\partial g^n}{\partial c_2}$$

The expression for the change of Δg (Gibbs potential per atom of the nucleus), after applying the parallel tangent rule, will be as follows:

$$\Delta g = g^n(c_1^n, c_2^n) - g^o(c_1^o, c_2^o) + (c_1^o - c_1^n)\partial g^o/\partial c_1 + (c_2^o - c_2^n)\partial g^o/\partial c_2 \quad (9.59)$$

Taking into account the aforementioned approximations, the formula for ΔG was obtained as

$$\Delta G = \pi R_2^2 n \left\langle \frac{4}{3} R_1 \left[\Delta g_0 + \frac{1}{2}\alpha_1^n aa^2 + \frac{1}{2}\alpha_2^n cc^2 + \alpha_+^n aacc - kk \right.\right.$$

$$+ (c_1^o - c_{01}^n - aa)\,ee + (c_2^o - c_{02}^n - cc)\,gg \bigg] + \frac{4}{15} R_1^3 \left[\frac{1}{2}\alpha_1^n bb^2 \right.$$

$$+ \frac{1}{2}\alpha_2^n dd^2 + \alpha_+^n bbdd - oo + \left(\frac{1}{L_1} - bb\right) ff + \left(\frac{1}{L_2} - dd\right) hh \bigg]\bigg\rangle \quad (9.60)$$

$$+ \begin{cases} 4\sigma\pi \left(\dfrac{R_2^2}{2} + \dfrac{R_2 R_1^2}{2\sqrt{R_1^2 - R_2^2}} \arcsin\sqrt{1 - \left(\dfrac{R_2}{R_1}\right)^2} \right), & \dfrac{R_2}{R_1} < 1 \\[1em] 4\sigma\pi \left(\dfrac{R_2^2}{2} + \dfrac{R_2 R_1^2}{2\sqrt{R_2^2 - R_1^2}} \ln\left(\sqrt{\left(\dfrac{R_2}{R_1}\right)^2 - 1} + \dfrac{R_2}{R_1}\right) \right), & \dfrac{R_2}{R_1} > 1 \end{cases}$$

$\Delta g_0 = g_0^n - g_0^o$

$$aa = \frac{\alpha_+^n}{(\alpha_+^n)^2 - \alpha_1^n \alpha_2^n} \left[\left(\alpha_2^o - \alpha_+^o \frac{\alpha_2^n}{\alpha_+^n}\right)(c_2^o - c_{02}^o) + \left(\alpha_+^o - \alpha_1^o \frac{\alpha_2^n}{\alpha_+^n}\right)(c_1^o - c_{01}^o) \right]$$

$$bb = \frac{\alpha_+^n}{(\alpha_+^n)^2 - \alpha_1^n \alpha_2^n} \left[\frac{\alpha_2^o}{L_2} + \frac{\alpha_+^o}{L_1} - \frac{\alpha_2^n \alpha_1^o}{\alpha_+^n L_1} - \frac{\alpha_2^n \alpha_+^o}{\alpha_+^n L_2} \right]$$

$$cc = \frac{\alpha_1^o}{\alpha_+^n} (c_1^o - c_{01}^o) + \frac{\alpha_+^o}{\alpha_+^n} (c_2^o - c_{02}^o) - \frac{\alpha_1^n}{(\alpha_+^n)^2 - \alpha_1^n \alpha_2^n} \left[\left(\alpha_2^o - \alpha_+^o \frac{\alpha_2^n}{\alpha_+^n} \right) \right.$$

$$\left. \times (c_2^o - c_{02}^o) + \left(\alpha_+^o - \alpha_1^o \frac{\alpha_2^n}{\alpha_+^n} \right) (c_1^o - c_{01}^o) \right]$$

$$dd = \frac{\alpha_1^o}{\alpha_+^n L_1} + \frac{\alpha_+^o}{\alpha_+^n L_2} - \frac{\alpha_1^n}{(\alpha_+^n)^2 - \alpha_1^n \alpha_2^n} \left[\frac{\alpha_2^o}{L_2} + \frac{\alpha_+^o}{L_1} - \frac{\alpha_2^n \alpha_1^o}{\alpha_+^n L_1} - \frac{\alpha_2^n \alpha_+^o}{\alpha_+^n L_2} \right]$$

$$ee = \alpha_1^o (c_1^o - c_{01}^o) + \alpha_+^o (c_2^o - c_{02}^o) ; ff = \left(\frac{\alpha_1^o}{L_1} + \frac{\alpha_+^o}{L_2} \right)$$

$$kk = \frac{1}{2} \alpha_1^o (c_1^o - c_{01}^o)^2 + \frac{1}{2} \alpha_2^o (c_2^o - c_{02}^o)^2 + \alpha_+^o (c_1^o - c_{01}^o)(c_2^o - c_{02}^o)$$

$$oo = \left(\frac{\alpha_1^o}{2L_1^2} + \frac{\alpha_2^o}{2L_1^2} + \frac{\alpha_+^o}{L_1 L_2} \right)$$

$$L_1 = 1 \Big/ \left(\frac{\partial c_1^o}{\partial x} \right), \quad L_2 = 1 \Big/ \left(\frac{\partial c_2^o}{\partial x} \right)$$

$$\alpha_+^n = \frac{\partial^2 g^n}{\partial c_1 \partial c_2}, \quad \alpha_1^n = \frac{\partial^2 g^n}{\partial c_1^2}, \quad \alpha_2^n = \frac{\partial^2 g^n}{\partial c_2^2}$$

$$\alpha_+^o = \frac{\partial^2 g^o}{\partial c_1 \partial c_2}, \quad \alpha_1^o = \frac{\partial^2 g^o}{\partial c_1^2}, \quad \alpha_2^o = \frac{\partial^2 g^o}{\partial c_2^2}$$

where R_1 is the radius of the nucleus along the direction of diffusion fluxes, R_2 is the radius of the nucleus along the direction perpendicular to that of diffusion fluxes, and c_1^o and c_2^o are the concentrations of components 1, 2 in the parent phase, in the point of formation of the new phase nucleus center.

As it is seen from these approximations, we used a model for the ternary system forming a series of solid solutions of each of the three components and having one intermediate phase (Figure 9.25). Old is the point on the diagram whose composition corresponds to the minimum of g^o (c_1^o, c_2^o) surface for solutions. New is the point on the diagram which corresponds to the minimum of the g^n (c_1^n, c_2^n) surface for the intermediate phase. Figure 9.25 shows the projections onto the concentration triangle of the intersection of the surfaces g^o and g^n for the solutions and the intermediate phase (line I), and the line intersecting the region where parallel tangents to the surfaces g^o and g^n exist (line II).

When studying the process, the initial composition of the diffusion couple was chosen and the corresponding diffusion path was constructed. Diffusion path construction was carried out using the model of constant diffusivities [25, 41, 42]:

$$c_1(t, x) = A_1 + B_1 \, \text{erf} \frac{x}{2\sqrt{D_1 t}} + C_1 \, \text{erf} \frac{x}{2\sqrt{D_2 t}}$$

$$c_2(t, x) = A_2 + B_2 \, \text{erf} \frac{x}{2\sqrt{D_1 t}} + C_2 \, \text{erf} \frac{x}{2\sqrt{D_2 t}}$$

$$A_1 = \frac{c_{1L}^o + c_{1R}^o}{2}, \quad A_2 = \frac{c_{2L}^o + c_{2R}^o}{2} \tag{9.61}$$

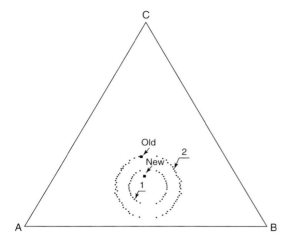

Figure 9.25 Projections onto the concentration triangle of the intersected lines of Gibbs potential for the solutions and the intermediate phase (line I), and of the line, bounding the region where parallel tangents to both surfaces exist (line II).

$$C_1 = c_{1R}^o - A_1 - B_1, \qquad C_1 = c_{2R}^o - A_2 - B_2, \qquad B_1 = \frac{(D_1 - D_{22})}{D_{21}} B_2$$

$$B_2 = \frac{1}{(D_1 - D_2)} \left[(D_2 - D_{22})(A_2 - c_{2R}^o) + D_{21}(c_{1R}^o - A_1) \right]$$

$$D_1 = \frac{1}{2}(D_{22} + D_{11}) + \sqrt{(D_{22} - D_{11})^2 + 4D_{12}D_{21}}$$

$$D_2 = \frac{1}{2}(D_{22} + D_{11}) - \sqrt{(D_{11} - D_{22})^2 + 4D_{12}D_{21}}$$

where D_{22}, D_{11} are the diffusivities describing the fluxes of elements 1(B) and 2(C) which are driven by their own concentration gradients; D_{21}, D_{12} are the diffusivities for the fluxes of one element driven by the concentration gradient of the other one; c_{1L}^o, c_{2L}^o characterize the initial composition of the diffusion couple for $x < 0$; and c_{1R}^o, c_{2R}^o characterize the initial composition of the diffusion couple for $x > 0$.

9.4.2
Algorithm and Results for the Model System

Varying the initial composition of the diffusion couple at fixed diffusivities $D_{22} = 1 \times 10^{-21}$ m² s⁻¹, $D_{11} = 2.5 \times 10^{-21}$ m² s⁻¹, $D_{12} = 2.5 \times 10^{-22}$ m² s⁻¹, $D_{21} = 1 \times 10^{-23}$ m² s⁻¹, we have obtained a set of diffusion paths (Figure 9.26).

The concentrations $c_{1L}^o = 0.1$, $c_{2L}^o = 0.2$ correspond to the point L, and c_{1R}^o, c_{2R}^o to points M, N, O, P, Q, R, S, T, U, and V. Further, $0.15 \leq c_{1R}^o \leq 0.95$ and $c_{2R}^o = 1 - c_{1R}^o$.

The next step was to investigate them according to the following algorithm:

1) One possible diffusion path was chosen.

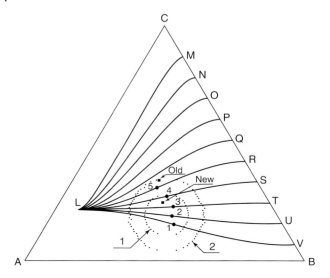

Figure 9.26 Diffusion paths for samples of various initial composition. Points 1–5 indicate the places of easiest possible nucleation of the intermediate phase in a solid solution for different diffusion couples.

2) For each point of the path, the dependence of $\Delta G(V)$ (advantage in Gibbs potential at formation of the new phase nucleus) on its volume was built. Thus, the possibility of formation of the intermediate phase nucleus was analyzed at the given point. In order to do this, the shape optimization for each of the nucleus' volumes in $\Delta G(V)$ was carried out. So, a ratio of R_2/R_1 for the given volume V was chosen, such that for it ΔG was minimal [37, 38]. After this, for each V the value of ΔG was calculated from Equation 9.60.

3) The maximum ΔG ($\Delta G_{\max}$) of the obtained dependence $\Delta G(V)$ was found. This maximum is the nucleation barrier for the intermediate phase in the given point of the diffusion path (the point for which the function $\Delta G(V)$ was built in step (2) of this algorithm). If the function $\Delta G(V)$ constantly increases, or its maximum $\Delta G_{\max}$ exceeds the critical nucleation barrier (taken as $60kT$ [43]), the nucleation of the intermediate phase in this point of diffusion path should be considered impossible.

4) Similarly, for all points, the whole diffusion path was examined. It was considered that formation of the intermediate phase nucleus will most probably occur in the region of the diffusion zone where the concentration composition corresponds to the point with minimal nucleation barrier. If there are no points with possible nucleation on the diffusion path, then the intermediate phase is not supposed to form during the annealing of the diffusion couple with the set initial composition (which the studied diffusion path corresponds to).

Parameters of the system under examination are $c_{1L}^o = 0.1$, $c_{2L}^o = 0.2$, $0.25 \leq c_{1R}^o \leq 0.95$, $c_{2R}^o = 1 - c_{1R}^o$, $\Delta g_0 = -1 \times 10^{-22}$ J, $\sigma = 0.05$ J/m², $c_{01}^n = 0.4$, $c_{02}^n = 0.23$,

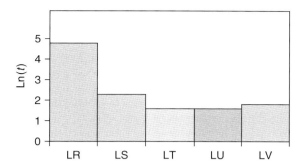

Figure 9.27 Incubation times of intermediate phase formation for diffusion couples with compositions LR, LS, LT, LU, and LV, corresponding to Figure 9.26.

$c_{01}^o = 0.33$, $c_{02}^o = 0.33$, $n = 1 \times 10^{29}$, $\alpha_+^n = 3 \times 10^{-18}$, $\alpha_1^n = 6 \times 10^{-18}$, $\alpha_2^n = 6 \times 10^{-18}$, $\alpha_+^o = 1 \times 10^{-18}$, $\alpha_1^o = 2 \times 10^{-18}$, and $\alpha_2^o = 2 \times 10^{-18}$. Having applied the described scheme, we obtained these results:

1) Formation of the intermediate phase (as should have been expected) is possible only for the systems with a diffusion path passing via the regions in the concentration triangle, bounded by lines 1 and 2 (Figure 9.26). In this case, the suitable diffusion couples are LR, LS, LT, LU, and LV. 1, 2, 3, 4, and 5 are the points with the minimal nucleation barrier ΔG_{max} of the intermediate phase.
2) It has turned out that even if the aforementioned condition is fulfilled, nucleation of the intermediate phase is possible only when the annealing time (diffusion time) exceeds a certain "critical" value (incubation time of nucleation) (Figure 9.27).
3) With time, the location of points with minimal nucleation barrier ΔG_{MAX} of the intermediate phase on the diffusion path changes within some narrow limits (Table 9.2 and Figure 9.28).
4) The shape of the intermediate phase nucleus at rather large concentration gradients $(\partial c_1/\partial x)$, $(\partial c_2/\partial x)$, observed at small annealing times, has the nonsphericity coefficient (aspect ratio) η (R_2/R_1), which is quite different from 1 (Table 9.3 and Figure 9.29).

9.4.3
Discussion

Similar to the case of a binary system, the theory for finding the incubation time of the intermediate phase formation was proposed for ternary systems. This time strongly depends not only on diffusion parameters of the system (diffusivities), but also on the initial composition of the samples which undergo isothermal diffusion. Unlike the binary system, the ternary initial composition influences not only the incubation time, but the result of the whole diffusion process.

Table 9.2 Data cited for diffusion couple with initial composition: $c_{1r} = 0.75$, $c_{2r} = 0.25$.

Number of points	Annealing time	Concentration composition of the point (region) with minimal nucleation barrier	
	Ln(t)	C_1	C_2
1	2	0.403	0.222
2	6.9	0.409	0.223
3	9.62	0.419	0.224
4	13.82	0.425	0.225

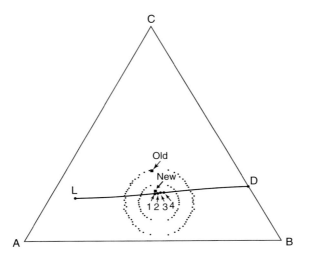

Figure 9.28 Time dependence of the point with minimal nucleation barrier in the diffusion path. The parameters of points 1, 2, 3, and 4 are given in Table 9.2. Calculations were made for the diffusion couple with the following initial composition: $c_{1L} = 0.1$, $c_{2L} = 0.2$, $c_{1R} = 0.75$, and $c_{2R} = 0.25$.

Changing the initial compositions, one can find the regions of the concentration triangle where the formation of the intermediate phase is impossible. Even if the new phase nuclei are formed, this will not necessarily lead to formation of the continuous layer of the phase. For example, the particles of the new phase may remain isolated, forming a two-phase zone.

If the system has several intermediate phases, then, by comparing their incubation times, we may find out which of the phases will appear first. This is essential for prediction of diffusion paths in multicomponent systems since the attempt to do so using balance equations for components written for interphase boundaries sometimes leads to nonsphericity (Section 9.3). Thus, as it follows from the theory, the system could simultaneously realize several diffusion paths. The analysis of

Table 9.3 Data cited for diffusion couple with initial concentrations: $c_{1r} = 0.75$, $c_{2r} = 0.25$.

Nucleus nonsphericity η	Annealing time t(s)	Ln($\Delta c_1 / \Delta x$)	Ln($\Delta c_2 / \Delta x$)
6.093	6.15	21.106	18.97
3.097	6.73	21.061	18.93
2.597	7.3	21.021	18.88
2.098	7.88	20.97	18.85
1.598	9.04	20.91	18.78
1.099	17.7	20.57	18.44

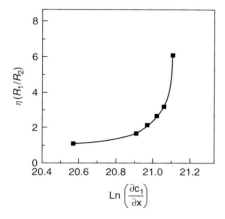

Figure 9.29 Nonsphericity coefficient dependence of the intermediate phase nucleus η on the concentration gradient of component 1 in the parent phase $\partial c_1 / \partial x$ for the diffusion path, given by Figure 9.28.

such a nonsphericity by comparing the incubation times for different possible phases, must remove the difficulties arising in such a case. At quite large gradients of the components' concentrations, a considerable nonsphericity of the new phase nuclei is observed.

References

1. Gusak, A.M. and Gurov, K.P. (**1982**) *Fizika Metallov i Metallovedeniye*, **53**, 842, 848 (in Russian).
2. Tu, K.N., Chu, W.K. and Mayer, J.W. (**1975**) *Thin Solid Films*, **25**, 403.
3. Goesele, U. and Tu, K.N. (**1982**) *Journal of Applied Physiology*, **53**, 3552.
4. Gurov, K.P., Kartashkin, B.A. and Ugaste, Yu.E. (**1981**) *Inter-diffusion in Multiphase Metallic Systems*, Nauka, Moscow (in Russian).
5. Poate, J. M. Tu, K. N. and Mayer, J. W. (eds) (**1978**) *Thin Films – Interdiffusion and Reactions*, John Wiley & sons.
6. d'Heurle, F.M., Gas, P., Philibert, J. and Zhang, S.L. (**2001**) *Diffusion and Defect Forum*, **194–199**, 1631.

7. Gurov, K.P. and Gusak, A.M. (1990) *Izvestiya AN SSSR, Metally*, **1**, 163 (in Russian).
8. Gusak, A.M., Bogaturev, O.O., Zaporogetz, T.V. et al. (2004) *Models of Solid State Reaction*, Cherkasy National University, Cherkasy.
9. Gusak, A.M. and Lyashenko, Yu.A. (1993) *Fizika i khimia obrabotky materialov*, **5**, 140 (in Russian).
10. Lyashenko, Yu.A. and Mangelinck, D. (2005) *Bulletin Cherkasy University*, **79**, 76.
11. Kalmukov, K.B. (1988) *Interaction Al, Mg with Cd alloys* Doct. Chem. Sc. Thesis, Moscow.
12. Gas, P. (1989) *Applied Surface Science*, **38**, 178.
13. Lavoie, C., d'Heurle, F.M., Detavernier, C. and Cabral, C. (2003) *Microelectronic Engineering*, **70**, 144.
14. Mangelinck, D., Gas, P., Gay, J.M. et al. (1998) *Journal of Applied Physics*, **84**, 2583.
15. Mangelinck, D., Dai, J.Y., Pan, J. and Lahiri, S.K. (1999) *Applied Physics Letters*, **75**, 1736.
16. Mangelinck, D. (2006) *Defect and Diffusion Forum*, **249**, 127.
17. www.crct.polymtl.ca/FACT (Ni-Si. Data from SGTE alloy data bases (revised 2004)).
18. Gulpen, J. (1996) *Reactive phase formation in the Ni-Si system*. Ph. D. Thesis, Eindhoven, Eindhoven University of Technology.
19. Kirkaldy, J.S. and Brown, L.C. (1963) *Canadian Metallurgical Quarterly*, **2** (1), 89.
20. Rönkä, K.J., Kodentsov, A.A., Kivilahti, J.K. and van Loo, F.J.J. (1997) *Defect and Diffusion Forum*, **143–147** 541.
21. Rapp, R.A., Ezis, A. and Yurek, G.J. (1973) *Metallurgical Transactions*, **4**, 1283.
22. van Loo, F.J.J., van Beek, J.A., Bastin, G.F. and Metselaar, R. (1985) The Role of thermodynamics, kinetics in multiphase ternary diffusion, in *Diffusion in Solids* (eds M.A. Dayananda and G.E. Murch), TMS, Warrendale, p. 231.
23. Gusak, A.M. (1990) *Jurnal fizicheskoy himiyi (Journal of Physical Chemistry)*, **64**, 510.
24. Gusak, A.M. and Lyashenko, Yu.A. (1991) *Metallofizika i Noveishiye Technologii (Metal Physics, Advanced Technologies)*, **13**, 91.
25. Tu, K.N. (1991) DD-91 International Conference on Diffusion and Defects in Solids Abstracts. Moscow, Part E–01.
26. Kornienko, S.V. and Gusak, A.M. (1994) *Inzhenerno-Fizicheskii Zhurnal (Journal of Physics, Engineering)*, **66**, 310 (in Russian).
27. Mokrov, A.P., Akimov, V.K. and Zaharov, P.N. (1975) *Zaschitnie Pokitiya v Metalah (Kiev)*, **9**, 10 (in Russian).
28. Kirkaldy, J.S. and Young, D.J. (1987) *Diffusion in the Condensed State*, The Institute of Metals, London.
29. Ziebold, T. and Ogilvie, R. (1967) *Transactions of the Metallurgical Society AIME*, **239**, 942.
30. van Loo, F.J.J., Smet, P.M., Rieck, P.M. and Verspui, G. (1982) *High Temperature-High Pressure*, **14**, 25.
31. Meschaninov, B.A. (1982) *Diffuzionie Processi v Metalah*, TPI, Tula, p. 67 (in Russian).
32. Nesbitt, J.A. and Heckel, R.W. (1987) *Metallurgical Transactions A.*, **18A**, 1987.
33. Gusak, A.M., Lyashenko, Yu.A., Kornienko, S.V. and Shirinyan, A.S. (1997) *Defect and Diffusion Forum*, **143–147**, 683.
34. Kornienko, S.V. and Gusak, A.M. (1998) *Metallofizika i Noveishiye Technologii (Metal Physics, Advanced Technologies)*, **20**, 28.
35. Gusak, A.M. (1990) *Ukrainskiy Fizicheskiy Jurnal (Ukrainian Physical Journal)*, **35**, 725.
36. Gusak, A.M., Dubiy, O.V. and Kornienko, S.V. (1991) *Ukrainskiy Fizicheskiy Jurnal (Ukrainian Physical Journal).*, **5**, 286 (in Russian).
37. Gusak, A.M. and Bogatyrev, A.O. (1994) *Metallofizika i Noveishiye Technologii (Metal Physics, Advanced Technologies)*, **16**, 28.
38. Desre, P.J. and Yavari, A.P. (1991) *Acta Metall. Mater.*, **39** (10), 2309.

39. Hodaj, F., Gusak, A.M. and Desre, P.I. (**1998**) *Philosophical Magazine A*, **77**, 1471.
40. Gusak, A.M., Hodaj, F. and Kovalchuk, A.O. (**1997**) *Bulletin of Cherkasy University*, **1**, 80.
41. Zakharov, P.N. and Mokrov, A.P. (**1978**) *Fizika Metallov i Metallovedeniye*, **46**, 431 (in Russian).
42. Gusak, A.M. and Lyashenko, Yu.A. (**1990**) *Inzhenerno-Fizicheskii Zhurnal (Journal of Physics, Engineering)*, **59**, 286 (in Russian).
43. Johnson, W.C., White, G.L., Marth, P.E. et al. (**1975**) *Met. Trans.*, **6A**, 911.

Further Reading

Sokolovskaya, E.M. and Guzej, L.S. (**1986**) *Metallohimiya*, Moscow University, Moscow.

10
Interdiffusion with Formation and Growth of Two-Phase Zones

Yuriy A. Lyashenko and Andriy M. Gusak

10.1
Introduction

Up to now, we have been treating the processes of interreactive and reactive diffusion with a diffusion zone consisting of sequential layers of solid solutions and intermediate phases. In ternary and multicomponent systems, the diffusion zone morphology may appear much more complicated. To be exact, diffusion is often accompanied by formation of two-phase zones with various topologies (isolated inclusions, percolation clusters, parallel phase connection, etc.). There was a certain break in this branch of diffusion science after the works of Wagner, Kirkaldy, and Morral; therefore, we attempt to continue these studies here. Further, we suggest some new ideas within a phenomenological approach regarding the concept of ambiguity and stochastization of the diffusion path.

To start with, we analyze here the general peculiarities of interdiffusion in ternary metallic systems. Theoretical and experimental investigations of interdiffusion in multicomponent systems present a challenge regarding the possibility of formation of two-phase regions as a result of diffusion interaction between alloys whose isothermal phase diagram contains two-phase regions. The simulation of interdiffusion in ternary systems with two-phase regions allows us to describe the growth of such regions, to determine the diffusion path and to estimate its stability. Using different simplifying approximations, we can present a theory and experimental investigation of some issues, which are highly essential for technology. In particular, we consider:

1) the influence of a third component added to a binary system in the diffusion-controlled phase growth at high temperatures, for example, the effect of suppression of technologically undesirable intermediate phases when producing semiconductors Nb_3Sn by bronze technology, when copper is added to the system Nb-Sn [1];
2) corrosion, thermal and mechanical stability of protective coatings, superalloys and other multicomponent alloys obtained and used in the process of oxidation, nitration, and so on;

Diffusion-Controlled Solid State Reactions. Andriy M. Gusak
Copyright © 2010 WILEY-VCH Verlag GmbH & Co. KGaA, Weinheim
ISBN: 978-3-527-40884-9

3) obtaining new materials of gradient type where, for instance, a metallic matrix of a holder and the working body consisting of intermetallics with high hardness have a contact zone not in the form of a planar interface but instead, penetrate into each other via a spatially extended two-phase region with a smoothly changing fraction of the second phase. At that, inside the two-phase region the concentration gradient refers to the gradient of volume fraction of phases rather than the gradient of chemical potential gradient and thus the thermal stability of such materials is provided (if coarsening is not essential).

One of the effective methods of affecting the kinetics and results of solid-state reactions under diffusion in metals is the addition of a third component, which can suppress certain phases and help the others to grow (see Chapter 9 and Sections 4.1–4.6 in Chapter 4). The influence of the third component can be connected either with its limited solubility in intermetallics, or with the change of the diffusion coefficient of the main component because of segregation of the impurity on the grain boundaries. At that, flux balance conditions at the moving interface may lead to the nonplanarity of diffusion transformation front. Thus, the interface can become smeared and the formation of two-phase regions is possible. Unlike binary systems (without external fields), the loss of diffusion front stability in ternary systems is possible even at later stages when parabolic growth dominates.

Today we have a well-developed theory of a particular case concerning the growth of two-phase regions, the case of internal oxidation. In the general case, we face a fundamental problem involving the self-consistent description of nucleation, phase growth, and coarsening in open multicomponent systems.

10.2
Peculiarities of the Diffusion Process in Ternary Systems

10.2.1
Notations

To exclude possible confusion, let us first fix the notations used in this chapter. The spatial region (in real space) containing both phases is called *two-phase zone*. The part of the concentration triangle between single-phase regions is called *two-phase region*. The interface between two phases in real space is called *interphase boundary (planar or nonplanar)*. The line in the concentration triangle, which is the margin of some single-phase composition region, is called *phase boundary*. Lines that connect two-phase boundaries in two points with compositions, which are in equilibrium with each other, are called *conodes* (or tie-lines). Conodes intersect the two-phase regions.

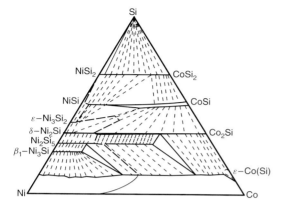

Figure 10.1 Isothermal section of Co-Ni-Si phase diagram at 800 °C with corresponding diffusion path (black line) [2].

10.2.2
Thermodynamic Peculiarities

Phase diagrams of ternary systems usually contain two-phase regions in the solid state (see e.g., [2] and Figure 10.1). The diffusion mass transfer in ternary and multicomponent systems is essentially different from the case of a binary system in quasiequilibrium as there exists the possibility of two-phase zone formation in the diffusion process. Though two-phase formation is connected with the thermodynamic disadvantage of interphase boundaries formation, there are cases when any other diffusion mode is impossible. Formation of two-phase regions may also proceed at high reaction rates at interfaces, that is, the assumption of quasiequilibrium of the interdiffusion process is imposed. Subsequently, we can apply the apparatus of linear thermodynamics for irreversible processes [3–5].

The quasiequilibrium condition implies the fulfillment of Gibbs phase rule when, at T and $P = $ const, the number of the degrees of freedom of the system, f, equals the number of components minus the number of phases. Applying this rule to binary systems we find that a two-phase region is unable to grow (under quasiequilibrium condition, $f = 2 - 2 = 0$) while its formation becomes possible starting from ternary systems: $f = 3 - 2 = 1$. The condition of local quasiequilibrium means that leveling of chemical potentials over each of three components in the given "physically small volume" (both between the grains of different phases and inside the grains) proceeds faster than the change of these chemical potentials due to macroflux divergence (provided that the "physically small volume" contains at least several grains of both phases). In this case, there is one degree of freedom in the two-phase zone. It can be, for example, the chemical potential of one of the species, or one of the concentrations in one of the two phases. This parameter changes along the two-phase zone, and its gradient is related to the driving force of interdiffusion and two-phase zone growth.

10.2.3
Diffusion Peculiarities

Unlike binary systems, in ternary and multicomponent systems there exists the problem of ambiguity of the diffusion path. The point is that the diffusion path linking initial compositions in single-phase regions on the concentration triangle is a curved line (S shaped) along which the composition change occurs [3, 5]. The curvature of the diffusion path arises from the number of degrees of freedom. At constant T and p within the single-phase region there are two degrees of freedom, for example, two independent concentrations $c_i = c_i(t, x)$, $i = 1, 2$. Formally speaking, the diffusion path may be a straight line, but this case is degenerate and needs very special interrelations between diffusivities. In general, the diffusion path is curved. Owing to conservation of matter, it cannot be one-sided (concave or convex), deviation to one side should be compensated by opposite deviation in the other part of the path. This yields the S-curved shape. In a two-phase zone (under quasiequilibrium conditions) only one degree of freedom remains. It means that the changes of two concentrations along the diffusion path are no longer independent. In this case the path may be continuous, it can contain straight intervals, zigzags and horns as well.

One of the peculiarities of the diffusion path [6–10] is that the concentration of one component in the diffusion zone layer may exceed the corresponding concentrations in the initial couple (uphill diffusion). Kirkaldy [3] argued about such redistribution of components using the main principles of Onsager's linear nonequilibrium thermodynamics.

Note, that two-phase formation is unfavorable since additional interphase boundaries must appear and this leads to additional energy dissipation. Perhaps, this point causes the biased attitude toward two-phase regions in metallic systems and has not been investigated properly. Moreover, two-phase regions are difficult to investigate experimentally as there are local resolution problems in EPMA (electron probe X-ray microanalysis).

However, in some cases when the structure of the initial samples is already defective (fine-grained structure or a large number of radiation defects), the possibility for a two-phase zone to form in a ternary system increases. If, in addition, the diffusion parameters of the initial alloys imply no other diffusion modes (for example, interphase boundary movement with the concentration jump on it (along tie-line) or growth of intermediate phases), then the system has to relax to equilibrium through the inevitable formation of a two-phase zone.

Diffusion processes in ternary metallic systems with two-phase regions and problems of their formation are discussed in [6–15]. Though there are no available experimental data on diffusion characteristics inside the grains of two-phase zones, it still cannot be doubted that two-phase zones are formed as a result of diffusion processes in such systems as V-Nb-Cr [14], Fe-Ni-Al [15, 16]. The most profound analysis regarding the regularities of diffusion formation of two-phase zones was made for the case of internal oxidation (nitration, etc.) of binary metallic systems [9, 17–29]. The work [9] treats possible variants of diffusion path as lines on the concentration triangle linking boundary compositions of phases. Having analyzed the solutions obtained from balance equations for fluxes at interphase boundaries, the authors made the following conclusions:

1) the number of phases in the diffusion zone of a ternary system depends not only on the initial conditions but also on the ratio between the diffusion coefficients in the phases;
2) there is a possibility of growth of internal phases whose concentrations exceed that of a given element in the end members of diffusion couple.

The morphology of the internal oxidation zone and stability of transformation front were studied in [29]. Solid-state reactions with formation of two-phase zones were analyzed in [3, 30, 31].

10.2.4
Types of Diffusion Zone Morphology in Three-Component Systems

Let us analyze possible types of diffusion zone morphology using some model systems. The simplest variant of diffusion interaction between three-component alloys and two-phase zones (see Figure 10.2) is mode I with the movement of a planar interphase boundary, which is stable with respect to perturbations. In this case, the diffusion path approaches the interphase boundary in points S and H. On the concentration triangle these points belong to the ends of one conode linking the compositions in equilibrium at the boundaries of the two-phase region (where the condition of equality of chemical potentials for each component in both phases is fulfilled). In real space, in the coordinate system with the x axis perpendicular to the interface, the two-phase zone is not formed (see spatially superposed points S and H for the mode I in Figure 10.2) and thus the movement of a planar interphase boundary is realized. At that, the conditions of flux balance are fulfilled at interfaces and this allows one to determine, unambiguously, the direction and rate of interphase boundary movement at given diffusion parameters in both phases, given initial and boundary conditions.

A two-phase zone is formed if in some layer $V_{\alpha\beta}$ (see illustration for mode III in Figure 10.2) the particles of both phases are in equilibrium. Here we should introduce the notion of volume fraction of phases, which is the main parameter when describing modes with formation of two-phase zones. However, to fully describe the case, one needs to know both the volume fraction of phases and the contact zone morphology, which cannot be predetermined. So far, there are no criteria for choosing the morphology of the two-phase zone in the general case.

Consider possible types of the two-phase zone morphology. The simplest way of the two-phase zone formation (provided the balance equations at interphase boundaries are maintained) can be the instability of the interaction front with respect to small perturbations of the latter resulting in its growing distortion (see mode II in Figure. 10.2). Similar instabilities can be described using the ideas proposed by Mullins and Sekerka [32, 33], further expanded for three-component systems by Kirkaldy (Ref. 3 in Chapter 12).

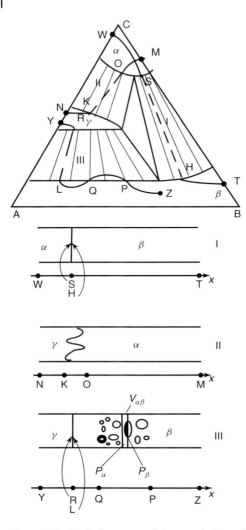

Figure 10.2 Typical patterns of the morphology of the two-phase zone: I – mode with the interphase boundary movement without crossing the conodes in the two-phase zone; II – mode of the two-phase zone formation with the loss of interphase boundary stability; III – mode with formation of the precipitate zone PQ and interphase boundary movement.

Another example of the two-phase zone morphology can be realized at the formation of the zone of isolated precipitates (see mode III in Figure 10.2). At that the diffusion path between the initial compositions of alloys Y and Z first goes through the two-phase zone with the jump along the conode (points R and L) (this is what the planar interphase boundary corresponds to), and then after passing the β-phase (between points L and Q) and intruding into the two-phase zone it crosses the conodes (between points Q and P), and this corresponds to the formation of precipitates. Such morphology was experimentally obtained and described by Kirkaldy [3, 6]. The model regarding the growth and interaction between the precipitates in the mean-field approximation [34], based on Cahn–Hilliard equations, assuming the diffusion in precipitates to be frozen, was outlined first in [35].

The third morphology type [11] corresponds to mode IV in Figure 10.3 of the interaction between the alloys with initial compositions, presented by points, say, N and M. At that, the two-phase zone grows between L and P, and volume fractions of phases "jump" (change step-wise) when crossing the boundary of the two-phase zone. Another type of morphology results from mode V in Figure 10.3. Here, some successive layers of intermediate phases and two-phase zones with different morphologies may grow.

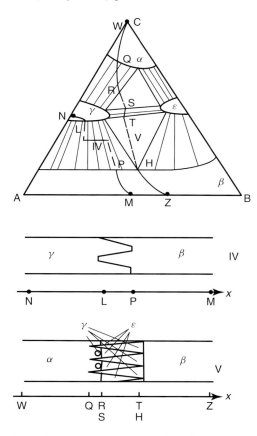

Figure 10.3 Typical patterns of the diffusion zone morphology: IV – mode with parallel phase connection in the two-phase zone; V – complex mode with parallel phase connection ST and precipitation zone QR.

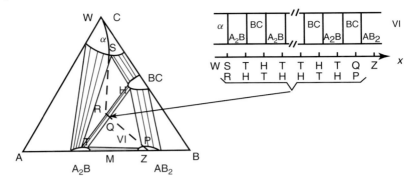

Figure 10.4 Typical patterns of the diffusion zone morphology: VI – mode with successive periodic phase formation in the two-phase zone, when the diffusion path demonstrates multiple repeating jumps within the segment RQ between points T and H of the phase diagram.

Systematic experimental investigations of such systems were carried out for the case of successive layers growth in [20, 36], and involving aggregative particles in the diffusion zone [20, 21]. In the model presentation of mode V (Figure 10.3), between Q and R there is a zone of γ-phase particles precipitation in the α-phase matrix, and between S and T there is the two-phase zone with volume phase fractions changing continuously. The description of the two-phase zone for mode V, in the general case presents some challenges within the phenomenological approach, since the volume fractions of the phases are considerable and one cannot neglect diffusion processes in one of the phases (for more details see [37, 38]).

Mode VI as depicted in Figure 10.4, is the most mysterious and theoretically unexamined one. It was first experimentally described in [39] and studied in different systems in [40–44]. Diffusion processes in each layer of the formed periodic structure with matching of boundary conditions are known [45], but kinetic peculiarities for new layers formed in the two-phase zone and the details of the formation of the two-phase zone periodic structures (sizes of layers and the succession of phases in the diffusion zone) are not fully described yet. It is important to note that the formation of such periodic structures is similar to the Liesegang effect [46] but cannot be reduced to it [43].

10.3
Models of Diffusive Two-Phase Interaction

Because of the formation and growth of two-phase zones, the description of diffusion interaction in ternary and multicomponent systems is more complicated when compared to binary ones, in which the diffusion path in quasiequilibrium conditions is unambiguous and the resulting diffusion zone does not contain any two-phase regions. Diffusion processes in two-phase zones are difficult to describe. In order to successfully overcome these difficulties we must introduce some simplifying assumptions. Looking at peculiarities of diffusion in ternary systems with two-phase zones applied to different models, we analyze now, the employed assumptions.

10.3.1
Model Systems

One of the first detailed formulations of the problem of the theoretical description of the diffusion processes in ternary systems leading to the formation of two-phase zones is given in [10]. It was pointed out that in order to describe the state of one layer of the two-phase zone it is sufficient to know the volume fractions of both phases, p_α and p_β (Figure 10.2), and the concentrations of the components in each phase.

The volume fractions are specified by the simple expressions

$$p_\alpha = \frac{V_\alpha}{V}, \quad p_\beta = \frac{V_\beta}{V}, \quad p_\beta = 1 - p_\alpha \tag{10.1}$$

where V_α and V_β are the volumes of the different phases. The layer under investigation is characterized by the averaged concentrations

$$c_i = p_\alpha c_i^\alpha + p_\beta c_i^\beta \tag{10.2}$$

It is possible to describe the growth of the two-phase zone if one knows the spatial and time change of the volume fractions of the phases and the concentrations of the components.

A phenomenological theory of these processes was proposed in [10]. Its main assumption is this: one of the averaged concentrations is taken as the independent variable, corresponding to the existing single degree of freedom. The evolution equation

$$\frac{\partial c_1}{\partial t} = \frac{\partial}{\partial x}\left(D_{11}\frac{\partial c_1}{\partial x} + D_{12}\frac{\partial c_2}{\partial x}\right) \tag{10.3}$$

together with the two equations

$$c_1 + c_2 + c_3 = 1 \tag{10.4}$$

$$c_2 = f(c_1) \tag{10.5}$$

fully describe the diffusion process between two-phase alloys. This approach is complicated by the necessity to determine the function $f(c_1)$, which is an additional separate problem. It is suggested to define the dependence $c_2 = f(c_1)$ after the profiles that have been averaged over concentration layers are experimentally obtained.

To solve the equations one needs to determine the initial and boundary conditions. This involves some problems. If in a binary case of semi-infinite samples, the boundary concentrations are constant at the interfaces, in the case of three-component systems we cannot say for sure whether or not it is so. In [10] it is indicated that the boundary concentrations change with the evolution of the diffusion path. For an adequate description of diffusion in the two-phase zone it is necessary to know both the positions of the phase boundaries (in the concentration triangle) and the conodes. In such a case, we may derive the self-contained system of equations, which allows us to describe diffusion processes in two-phase zones. Moreover, the problem of how the diffusion path approaches the two-phase region remains unsolved. If at the beginning of annealing the diffusion couple consists of pure phases, it is impossible to predict (at least from the phase diagram) the points on the phase boundaries (in the concentration triangle) that the diffusion path will reach. If they happen to be the ends of one conode, no two-phase zone will develop, and the mode with planar interphase boundary movement will take place.

The problem of the diffusion path going through the two-phase region between the single-phase alloys was discussed in [11] in terms of real, virtual, and complex diffusion paths. The real diffusion path fails to completely reflect the structure of the formed two-phase zone, since it does not cross it, but passes along the boundaries of individual particles of different phases in real space. In this case, only concentrations at the phase boundaries are indicated and the information about volume fractions of phases is lost. Kirkaldy and Brown [6] treating the diffusion couple in a pseudobinary approximation have introduced the notion of a virtual diffusion path. So, it is assumed that the interphase boundary remains planar and its movement is parabolic, though the diffusion path may virtually get into the two-phase region.

When describing the intrusion of the virtual diffusion path into the two-phase region it is necessary to put down the diffusion equations using interdiffusion coefficients in pure phases. It is also predicted that the two-phase zone is not formed, and diffusion takes place in supersaturated solid solutions. Generally, virtual diffusion paths obtained experimentally and those using the pseudobinary approximation, may vastly differ. They coincide when the diffusion path in the studied system comes to the ends of the one conode, jumping over the two-phase region, that is, the two-phase zone does not appear. If real experiment proves that the two-phase zone is formed either due to the growth of precipitates or due to the loss of the planar structure by the interphase boundary, the experimental diffusion path and the unstable virtual one do not coincide. It results from the violation of the pseudobinary conditions (i.e., the existence of only planar interphase boundaries) in the real system. Similar to virtual diffusion paths, complex ones contain information both about concentrations on interphase boundaries and phase fractions, and in addition reflect the real microstructure of the formed two-phase zone. These paths were introduced by Roper and Whittle [11].

To distinguish between virtual and complex diffusion paths let us consider the model system depicted in Figure 10.5. Diffusion occurs between α and β alloys with initial compositions illustrated by points X and Y on the concentration triangle. If we proceed from the approximation of a planar interphase boundary (Kirkaldy and Brown), the diffusion zone morphology depends on where the diffusion path comes after intrusion into the two-phase region. If it comes back to the same single-phase region, this implies just a precipitation zone of isolated precipitates

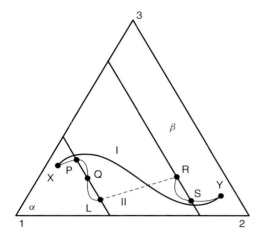

Figure 10.5 Model phase diagram: I – virtual diffusion path between two single-phase alloys X and Y (pseudobinary approximation); II – complex diffusion path. PQ – precipitate zone; LR – jump along the conode; RS – nonplanar interphase boundary.

(segment PQ) and does not violate the condition of planar boundary. If the diffusion path after intrusion into the two-phase zone enters another single-phase region (β-phase), then a nonplanar interphase boundary must appear (perhaps, together with precipitates), that is, the two-phase zone must arise. Here the dotted line corresponds to a planar interphase boundary, and the segment RS to its distortion with the formation of the two-phase zone. This is the segment PS of the diffusion zone that determines the unstable part, which is responsible for the path's virtuality. Therefore, from the pseudobinary approximation, when there is no concentration gradient in the equilibrium two-phase zone, that is, there is a jump over the conode, the boundaries remain planar and the two-phase zone does not form. If the mentioned condition fails, the two-phase zone appears and is described by the virtual diffusion path. If, from the very beginning, the diffusion couple consists of single- and two-phase alloys, we cannot assume the diffusion path to be virtual in any case. Indeed, the diffusion path must pass the two-phase region, that is, there is a concentration gradient in it. Hereinafter, the term *diffusion* path will denote a complex diffusion path.

References [12, 13] treat the problem of diffusion path in a multicomponent diffusion couple. For the model of a three-component system [12], a zigzag diffusion path in the diffusion zone was obtained (Figure 10.6). The matrix formalism applied in [47] for the transformation of coordinates and kinetic coefficients in single-phase regions leads to solutions with standard lacet-like diffusion paths. In case of a two-phase zone, the diffusion path crosses the conodes along the principle value vectors of the matrix of constant interdiffusion coefficients (segments XL and RY in Figure 10.6) with the break along one conode (segment LR in Figure 10.6).

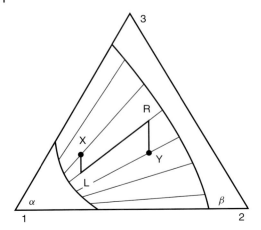

Figure 10.6 Zigzag diffusion path in the two-phase zone: X and Y are the initial compositions; LR is the jump over the conode.

In the model calculations [13], the effective diffusion coefficients in the two-phase zone were used. It was established that diffusion coefficients were constant along the conode in case of a phase diagram with parallel conodes. Experimentally obtained zigzag diffusion paths in the two-phase zone of an Al-Cr-Ni system were also presented. The most detailed development of the theoretical approach [12, 13] for the investigation of diffusion interaction between the α phase of the solid solution and the $\alpha - \beta$ two-phase alloy was performed in [48]. The authors have obtained the dependencies of the contact zone microstructure on the initial conditions of diffusion annealing. The model is based on the following assumptions:

- In the model phase diagram, the phase boundaries have the form of segments with conodes orthogonal to them.
- In the single-phase region, the description of diffusion is typical for three-component systems, that is, using Fick's equations and the matrix of constant interdiffusion coefficients.
- In the two-phase zone, averaged concentrations are used, and the matrix of diffusion coefficients and concentration gradients in the α phase (in the β phase the diffusion is frozen) are taken into account in Fick's equation. The diffusion equations in the two-phase zone are then rewritten using the gradients of the averaged concentrations and the matrix of effective diffusion coefficients, taking into account the peculiarities of constructed phase diagrams.
- The obtained simplified systems of equations are solved analytically, and then balance equations for fluxes at the moving boundary of diffusion transformation are written down using Boltzmann's substitution.
- The obtained set of equations allows one calculating the diffusion path, to define the value and direction for the velocity of the transformation front depending on the initial compositions of diffusion couples and the values of the interdiffusion coefficient matrix.

- It is possible to make up microstructure maps, which separate the regions of initial compositions leading to different final states (resulting single-phase alloy or a two-phase zone) in the phase diagram.

Another development of the approaches [12, 13, 48] is given in [35], as it describes diffusion interaction between two two-phase alloys with different volume fractions of phases, grounding upon the phase field model, which employs generalized Cahn–Hilliard equations [34]. The model treats different components' mobilities in different phases, and accounts for the surface energy between the particles of phases, which is connected with gradient terms from Cahn–Hilliard equations. This approach enables analysis of the Kirkendall effect in the contact zone. As a result of 2D computer simulations, different types of diffusion zone microstructure were obtained. The influence of Kirkendall effect on the behavior of the zigzag diffusion path is treated, and the possible type of the resulting diffusion zone between two-phase alloys without particles of the other phase is obtained. We concluded that the change of the rotation angle of the principle value vector of the effective diffusion coefficients matrix and the concentration dependence of mass transfer coefficients lead to deviations of the averaged concentrations in the two-phase zone from the directions of the principle value vector obtained in the models developed earlier [12, 13, 48].

10.3.2
Phenomenological Approach to the Description of Interdiffusion in Two-Phase Zones

In [37, 38], a phenomenological scheme for interdiffusion in the two-phase zone at a quasiequilibrium process, taking into account diffusion fluxes in both phases of the diffusion zone, was proposed. The analysis was performed for a model pseudobinary system (Figure 10.7), which, however, does not impose any limitations with regard to the generality of the developed phenomenological method.

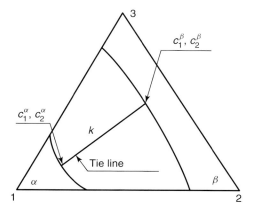

Figure 10.7 Phase diagram for a pseudobinary ternary system.

Local equilibrium between α and β phases in each physically small volume (cell) is determined by three equations:

$$\mu_i^\alpha \left(c_1^\alpha, c_2^\alpha\right) = \mu_i^\beta \left(c_1^\beta, c_2^\beta\right), \quad i = 1, 2, 3 \tag{10.6}$$

Here $c_1^\alpha, c_2^\alpha, c_1^\beta, c_2^\beta$ are the boundary phase concentrations, and

$$c_3^{\alpha,\beta} = 1 - c_1^{\alpha,\beta} - c_2^{\alpha,\beta} \tag{10.7}$$

The conditions, (Equation 10.6) are imposed at these four values and only one of them remains independent. More exactly speaking, all the four boundary concentrations are defined with one parameter k: $c_1^\alpha(k), c_2^\alpha(k), c_1^\beta(k), c_2^\beta(k)$.

We denote k as the "the number (parameter) of the conode" keeping in mind that k is undergoing constant change from one conode to another. The meaning of the conode is characterized by the following:

1) In the diagram, the conode links the phase points with equal chemical potentials.
2) States of cells, located at the same conode (i.e., with equal chemical potentials) differ only in α- and β-phase fractions.
3) If the diffusion path comes out from both sides onto the same conode, then a two-phase region does not form (the movement of planar interphase boundaries with concentration jumps along the conode takes place).

As the "conode parameter," any boundary concentration on the ends of the conode can be chosen.

Under the condition of local quasiequilibrium, the chemical potentials of the components (which are equal for both phases in each cell) are determined only by one conode parameter k; this conode contains the corresponding point of the concentration triangle with the average concentrations $\bar{c}_i$,

$$\bar{c}_i = c_i^\alpha p_\alpha + c_i^\beta p_\beta, \quad i = 1, 2 \tag{10.8}$$

where p_α, p_β are the phase fractions, and $p_\alpha + p_\beta = 1$ holds.

The chemical potentials of all the cells are equal if their compositions are on the same conode. States of these cells differ only by the fractions of the phases $p_\alpha(x)$ and not by $\mu(x)$. That is why all the cells with the concentrations on the same conode are balanced and do not cause any interdiffusion. It means that the first Fick's law is not valid for diffusion in the two-phase zone: the concentration gradient does not generate the diffusion flux. Instead, the true driving force is the gradient of the chemical potential. The chemical potential does not change along the conodes but changes across them. Thus, to form the two-phase zone, the diffusion path should cross the conodes rather then "touch" them. In other words, a continuous path is possible only across the conodes (with a changing conode parameter k), and along the conodes the path can only jump.

According to the general phenomenological equations of nonequilibrium thermodynamics in the laboratory reference frame (in which $J_1 + J_2 + J_3 = 0$), independent fluxes of components 1 and 2 (the third is taken as the solvent) are being given by the expressions

$$J_m = -\sum_{n=1}^{2} \tilde{L}_{mn} \frac{\partial(\mu_n - \mu_3)}{\partial x}, \quad m = 1, 2 \tag{10.9}$$

where $\tilde{L}_{mn}$ are Onsager's coefficients. Taking into account that all μ_i depend eventually on the parameter k alone, we will get an equation, formally analogous to the first Fick's law

$$J_i = \tilde{M}_i \frac{\partial k}{\partial x} \tag{10.10}$$

where

$$\tilde{M}_i = \tilde{L}_{ii} \frac{d(\mu_j - \mu_3)}{dk} + \tilde{L}_{ij} \frac{d(\mu_j - \mu_3)}{dk}, \quad i,j = 1, 2, \quad i \neq j \tag{10.11}$$

Let us now employ the mass conservation law written in an hydrodynamic scale:

$$\frac{\partial}{\partial t}\left(c_i^{\alpha} p_{\alpha} + c_i^{\beta} p_{\beta}\right) = -\frac{\partial J_i}{\partial x} = \frac{\partial}{\partial x}\left(\tilde{M}_i \frac{\partial k}{\partial x}\right), \quad i = 1, 2 \tag{10.12}$$

Taking into account the quasiequililibrium conditions, one will have two equations for two functions $k(t,x)$ and $p_{\alpha}(t,x)$

$$\left(p_{\alpha} \frac{\partial c_i^{\alpha}}{\partial x} + p_{\beta} \frac{\partial c_i^{\beta}}{\partial x}\right) \frac{\partial k}{\partial t} + \left(c_i^{\alpha} - c_i^{\beta}\right) \frac{\partial p_{\alpha}}{\partial t} = \frac{\partial}{\partial x}\left(\tilde{M}_i(k, p_{\alpha}) \frac{\partial k}{\partial x}\right), \quad i = 1, 2 \tag{10.13}$$

There are two ways of defining $\tilde{M}_i$ in the framework of the employed phenomenological approach. The first one means expressing $\tilde{M}_i$ in terms of diffusion parameters of pure phases. On the boundary between two- and one-phase regions, such a relation is given by the equations:

$$\tilde{M}_i^{\alpha,\beta}(k, p_{\alpha} = 1) = \tilde{D}_{ii}^{\alpha,\beta}\left(c_1^{\alpha,\beta}(k), c_2^{\alpha,\beta}(k)\right) \frac{dc_i^{\alpha,\beta}}{dk}$$
$$+ \tilde{D}_{ij}^{\alpha,\beta}\left(c_1^{\alpha,\beta}(k), c_2^{\alpha,\beta}(k)\right) \frac{dc_j^{\alpha,\beta}}{dk}, \quad i,j = 1, 2, \quad i \neq j \tag{10.14}$$

The general expression for the interdiffusion coefficients within the $\alpha-\beta$ region has not been found yet. So far, different approaches [26] of introducing the transport parameters in a locally inhomogeneous system are used. But in many cases (if the diffusivity of α and β phases does not differ that much), "the model of phase parallel connection" is considered a reasonable approximation

$$\tilde{M}_i(k, p_{\alpha}) = p_{\alpha} \tilde{M}_i^{\alpha}(k) + p_{\beta} \tilde{M}_i^{\beta}(k) \tag{10.15}$$

where the coefficients $\tilde{M}_i^{\alpha,\beta}$ are given with the boundary conditions of conjugation with pure phases, that is, via the relations (Equation 10.14).

The second way is expressing $\tilde{M}_i$ in terms of tracer diffusivities D_i^* in a two-phase alloy [38]. Assuming the Kirkendall effect is suppressed in the hydrodynamics scale, we finally have

$$\tilde{M}_i = \frac{\left(c_1^\alpha p_\alpha + c_1^\beta p_\beta\right)}{kT} \left(\frac{d\mu_i}{dk} - \frac{\sum_{n=1}^{3}\left(c_n^\alpha p_\alpha + c_n^\beta p_\beta\right) D_n^* \frac{d\mu_n}{dk}}{\sum_{n=1}^{3}\left(c_n^\alpha p_\alpha + c_n^\beta p_\beta\right) D_n^*} \right) \qquad (10.16)$$

In general, the problem must be solved by numerical methods. However, from the general thermodynamic considerations some peculiarities of solutions can be predicted.

10.3.3
Choice of the Diffusion Interaction Mode

In [49, 50], the conditions under which the two-phase zone forms at interdiffusion in ternary metallic systems, at constant temperature and pressure, are analyzed. In the case of a three-component diffusion couple, when the quasiequilibrium condition is being maintained, three fundamentally different situations are possible: first is movement of the interphase boundary in the systems with limited solubility, second is growth of one or several intermediate phases, and third is formation of the two-phase zone as a result of the diffusion process. It is presupposed that the two-phase zone is formed at quasiequilibrium diffusion only if the diffusion mode is impossible without it; in other words, if the corresponding set of equations for diffusion in two single phases with the flux balance boundary condition at the moving interface has no physically reasonable solutions. The analysis is based on the phenomenological theory of diffusion processes in a two-phase zone. It is assumed that the boundary will move without formation of the two-phase zone and without corresponding increase of surface energy while it is kinetically possible.

The first possibility is considered in the quasibinary system $A - (B_{1-X}C_X)$ with limited solubility, but without intermediate phases. Analytical and numerical methods were used to find the solutions, which correspond to the case of a diffusion path going through the conode (at the interphase boundary, the concentrations jump between two edge points of the conode). At that the interdiffusion in the couple A, solid solution (BC) was treated in the approximation of parabolic movement of the boundary. On the basis of diffusion equations for α and β phases and two equations for flux balance (for A and B components), one has

$$\left(c_i^\alpha - c_i^\beta\right)\frac{dy}{dt} = J_i^\alpha - J_i^\beta, \quad i = A, B \qquad (10.17)$$

The range of values for the parameters at which the system of differential equations has no solutions (negative concentrations are obtained) and thus diffusion without the two-phase zone is impossible, was found. Relying upon the results of

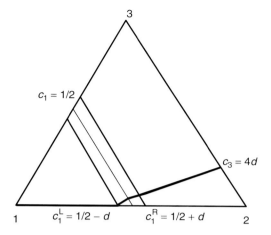

Figure 10.8 Model system with the stretched intermediate phase (1, 2, and 3 correspond to components A, B, and C); d – concentration width of the intermetallide.

the analysis, it was concluded that when the interdiffusion in solution (BC) is too slow when compared to that in the α phase, the diffusion path in the two-phase zone cannot go along the conode, and the two-phase zone must appear.

Another alternative is connected with the possibility of the intermediate phase to grow in the quasibinary system, which can be more favorable than the formation of the two-phase zone. So, the growth of an intermediate γ phase was also analyzed at zero solubility of B and C in A, and A in alloy BC (Figure 10.8).

While carrying out the numerical solutions of the corresponding system of nonlinear equations using Newton's method and methods of minimization, the constant of the growth rate for the intermediate phase was calculated. The growth constant decreases with the initial concentration $c_3^\beta(\infty)$, and at $c_3^\beta(\infty) < 1 - 4d$ becomes negative. From the indicated calculations we may conclude that the slow-diffusing component in the γ phase may reduce or even prevent the growth of the continuous layer of the intermetallide.

At that boundary, the composition of the alloy BC shifts along the BC side. Indeed, if the intermetallide requires different quantities of B and C, the initial alloy BC will become depleted of these components to a certain extent. This, in particular, leads to homogeneity violation of the initial alloy of the BC couple, which in turn, causes fluxes in this part of the diffusion couple, and this influences the kinetics of the boundary movement. When there are several intermediate phases on the phase diagram, the mentioned effect may cause competition between them. The failure of the two indicated modes may lead to the formation of the two-phase zone in the ternary system during the diffusion process; the boundary concentration may appear to lie on different conodes and the diffusion path will not be able to bypass them.

10.4
Results of Modeling and Discussion

10.4.1
One-Dimensional Model of Interdiffusion between Two-Phase Alloys

It is necessary to find, by numerical methods, the diffusion path and profiles $c_1(t, x)$, $c_2(t, x)$, $p(t, x)$, and $k(t, x)$ if the phase diagram and the dependence $\widetilde{M}_i(k, p_\alpha)$ are known [51]. Here we proceed as follows.

The diffusion couple is supposed to consist of two-phase alloys, $L(c_1^L, c_2^L)$ and $R(c_1^R, c_2^R)$. In this simplified model we shall consider the diffusion of B and C components to proceed independently in both α and β phases. The coefficients $D_{ij}^{\alpha\beta}$ are given. We employ further the approximation Equations 10.14 and 10.15 to describe the dependencies $\widetilde{M}_i(k, p_\alpha)$.

The form of the two-phase region in the concentration triangle and the exact geometry of conodes are not that important for the investigation of basic regularities. That is the reason why α and β phase boundaries in the concentration triangle are chosen as straight lines parallel to the side 1-2 (Figure 10.9)

$$c_1^\alpha + c_2^\alpha = e_\alpha, \qquad c_1^\beta + c_2^\beta = 1 - e_\beta > e_\alpha \tag{10.18}$$

We consider that all the conodes continue and meet each other in the apex 3 of concentration triangle, so that for the equal phase boundaries points (lying on one and the same conode) the following equations are valid.

$$\frac{c_1^\alpha}{e_\alpha} = \frac{c_1^\beta}{1 - e_\beta}, \qquad \frac{c_2^\alpha}{e_\alpha} = \frac{c_2^\beta}{1 - e_\beta} \tag{10.19}$$

Let us choose c_1^β as the conode parameter k

$$c_1^\alpha(k) = e_\alpha / (1 - e_\beta) \, k, \qquad c_1^\beta(k) = k,$$
$$c_2^\alpha(k) = e_\alpha / (1 - k/(1 - e_\beta)), \qquad c_2^\beta(k) = 1 - e_\beta - k \tag{10.20}$$

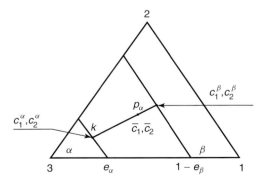

Figure 10.9 Phase diagram of the model system.

We use the finite-difference method for the determination of the profiles c_1, c_2 and the diffusion path $f(c_1, c_2)$:

$$c_i(j) = c_i(j) + \frac{\Delta t}{h^2}\left\{\tilde{M}_i(i+1)(k(j+1) - k(j)) - \tilde{M}_i(j-1)(k(j) - k(j-1))\right\}, \quad i = 1, 2 \quad (10.21)$$

To check the correctness of the obtained solutions we used the semi-discrete method of numerical solving of two differential equations in specific derivatives. The point of the method lies in modeling the finite-difference scheme for the space axis, the variable t being considered as continuous. In this case, the $2N$ ordinary differential equation system, which can be solved, for example, by the Runge–Kutta–Felberg's method with automatic choice of step along t, is obtained:

$$\frac{dc_i(j)}{dt} = \frac{1}{2h^2}\left[\left(\tilde{M}_{i,j+1} + \tilde{M}_{i,j}\right)(k_{j+1} - k_j) - \left(\tilde{M}_{i,j-1} + \tilde{M}_{i,j}\right)(k_j - k_{j-1})\right] \quad (10.22)$$

where i corresponds to the component's number and j is the number of the spatial point. In this case, N denotes the number of points on the axis x. This method provides avoiding the accumulation of errors with time (such errors being difficult to avoid while using the usual finite differences along t).

The initial conditions are being given by a step function. Under this condition, the net approximations, if not modified, can give great errors. That is why instead of the break point we have a narrow transition zone consisting of some net steps. This boundary layer correction is searched from the stationary case of the system of equations (Equation 10.12)

$$\nabla\left(\tilde{M}_1 \nabla k\right) = 0, \quad \nabla\left(\tilde{M}_2 \nabla k\right) = 0 \quad (10.23)$$

$$\implies \tilde{M}_1 \nabla k = \text{const}, \quad \tilde{M}_2 \nabla k = \text{const}$$

Thus, in this case the boundary layer correction in the boundary layer refers to the function connecting the points to the left and the right of the break.

The numerical solutions of diffusion paths and concentration profiles were obtained for different matrices of the diffusion coefficients in the α and β phases. All the parameters here are reduced to a dimensionless form. Typical examples of choice of parameters are:

$D_{11}^{\alpha} = 3, \quad D_{12}^{\alpha} = -2.4, \quad D_{21}^{\alpha} = 1.6, \quad D_{22}^{\alpha} = 2, \quad e_\alpha = 0.3,$

$D_{11}^{\beta} = 3, \quad D_{12}^{\beta} = 0.4, \quad D_{21}^{\beta} = -1.2, \quad D_{22}^{\beta} = 1.5, \quad e_\beta = 0.25,$

$c_1^L = e_\alpha \times 10^{-2}, \quad c_2^L = e_\alpha \times 1.6, \quad c_1^R = (1 - e_\beta) \times 0.6, \quad c_2^R = (1 - e_\beta) \times 10^{-2}$

It was found out that the solution of the given system of equations depends strongly on the initial conditions. The solution behavior is just the same with all testing compositions of diffusion parameters and it differs only in the rate of the process. Thus, if k_L differs greatly from k_R (Figure 10.10) of the initial diffusion couple, the solution has a smooth character, at first. But in the course of time, the functions

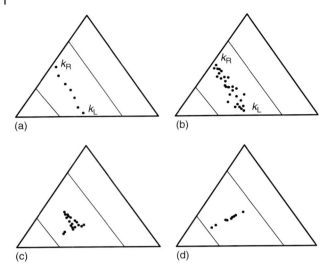

Figure 10.10 Diffusion path evolution between two-phase alloys: (a) initial stage (S shape, $t \geq 0$); (b) oscillation mode; (c) "stochastization" of diffusion path; (d) state of final equilibrium along the conode ($t \to \infty$).

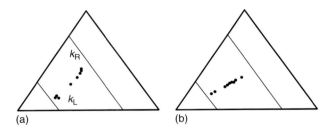

Figure 10.11 Calculation of the diffusion path at close values of k_L and k_R but with different volume fractions: (a) initial stage; (b) stratifications of the final path.

$c_1(x)$ and $c_2(x)$ and the diffusion path in the concentration triangle become broken (scattered).

The second important case is observed when the initial value of k_L is given close to k_R but with different phase fractions (Figure 10.11). In this case, the diffusion path comes out onto one and the same conode just in different points.

10.4.2
The Problem of Indefiniteness of the Final State

Let the initial single-phase alloy consist of two regions of volumes V_L and V_R with concentrations c_1^L, c_2^L and c_1^R, c_2^R. It can be predicted, definitely and independent of the concrete solution of equations, that in the final state the homogeneous

F-composed alloy is formed:

$$c_i^F = \frac{V_L c_i^L + V_R c_i^R}{V_L + V_R}, \quad i = 1, 2 \tag{10.24}$$

The situation becomes more complicated with two-phase alloys.

Consider, the points c_L and c_R in the two-phase region of the concentration triangle. In the final state, the average concentration $\bar{c}_F \equiv (\bar{c}_1, \bar{c}_2)$ is determined definably by Equation 10.24, but now the system in its final state cannot simply be a homogeneous one. For example, it may consist of two regions 1 and 2, which are in equilibrium, the average concentrations of which are on one and the same conode with the point of average state and the volume fractions are bound by the correlation

$$p(\bar{F}) = \frac{V_L p(1) + V_R p(2)}{V_L + V_R} \tag{10.25}$$

Apparently, in the final state (if coarsening is excluded!) stratification is possible into an arbitrary number of regions on one and the same conode, with the normalizing condition maintained, that is, Equation 10.24.

When all the final states are balanced, they correspond to the identical average over all volume concentrations and in this sense they are equal. The main difference from the single-phase case is just this indefiniteness. Naturally we should expect that the indefiniteness of the final balanced state would influence the specific solutions of diffusion equations in the two-phase region, describing the system relaxation to the equilibrium.

10.4.3
Diffusion Path Stochastization in the Two-Phase Region

The presence of the jump of p_α [51], which is equivalent to diffusion path breaks along one of the conodes can be described using the balance flux equation on the ordinate y:

$$\Delta c_i \frac{dy}{dt} = \Delta J_i, \quad i = 1, 2 \tag{10.26}$$

Employing Equation 10.10 we obtain

$$\left(c_i^\alpha - c_i^\beta\right) \Delta p_\alpha \frac{dy}{dt} = \Delta J_i = -\Delta \left(\tilde{M}_i \nabla k\right) \tag{10.27}$$

Let M_1 and M_2 be constant (they do not change at the p jump). In this case for the compatibility of Equation 10.27 the following conditions

$$\frac{dy}{dt} = 0, \quad \Delta \frac{\partial k}{\partial x} = 0$$

have to be fulfilled, otherwise dividing Equation 10.27 by Equation 10.26 we will come to the equality

$$\frac{c_1^\alpha - c_1^\beta}{c_2^\alpha - c_2^\beta} = \frac{M_1}{M_2} \tag{10.28}$$

which is hardly probable. Thus, a jump of p is possible and the derivative $(\partial k/\partial x)$ has no breaks. If M_1 and M_2 depend on p (for example, in the model of parallel connection Equation 10.15), then fluxes are unbroken:

$$\Delta(\widetilde{M}_i \nabla k) = 0$$

and the derivative $(\partial k/\partial x)$ occurs to be broken. The number of the diffusion path breaks (and concentrational profiles) may be different. A situation is possible when the solution is the function $p(t, x)$ broken in every point and the function $k(t, x)$ is continuous but with the derivative $(\partial k/\partial x)$ broken in every point.

These results (the diffusion path "stochastization" in the concentration space) are unusual and need some interpretation. It is similar to the break along the zigzag diffusion path obtained in [12, 13], but in this model the breaks are possible in every point of the diffusion path in the two-phase zone. As mentioned above, it is connected with the ambiguity of the final equilibrium state of the two-phase alloy.

Let us consider another interpretation: interdiffusion in a two-phase zone of a ternary system can be formally defined with the help of ordinary Fick's equation system with a 2×2 matrix of diffusion coefficients D_{ij}. But here the coefficients are not independent [37] (it means that there is only one thermodynamic degree of freedom at fixed T and p). Here det $D_{ij}=0$ or $\widetilde{D} = 0$, that is, the matrix of interdiffusion coefficients is degenerated. It should be mentioned that the condition det $D_{ij}=0$ (or $\widetilde{D} = 0$ for the binary alloy) is inherent for the spinodal, which is also characterized by the "stochastization" and the final state indefinability.

10.4.4
Invariant Interdiffusion Coefficients in the Two-Phase Zone

The above suggested scheme of interdiffusion in the two-phase zone can be formulated also in terms of "invariant coefficients" [37, 38, 52] (see also [53–57]). The explicit 2×2 matrix of interdiffusion coefficients in a ternary system depends on the solvent chosen, but the matrix eigenvalues $D^{(1)}$, $D^{(2)}$ remain invariant at such transformation in the concentration space. These values are the elements of the diagonalized matrix $(D^{(0)} = \widehat{a}D\widehat{a}^{-1})$ and the diagonalizing matrix $\widehat{a}$ gives linear combinations of concentrations

$$\delta u_1 = a_{11}\delta c_1 + a_{12}\delta c_2, \qquad \delta u_2 = a_{21}\delta c_1 + a_{22}\delta c_2$$

which behave independently both at fluctuations and macro-nonequilibrium processes (the behavior of u_1 is determined only by its "own" coefficient $D^{(1)}$ and space distribution).

Formally, one can put down the usual expressions for fluxes in the two-phase zone as well [10]:

$$\bar{J}_1 = -D_{11}\frac{\partial \bar{c}_1}{\partial x} - D_{12}\frac{\partial \bar{c}_2}{\partial x}$$
$$\bar{J}_2 = -D_{21}\frac{\partial \bar{c}_1}{\partial x} - D_{22}\frac{\partial \bar{c}_2}{\partial x} \qquad (10.29)$$

But unlike the one-phase case, the coefficients D_{ik} are no longer independent. The point is that we may rewrite Equation 10.29 taking into account

$$\frac{\partial \bar{c}_i}{\partial t} = \left(\frac{dc_i^\alpha}{dk} p_\alpha + \frac{dc_i^\beta}{dk} p_\beta \right) \frac{\partial k}{\partial x} + \left(c_i^\alpha - c_i^\beta \right) \frac{\partial p_\alpha}{\partial x} \qquad (10.30)$$

as follows:

$$J_i = - \left[D_{i1} \left(\frac{dc_1^\alpha}{dk} p_\alpha + \frac{dc_1^\beta}{dk} p_\beta \right) + D_{i2} \left(\frac{dc_2^\alpha}{dk} p_\alpha + \frac{dc_2^\beta}{dk} p_\beta \right) \right] \frac{\partial k}{\partial x}$$
$$- \left[D_{i1} \left(c_1^\alpha - c_1^\beta \right) \right] \frac{\partial p_\alpha}{\partial x} \qquad (10.31)$$

In the two-phase zone at quasiequilibrium diffusion, the fluxes cannot depend on $\partial p_\alpha / \partial x$. Therefore, the coefficients in Equation 10.31 at $\partial p_\alpha / \partial x$ are equal to zero. Hence we obtain the relations

$$D_{12} = -D_{11} \frac{c_1^\alpha - c_1^\beta}{c_2^\alpha - c_2^\beta}, \quad D_{21} = -D_{22} \frac{c_2^\alpha - c_2^\beta}{c_1^\alpha - c_1^\beta} \qquad (10.32)$$

$$D_{11} D_{22} - D_{21} D_{12} = 0 \qquad (10.33)$$

From the determinant of the matrix $\widehat{D}$ being equal to zero one gets

$$D^{(1)} + D^{(2)} = D_{11} + D_{22}, \quad D^{(1)} D^{(2)} = 0$$

and then

$$D^{(1)} = D_{11} + D_{22}, \quad D^{(2)} = 0 \qquad (10.34)$$

$$\delta u_1 = \delta c_1 - \frac{c_1^\alpha - c_1^\beta}{c_2^\alpha - c_2^\beta} \delta c_2, \quad \delta u_2 = \frac{c_2^\alpha - c_2^\beta}{c_1^\alpha - c_1^\beta} \frac{D_{22}}{D_{11}} \delta c_1 + \delta c_2 \qquad (10.35)$$

It is easy to make sure that δu_1 keeps close to zero while moving along the conode, when

$$\delta c_i = \left(c_i^\alpha - c_i^\beta \right) \delta p_\alpha$$

Thus, the invariant coefficient $D^{(1)} = D_{11} + D_{22}$ describes the diffusion "perpendicular" to the conode. When the coefficient $D^{(2)}$ equals to zero it means that the corresponding value of u_2 does not evolve with time. The equality $\delta u_2 = 0$ would specify the diffusion path,

$$\frac{dc_2}{dc_1} = -\frac{c_2^\alpha - c_2^\beta}{c_1^\alpha - c_1^\beta} \frac{D_{22}}{D_{11}}$$

if we could consider the corresponding coefficients in Equation 10.31 as constant.

10.4.5
Conclusions

The complexity of describing the diffusion interaction in ternary systems with two-phase zones leads to the consideration of models based on different assumptions. Many models involve diffusion equations in the two-phase zone at frozen diffusion in one of the phases, which imposes limitations on the model's applicability to systems with considerable fractions of both phases. The problems of precipitation and coarsening in the diffusion zone of ternary systems with two-phase zones, the influence of Kirkendall effect on phase separation and movement of inclusions, and description of periodic structures' diffusion growth, are not completely solved. The issues of diffusion mode choice, and, thus, the criterion of the two-phase zone formation, and the final two-phase zone morphology, are still of high importance. Additional use of thermodynamic data for the studied system allows self-consistent describing the diffusion and thermodynamic properties of definite ternary systems [58]. The offered phenomenological approach enables the description of two-phase zones formation and evolution taking into account the diffusion fluxes in both phases in terms of diffusion path peculiarities, in particular, stability or instability with respect to breaks, indefiniteness of the path, and final state. To build the closed system of equations for finding the parameters of formation and growth of two-phase zones, it is possible to use additional thermodynamic principles.

References

1. Cheng, C.C. and Verhoeven, J.D. (**1988**) *Journal of the Less-Common Metals*, **139**, 15.
2. Rijnders, M.R., Oberndorff, P.J.T.L., Kodentsov, A.A. and van Loo, F.J.J. (**2000**) *Journal of Alloys and Compounds*, **297**, 137.
3. Kirkaldy, J.S. and Young, D.J. (**1987**) *Diffusion in the Condensed State*, The Institute of Metals, London.
4. De Groot, S.R. and Mazur, P. (**1962**) *Nonequilibrium Thermodynamics*, North-Holland, Amsterdam.
5. Borovskii, I.B., Gurov, K.P., Marchukova, I.D. and Ugaste, Yu.E. (**1971**) *Interdiffusion Processes in Metals*, Nauka, Moscow (in Russian).
6. Kirkaldy, J.S. and Brown, L.C. (**1963**) *Canadian Metallurgical Quarterly*, **2**(1), 89.
7. Mecshaninov, B.A. (**1982**) *Diffusion Processes in Metals*, TPI, Tula (in Russian).
8. Kirkaldy, J.S. (**1966**) *Transactions of ASM*, **51**, 218.
9. Borisov, V.T., Golikov, V.M. and Scherbedinskiy, G.M. (**1968**) *Protective Metal Coating*, Kiev (in Russian).
10. Gurov, K.P., Kartashkin, B.A. and Chadov, A.N. (**1980**) *Diffusion Processes in Metals*, TPI, Tula (in Russian).
11. Roper, G.W. and Whittle, D.P. (**1981**) *Metal Science*, **15**, 148.
12. Hopfe, W.D. and Morral, J.E. (**1994**) *Acta Metallurgica Materialia*, **42**(11), 3887.
13. Chen, H. and Morral, J.E. (**1999**) *Acta Materialia*, **47**(4), 1175.
14. Kodentzov, A.A., Dunaev, S.F. and Slusarenko, E.M. (**1987**) *Journal of the Less-Common Metals*, **135**, 15.
15. Nesbitt, J.A. and Heckel, R.W. (**1987**) *Metallurgical Transactions A*, **18A**, 1987.
16. Dayananda, M.A. (**2000**) *Solid State Phenomena*, **72**, 123.

17. van Loo, F.J.I. (1990) *Progress in Solid State Chemistry*, **20**, 47.
18. Wagner, C. (1938) *Zeitschrift für Anorganische und Allgemeine Chemie*, **236**, 320.
19. Hauffe, K. (1963) *Solid State and Surface Reaction*, Moscow (in Russian).
20. Rapp, R.A., Ezis, A. and Yurek, G.J. (1973) *Metallurgical and Materials Transactions B*, **4**, 1283.
21. Yurek, G.J., Rapp, R.A. and Hirth, J.P. (1973) *Metallurgical and Materials Transactions B*, **4**, 1293.
22. Scherbedinskiy, G.M., Isakov, M.G. and Abramov, G.S. (1977) *Diffusion Processes in Metals*, TPI, Tula (in Russian).
23. Shatynski, G.S., Hirth, J.P. and Rapp, R.A. (1979) *Metallurgical Transactions A*, **10A**, 591.
24. Tangchitvittaya, C., Hirth, J.P. and Rapp, R.A. (1982) *Metallurgical Transactions A*, **13A**, 585.
25. Isakov, M.G. (1978) *Diffusion Processes in Metals*, TPI, Tula (in Russian).
26. Scherbedinskiy, G.M. and Abramov, G.S. (1978) *Diffusion Processes in Metals*, TPI, Tula (in Russian).
27. Isakov, M.G. and Tolpigo, V.K. (1987) *Diffusion Processes in Metals*, TPI, Tula (in Russian).
28. Mchedlov-Petrossyan, P.O. and Sodin, S.L. (1988) *Journal Technical Physics*, **58**, 652.
29. Mchedlov-Petrossyan, P.O. (2002) *Metallofizika i Noveishie Tekhnologii*, **24**(1), 25.
30. Nekrasov, E.A. (1988) *Phenomenological theory of inter-diffusion in non single phase region of multicomponent metallic systems*. Tomsk, VINITI, N5368-B88).
31. Sinclair, C.W., Purdy, C.W. and Morral, C.W. (2000) *Metallurgical Transactions A*, **31A**, 1187.
32. Mullins, W.W. and Sekerka, R.F. (1963) *Journal of Applied Physics*, **34**, 323.
33. Mullins, W.W. and Sekerka, R.F. (1964) *Journal of Applied Physics*, **35**, 444.
34. Chen, L.Q. (1994) *Acta Materialia*, **42**, 3503.
35. Wu, K., Morral, J.E. and Wang, Y. (2001) *Acta Materialia*, **49**, 3401.
36. Clark, J.B. and Rhines, F.N. (1959) *Transactions of ASM*, **51**, 199.
37. Gusak, A.M. (1990) *Journal of Physical Chemistry*, **64**, 510 (Moscow, in Russian).
38. Gusak, A.M. and Lyashenko, Yu.A. (1988) *Problem of the Two-phase Zones Formation at Interdiffusion Description*, VINITI, Kiev, N992-B89 (in Russian).
39. Osinski, K., Vriend, A.W., Bastin, G.F. and van Loo, F.J.J. (1982) *Zeitschrift fur Metallkunde*, **73**, 258.
40. van Loo, F.J.J., Vriend, A.W. and Osinski, K. (1990) in *Fundamentals and Applications of Ternary Diffusion* (ed. G.R. Purdy), Pergamon Press, Oxford.
41. Rijnders, M.R. and van Loo, F.J.J. (1995) *Scripta Metallurgica et Materialia*, **32**(12), 1931.
42. Rijnders, M.R., van Beek, J.A., Kodentsov, A.A. and van Loo, F.J.J. (1996) *Zeitschrift fur Metallkunde*, **87**, 732.
43. Kodentsov, A.A., van Dal, M.J.H., Cserhati, C. et al. (2001) *Defect and Diffusion Forum*, **194-199**, 1491.
44. Kodentsov, A.A., Rijnders, M.R. and van Loo, F.J.J. (1998) *Acta Materialia*, **46**, 6521.
45. Kao, C.R. and Chang, Y.A. (1993) *Acta Materialia*, **43**, 3463.
46. Liesegang, R.E. (1896) *Naturwissenschaftliche Wochenschrift*, **11**, 353.
47. Thompson, M.S. and Morral, J.E. (1986) *Acta Metallurgica Materialia*, **34**, 2201.
48. Boettinger, W.J., Coriell, S.R., Campbell, C.E. and McFadden, G.B. (2000) *Acta Materialia*, **48**, 481.
49. Gusak, A.M. and Lyashenko, Yu.A. (1991) *Metallofizika i Noveishiye Technologii*, **13**(4), 48 (in Russian).
50. Gusak, A.M., Lyashenko, Yu.A., Kornienko, S.V. and Shirinyan, A.S. (1997) *Defect and Diffusion Forum*, **143-147**, 683.
51. Gusak, A.M. and Lyashenko, Yu.A. (1990) *Inzhenerno - Fizicheskij Journal*, **59**(2), 286 (in Russian).
52. Desre, P.J. and Gusak, A.M. (2001) *Philosophical Magazine A*, **81**(10), 2503.
53. Gusak, A.M. and Zakharov, P.N. (1979) *Journal of Physical Chemistry*, **53**(6), 1573 (in Russian).
54. Krishtal, M.A., Zakharov, P.N. and Mokrov, A.L. (1971) *FTT*, **13**(5), 1332 (in Russian).

55. Zaharov, P.N. (1982) *Diffusion Processes in Metals*, TPI, Tula (in Russian).
56. Zaharov, P.N. and Mokrov, A.L. (1978) *Fizika Metallov i Metallovedeniye* **46**(2), 431 (in Russian).
57. Gusak, A.M. and Desre, P.J. (2001) *Defect and Diffusion Forum*, **194-199**, 201.
58. Lyashenko, Yu.A. (2003) *Uspekhi Fiziki Metallov*, **4**(2), 81 (in Russian).

Further Reading

Lyashenko, Yu.A. (2002) *Bulletin of Cherkasy University, Diftrans 2001*, **37-38**, 69 (in Russian).

Lyashenko, Yu.A. and Kornienko, S.V. (2000) *Solid State Phenomena*, **72**, 135.

Lyashenko, Yu.A. and Mukovoz, T.P. (1994) *Metallofizika*, **16**(6), 23 (in Russian).

11
The Problem of Choice of Reaction Path and Extremum Principles
Andriy M. Gusak and Yuriy A. Lyashenko

11.1
Introduction

Thermodynamics of irreversible processes in diffusion interaction and phase transformation models, based on the second law of thermodynamics and Onsager's correlations, enables one to describe the behavior of an isolated system near equilibrium states. While analyzing the choice of the evolutionary path through which an open nonequilibrium system comes to equilibrium, an additional thermodynamic principle should be used. In the present chapter, the role of entropy generation and its extreme rate is analyzed. We have also attempted to apply it to an unambiguous description of both, isolated diffusion-interactive metallic systems and systems that are in not uniform environments; that is, in the fields of concentration gradients or chemical potentials. The bases of application of maximum rate principles for Gibbs's potential release are given. It is performed for those systems for which the choice of the reaction path is ambiguous.

11.2
Principle of Maximal Entropy Production at Choosing the Evolution Path of Diffusion-Interactive Systems

Most systems used in material science are nonequilibrium ones: aging (supersaturated) alloys dissociate by initiation and coarsening of decay products. Grains start growing in nano- and polycrystalline materials, amorphous alloys crystallize, interdiffusion takes place in protective coatings and powder alloys, metals oxidize in the atmosphere irreversibly, and so on. All materials listed above are considered to be either metastable or absolutely unstable ones and it is just a matter of the time period required for relaxation to equilibrium, or, more commonly, to a less nonequilibrium state. The production of those materials following the chemical reactions, thermal treatment or mechanical operation is accompanied thus, by irreversible nonquasistatic processes.

Diffusion-Controlled Solid State Reactions. Andriy M. Gusak
Copyright © 2010 WILEY-VCH Verlag GmbH & Co. KGaA, Weinheim
ISBN: 978-3-527-40884-9

11 The Problem of Choice of Reaction Path and Extremum Principles

As it is well known, the second law of thermodynamics alone is not sufficient for forecasting the evolution of a nonequilibrium system. The law indicates only the general direction of the evolution of the mentioned system, the tendency of a closed system's entropy to reach the maximum, the tendency of free energy to be minimum at fixed volume and temperature, and the tendency of Gibbs's potential to reach minimum at fixed temperature and pressure [1–5]. From these reasons, there arise some fundamental questions that have important practical significance and go beyond the second law of thermodynamics:

1) At what rate and according to what law does the relaxation to equilibrium occur?
2) If more than one path to relaxation is possible, how is the choice made?
3) What does the evolution of an open system look like, if the system passes through matter and energy fluxes?

Let us discuss some concrete examples in this respect.

1) At the deposition of a thin film of nickel onto silicon and further annealing at high temperature, the process, basically, must end with total nickel dissolution since each substance has at least a very small solubility in any other substance. In fact, a total dissolution process is so long that the formation of nickel disilicide $NiSi_2$ can be considered to be the result of the reaction [6, 7]. But still, this product appears as a result of a solid-phase reaction succession, the formation of Ni_2Si occurs first. After pure nickel runs out, it will become a product for a new growing phase NiSi. That one, in its turn, having used up the previous phase Ni_2Si, itself becomes a product for silicide growth . The problem of succession and intermediate phase competition does not possess any full, generally accepted answer.
2) Furthermore, in solid-phase reactions of diffusion amorphization (e.g., aurum and lanthanum [8], nickel and zirconium [9]), a metastable amorphous phase, which is absent in an equilibrium state diagram, appears first. In multicomponent systems with two-phase zones, the problem of evolutionary path choice becomes even more complicated [10–16]. Therefore, the formulation of a heuristic principle of choice forecasting the evolutionary path based on a general thermodynamic background would be very important.
3) The reaction zone of the copper–liquid solder (e.g., stannum–lead) system is a typical example of an open system. The reaction between copper and solder at temperatures about 200 °C will lead to an intermetallide Cu_6Sn_5 formation and growth. Its grains are shaped like teeth, with the copper base separated by liquid solder with substrate. The reaction goes on due to the copper flux which comes in the way of facilitated diffusion between the teeth and is used for phase growth. In the course of time, the increase of the intermetallide volume and grain growth are observed simultaneously: large grains "devour" small ones. The rules of such growth in the external flux differ greatly from the normal grain growth. In particular, the full area of grains (and interphase energy) remains nearly constant [17–19].

To develop generalizations of thermodynamics, for irreversible processes it is convenient to use the second law of thermodynamics in Prigogine's formulation [1, 2, 4, 5, 10, 20]. Namely, for any system, the full change of its entropy is presented in the form of a sum

$$dS = d_e S + d_i S \tag{11.1}$$

The first item presents the entropy change of the system at the expense of its flux through the boundary and if the fluxes of substance through the boundary are absent, it is given by the Clausius equation:

$$d_e S = \oint \frac{d'Q}{T} = \oint \frac{J_n^Q d\Sigma}{T}$$

The second term corresponds to entropy production; that is, its change is not due to a redistribution of the parts of the system but due to different dissipative processes of chaotization (temperature equalizing, chemical potential equalizing, and chemical reaction processes). Entropy generation (unlike its full change) is always positive and approaches zero only at equilibrium. Usually, entropy generation is determined per unit volume and unit time:

$$\sigma = \frac{1}{V} \frac{d_i S}{dt} \tag{11.2}$$

Entropy generation is considered a key concept of nonequilibrium thermodynamics. Let us view entropy generation for the following cases: isolated systems, systems in a homogeneous thermostat, and systems in a nonhomogeneous environment (in the temperature gradient field, in chemical potential field, etc.). At that, let us divide systems into two types: weakly nonequilibrium (linear) and far from equilibrium (nonlinear).

11.3
Nonequilibrium Thermodynamics: General Relations

11.3.1
Isolated Systems

Let us identify isolated systems or those that are in a homogeneous environment (e.g., at constant volume and temperature or pressure and temperature) as closed. Let us consider first, an isolated system. Its state can be described by a set of scalar parameters $\eta_k, k = 1, 2, \ldots, M$ (e.g., parameters of a long-range order in an ordered alloy). At equilibrium, the magnitude of these parameters are such that the entropy reaches a maximum:

$$S^{eq} = S^{max} = S\left(\eta_1^{eq}, \ldots \eta_M^{eq}\right)$$

At small fluctuations, the entropy can be presented as a Taylor's series, taking into account those terms which are quadratic with respect to the fluctuation.

Here linear terms disappear, as all first-order derivatives at the maximum are equal to zero. Hence

$$S(\eta_1,\ldots\eta_M) = S^{\max} + \frac{1}{2}\sum_{i=1}^{M}\sum_{k=1}^{M}\left(\frac{\partial^2 S}{\partial\eta_i\partial\eta_k}\right)^{\mathrm{eq}}\left(\eta_k - \eta_k^{\mathrm{eq}}\right)\left(\eta_i - \eta_i^{\mathrm{eq}}\right)$$

$$= S^{\max} - \frac{1}{2}\sum_{i=1}^{M}\sum_{k=1}^{M}\beta_{ik}x_kx_i \tag{11.3}$$

where

$$\beta_{ik} = -\left(\frac{\partial^2 S}{\partial\eta_i\partial\eta_k}\right)^{\mathrm{eq}} = \beta_{ki}$$

is the positively defined matrix of thermodynamic coefficients, and $x_i = \eta_i - \eta_i^{\mathrm{eq}}$ is the magnitude of the deviation of the respective parameter from equilibrium.

For each parameter (or its fluctuation), the derivative of entropy with respect to it is denoted as the thermodynamic force. By analogy with Hooke's law in mechanics, it is a linear function of the deviation from equilibrium:

$$X_i = \frac{\partial S}{\partial\eta_i} = \frac{\partial S}{\partial x_i} = -\sum_{k=1}^{M}\beta_{ik}x_k$$

The rate of change of the scalar parameter is called *scalar flux* and given by

$$J_i = \frac{dx_i}{dt}$$

As long as the system is isolated, entropy production itself is the full change of entropy in a unit of time:

$$\frac{dS}{dt} = -\frac{1}{2}\sum_{i=1}^{M}\sum_{k=1}^{M}\beta_{ik}\left(x_k\frac{dx_i}{dt} + x_i\frac{dx_k}{dt}\right)$$

$$= -\sum_{i=1}^{M}\sum_{k=1}^{M}\beta_{ik}x_k\frac{dx_i}{dt} = \sum_{i=1}^{M}X_iJ_i \tag{11.4}$$

Thus, entropy production is the sum of thermodynamic forces multiplied by the corresponding fluxes [5].

In weakly nonequilibrium systems, fluxes and forces are connected by Onsager's linear correlations [21–23]:

$$J_i = \sum_{k=1}^{M}L_{ik}X_k \tag{11.5}$$

The matrix of Onsager's coefficients L_{ik}, is generally nondiagonal. Therefore, the flux is determined not only by means of one thermodynamic force, but also by taking into account the other forces. A classical example in this respect is the thermoelectric effect. Thermoelectric effects will be discussed in greater detail later, for the example of multicomponent diffusion. Onsager also proved that the matrix L_{ik} is symmetrical, and this is a direct consequence of the microscopic reversibility of motion.

As far as the relation between forces and order parameter fluctuation is linear, one can express the fluxes directly through these fluctuations:

$$J_i = -\sum_{n=1}^{M} \lambda_{in} x_n$$

where

$$\lambda_{in} = \sum_{k=1}^{M} L_{ik} \beta_{kn} \qquad (11.6)$$

Note that the matrix of coefficients λ_{in} (matrix product of kinetic and thermodynamic coefficients) is generally nonsymmetrical, unlike the L_{in} and β_{nk} matrices.

11.3.2
System in a Thermostat

If a system C is located in a thermostat T, for example, at fixed temperature and pressure, the maximum entropy principle can be applied only to the whole system "C + thermostat". It can be considered isolated. If the system C is macroscopic in all directions, the entropy of the whole system can be considered to be additive:

$$S^{\text{total}} = S^T + S = S^T \left(U^{\text{total}} - U, V^{\text{total}} - V \right) \qquad (11.7)$$

Here, $(U^{\text{total}} - U)$ is the energy of the thermostat, equal to the difference of energies of the whole system and C, $(V^{\text{total}} - V)$ is the corresponding difference of volumes. The system C can be in any nonequilibrium state. At that, the thermostat is supposed to be in quasi-equilibrium at the energy and volume the system C reserved for it. In this case, the thermodynamic identity can be applied to the thermostat. The entropy of the thermostat is preliminarily expanded with respect to the small parameters U and V:

$$S^T \left(U^{\text{total}} - U, V^{\text{total}} - V \right) + S = S^T_{\text{eq}} \left(U^{\text{total}}, V^{\text{total}} \right) - U \frac{\partial S^T}{\partial U^T} - V \frac{\partial S^T}{\partial V^T} + S$$

$$= S^{\text{total}}_{\text{max}} - \frac{U}{T} - \frac{Vp}{T} + S = S^{\text{total}}_{\text{max}} - \frac{G}{T} \qquad (11.8)$$

Here, $G = U + pV - TS$ is the Gibbs' potential of the system C, and $S^{\text{total}}_{\text{max}}$ is the equilibrium value of the entropy of the whole system, C + T. Thus, if a system C is located in a thermostat T at fixed temperature and pressure, then the entropy production in the whole system C + T is equal to the rate of release of Gibbs' potential of the system C divided by temperature:

$$\frac{dS^{\text{total}}}{dt} = -\frac{1}{T} \frac{dG}{dt} \qquad (11.9)$$

A similar Taylor's series expansion may be applied to Gibbs' potential of the system C (here for the vicinity of the minimal equilibrium value G_{min}) as a quadratic form of the deviations of the state parameters from equilibrium.

If the system C has not only thermal and mechanical but also diffusional contact with the thermostat (with fixed chemical potential μ_i of each component), we may prove similarly that

$$\frac{dS^{total}}{dt} = -\frac{1}{T}\frac{d\left(G - \sum_{i=1}^{n}\mu_i N_i\right)}{dt} \qquad (11.10)$$

It is evident that at the equilibrium state, all time derivatives are equal to zero and moreover,

$$G_{eq} = \sum_{i=1}^{n}\mu_i N_i^{eq}$$

11.3.3
Inhomogeneous Systems: Postulate of Quasi-Equilibrium for Physically Small Volumes

In most cases, nonequilibrium systems are spatially inhomogeneous. For the entropy of such systems to be determined, the approximation of local quasi-equilibrium [2, 4] should be realized as the basic one. To be exact, divide the inhomogeneous system into physically small volumes (sites) having the following properties:

1) The size of the site Δx is less than the characteristic inhomogeneity length over the considered property (for example, $\Delta x < T/|\mathrm{grad}\,T|$).
2) The relaxation time of the site to equilibrium at external parameters being fixed is less than the characteristic time for a nonequilibrium process (for example, $\tau < T/|\partial T/\partial t|$).
3) The site contains many more bulk atoms when compared to the number of surface atoms; this separates it from other sites (which provides additivity of energy and entropy).

These three requirements (properties) can be met simultaneously if the system's degree of inhomogeneity is not very high. If so, each physically small volume can be considered to be in quasi-equilibrium and its entropy can be determined from the relations of equilibrium thermodynamics.

The whole system's entropy is defined here as simply the sum of entropies of all sites. If the characteristic inhomogeneity of the system is of the order of several nanometers, other methods should be applied. This refers, for example, to spinodal decomposition of supersaturated solid solutions, to nucleation of new phases, and to nano-crystalline alloy behavior. The simplest modification of local quasi-equilibrium approximation for strongly inhomogeneous systems implies adding the squared order parameter gradient into the set of local thermodynamic parameters (van der Waals [24, 25], Ginzburg–Landau [26, 27], Hillert [28, 29], Cahn–Hilliard [30, 31], Khachaturyan [32]).

The description of the case of inhomogeneous systems directly results from the general scheme developed above. Indeed, consider an alloy with an inhomogeneous distribution of temperature $T(\vec{r})$ and chemical potentials $\mu_i(\vec{r})$ for each of the n components. Let us divide the whole volume of an alloy into a number (M) of physically small volumes, such that each of them can be considered to be in equilibrium at any time, and have an internal energy $U(m)$ and number of atoms $N_i(m)$ pertaining to the i-th component. We will treat the local parameters $U(m)$ and $N_i(m)$ as independent scalar parameters η_i to describe the system's state within Onsager's scheme. Also remember that, at sources/sinks of heat and when the substance is absent, the chosen parameters fit local conservation laws

$$\frac{dS}{dt} = \sum_{m=1}^{M} \frac{dS(m)}{dt} = \sum_{m=1}^{M} \left(\frac{\partial S(m)}{\partial U(m)} \frac{dU(m)}{dt} + \sum_{i=1}^{n} \frac{\partial S(m)}{\partial N_i(m)} \frac{dN_i(m)}{dt} \right) \quad (11.11)$$

Further, we employ the following considerations:

1) the approximation of local equilibrium, and hence, thermodynamic identity for each volume so that

$$\frac{\partial S(m)}{\partial U(m)} = \frac{1}{T(m)}, \quad \frac{\partial S(m)}{\partial N_i(m)} = -\frac{\mu_i(m)}{T(m)}$$

2) thermal energy conservation law (in cases of local sources of energy being absent),

$$\frac{dU(m)}{dt} = -\Delta V(m) \cdot \text{div} \vec{J}_Q$$

the rate of internal energy change in a nonmoving site of $\Delta V(m)$ volume is equal to the difference between inside- and outside-directed heat fluxes, and thus, proportional to the divergence of heat flux $\vec{J}_Q$ density;

3) law of conservation of matter for each component (in the absence of chemical and nuclear reactions),

$$\frac{dN_i(m)}{dt} = -\Delta V(m) \cdot \text{div} \vec{J}_i(m)$$

where $\vec{J}_i$ is the flux density of i-th component;

4) a relation well known from vector analysis:

$$\varphi(\vec{r}) \text{div} \vec{A}(\vec{r}) = \text{div} \left(\varphi(\vec{r}) \vec{A}(\vec{r}) \right) - \vec{A} \cdot \text{grad} \varphi(\vec{r})$$

Then, we get (in coarse-grained spatial roughening over physically small volumes):

$$\frac{dS}{dt} = \iiint dV \left[-\text{div} \left(\frac{\vec{J}_Q}{T} - \sum_{i=1}^{n} \frac{\vec{J}_i \mu_i}{T} \right) + \vec{J}_Q \text{grad} \frac{1}{T} + \sum_{i=1}^{n} \vec{J}_i \left(-\text{grad} \frac{\mu_i}{T} \right) \right]$$

$$= \iiint dV \left[-\text{div} \left(\vec{J}_S \right) + \sigma \right] \quad (11.12)$$

So, the density of total entropy change contains, as was expected, two summands: the divergence of density of entropy flux (describing the redistribution of already available entropy between the sites), and the density of entropy production including the sum of products of "fluxes" and corresponding "forces". Apparently, for the order parameters obeying the conservation law in the inhomogeneous system, it is more convenient to use vector quantities as pairs of fluxes and forces, heat flux density and reverse temperature gradient, component flux density and minus gradient of its chemical potential, divided by temperature.

In the more generalized case, taking into account the convective motion at a certain velocity $\vec{V}$ and the corresponding viscosity, external forces $\vec{F}_i$ acting on each component (for example, electric field, chemical reactions), the density of entropy flux, and the density of entropy production become [1]:

$$\vec{J}_s = \frac{\vec{J}_Q}{T} - \sum_{i=1}^{n} \frac{\vec{J}_i^{dif} \mu_i}{T} + s \cdot \vec{V} \tag{11.13}$$

$$\sigma = \vec{J}_Q \cdot \text{grad}\frac{1}{T} + \sum_{i=1}^{n} \vec{J}_i \cdot \left(-\text{grad}\frac{\mu_i}{T} + \frac{\vec{F}_i}{T}\right) + \frac{p_{\alpha\beta}}{T} \frac{\partial V_\alpha}{\partial x_\beta} + \sum_{\rho=1}^{n} \frac{w_\rho A_\rho}{T}$$

Here, the chemical affinity A_ρ for the reactions of ρ type is defined as the change of the Gibbs potential per structural unit ($A_\rho = \sum v_{\rho i} \mu_i$), w_ρ is the rate of this reaction (in moles per unit time in a unit of volume), $p_{\alpha\beta}$ is the dissipative component of the pressure tensor (minus stress tensor). If the deviation of a system from equilibrium is minor, vector fluxes and forces must be connected by linear relations, like in the case of the scalar forces and fluxes examined above.

11.3.4
Extremum Principles

Any system that does not pass the directional fluxes of substance and energy through itself tends to equilibrium. For isolated systems, the equilibrium (that is, the stable state the system is unable to leave by itself if one neglects fluctuations,) corresponds to the maximum entropy, S. For systems in a thermostat at fixed temperature T and pressure p (the most common case for diffusion processes in alloys), the equilibrium corresponds to a minimum of Gibbs thermodynamic potential, $G = U - TS + pV$. The change to equilibrium is often accompanied by the formation of steady (metastable) states for some time (quite long in some cases). At these points, local minimums of F or G, being the functions of the system's parameters, correspond to these states. If the duration of the stay of the system in a metastable state is by far larger than the time of the experiment, this state can be regarded as stable and thus, equilibrium thermodynamic laws may be applied to it.

If the system is regarded to be under nonhomogeneous external conditions (e.g., different temperatures or chemical potentials at different places of the external boundary), then it lets energy and substance fluxes pass through itself and thus, it cannot pass on to equilibrium. On the other hand, external nonhomogeneous conditions remaining unchangeable, the system must transfer to equilibrium. It is often said that at the transition to such a state, the entropy production in the system tends to a minimum. This statement is referred to as *Prigogine's principle*, the principle of minimum entropy production [1, 2, 4, 5, 10, 20]. It should be taken into

11.3 Nonequilibrium Thermodynamics: General Relations

consideration that the principle formulated in such a form is correct within a fairly limited interval of parameters. In particular, for the principle to not fail, Onsager's coefficients must be constants but this is never realized, especially for interdiffusion and heat transfer processes. More precisely, Onsager's coefficients practically do not depend on thermodynamic forces (such as temperature gradients), and in this sense, the linearity is realized but at the same time, they always depend on local parameters of state (temperature and components concentration). Since local parameters of state change from one place to another and from one moment of time to another, in all real cases, Progigine's principle appears to fail.

However, Prigogine's group also established a more general statement which does not require Onsager's coefficients to be constant; that is, it is correct even in nonlinear thermodynamics [2]. To be precise, we can always present the derivative of entropy production with respect to time, in the form of two summands:

$$\frac{d\sigma}{dt} = \frac{d}{dt}\sum X_i J_i = \sum \frac{dX_i}{dt} J_i + \sum X_i \frac{dJ_i}{dt} = \frac{d_X\sigma}{dt} + \frac{d_J\sigma}{dt} \quad (11.14)$$

It has been proven that the first summand (that part of the rate of entropy production change connected with the change of forces) is nonpositive: $d_X\sigma/dt \leq 0$. In other words, thermodynamic force change always aims at decreasing entropy production. We may say that at constant (and symmetric) Onsager's coefficients, we have

$$\frac{d_J\sigma}{dt} = \frac{d_X\sigma}{dt} = \frac{1}{2}\frac{d\sigma}{dt} \leq 0$$

so the minimum entropy production principle presents a particular case of a more general evolution criterion.

Onsager principle is the one of the least energy dissipation [1, 5, 21, 22]: if we treat the difference $(\sigma - \Phi)$ between entropy production $\sigma = \sum J_i X_i$ and a so-called dissipation potential

$$\Phi = \frac{1}{2}\sum\sum L_{ik}^{-1} J_i J_k$$

as a function of local fluxes, with thermodynamic forces being fixed, the evolution path of the weak nonequilibrium (linear) system has chosen to correspond to the maximum of the indicated value. Gyarmati proposed an alternative principle [5]: if we consider the difference between entropy production $\sigma = \sum J_i X_i$ and $1/2 \sum\sum L_{ik} X_i X_k$ as a function of local values of thermodynamic forces, at fixed fluxes, the evolution path for a weak nonequilibrium (linear) system corresponds to a maximum of the indicated difference. There are also the so-called Biot principle [33, 34] (minimum dissipation) and Ziegler principle [35] (maximal rate of entropy production or maximal rate of work dissipation).

All principles, cited above, are equivalent to one another and to Onsager's equations, connecting forces and fluxes by a symmetrical matrix of kinetic coefficients. They differ in the parameters chosen as variable and fixed. In this sense, extremum principles do not present any peculiarly new results. However, there exists at least one extremum principle, which, though not proven strictly so far, long ago became a powerful heuristic means of forecasting the system's evolution in material science. This is the principle of maximal rate of Gibbs potential release for the systems in which the evolution path cannot be chosen in a unique way [36–45, Ivanov, M.A. personal communication, 2003]. Namely, it is assumed that among all possible ways to reaching equilibrium (under homogeneous external conditions) or at a stationary state (under nonhomogeneous conditions), nature is likely to choose a path at which the value of $-dG/dt$ is maximal (if temperature and pressure are fixed).

In order to emphasize the physical meaning of the difference between this principle and variational principles established by Onsager, Gyarmati, Biot, Ziegler, and others, we use the following analogy. Each evolutionary path of the system can be presented as a groove leading to, for example, a minimum of Gibbs potential. Onsager's equations determine the definite trajectory inside the groove. Minor force or flux variations slightly (in the second order of smallness) deflect the indicated trajectory, keeping it inside the groove. But nature often suggests a more interesting development; when starting from the same state, it is possible to go down via different grooves, separated by quite high walls which are easy to jump over only at the initial stage, but become increasingly difficult with time. Arrival at each groove needs, as a rule, overcoming a certain barrier (the nucleation one, for instance). After having overcome different barriers, initially identical system samples find themselves in different states, at different groove bottoms, with different sets of kinetic coefficients (therein lies the nonlinearity; that is, the dependence of kinetic coefficients on state parameters).

The principle of maximum for $-dG/dt$ shows a better groove for descent without analyzing the complex problem concerning competitive overcoming of different barriers during the selection stage. To this problem class, we can refer competitive phase transformations, diffusion paths choosing at interdiffusion in three- and multicomponent systems with limited solubility, formation, and evolution of two-phase zones, defining the lamellar structure period at cellular precipitation and spinodal decomposition.

For the time being, as far as we are concerned, the principle of maximal Gibbs free energy release (in the aforementioned sense) is not rigorously proven. Moreover, we cannot assert it to be always valid. However, being a heuristic principle, it is employed successfully in many problems. One can suppose that in the transformation zone in nanoparticle and nanograin ensembles, with barriers between grooves being low, and at Gibbs potential surface, some split of paths will be observed rather than a single determined unique evolutionary path.

11.4
Application of the Principles of Thermodynamics of Irreversible Processes: Examples

11.4.1
Criterion of First Phase Choice at Reaction–Diffusion Processes

Formation and phase-growth processes observed in the diffusion zone at interdiffusion in a binary metal system often turn out to be different from the ideal case.

11.4 Application of the Principles of Thermodynamics of Irreversible Processes: Examples

The latter implies that all phases, existing on the phase diagram at the given temperature, appear in the diffusion zone and grow according to a parabolic law. There are different theories that aim at explaining these peculiarities (limited reaction rates at interfaces, nucleation in the concentration gradient field, and diffusion phase competition at the nucleation stage) [16, 46–52]. All these theories are based on certain model assumptions. Let us check whether it is possible or not to explain these peculiarities using general thermodynamic arguments, without introducing additional presuppositions regarding the nucleation mechanisms or the kind of interface transition experienced by atoms.

Let us consider a solid-state reaction between two almost mutually insoluble materials A and B, at which we get two or more intermediate phases to form and grow. If for some reason, at a certain stage of a process, the layer of only one intermediate phase "i" starts growing with an average concentration (molar fraction) of the B component $\bar{c}_B = \bar{c}_i$, the growth rate of the thickness ΔX_i is defined as the product of the average interdiffusion coefficient $\overline{D}_i$ in this phase and the homogeneity interval width of the phase, Δc_i:

$$\frac{d\Delta X_i}{dt} = \frac{1}{c_i(1-c_i)} \frac{\overline{D}_i \Delta c_i}{\Delta X_i + \lambda_i} \tag{11.15}$$

Here λ_i is characteristic length, connected with the finite reaction rate at interfaces, and

$$\overline{D}_i = \frac{1}{\Delta c_i} \int_{\Delta c_i} \tilde{D}(c) dc$$

At that, the product $\overline{D}_i \Delta c_i$, defined as the integral of interdiffusion coefficient over the homogeneity interval, $\int_{\Delta c_i} \tilde{D}(c) dc$, is called *Wagner's integral coefficient* [53]. If the phase thickness is quite large, $\Delta X_i \gg \lambda_i$ (such that the time required for atoms to diffuse through the layer is much larger than the time of their delay at the boundary), the phase layer growth becomes parabolic:

$$\frac{d\Delta X_i}{dt} = \frac{1}{c_i(1-c_i)} \frac{\overline{D}_i \Delta c_i}{\Delta X_i}, \quad \Delta X_i \cong \sqrt{\frac{2\overline{D}_i \Delta c_i}{c_i(1-c_i)} t} \tag{11.16}$$

There exists a widespread erroneous interpretation of solid-phase reactions. It is connected with the idea that intermediate chemical compounds with a very small (or, to be more exact, unobserved during the direct experiment) concentration zone of existence ($\Delta c \to 0$) can grow just by generating and transferring nonequilibrium defects or even via some nondiffusion way. Such ideas are wrong and present, in particular, the inconvenience when the concentration c is used as a state parameter for the intermediate phases. Many things become clear when instead of concentration, the chemical potential is used. To be more exact, we use the difference of chemical potentials, which is equal to the derivative of Gibbs free energy (per atom) with respect to atomic concentration; that is,

$$c_B \left(\frac{\partial g}{\partial c_B} \right) = \mu_B - \mu_A = \tilde{\mu}_B$$

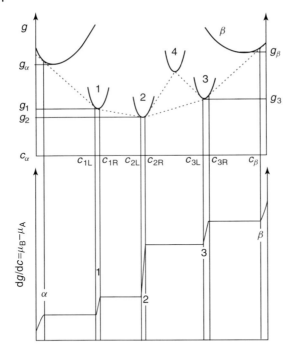

Figure 11.1 Phase diagram, phase equilibriums, and concentration profile in a binary multiphase system: α and β are solid solutions; 1, 2, 3 are stable; and 4 is a metastable intermediate phase.

the reduced chemical potential. In this scale, the interval of existence of chemical compounds, the compounds being considered as intermediate phases between A and B, is not at all short and cannot be reduced to a point.

Figure 11.1 shows the curves of the dependence of the Gibbs potential on concentration for two solid solutions α, β, and three stable (1, 2, 3) and one metastable (4) intermediate phases. The boundaries of the stable existence zone for each compound are established by the points of common tangents for the curves of neighboring phases. We denote the minimum of the $g(c)$ curve for the i-th phase as g_i. The value of the flux through the phase layer is directly determined by the chemical potential difference μ between the boundaries. As is seen from Figure 11.1, this difference is mainly influenced by the difference between slopes of common tangents rather than by the small concentration width of curves $g(c)$. The aforementioned slopes of common tangents, in their turn, are determined by the difference between "well" depths:

$$\left.\frac{\partial g}{\partial c}\right|_{i,i+1} \simeq \frac{g_{i+1} - g_i}{c_{i+1} - c_i} \tag{11.17}$$

where c_i is the concentration corresponding to the minimum of i-th phase Gibbs energy. This concentration, as a rule, is close to a certain stoichiometric one.

11.4 Application of the Principles of Thermodynamics of Irreversible Processes: Examples

Thus, the difference between chemical potentials is determined by a simple approximate equation which is nominally similar to a finite-difference form of the second derivative in the concentration space:

$$\Delta(\mu_B - \mu_A) = \left.\frac{\partial g}{\partial c}\right|_{i,i+1} - \left.\frac{\partial g}{\partial c}\right|_{i-1,i} \simeq \frac{g_{i+1} - g_i}{c_{i+1} - c_i} - \frac{g_i - g_{i-1}}{c_i - c_{i-1}} \quad (11.18)$$

This difference is not small at all if compared, for example, with kT, and it can cause significant diffusion fluxes, $-\tilde{L}\frac{\Delta(\tilde{\mu}/kT)}{\Delta X}$, without the assistance of nonequilibrium defects. Usually, nonequilibrium defects appear at interfaces, for example, but then the phase growth will differ from the parabolic one which is actually observed in many cases. At not too small times, it is the parabolic phase growth that becomes an evidence of the act of equilibrium difference between chemical potentials in the growing layers of chemical compounds.

The given description, in terms of chemical potentials and Onsager's coefficients, completely agrees with the description within Fick's first law. Indeed, Wagner's integral coefficient, introduced above, can be transformed after applying Darken's relation, expressing the interdiffusion coefficient in a binary system through diffusion coefficients of components' marked atoms and the second derivative of Gibbs potential (per atom) with respect to concentration:

$$\tilde{D} = (cD_A^* + (1-c)D_B^*)\frac{c(1-c)}{kT}\frac{\partial^2 g}{\partial c^2} \quad (11.19)$$

Having substituted this expression into Wagner's integral coefficient and taking into account the small width of the homogeneity zone in a phase, we get

$$\int_{\Delta c_i} \tilde{D}(c)dc = \overline{D_i^*}\frac{c_i(1-c_i)}{kT}\left(\left.\frac{\partial g}{\partial c}\right|_{i,i+1} - \left.\frac{\partial g}{\partial c}\right|_{i-1,i}\right)$$

$$\cong \overline{D_i^*}\frac{c_i(1-c_i)}{kT}\left(\frac{g_{i+1}-g_i}{c_{i+1}-c_i} - \frac{g_i-g_{i-1}}{c_i-c_{i-1}}\right)$$

$$= \overline{D_i^*}\frac{c_i(1-c_i)(c_{i+1}-c_{i-1})}{kT(c_{i+1}-c_i)(c_i-c_{i-1})}\Delta g_i(i+1, i-1 \to i) \quad (11.20)$$

Here $\overline{D_i^*}$ is the diffusion coefficient of marked atoms averaged over the phase

$$\overline{D_i^*} \equiv c_i\overline{D_A^*} + (1-c_i)\overline{D_B^*}$$

and $\Delta g_i(i+1, i-1 \to i)$ is the thermodynamic driving force (per atom) for phase "i"-formation from neighboring "$i-1$" and "$i+1$" phases.

If, during the reaction, there grows a layer consisting of one phase only (i.e., phase "$i-1$" is almost pure A, and phase "$i+1$" is almost pure B), the expressions become very simple and physically clear:

$$\int_{\Delta c_i}^{\tilde{D}}(c)dc = \overline{D_i^*}\frac{\Delta g_i(A+B \to i)}{kT} \quad (11.21)$$

Thus, the growth rate of a single intermediate phase is determined just by its atoms' mobility and the thermodynamic driving force for the phase formation.

At that, the rate of Gibbs free energy release per unit of the layer's area is defined as

$$-\frac{dG}{dt} = \frac{d}{dt}\left(\frac{\Delta X_i}{\Omega}\Delta g_i\right) = \frac{1}{\Omega}\frac{\overline{D}_i^*(\Delta g_i)^2}{c_i(1-c_i)\Delta X_i} = \frac{1}{\Omega}\sqrt{\frac{\overline{D}_i^*(\Delta g_i)^3}{2tc_i(1-c_i)}} \qquad (11.22)$$

The given expression presents the basis for understanding the thermodynamic reasons for solid-phase amorphization, and of solving "the first phase problem" at reaction–diffusion in general. The first phase, to be formed in the diffusion zone, must have a maximal product of mobility and free formation energy cubed. In case of amorphous layer formation, it means that the increased mobility of atoms in the disordered phase compensates for the low transformation driving force.

The principle of maximal Gibbs potential release allows one an understanding (from the viewpoint of thermodynamics) of why only one phase grows in the majority of cases at the initial stage of reaction–diffusion. Consider, for clarity, the case of two intermediate phases on the phase diagram. If the concentration homogeneity ranges for both phases are narrow and close to stoichiometric concentrations c_1, c_2, the rate of their growth is defined from the following system of equations [46]:

$$\frac{d\Delta X_i}{dt} = \sum_{k=1}^{2} a_{ik}\frac{D_k \Delta c_k}{\Delta X_k} \qquad (11.23)$$

where

$$||a|| = \begin{pmatrix} a_{11} & a_{12} \\ a_{21} & a_{22} \end{pmatrix} = \frac{1}{c_2 - c_1}\begin{pmatrix} \frac{c_2}{c_1} & -1 \\ -1 & \frac{1-c_1}{1-c_2} \end{pmatrix}$$

As discussed previously, Wagner's integral coefficients for both phases can be expressed through the mobilities of marked atoms and the formation driving forces of each phase. It is important that in the case of two phases growing, the driving forces for each of them are lower than in the case of one phase growth, the phase being formed from pure initial components. For example, the driving force of formation for nickel monosilicide between Si and Ni_2Si is considerably less than that between Si and Ni (see Figure 11.2) [6, 7, 53]:

Consider, for the sake of concreteness, the case where it is possible to form two phases in a binary system. The phases have average concentrations such as $c_1 = 1/3$, $c_2 = 2/3$ which are equal to the formation driving forces from pure A and B,

$$\Delta g_1(A, B \to 1) = \Delta g_2(A, B \to 2) \equiv \Delta g_0$$

At that, the driving forces of phase 1 formation from A and phase 2, and phase 2 formation from phase 1 and B, are also equal, and as can be easily checked, amount to a half of the formation driving force from pure components (Figure 11.3):

$$\Delta g_1(A, 2 \to 1) = \Delta g_2(1, B \to 2) = \Delta g_0/2 \equiv \Delta g'$$

11.4 Application of the Principles of Thermodynamics of Irreversible Processes: Examples

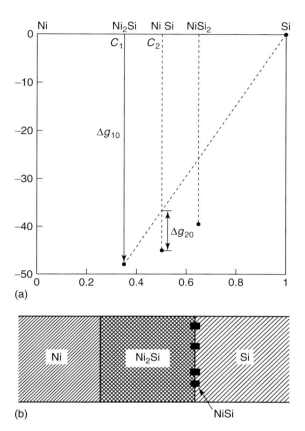

Figure 11.2 (a) Phase formation scheme and driving forces for reaction between Ni and Si; (b) contact zone morphology for the system Ni–Si.

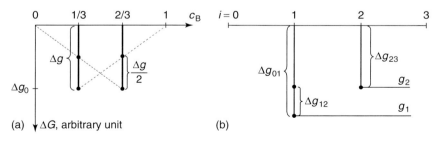

Figure 11.3 (a) Scheme of phase formation and reaction driving forces with and without formation of phase 2. (b) Scheme of phase formation with different driving forces for the formation of two intermediate phases.

We get

$$D_1 \Delta c_1 = \overline{D_1^*} \frac{\frac{1}{3}\left(1-\frac{1}{3}\right)\left(\frac{2}{3}-0\right)}{kT\left(\frac{2}{3}-\frac{1}{3}\right)\left(\frac{1}{3}-0\right)} \Delta g' = \frac{2}{3} \frac{\overline{D_1^*}\Delta g_0}{kT} \quad (11.24)$$

$$D_2 \Delta c_2 = \frac{2}{3} \frac{\overline{D_2^*}\Delta g_0}{kT}$$

As the next step, we analyze two cases for different relations between diffusion characteristics of the phases involved. In the first case, we take $\overline{D_1^*} = \overline{D_2^*}$. In the second $\overline{D_1^*} \gg \overline{D_2^*}$.

1) In the first case $(\overline{D_1^*} = \overline{D_2^*})$, the situation is absolutely symmetrical, the growth rates are equal for both phases and one can easily define them from the aforementioned system of equations (Equation 11.23):

$$\frac{d\Delta X_1}{dt} = \frac{d\Delta X_2}{dt} = \sqrt{2\overline{D_1^*}\Delta g'}\,\frac{1}{\sqrt{t}} = \sqrt{\overline{D_1^*}\Delta g_0}\,\frac{1}{\sqrt{t}} \quad (11.25)$$

At that point, the general expression for the rate of Gibbs potential release per unit of area at simultaneous growth of both phases equals

$$-\frac{dG^{(1+2)}}{dt} = \frac{d}{dt}\left(2\frac{\Delta X_1}{\Omega}\Delta g_0\right) = \frac{2\Delta g_0}{\Omega}\sqrt{\frac{\overline{D_1^*}\Delta g_0}{t}} \quad (11.26)$$

If there is only phase 1 that grows, then, according to the above formula,

$$-\frac{dG^{(1)}}{dt} = \frac{d}{dt}\left(\frac{\Delta X_1}{\Omega}\Delta g_0\right) = \frac{\left(\frac{3}{2}\right)\Delta g_0}{\Omega}\sqrt{\frac{\overline{D_1^*}\Delta g_0}{t}} \quad (11.27)$$

As observed, according to the maximum principle $-dG/dt$ in case of similar thermodynamic and diffusion properties, two phases are most likely to grow simultaneously $(2 > 3/2)$.

2) Let, now, $\overline{D_1^*} \gg \overline{D_2^*}$. Here, phase 2 grows far more slowly than phase 1 (in a quasi-stationary way), so the solution of the system can be simplified using the quasi-stationary condition

$$\frac{d\Delta X_2}{dt} \approx 0$$

Then,

$$\frac{d\Delta X_1}{dt} = \sqrt{3\overline{D_1^*}\Delta g'}\,\frac{1}{\sqrt{t}} = \sqrt{\frac{3}{2}}\sqrt{\overline{D_1^*}\Delta g_0}\,\frac{1}{\sqrt{t}}$$

that is, the growth rate for phase 1 is still less than in the case of absolute absence of phase 2. Consequently, the rate of Gibbs potential release is less as well ($\sqrt{3/2} < 3/2$), since the driving force of the A + 2 → 1 reaction is less than that of A + B → 1 reaction. Thus, in case 2 it is more favorable when phase 2 does not simply grow slowly but is completely absent. In this manner, the suppression of slow-growing phases by fast-growing ones is thermodynamically favorable.

11.4 Application of the Principles of Thermodynamics of Irreversible Processes: Examples

It is clear that the proposed analysis is correct only if applied to the initial stage. With time, when the first phase while extending lessens its growth rate, at a certain point of time it becomes favorable that phase 2 appears and the suppression no longer takes place. The conclusions cited above are matched to the results of a more complicated and detailed theory of phase competition, based on nucleation analysis in the field of fluxes and concentration gradients [52].

Next, we calculate and compare the criteria for the beginning of the growth of phase 1, suppressed by phase 2, on the basis of a kinetic approach [46, 47, 50], and using the criterion of maximal entropy production. In order to complete this, consider the binary system described above, with the driving forces of formation of the two intermediate phases being different. We will use the following relations between Gibbs potentials of the phases (Figure 11.3b, where i is the phase number):

$$g_0 = 0, \quad g_1 = -\Delta g_{01}, \quad g_2 = -\Delta g_{23}, \quad g_3 = 0,$$
$$g_1 - g_2 = -\Delta g_{12}, \quad \Delta g_{23} = \Delta g_{01} - \Delta g_{12} = \Delta g_{01}(1-\lambda) \tag{11.28}$$

where

$$\lambda = \frac{\Delta g_{12}}{\Delta g_{01}}$$

and in case of $\lambda = 0$, the relation $\Delta g_{01} = \Delta g_{23}$ is fulfilled.

For the given diffusion problem, the equations for the growth rates of phases 1 and 2 are the following [46]:

$$\left(\frac{d\Delta X_1}{dt}\right)^* = \frac{1}{c_2 - c_1}\left(\frac{c_2}{c_1}\frac{\overline{D}_1 \Delta c_1^*}{\Delta X_1} - \frac{\overline{D}_2 \Delta c_2^*}{\Delta X_2}\right) \tag{11.29}$$

$$\left(\frac{d\Delta X_2}{dt}\right)^* = \frac{1}{c_2 - c_1}\left(-\frac{\overline{D}_1 \Delta c_1^*}{\Delta X_1} + \frac{1-c_1}{1-c_2}\frac{\overline{D}_2 \Delta c_2^*}{\Delta X_2}\right)$$

Keeping to the results of [46], we get the growth condition for phase 1 and suppression condition of phase 2

$$\left.\frac{d\Delta X_1}{dt}\right|_{l_{cr}} > 0, \quad \left.\frac{d\Delta X_2}{dt}\right|_{l_{cr}} < 0 \tag{11.30}$$

l_{cr} standing for the critical size of phase nuclei. In accordance with [46, 47, 50], phase 2 is suppressed by the growing phase 1, and phase 2 will be able to start growing only when phase 1 reaches the thickness

$$\Delta X_1|^{GG} = \frac{1-c_2}{1-c_1}\frac{\overline{D}_1 \Delta c_1}{\overline{D}_2 \Delta c_2}l_{cr} \tag{11.31}$$

Further, we transform Equation 11.31 using Equation 11.20:

$$\Delta X_1|^{GG} = \frac{1-c_2}{1-c_1}\frac{\overline{D}_1^* \frac{c_1(1-c_1)}{kT}\left(\frac{0-g_1}{1-c_1} - \frac{g_1-0}{c_1-0}\right)}{\overline{D}_2^* \frac{c_2(1-c_2)}{kT}\left(\frac{0-g_2}{1-c_2} - \frac{g_2-0}{c_2-0}\right)}l_{cr}$$

$$= \frac{1-c_2}{1-c_1}\frac{\overline{D}_1^*}{\overline{D}_2^*}\frac{l_{cr}}{(1-\lambda)} \tag{11.32}$$

which allows us to determine the critical thickness of phase 1. After this thickness is reached, the growth of phase 2 becomes possible from the viewpoint of kinetics. In terms of the given description, the continuous layer of phase 2 nuclei must exist in the diffusion zone between phase 1 and pure B. In particular, when the condition

$$\Delta g_{12} = \Delta g_{23} \quad (\lambda = 0, c_1 = 1/3, c_2 = 2/3)$$

(symmetrical case) is fulfilled, we obtain the critical thickness of phase 1:

$$\Delta X_1|^{GG} = \frac{1}{2} \frac{\overline{D}_1^*}{\overline{D}_2^*} l_{cr} \qquad (11.33)$$

In the second case, let us consider whether the suppression condition for the growth of phase 2 is fulfilled or not. For that, we use the criterion of maximal rate of free energy release. Let us write down the expression for the rate of free energy release at phase 1 growing and the nuclei of phase 2 being present in the diffusion zone:

$$-\left(\frac{dG}{dt}\right)^* = \Delta g_{01} \left(\frac{d\Delta X_1}{dt}\right)^* + \Delta g_{23} \left(\frac{d\Delta X_2}{dt}\right)^* = \frac{\Delta g_{01}}{c_2 - c_1}$$
$$\times \left[\left(\frac{c_2}{c_1} \frac{\overline{D}_1 \Delta c_1^*}{\Delta X_1} - \frac{\overline{D}_1 \Delta c_2^*}{l_{cr}} \right) + (1 - \lambda) \left(-\frac{\overline{D}_1 \Delta c_1^*}{\Delta X_1} + \frac{1 - c_1}{1 - c_2} \frac{\overline{D}_2 \Delta c_2^*}{l_{cr}} \right) \right]$$
$$(11.34)$$

Similarly, the expression for the rate of free energy release during the growth process of the intermediate phase 1, without the nuclei of phase 2 in the diffusion zone, is as follows:

$$-\left(\frac{dG}{dt}\right) = \Delta g_{01} \frac{\overline{D}_1 \Delta c_1}{\Delta X_1} \frac{1}{c_1 (1 - c_1)} \qquad (11.35)$$

Suppose that phase 2 starts increasing when the growing nuclei lead to the fulfillment of relation:

$$-\left(\frac{dG}{dt}\right)^* \geq -\left(\frac{dG}{dt}\right) \qquad (11.36)$$

Then, from Equation 11.36, taking into account Equations 11.34 and 11.35, one obtains

$$\left[\frac{1}{(1 - c_2)} - \frac{\lambda (1 - c_1)}{(c_2 - c_1)(1 - c_2)} \right] \frac{\overline{D}_2 \Delta c_2^*}{l_{cr}}$$
$$\geq \left[\frac{\Delta c_1}{c_1 (1 - c_1)} - \frac{c_2 - (1 - \lambda) c_1}{(c_2 - c_1) c_1} \Delta c_1^* \right] \frac{\overline{D}_1}{\Delta X_1} \qquad (11.37)$$

where Δc_1^* corresponds to the concentration width of phase 1 in the diffusion zone in the presence of competitive nuclei of phase 2, and Δc_1 denotes the concentration width of phase 1, when the nuclei of phase 2 are absent in the diffusion zone.

11.4 Application of the Principles of Thermodynamics of Irreversible Processes: Examples

To interrelate Δc_1^* and Δc_1, consider the following relation:

$$\frac{(\overline{D}_1 \Delta c_1)^*}{\overline{D}_1 \Delta c_1} = \frac{\overline{D}_1^* \frac{c_1(1-c_1)}{kT} \left(\frac{g_2 - g_1}{c_2 - c_1} - \frac{g_1 - 0}{c_1 - 0} \right)}{\overline{D}_1^* \frac{c_1(1-c_1)}{kT} \left(\frac{0 - g_1}{1 - c_1} - \frac{g_1 - 0}{c_1 - 0} \right)} = \frac{c_2 - (1-\lambda)c_1}{(c_2 - c_1)} (1 - c_1) \quad (11.38)$$

Here, we neglect the fact that at abrupt dependence of $D(c)$, the values $\overline{D}$ for different Δc can vary. Thus,

$$\Delta c_1^* = \frac{c_2 - (1-\lambda)c_1}{(c_2 - c_1)} (1 - c_1) \Delta c_1 \quad (11.39)$$

which allows us to rewrite Equation 11.37 as

$$\frac{\overline{D}_2 \Delta c_2^*}{l_{cr}} \geq \frac{\overline{D}_1 \Delta c_1}{\Delta X_1} \frac{(1-c_2)\left((c_2-c_1)^2 - (1-c_1)^2(c_2-(1-\lambda)c_1)^2\right)}{c_1(1-c_1)(c_2-c_1)(c_2-c_1-(1-c_1)\lambda)} \quad (11.40)$$

When the critical thickness of phase 1 is reached, the growth of phase 2 becomes favorable. This thickness is defined from the relation

$$\Delta X_1 \left| \frac{dG}{dt} \right. = \frac{\overline{D}_1^*}{\overline{D}_2^*} \frac{(1-c_2)\left((c_2-c_1)^2 - (1-c_1)^2(c_2-(1-\lambda)c_1)^2\right)}{c_1 c_2 (1-c_1)(c_2-c_1-(1-c_1)\lambda)^2} l_{cr} \quad (11.41)$$

Equation 11.41 in the model approximations

$$\Delta g_{12} = \Delta g_{23} \qquad (\lambda = 0, c_1 = 1/3, c_2 = 2/3)$$

and Equation 11.28 are reduced to

$$\Delta X_1 \left| \frac{dG}{dt} \right. = \frac{5}{4} \frac{\overline{D}_1^*}{\overline{D}_2^*} l_{cr} \quad (11.42)$$

which is 2.5 times larger than that calculated from Equation 11.33.

The obtained result implies that at phase thickness increase up to the value indicated by Equation 11.33, first comes the kinetic possibility of the growth of phase 2 though it becomes thermodynamically favorable only when the thickness of phase 1, defined from Equation 11.41, is reached. It can be explained in the following way. Though the growth of phase 2 becomes possible when the condition Equation 11.32 is fulfilled, the relaxation rate of the system falls abruptly for phase 2 and cannot grow fast enough; the rate of free energy release by the first phase noticeably drops due to the driving force decrease. Only after phase 1 has reached the thickness given by Equation 11.41, both growing phases can reduce free energy of the system more efficiently than in case of phase 1 growth in the presence of the nuclei of phase 2. Thus, one more conclusion becomes apparent: until the critical thickness of phase 1 (defined from Equation 11.41) is reached, the nuclei of phase 2, forming a continuous layer, are not favorable in the diffusion zone from the viewpoint of maximal rate of decrease of the system's free energy.

In Figure 11.4, one gets a calculated dependence of critical thickness of phase 2 on the ratio of the thermodynamic driving forces for the growth of intermediate

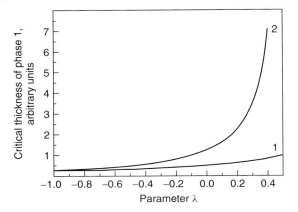

Figure 11.4 Dependence of critical thickness of phase 1 (from which the growth of phase 2 begins) on the parameter λ: 1—according to Equation 11.32; 2—according to Equation 11.41.

phases λ in the range of their stable coexistence, when

$$\Delta g_{23}/(1-c_2) > \Delta g_{01}/(1-c_1), \qquad \Delta g_{01}/c_1 > \Delta g_{23}/c_2$$

which in the given case corresponds to the range $-1 \leq \lambda \leq 0.5$. As observed in any relation between thermodynamic driving forces of the growth of the two phases, the kinetic stage of suppression of phase 2 and beginning of its growth, in accordance with Equation 11.32 occurs earlier (especially at $\lambda \to 0.5$), which is determined by the principle of maximal rate of free energy release according to Equation 11.41.

11.5
Conclusions

As the revisionist Bernstein once said, "Goal is nothing, movement is everything." Actually, we are interested not in the final result of the process (minimum of Gibbs potential in case of isothermal and isobaric external conditions), but in the way of evolution of how this result is achieved. There are a number of problems where more than one way is possible and therefore, a problem of choice arises. When analyzing the choice of the evolutionary path for a nonequilibrium system reaching the equilibrium state, one must use an additional thermodynamic principle. The chapter presents arguments for the principle of maximal rate of Gibbs potential release.

In the chapter, the peculiarities of phase formation in binary metal systems with competitive formation of intermediate phases resulting from diffusion interaction are analyzed. The previous kinetic analysis of diffusion phase interaction [46, 47] has shown that only after a suppressing phase reaches a certain critical thickness, the growth rate of the suppressed phase nucleus becomes positive. Thus, the kinetic criterion for the beginning of the suppressed phase growth was provided. However, as shown above, such kinetic permission still does not guarantee the beginning of growth, as it can conflict with the maximal rate of free energy release. The reason is as follows: the formation of the second phase leads to essential lessening of the first phase growth rate (because of the decrease in the transformation driving force), and the second phase is unable to perform a sufficiently fast growth. It is shown that some additional time is required (and additional growth of the suppressing phase) to reach thermodynamic profitability of the growth of two phases as compared to the growth of one phase. During this intervening period, the growth of the second phase is both possible and favorable (from the viewpoint of equilibrium thermodynamics) but is not the most favorable one. Such a process can be referred to as a *metastable regime*. Transition from "metastable" to the "stable regime" (when the rate of free energy release is maximal), can be called the *thermodynamic criterion* of the regime's stability. After the old regime suffers "thermodynamic loss of stability," the system becomes able to reduce its Gibbs potential, that is to move to an equilibrium state, at an even higher rate than previously.

References

1. de Groot, S.R. and Mazur, P. (**1962**) *Non-equilibrium Thermodynamics*, North-Holland, Amsterdam.
2. Glansdorff, P. and Prigogine, I. (**1971**) *Thermodynamic Theory of Structure, Stability and Fluctuations*, John Wiley & Sons, Ltd, New York.
3. Landau, L.D. and Lifshitz, E.M. (**1964**) *Statistical Physics, Part 1*, Nauka, Moscow (in Russian).
4. Gurov, K.P. (**1978**) *Phenomenological Thermodynamics of Irreversible Processes (Physical Foundations)*, Nauka, Moscow (in Russian).
5. Gyarmati, I. (**1970**) *Nonequilibrium Thermodynamics*, Springer, Berlin.
6. van Gulpen, J. (**1996**) *Reactive phase formation in the Ni-Si system*. Ph. D. thesis. Eindhoven University of Technology, Eindhoven.
7. Gas, P. (**1989**) *Applied Surface Science*, **38**, 178.
8. Schwarz, R.B. and Johnson, W.L. (**1988**) *Journal of the Less-Common Metals*, **140**, 1.
9. Vredenberg, A.M. (**1986**) *Journal of Materials Research*, **1**(6), 773.
10. Kirkaldy, J.S. and Young, D.J. (**1987**) *Diffusion in the Condensed State*, The Institute of Metals, London.
11. Kirkaldy, J.S. and Brown, L.C. (**1963**) *Canadian Metallurgical Quarterly*, **2**(1), 89.
12. Gusak, A.M. (**1990**) *Zhurnal Fizicheskoi Khimii (Journal of Physical Chemistry)*, **64**, 510 (Moscow, in Russian).
13. Gusak, A.M. and Lyashenko, Yu.A. (**1990**) *Ingenerno - Fizicheskij Journal*, **59**(2), 286 (in Russian).
14. Hopfe, W.D. and Morral, J.E. (**1994**) *Acta Metallurgica Materialia*, **42**(11), 3887.
15. Chen, H. and Morral, J.E. (**1999**) *Acta Materialia*, **47**(4), 1175.
16. Gusak, A.M., Bogatyrev, O.O., Zaporozhetz, T.V. et al. (**2004**) *Models of Solid State Reactions*, Cherkasy National University, Cherkasy.
17. Frear, D., Gravas, D. and Morris, J.W. (**1986**) *Journal of Electronic Materials*, **16**, 181.
18. Kim, H.K., Liu, H.K. and Tu, K.N. (**1995**) *Applied Physics Letters*, **66**, 2337.
19. Gusak, A.M. and Tu, K.N. (**2002**) *Physical Review B*, **66**, 115403.
20. Prigogine, I. (**1961**) *Thermodynamics of Irreversible Processes*, Interscience, New York.
21. Onsager, L. (**1931**) *Physical Review*, **37**(II), 1404.

22. Onsager, L. (1931) *Physical Review*, **38**(II), 2265.
23. Kubo, R. (1959) in *Lectures in Theoretical Physics*, (eds W.E., Brittin and L.G., Dunham), Interscience, New York, Vol. **1**, 120.
24. van der Waals, J.D. and Kohnstamm, J.D. (1908) *Lehrbuch der Thermodynamik*, Maas and Suchtelen, Leipzig.
25. Baidakov, V.G. (2002) *Nucleation Theory and Applications*, JINR, Dubna.
26. Patashinskiy, A.Z. and Pokrovskiy, V.L. (1975) *Fluctuation Theory of Phase Transformations*, Nauka, Moscow (in Russian).
27. Devyatko, Yu.N., Rogogkin, S.V., Musin, R.N. and Fedotov, B.A. (1993) *Journal of Experimental and Theoretical Physics*, **103**(1), 285.
28. Hillert, M. (1970) *Phase Transformations*, American Society of Metals, Metals Park, Ohio.
29. Hillert, M. (1998) *Phase Equilibria, Phase Diagram, and Phase Transformations - Their Thermodynamic Basis*, Cambridge University Press, Cambridge.
30. Cahn, J.W. and Hilliard, J.E. (1959) *Journal of Chemical Physics*, **31**, 1965.
31. Christian, J.W. (1975) *The Theory of Transformation in Metals and Alloys*, Oxford University Press, Oxford.
32. Wang, Y., Chen, L. and Khatchaturyan, A. (1994) in *Proceedings of an International Conference on Solid - Solid Phase Transformations* (eds W.C., Johnson, J.M., Howe, D.E., Laughlin and W.A., Soffa), TMS, USA.
33. Biot, M.A. (1955) *Physical Review*, **97**, 1463.
34. Biot, M.A. (1970) *Variational Principles in Heat Transfer*, Clarendon, Oxford.
35. Ziegler, H. (1963) in *Progress in Solid Mechanics*, Vol. 4, Ch. 2 (I.N., Sneddon and R., Hill), North-Holland, Amsterdam.
36. Bene, R.W. (1987) *Journal of Applied Physics*, **61**, 1826.
37. Goesele, U. and Tu, K.N. (1982) *Journal of Applied Physics*, **53**, 3252.
38. Bogel, A. and Gust, W.A. (1988) *Zeitschrift fur Metallkunde*, **79**, 296.
39. Martyushev, L.M., Seleznew, V.D. and Kuznetzova, I.E. (2000) *Journal of Experimental and Theoretical Physics*, **118**(1), 149 (Moscow, in Russian).
40. Martyushev, L.M. and Seleznew, V.D. (2006) *Physics Reports*, **426**, 1.
41. Podolskiy, S.E. (1996) *Metallofizika i Noveishiye Technologii*, **18**(1), 18 (in Russian).
42. Hillert, M. and Agren, J. (2006) *Acta Materialia*, **54**, 2063.
43. Lyashenko, Yu.A. (2004) *Technical Physics Letters*, **30**(2), 109; (Translated from (2004) *Pis'ma v Zhurnal Teknihеskoi Fiziki*, **30**(3), 54.
44. Lyashenko, Yu.A., Gusak, A.M. and Shmatko, O.A. (2005) *Metallofizika i Noveishiye Technologii*, **27**(7), 873 (in Russian).
45. Lyashenko, Yu.A. and Gusak, A.M. (2006) *Defect and Diffusion Forum*, **249**, 81.
46. Gurov, K.P., Kartashkin, B.A. and Ugaste, Yu.E. (1981) *Interdiffusion in Multiphase Metallic Systems*, Nauka, Moscow (in Russian).
47. Gusak, A.M., Gurov, K.P. (1982) *Fizika Metallov i Metallovedeniye*, **53**, 842, 848 (in Russian).
48. Tu, K.N., Chu, W.K. and Mayer, J.W. (1975) *Thin Solid Films*, **25**, 403.
49. d'Heurle, F.M., Gas, P., Philibert, J. and Zhang, S.L. (2001) *Diffusion and Defect Forum*, **194-199**, 1631.
50. Gurov, K.P. and Gusak, A.M. (1990) *Izvestiya AN SSSR, Metally*, 1, 163 (in Russian).
51. Gurov, K.P. and Gusak, A.M. (1985) *Fizika Metallov i Metallovedeniye*, **59**, 1062 (in Russian).
52. Gusak, A.M., Hodaj, F. and Bogatyrev, A. (2001) *Journal of Physics. Condensed Matter*, **13**(12), 2767.
53. van Loo, F.J.I. (1990) *Progress in Solid State Chemistry*, **20**, 47.

12
Choice of Optimal Regimes in Cellular Decomposition, Diffusion-Induced Grain Boundary Migration, and the Inverse Diffusion Problem

Yuriy A. Lyashenko

12.1
Introduction

In the previous chapter, we considered the problem of selection of the path for the reactive diffusion when the choice is made among a finite number of phase formation modes. Let us now treat the problems allowing for an infinite set of solutions, all of which are compatible with the matter balance equations. These are mainly the problems of morphology choice when different modes are possible, and the task is to find the optimal one to be realized in practice. In the first two sections of the chapter, we consider the issue of low-temperature phase diffusion transformations, namely discontinuous precipitation and DIGM. At that, the evolution equations based on matter conservation laws allow an infinite number of solutions corresponding to different thicknesses of the phase formed and different velocities of phase transformation front movement.

First, we treat a model of discontinuous precipitation of binary polycrystalline supercooled alloys at low temperatures as a result of DIGM. In the proposed approach, we independently determine the main parameters: interlamellar distance, maximum velocity of the phase transformation front, and residual supersaturation at the front. This is achieved by using a set of equations for

1) diffusion mass transfer in the moving interphase boundary (coinciding with the transformation front) including a triple product of interdiffusion coefficient in the interphase boundary, segregation coefficient, and interphase boundary thickness;
2) balance of entropy fluxes at the phase transformation front;
3) maximum rate of free energy release.

We will determine the concentration profile in the precipitation lamella and compare it with experimental results for Ni–In and Pb–Sn systems at different supersaturations.

We also describe the model of alloying in the three-layer thin film system at low temperatures. In such cases, solid solution formation takes place as a

Diffusion-Controlled Solid State Reactions. Andriy M. Gusak
Copyright © 2010 WILEY-VCH Verlag GmbH & Co. KGaA, Weinheim
ISBN: 978-3-527-40884-9

result of the DIGM. The unknown parameters are determined from the set of equations for:

1) grain boundary diffusion along the moving planar phase boundary;
2) the entropy balance in the region of the phase transformation moving at constant velocity;
3) the maximum rate of the free energy release.

We consider a model system with complete solubility of the components. The main parameters are self-consistently determined using a thermodynamic and kinetic description in the frame of the regular solution model. The model allows one to determine the concentration distribution along the planar moving phase boundary, its velocity, the thickness of the forming solid solution layer, and the limiting average concentration in this layer.

The third section of the chapter features the use of the principle of minimum entropy production at solving the inverse diffusion problem. Generally, the interdiffusion coefficients are found by experimental methods, for example, by the Matano–Boltzmann method. This method is convenient for the determination of the interdiffusion coefficient in binary systems provided the Boltzmann substitution can be used; however, it is difficult to adapt it to other boundary conditions or multicomponent systems. There exists an alternative approach, in which diffusion coefficients are chosen because of coincidence of the calculated and experimentally obtained concentration profiles. Such ill-posed problems can be solved by the Tikhonov regularization methods when a specially built discrepancy functional is minimized. The task is complicated by the necessity to solve the inverse diffusion problem using experimental concentration profiles measured with rather considerable errors. This makes the minimization problem (already unstable with respect to perturbations) diverge and not give the correct results. By the example of the solution of the inverse diffusion problem in the model binary system, we demonstrate that including the entropy production into the procedure of discrepancy functional minimization makes the solution stable even at noticeable distortions of the concentration profile. The choice of the entropy production as a smoothing part of the discrepancy functional is physically well founded and can be used while solving arbitrary inverse problems concerning heat and mass transfer.

12.2
Model of Self-Consistent Calculation of Discontinuous Precipitation Parameters in the Pb–Sn System

The problems of the choice of the diffusion evolution path for nonequilibrium metallic systems at phase transformations when different energetically favorable modes of mass transfer at the same initial and boundary conditions are possible have been intensively investigated both theoretically and experimentally. An actual topic of material science is the problem of the choice of diffusion evolution path for nonequilibrium metallic systems at phase transformations under the conditions of ambiguity when different modes are possible at the same initial and boundary conditions and these modes described by, say, mass transfer equations. One of these problems, which is formulated in terms of the invariant that is equal to the product of the squared structure period and the growth velocity, is the problem of discontinuous precipitation of a supersaturated binary alloy. The most thorough theoretical and experimental reviews of discontinuous precipitation peculiarities are given in [1–4].

The process is peculiar because of the necessity to treat the substance fluxes in two dimensions; here, the diffusion grain boundary fluxes, which leads to the removal of the initial supersaturation of the binary alloy and the formation of the lamellae and go in the direction perpendicular to the transformation front. It is a generally accepted point of view that at low temperatures the complete removal of the supersaturation is not achieved; that is, behind a moving transformation front the system does not reach an equilibrium state. Therefore, the second and even the third coarsening stage of a cellular structure follow consecutively and a much smaller growth rate is possible. The saturation reached because of discontinuous precipitation reaction can be arbitrary. At low rates of phase transformation (thus, large sizes of the interlamellar distance), a high degree of component redistribution and supersaturation removal is reached. Here, the system reaches almost equilibrium; thus, the decaying phase concentration approaches the composition determined by the tangent at the binary phase diagram.

Cahn [5] found a solution of the problem of the diffusive redistribution of components in the moving grain boundary, which depends on the combination of two parameters: interlamellar distance and growth rate. Moreover, Cahn used the following assumptions: (i) linear relation between growth rate and driving force introducing one more kinetic parameter, the boundary mobility; (ii) the maximum free energy release principle. This set of equations allowed the determination of the interlamellar distance, the rate of transformation boundary movement, and concentration profile along the lamellae at the known triple product of grain boundary diffusion coefficient, segregation coefficient, and grain boundary thickness.

In [6, 7], another set of equations and the algorithm for solving the problem of discontinuous precipitation in the case of the most mobile grain boundary are analyzed. There, the principle of entropy balance and principle of maximum entropy production are additionally used. In this approach, we neglect the energy dissipation by resistance to the boundary motion (the grain boundary mobility tends to infinity), so that the total energy dissipation is caused by the grain boundary diffusion. Both in Cahn's approach and in the model [6] for hypothetical systems, coincidence was obtained of interlamellar distance with the experimentally determined one within a reasonable error interval; the calculated magnitude of transformation front velocity was by 2–3 orders larger than the experimental one. Therefore, we face with a question as to whether it is appropriate to use the principle of maximum entropy production for unambiguous and reliable determination of discontinuous precipitation parameters.

The approach [6] employs a simplified expression for the concentration dependence of the Gibbs potential, namely, a quadratic concentration dependence without logarithmic terms connected with mixing entropy. Reference [8] provides the solution of the discontinuous precipitation problem in a binary Pb–Sn system using a thermodynamic description based on the regular solution model taking into account the energy dissipation as a result of grain boundary diffusion only. However, the estimated transformation rate calculated by both approaches remains too high, especially at small supersaturations, which necessitates the search for other mechanisms of energy dissipation. There, accounting for additional ways of energy dissipation results in the reduction of the calculated rate of transformation front motion and, thus, in a higher degree of component redistribution.

In the model considered below, the role of both grain boundary and bulk diffusion in the transformation front and close to it, respectively, is analyzed within the problem of unambiguous determination of the discontinuous precipitation parameters in the binary Pb–Sn system at room temperature [9]. In order to complete this, we use the principle of maximum rate of free energy release and balance of entropy fluxes for the description of discontinuous precipitation kinetics for binary polycrystalline alloys and independent determination of three basic parameters: interlamellar distance, rate of phase transformation front, and concentration profile close to the transformation front. While solving the problem, we also find the optimal concentration distribution of components both along the precipitation lamella behind the transformation front and close to it, as well as the degree of the components' separation.

12.2.1
General Description of the Model Systems

We analyze now the formation of cells in the grains of an alloy α_0 supersaturated at the aging temperature with a concentration c_0. The configuration of the system at the formation of a single cell, with depleted phase α and phase β being in equilibrium with it, is depicted in Figures 12.1b and c. We introduce here the following notations: $c^{\alpha/\beta}$ is the equilibrium concentration at the junction of α and β phases in the α phase; c^β is the equilibrium concentration in the β phase lamella; and c_1^{max} is the maximum concentration in the α phase, reached inside the α lamella (along the z axis). The transformation region R coincides with the interphase boundary between α_0 and α phases, and is of the following sizes: b is the height, h is the width, and Δz is the length.

The thermodynamic driving force of the discontinuous precipitation, as for any process at constant temperature and pressure, is the decrease of the system's Gibbs potential, which is schematically illustrated in the model phase diagram (Figure 12.2).

The dependence of the concentration in the α phase on the coordinate z along the transformation boundary is described by a steady-state grain boundary diffusion

12.2 Calculation of Discontinuous Precipitation Parameters in the Pb–Sn System

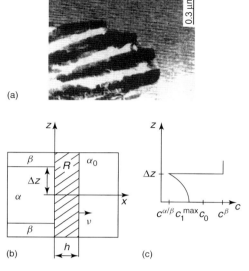

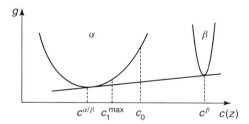

Figure 12.1 (a) Image of the discontinuous precipitation front in the Co32W alloy [10] (for model illustration). Model system: (b) configuration in the x–z-plane; (c) concentration distribution in and behind the transformation front at the supersaturated solution's concentration being equal to c_0 close to the transformation front.

Figure 12.2 Model phase diagram and phase equilibria in the Pb–Sn system.

equation for a moving boundary with the constant velocity of the boundary v [5]:

$$s\tilde{D}\frac{d^2 c(z)}{dz^2} + \frac{c_0 - c(z)}{h} v = 0 \tag{12.1}$$

where s is the segregation coefficient, $\tilde{D}$ is the interdiffusion coefficient in the boundary R, $c(z)$ is the local concentration in the boundary R, and h is the boundary width. The Cahn solution of this equation is as follows:

$$c(z) = c_0 - (c_0 - c^{\alpha/\beta})\frac{\cosh(z/L)}{\cosh(\Delta z/L)} \tag{12.2}$$

where the coordinate z varies from zero (in the middle of the α-phase lamella) to Δz at the α–β interface. Accordingly, $c(z)$ changes from c_1^{max} to $c^{\alpha/\beta}$ over a half-length of the cell, depending on the combination of parameters $\Delta z/L$. The interlamellar distance Δz, the kinetic coefficient $L = \sqrt{s\tilde{D}h/v}$, and the triple product $s\tilde{D}h$ are related in the models of

1) Turnbull [11] via

$$sh\tilde{D} = \frac{c_0}{c_0 - c_1} \lambda^2 v \qquad (12.3)$$

where $\lambda \approx \lambda_\alpha = 2\Delta z$ and c_1 is the average concentration in the α phase;

2) Cahn [5] via

$$sh\tilde{D} = \frac{1}{C}\Delta z^2 v \qquad (12.4)$$

where Cahn's parameter C is found from the expression

$$W = \frac{c_0 - c_1}{c_0 - c^{\alpha/\beta}} = \frac{2}{\sqrt{C}} \text{th}\left(\frac{\sqrt{C}}{2}\right) \qquad (12.5)$$

3) Petermann and Hornbogen [12] via

$$sh\tilde{D} = \frac{RT}{-8\Delta G}\Delta z^2 v \qquad (12.6)$$

where ΔG is the change of Gibbs free energy due to reaction ($\Delta G < 0$);

4) Bogel and Gust [13] via

$$sh\tilde{D} = \frac{RT}{C\Delta G}\left(\frac{c^\beta - c_0}{c^\beta - c_1}\right)^2 \Delta z^2 v. \qquad (12.7)$$

In all the models presented, the product $\Delta z^2 v$ enters the solution as a single parameter (invariant) which is indivisible. This makes it impossible to solve the inverse problem of prediction of the morphology parameters in a two-phase system at discontinuous precipitation.

Each kinetically possible solution for an arbitrary transformation rate and the corresponding value of the parameter L gives a certain interlamellar distance Δz. For $\Delta z/L \to 0$, the maximum concentration c_1^{max} in the α phase (at the center of the α-phase lamella and along the lamella) tends to the equilibrium value $c^{\alpha/\beta}$. As the $\Delta z/L$ ratio increases, the degree of redistribution of the components decreases, the interphase boundary velocity increases, and the system remains substantially in nonequilibrium. Thus, a single equation, (Equation 12.1) with the solution Equation 12.2 cannot unambiguously describe the kinetics of the discontinuous precipitation and does not provide an independent determination of Δz and L. In order to eliminate this uncertainty, Cahn employed the principle of maximum free energy release rate ΔF during the discontinuous precipitation reaction. This was achieved by considering a linear relationship between ΔF and the boundary velocity [5] and by introducing the second kinetic coefficient representing the boundary mobility. However, this linear relationship is not always confirmed in experiment [14].

The main idea of our approach is in properly taking into account the sources of energy appearance and degradation in the transformation boundary moving at constant velocity. We consider that the total free energy reduction must balance with all ways of energy degradation both at the expense of diffusion along the transformation region and because of the kinetics of atomic interphase jumps $\alpha_0 - \alpha$, determining the mobility of the boundary. To independently find the basic kinetic transformation parameters, we use the principle of maximum rate of free energy release, balance equations for entropy fluxes, and Cahn's solution for the equation of mass transfer (Equation 12.2).

12.2.2
Model Based on the Balance and Maximum Production of Entropy

While analyzing the above-presented models, one realizes that the problem of mode choice cannot be unambiguously solved within the solution of mass transfer equation. This makes it necessary to consider thermodynamic or kinetic approaches to the analysis of transformation front stability and to choose a certain contact zone morphology. From the point of view of kinetics, the interphase boundary instability may be caused either by instability with respect to fluctuations of the boundary shape [15–17] or by the failure of balance equations for fluxes at the moving boundaries [16]. From a thermodynamic viewpoint, the problem of choice of one kinetically allowed mode can be solved using the variation principles of nonequilibrium thermodynamics [18–29].

In [6–9], we presented an analysis of the balance of the free energy release and dissipation in the transformation boundary at a constant velocity: namely, the moving reaction front (grain boundary) is an open system in a steady-state regime. Its entropy should be constant. It means that the entropy production inside this boundary due to diffusion should be compensated by in and out entropy fluxes. The difference of these fluxes is just the free energy release rate due to the decomposition. We calculate the entropy production rate according to the Onsager scheme. On the other hand, we calculate the free energy release rate in the reaction using the explicit form of free energy of a regular solid solution and velocity of front propagation.

Both the entropy production and the divergence of entropy fluxes are functions of two variables (say, interlamellar distance and L). Each of these functions independently has no maximum. Yet, at the subset of variables corresponding to the constraint of steady-state balance of entropy, the maximum of entropy production does exist and should correspond to the most probable evolution path.

We described the model for independent determination of the main parameters Δz and L at known thermodynamic parameters of the system, triple product $s\tilde{D}h$ for the interphase boundary, and initial concentration c_0 of the supersaturated solution. The employed approach allows us to find the concentration distribution $c(z)$ along the α-phase lamella and, therefore, the averaged concentration inside the α-phase lamella. The main approximation is that we assume diffusion redistribution to take place only in the planar interphase boundary moving at constant velocity (Figure 12.1).

Let us assume that a change in the entropy $d_e S$ during the elementary time interval dt due to the phase transformation in the region R shifting by its width is equal to the change in the entropy $d_i S$ due to the diffusive redistribution of components in the same region R. We consider a steady-state process obeying the condition [18]

$$\frac{dS}{dt} = \frac{d_i S}{dt} + \frac{d_e S}{dt} = 0 \tag{12.8}$$

which means that the total entropy change in the moving open system is zero (this condition is valid at a constant transformation front velocity v). In order to pass from the rate of entropy change to the rate of free energy release, we can use the relation (valid at a constant temperature and pressure)

$$\frac{dS_{i,e}}{dt} = -\frac{1}{T}\frac{dG_{i,e}}{dt}$$

In this case, the rate of the free energy release is [18, 19]

$$\Psi \equiv T\frac{d_i S}{dt} = T\int_V \sigma \, dV = -\frac{d_i G}{dt} = \frac{d_e G}{dt} > 0 \tag{12.9}$$

12.2.2.1 Phase Transformations and Law of Conservation of Matter

Let us determine the change in the Gibbs potential due to the transformation of the α_0 phase in an element $dz + dz'$ at the point z of the interphase boundary. For this element, transformed into a section of the α phase of length dz and a section of the β phase of length dz', conservation of matter yields (Figure 12.3)

$$c_0(dz + dz') = c(z)dz + c^\beta dz' \tag{12.10}$$

The change in the Gibbs potential can be written as

$$\Delta G(z) = g(c(z))dz + g^\beta dz' - g(c_0)(dz + dz') \tag{12.11}$$

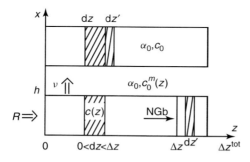

Figure 12.3 Model scheme of concentration redistribution accounting for mass conservation in the reaction front R. NGb indicates the newly formed grain boundary between the α and β phases.

where $g(c(z))dz$ is the Gibbs potential of the α-phase band of length dz at the point z with the concentration $c(z)$ in the region R, g^β is the Gibbs potential of the β phase, and $g(c_0)$ is the Gibbs potential in the supersaturated α_0 phase in front of the transformation boundary.

12.2.2.2 Calculation of the Driving Force

The rate of the free energy release due to the phase transformation in the region R can be written as

$$\Psi_{\text{release}} = \frac{d_e G}{dt} = -\frac{\upsilon b}{\Delta z} \int_0^{\Delta z} \Delta G(c(z))dz - g^{\alpha/\beta} b \upsilon \qquad (12.12)$$

Here, the first term in the right-hand side is the energy gained per unit volume and the second term corresponds to the formation of a new boundary region of length h between α_0 and β phases,

$$\Delta G(z) = \left[g(c(z)) + g^\beta \frac{c_0 - c(z)}{c^\beta - c_0} - g(c_0) \frac{c^\beta - c(z)}{c^\beta - c_0} \right] dz^\alpha \qquad (12.13)$$

Integration of Equation 12.12 is performed only for the part of the boundary that is located between α and α_0 phases. The second item in Equation 12.12 corresponds to the loss in the change of total free energy and to the formation of the new part of the boundary between α and β phases, which is of length h. Therefore, for a unit volume we get $g^{\alpha/\beta} = \frac{2\gamma}{\Delta z}$, where γ is the free energy per unit area of the α–β boundary.

12.2.2.3 Calculation of Energy Dissipation in the Transformation Front along the Precipitation Lamella

The entropy production as a result of the diffusive redistribution of components in the region R can be written as [16, 18, 19]

$$\Psi_{DZ} = \frac{hb}{\Delta z} \int_0^{\Delta z} IX dz = \frac{hb}{\Delta z} \int_0^{\Delta z} -s\tilde{D} \frac{\partial c(z)}{\partial z} \left(-\frac{\partial \tilde{\mu}(z)}{\partial z} \right) dz \qquad (12.14)$$

where I is the generalized flux along the z axis and correspondingly along the transformation front R, X is the driving force causing the flux along the z axis, Δz is half the size of the α-phase cell along the region R, and $\tilde{\mu}(z)$ is the generalized chemical potential at the point z of region R, constant along the boundary thickness, h.

Model 1: Quadratic Dependence of Gibbs Potential and Relatively High Mobility of the Boundary Consider the case of diffusion-controlled reaction at decomposition in the approximation of rather high mobility of the boundary. Expanding Gibbs potential into a Taylor series in α and α_0 phases with respect to $g(c^{\alpha/\beta})$ up to the second order inclusive, which is acceptable for regular solutions with high intermixing energy, we obtain

$$\Delta G = \frac{1}{2} \left[\left(c(z) - c^{\alpha/\beta} \right)^2 - \frac{c^\beta - c(z)}{c^\beta - c_0} \left(c_0 - c^{\alpha/\beta} \right)^2 \right] g''|_{c^{\alpha/\beta}} = f(c(z))k \qquad (12.15)$$

where $f(c(z)) < 0$ and $k = g''|_{c^{\alpha/\beta}} > 0$ is the Gibbs potential curvature in the α-phase in the point $c^{\alpha/\beta}$. For the driving force one has

$$X = -\frac{\partial \tilde{\mu}}{\partial z} = -\frac{\partial c(z)}{\partial z} k$$

where

$$\tilde{\mu} = \frac{\partial \Delta G}{\partial c(z)} = \left[(c(z) - c^{\alpha/\beta}) - \frac{(c_0 - c^{\alpha/\beta})^2}{2(c^\beta - c_0)} \right] k \qquad (12.16)$$

Here $\tilde{\mu}$ is a reduced chemical potential per unit volume.

Using Equations 12.2, 12.4, 12.5, and 12.12–12.16 and integrating with respect to z, we get

$$\Psi_{DZ} = \frac{d_e G}{dt} = -\frac{kL\upsilon}{16\Delta z}(c_0 - c^{\alpha/\beta})^2$$
$$\times \left\{ \frac{\Delta z}{L} \text{sch}^2\left(\frac{\Delta z}{L}\right) - \frac{3c^\beta - c_0 - 2c^{\alpha/\beta}}{(c^\beta - c_0)} \text{th}\,(\Delta z/L) \right\} - \frac{2\gamma \upsilon}{\Delta z}$$
$$= -\frac{d_i G}{dt} = \frac{kL\upsilon}{2\Delta z}(c_0 - c^{\alpha/\beta})^2 \left\{ \text{th}\,(\Delta z/L) - \frac{\Delta z}{L}\text{sch}^2\left(\frac{\Delta z}{L}\right) \right\} \qquad (12.17)$$

which enables us to determine the first relation between Δz and L:

$$kL(c_0 - c^{\alpha/\beta})^2 = 8\gamma \left\{ \frac{\Delta z}{L}\text{sch}^2\left(\frac{\Delta z}{L}\right) - \frac{2c^{\alpha/\beta} - c_0 - c^\beta}{c^\beta - c_0}\text{th}\left(\frac{\Delta z}{L}\right) \right\}^{-1} \qquad (12.18)$$

Having established the relation between Δz and L, we use the principle of maximum rate of free energy release to find the optimal Δz from the condition

$$\frac{d\Psi}{d(\Delta z)} = 0 \qquad (12.19)$$

Let us apply now numerical methods to find the maximum of the expression 12.19 with respect to Δz provided Equation 12.18 is fulfilled. Ultimately, we find the solution of the set of equations 12.18 and 12.19, determining the parameters $\Delta z = \Delta z^{SOL}$ and L at a given triple product $sh\tilde{D}$ and equilibrium concentration $c^{\alpha/\beta}$ (Figure 12.4).

To compare the results with experiment, we use the data for the system Ni-1.4 at.% In at 703 K [1, 13] at the following parameters: $k = 1.075 \times 10^{11} \text{J m}^{-3}$, $\gamma = 0.5 \text{J m}^{-2}$, $c_0 = 1.4$ at.%, $c^{\alpha/\beta} = 0.43$ at.%, $c^\beta = 25$ at.%, and $s\tilde{D}h = 2.12 \times 10^{-24} \text{m}^3 \text{s}^{-1}$. As a result of calculations, we get the decomposition parameters which coincide with experimental ones: averaged concentration inside the cell $c_1 = 0.82$ at.%, interlamellar distance $\Delta z^{SOL} = 0.134 \times 10^{-6}$ m, and Cahn's parameter $C = 9.2$. The rate of the transformation boundary is derived from $\upsilon = s\tilde{D}h/L^2 = 2.7 \times 10^{-10} \text{m s}^{-1}$. Also calculated is the change of the Gibbs free energy at the given parameters $\Delta G = -14.6$ J mol^{-1}, by integrating Equation 12.15 and taking into account the surface energy (according to Equation 12.12). The obtained value of the total change of free energy at discontinuous decomposition confirms the experimentally found one [1, 13].

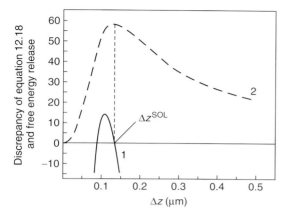

Figure 12.4 Dependences for (i) the discrepancy of Equation 12.18 from the half-period of cellular structure; (ii) the free energy release rate in the transformation region. The optimal solution corresponds to Δz^{SOL}.

Thus, using the balance equation for entropy fluxes allows us to obtain a new scheme of determining the main parameters of discontinuous precipitation without Cahn's assumption about linear dependence of the driving force and the velocity of transformation boundary (the form of the dependence can be established in the process of calculation). The scheme may take into account some additional ways of energy dissipation in the transformation region, which result in the reduction of the boundary's mobility. Also, the system of problem solving may contain a more detailed description of the thermodynamic parameters and take into account additional kinds of driving forces.

Model 2: Peculiarities of Discontinuous Precipitation in the Pb–Sn System Below we describe the solution of the discontinuous precipitation problem for the binary system Pb–Sn using a thermodynamic description based on the regular solution model [8], since the approach [6] employed a simplified expression for the concentration dependence of the Gibbs potential (quadratic concentration dependence without taking into account logarithmic terms connected with the mixing entropy). Here, we apply the principle of maximum rate of free energy change and balance of entropy fluxes to the description of the kinetics of discontinuous precipitation for binary polycrystalline supersaturated alloys and independent determination of two chief parameters: the interlamellar distance, velocity of phase transformation boundary, and their interrelation with the triple product. In the process of solving, we find the supersaturation close to the transformation front and the degree of the components' separation. A comparison with the experimental results for the system Pb–Sn will be made.

The main approximation employed here is the assumption of diffusion redistribution of components both inside the planar interphase boundary moving at constant velocity (Figure 12.1b and c) and close to the transformation front because of bulk diffusion. We demonstrate that, despite that the distance over which the concentration profile undergoes changes close to the transformation front due to bulk diffusion can be small, the corresponding contribution to energy dissipation may turn out to be significant.

The Gibbs potentials in the α and α_0 phases and in the grain boundary phase (per unit volume) are calculated within the regular solution model:

$$G(c(z)) = g_0 + g_1 c(z) - g_2 c(z)^2 + \frac{RT}{V_m} \left[c(z) \ln(c(z)) + (1-c(z)) \ln(1-c(z)) \right] \quad (12.20)$$

where

$$g_0 = \frac{G_{Pb}^0}{V_m}, \quad g_1 = \frac{G_{Sn}^0 - G_{Pb}^0 + G^{xs}}{V_m}, \quad g_2 = \frac{G^{xs}}{V_m}$$

and G_{Pb}^0 and G_{Sn}^0 are the Gibbs free energy for the pure components, G^{xs} is the mixing energy, and $V_m = 1.92 \times 10^{-5}$ m^3 mol^{-1} is the molar volume of the alloy Pb–Sn.

The parameters of the regular solution model for the system Pb–Sn [30] for an fcc phase of solid solution Sn in Pb (in J mol^{-1}) are the following:

$$G_{Pb}^0 = -7650.085 + 101.715188T - 24.5242231 T \ln(T)$$
$$-0.00365895 T^2 - 2.4395 \times 10^{-7} T^3$$

$$G_{Sn}^0 = 4150 - 5.2T + G_{Sn}^{SER}$$

where

$$G_{Sn}^{SER} = -5855.135 + 65.427891T - 15.961 T \ln(T)$$
$$-0.0188702 T^2 - 3.121167 \times 10^{-6} T^3 - 61.96/T$$

$$G^{XS} = 5132.4154 + 1.5631T \quad (12.21)$$

For the β phase, the value G_{Sn}^{SER} is taken. As a result of the phase equilibria calculation (Figure 12.5), we obtain the equilibrium concentration $c^{\alpha/\beta} = 2.043$ at. %, which coincides with the experimentally determined value [31]. So,

$$\tilde{\mu} = \frac{\partial G}{\partial c(z)} = \mu_B - \mu_A = g_1 - 2g_2 c(z) + \frac{RT}{V_m} \ln \frac{c(z)}{1-c(z)} \quad (12.22)$$

Finally, for the driving force we have

$$X = -\frac{\partial \tilde{\mu}}{\partial z} = -\left[-2g_2 + \frac{RT}{V_m} \frac{1}{c(z)(1-c(z))} \right] \frac{\partial c(z)}{\partial z} \quad (12.23)$$

Now, for the free energy release rate (Equation 12.14), using Equation 12.2, one may write down

$$\Psi_{DZ} = -\frac{d_i G}{dt} = \frac{hbs\tilde{D}}{\Delta z} \int_0^{\Delta z} \left[-2g_2 + \frac{RT}{V_m} \frac{1}{c(z)(1-c(z))} \right] \left(\frac{\partial c(z)}{\partial z} \right)^2 dz$$

$$= \frac{L\upsilon}{\Delta z} \left[-g_2 A \left(\mathrm{sh}\left(\frac{\Delta z}{L}\right) \mathrm{ch}\left(\frac{\Delta z}{L}\right) - \frac{\Delta z}{L} \right) \right. \quad (12.24)$$

$$\left. + \frac{RT}{V_m} \left(-\frac{\Delta z}{L} + 2B \, \mathrm{ath}(\mathrm{arg}_1) + 2C \, \mathrm{ath}(\mathrm{arg}_2) \right) \right]$$

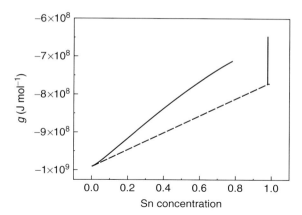

Figure 12.5 Pb–Sn phase diagram and phase equilibria calculated from Equations 12.20 and 12.21.

where

$$A = \frac{(c_0 - c^{\alpha/\beta})^2}{ch^2(\Delta z/L)}, \quad A_1 = c_0 - A, \quad A_2 = c_0 + A$$

$$B = \sqrt{\frac{A_1}{A_2}}, \quad C = \sqrt{\frac{1-A_1}{1-A_2}}, \quad \arg_1 = \sqrt{\frac{A_2}{A_1}}\,\text{th}\left(\frac{\Delta z}{2L}\right)$$

$$\arg_2 = \sqrt{\frac{1-A_2}{1-A_1}}\,\text{th}\left(\frac{\Delta z}{2L}\right)$$

12.2.2.4 Calculation of Energy Dissipation Close to the Transformation Front

To calculate the energy dissipation close to the phase transformation front due to bulk diffusion, we search for the concentration profile close to the front [9]. In order to do this, let us put down a steady-state equation of flux balance in the element of region R (Figure 12.6a) moving at constant velocity of the boundary v, in the reference frame of the transformation boundary:

$$\frac{\partial c}{\partial t} = \left[I_x^*(x,z) - I_x^*(x+dx,z)\right]dz \cdot b + \left[I_z^*(x,z) - I_z^*(x,z+dz)\right]dx \cdot b = 0 \tag{12.25}$$

which, on transition from the discrete scheme to the continuous one, changes into

$$\frac{\partial I_x^*}{\partial x} + \frac{\partial I_z^*}{\partial z} = 0 \tag{12.26}$$

where I_x^* and I_z^* are fluxes in the reference frame of the boundary along the axes x and z, respectively.

Let us establish the interrelation of fluxes in the boundary and laboratory reference frames, all the equations of energy dissipation being written in the latter:

$$I_x^* = I_x^* - c(x,z)v, \quad I_z^* = -s\tilde{D}\frac{\partial c(x,z)}{\partial z} \tag{12.27}$$

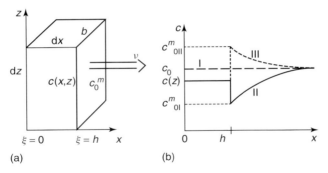

Figure 12.6 (a) Element of the transformation region R and its parameters. (b) Lines I–III illustrate the possible concentration distributions close to the transformation front; mode I – without taking into account bulk diffusion; II and III – taking into account bulk diffusion.

where I_x^* is the flux in the laboratory reference frame. Here, we consider the concentration $c(z)$ to be set at the left side of the chosen element R of the boundary, and a certain unknown concentration c_0^m at the right side. The latter generally can differ from the concentration of the supersaturated alloy c_0 (Figure 12.6b).

The balance equation for the fluxes in the volume $V = dz \cdot b \cdot h$ through diffusion fluxes in the laboratory reference frame is

$$dz \cdot b \cdot \int_0^h \left[\frac{\partial I_x}{\partial x} - v \frac{\partial c}{\partial x} + \frac{\partial I_z^*}{\partial z} \right] dx = 0 \quad (12.28)$$

where we use the following approximation while integrating the flux divergence $(\partial I_z^*/\partial z)$ over the boundary thickness:

$$\int_0^h \frac{\partial I_z^*}{\partial z} dx = -s\tilde{D}\frac{\partial^2}{\partial z^2} \int_0^h c(x,z) dz = -s\tilde{D}h \frac{\partial^2 c(z)}{\partial z^2} \quad (12.29)$$

Thus, Equation 12.28 transforms into

$$-hs D_b \frac{d^2 c(z)}{dz^2} - v(c_0^m - c(z)) + I_x(h,z) = 0 \quad (12.30)$$

where we neglect the flux through the left boundary of the volume element in the laboratory reference frame, which corresponds to the frozen bulk diffusion behind the transformation front.

Now, let us write down the balance equation of diffusion fluxes at the right boundary of the chosen volume element in the region R in the laboratory reference frame:

$$I_x(h,z) = -D_v \left. \frac{dc}{dx} \right|_{x=h} \quad (12.31)$$

To find the concentration profile close to the transformation front, we use both Equation 12.31 and the solution of Fick's equation for bulk diffusion close to the

12.2 Calculation of Discontinuous Precipitation Parameters in the Pb–Sn System

transformation front taking into account the steady-state mode of the transformation boundary movement:

$$D_v \frac{d^2 c(x)}{dx^2} + v \frac{dc(x)}{dx} = 0 \quad (12.32)$$

which brings to interrelation at the boundary $x = h$:

$$\left. \frac{dc}{dx} \right|_{x=h} = -\frac{I_x(h, z)}{D_v} = a \exp(-kx) \quad (12.33)$$

where a is the integration constant and $k = v/D_v$.

To find the integration constant a, we use the matter conservation law for the concentration profile close to the transformation front (Figure 12.6b):

$$-vc_0^m + I_x(h, z) = -v(c_0^- c(z)) \quad (12.34)$$

which gives

$$a = \frac{v(c_0^- c_0^m)}{D_v} \cdot \exp(kh) \quad (12.35)$$

Now, let us find the concentration distribution close to the transformation front at $x \geq h$ from the solution of Equation 12.32 and taking into account the boundary condition Equation 12.33:

$$c(x) = c_0^m - (c_0^- c_0^m) \exp(k(h - x) - 1) = c_0^- (c_0^- c_0^m) \exp(k(h - x)) \quad (12.36)$$

which meets boundary conditions at $x = h$ and $c(x) = c_0$ at $x \to \infty$.

To determine the free energy release rate due to diffusion redistribution of components close the moving front of discontinuous precipitation reaction, we can put down similar to Equation 12.14:

$$\Psi_{DX} = \frac{1}{\Delta z} \int_0^{\Delta z} \int_h^\infty I_x X_x dx dz = \frac{1}{\Delta z} \int_0^{\Delta z} \int_h^\infty (-D_v) \frac{\partial c(x, z)}{\partial x} \left(-\frac{\partial \tilde{\mu}(x, z)}{\partial x} \right) dx dz \quad (12.37)$$

where I_x is the generalized diffusion flux along the x axis close to the reaction front, X_x is the corresponding driving force, and $\tilde{\mu}(x, z)$ is the generalized chemical potential in the point (x, z) close to the transformation front.

For the driving force of diffusion redistribution close to the front R, we put down the expression

$$X_x = -\frac{\partial \tilde{\mu}}{\partial x} = -\left[-2g_2 + \frac{RT}{V_m} \frac{1}{c(x,z)(1 - c(x,z))} \right] \frac{\partial c(x, z)}{\partial x} \quad (12.38)$$

Assuming that just close the transformation front the concentration along the cell is $c(z, x = h) = c_0^m$, the free energy release rate Equation 12.37, using Equation 12.2, we may obtain

$$\Psi_{DX} = D_v k^2 (c_0 - c_0^m)^2 \int_h^\infty e^{2k(h-x)} \left[-2g_2 + \frac{RT}{V_m} \frac{1}{c(x)(1 - c(x))} \right] dx$$

$$= v \left[g_2 (c_0 - c_0^m)^2 - \frac{RT}{V_m} \left((1 - c_0) \ln \left| \frac{1 - c_0}{1 - c_0^m} \right| + c_0 \ln \left| \frac{c_0}{c_0^m} \right| \right) \right] \quad (12.39)$$

From Equation 12.39, one can see that the free energy release rate close to the reaction front is independent of the interdiffusion coefficient. It depends only on the front velocity and the concentration difference between the supersaturated alloy c_0 (far from the front) and c_0^m (close to the front) (Figure 12.6b). The concentration distribution close to the front (Equation 12.36) depends on both the deviation of c_0^m from c_0 and the coefficient $k = v/D_v$. Therefore, in calculating such a kind of energy dissipation, the concentration c_0^m deviation at the front can be arbitrary, it can be determined from the optimization procedure, and the concentration profile close to the transformation front must self-consistently adjust to this arbitrary value c_0^m at the distance given by the coefficient $k = v/D_v$ (see Equation 12.36). In the marginal case, when $D_v \to 0$ and the velocity v does not tend to infinity, from Equation 12.36 we get the relation $c(x)|_{x=h} = c_0$, which implies insignificantly small energy dissipation due to bulk diffusion. At the increase of D_v, the profile may develop at the distance which is commensurate with the interatomic one, which will require the necessity to take into account the energy dissipation due to bulk diffusion without the developed profile of concentration distribution close to the front. On further increase of D_v, caused by, say, a temperature rise, we can obtain a developed profile close to the transformation front.

Calculation of Optimal Parameters To determine the optimal parameters Δz, c_0^m, and L, we use the principle of maximum rate of free energy release (analogously, maximum entropy production). In order to do this, we find the extremum over the parameters Δz and c_0^m from the solution of the variation problem $\frac{d\Psi_{DF}}{d(\Delta z)} = 0$ and $\frac{d\Psi_{DF}}{d(c_0^m)} = 0$ taking into account Equation 12.9. Therefore, the total rate of energy dissipation is written as $\Psi_{DF} = \Psi_{DX} + \Psi_{DZ}$. For determination of the optimal value L, let us use Equations 12.2, 12.12, 12.14, 12.20, 12.24, and 12.39. Having found L, one can calculate the growth rate of a cell, using the triple product $s\tilde{D}h$ (equal to the product of segregation coefficient, grain boundary diffusion coefficient, and boundary thickness).

Calculation Results To analyze the solutions of the model worked out, we use the following parameters for the system Pb–Sn: $\gamma = 0.1437$ J m^{-2} [32], $c^{\alpha/\beta} = 2.043$ at.%, $c^\beta = 100$ at.%, $s\tilde{D}h = 2.1 \times 10^{-20} \exp(-4760/RT)$m^3 s^{-1} [17], and $T = 293$ K.

To find the optimal values for Δz, L, and c_0^m, a special numerical procedure was applied. The peculiarity of the problem is the absence of global maxima on the surfaces $\Psi_{DF}(\Delta z, L, c_0^m)$, $\Psi_{DZ}(\Delta z, L, c_0^m)$, and $\Psi_{DX}(\Delta z, L, c_0^m)$. It can be visually illustrated by solving the problem without optimization over c_0^m. In such a case, one has to maximize the free energy release rate $\Psi_{DZ}(\Delta z, L)$ (which is similar to maximization of $\Psi_{DF}(\Delta z, L)$) taking into account the balance of entropy production: $\Psi_{DF}(\Delta z, L) = \Psi_{DZ}(\Delta z, L)$. The behavior of production rate and free energy release isolines in the vicinity of the solution found by the numerical procedure with $\Delta z^{sol} = 1.02 \times 10^{-7}$m and $L^{sol} = 7.81 \times 10^{-8}$m is given in Figure 12.7a–b. From Figure 12.7a, it is seen that at the increase of both Δz and L (the latter is analogous to the reduction of the transformation rate) the rate of free energy production due to

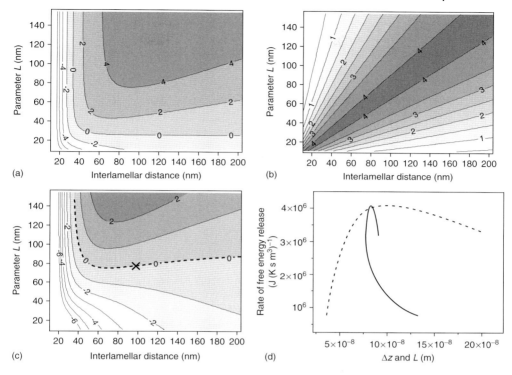

Figure 12.7 The isolines of rates: (a) production and (b) release of free energy depending on the interlamellar distance (x axis) and the parameter L (y axis) (the scale corresponds to the scale of the x axis of (d). (c) shows isolines of the differences between the rates of the free energy production and release (zero line corresponds to the line of optimal solution search). (d) presents the rate of free energy release along the zero line of (c): dotted line – dependence on Δz and solid line – dependence on L. Calculations are performed for $c_0 = 10$ at.%.

phase transformation just increases and does not have a global maximum. The rate of free energy release due to diffusion redistribution along the precipitation lamella (Figure 12.7b) has approximately one and the same maximum value along the line $\Delta z/L = $ const, which also corresponds to the absence of a global maximum. And, only taking into account the entropy balance $\Psi_{DF}(\Delta z, L) = \Psi_{DZ}(\Delta z, L)$ allows one unambiguously to solve the problem of optimum (the solution must be searched for along the line with zero level in Figure 12.7c). The dependence $\Psi_{DZ}(\Delta z, L)$ along the indicated line is presented in Figure 12.7d, where both interlamellar distance and the parameter L are depicted along the x axis. As can be seen, the maxima exist along the line of entropy production balance at $\Delta z = \Delta z^{sol}$ and $L = L^{sol}$.

After the calculations were carried out for different supersaturations, it was established that within the range of $c_0 \leq 20$ at.% the maximum rate of free energy release corresponds to the profile close to the front with $c_0^m = c_0$, and for supersaturations of $c_0 = 25$ and $c_0 = 30$ at.% to that with $c_0^m = 22.5$ at.%.

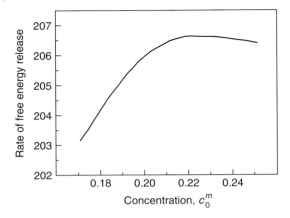

Figure 12.8 Dependence of free energy release rate on the concentration c_0^m at decomposition of alloy with $c_0 = 25$ at.%.

In Figure 12.8, the dependence of free energy release rate on the concentration c_0^m is plotted for decomposition of the supersaturated alloy with concentration $c_0 = 25$ at.%. Thus, one may conclude that energy dissipation due to bulk diffusion close to the front of discontinuous precipitation at small supersaturations is insignificant (within the approximations accepted at the introduction of Equation 12.33), which conforms to the generally accepted idea that only discontinuous precipitation takes place in the Pb–Sn system [32, 33]. In this case, the parameters of discontinuous precipitation depend on the shape of Gibbs potential surface energy and of phase boundaries and grain boundary diffusion parameters. At large supersaturations, the mentioned parameters depend on energy dissipation caused by the bulk diffusion close to the transformation front.

Let us also compare the parameters of discontinuous precipitation in the Pb–Sn system at 293 K, obtained in [34] and determined from the model calculation. In [34], an X-ray study of aging of Pb–Sn alloys at high (up to 29 at.%) Sn concentrations was realized. This work was significant in providing the dependence of the tin fraction W (Equation 12.5), which precipitates at the first stage of aging, on the tin concentration c_0 in the supersaturated α phase (Figure 12.9a). At average supersaturations, the experiments reported $W \approx 60$–65% (for example [17, 32]) at discontinuous precipitation. Here, we face an issue on the physical reason for the increase of the tin fraction W at low and high supersaturations of the initial alloys c_0.

From the model calculations, it is clear that, by taking into account the energy dissipation induced by grain boundary diffusion in the transformation front and bulk diffusion close to it, the value of the tin fraction W is close to the experimental ones at average and large supersaturations. The obtained concentration profile along the α-phase cell is presented in Figure 12.9b. Let us compare concentration dependencies for experimental and model values of interlamellar distance

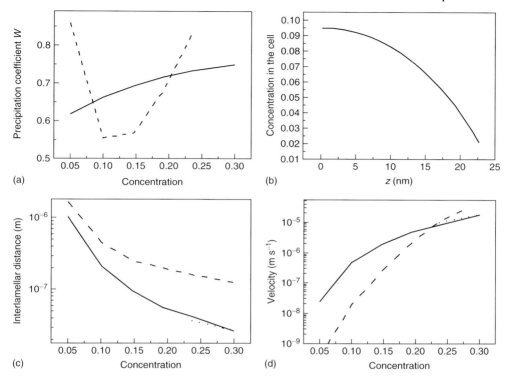

Figure 12.9 (a) experimental (dotted line) and model (solid line) concentration dependences of the precipitation coefficient W (Equation 12.5) in the Pb–Sn system; (b) calculated concentration change along the lamella between the middle of the α-phase lamella and the joint of α and β phases at $c_0 = 15$ at.%; (c–d) experimental (dashed line) and model (solid and dotted (taking into account interdiffusion) lines) dependences of interlamellar distance and boundary velocity on the concentration of the supersaturated alloy c_0.

(Figure 12.9c) and transformation front rate (Figure 12.9d). On analyzing the given dependence, we conclude that model values of interlamellar distance have the same functional dependence on the concentration at different initial supersaturations as experimentally obtained values, though they are slightly underestimated. Taking into account the energy dissipation due to grain boundary diffusion close to the transformation front leads to minor corrections, which confirms the hypothesis on the chief role of grain boundary diffusion and only discontinuous precipitation occurring in the Pb–Sn system at low temperatures. The deviations of experimental and model values of the rate of discontinuous precipitation front may be used while determining the concentration dependence of the grain boundary diffusion parameters on the concentration of moving transformation boundary.

So, the difference between experimental and model values can be explained either by the concentration dependence of the diffusivity in the grain boundary or by some additional mechanism of energy dissipation close to the phase transformation front not connected with the ordinary idea of bulk diffusion mechanism, or by kinetics of solid-state reactions at the transformation boundaries, when the rate of the process is influenced by both the rate of atomic transitions through the boundary and by the rate of new phase formation, which tends to infinity in our approach.

12.2.3
Calculation of Entropy Production Taking into Account Grain Boundary Diffusion and Atomic Jumps through the Grain Boundary

Entropy production due to diffusion redistribution of components in the region R can be written in the form of Equation 12.14. The driving force in this case can be obtained as the sum of the bulk chemical potential gradient X_b and the interphase boundary energy gradient X_σ. The first form of the driving force corresponds to the components' fluxes along the moving grain boundary, and in our assumptions it is taken to be equal to the gradient of the generalized chemical potential (Equation 12.23). Here, the rate of the grain boundary diffusion-induced free energy dissipation can be put down as in Equation 12.24.

Another kind of driving force, X_σ, is connected with the change of grain boundary energy along the reaction front. Here, it is presupposed that the interphase boundary energy depends on the concentration jump through the transformation boundary. First, let us calculate the grain boundary energy between two grains of the same phase but with different concentrations, that is, at the α_0/α interface, according to the approach outlined in [35]:

$$\sigma\left(c(z)\right) = \sigma_0 (c(z) - c_0)^2 + \text{const} \tag{12.40}$$

where

$$\sigma_0 = n_s z_s \frac{g_2 V_m}{z N_A}$$

n_s stands for the number of atoms per unit area of the interface, $z_s = 3$ is the coordination number of atoms in the grain boundary, $z = 12$ is the coordination number in the grain bulk, and N_A is the Avogadro number. To estimate the parameter n_s, we may use either the expression $n_s = \frac{4}{\sqrt{3a^2}}$, taken in the approximation of fcc structure of grain boundary, or the expression $n_s = \frac{4}{\sqrt{2a^2}}$, derived from the approach [36], where a stands for the interatomic distance. Here $n_s z_s$ is considered as the number of split interatomic bonds in the grain boundary as compared to the bond structure in the bulk. A certain mixing energy $\frac{g_2 V_m}{z N_A}$ per one atom corresponds to each bond, where $v_{at} = \frac{V_m}{N_A}$ is the atomic volume and the volume g_2 is taken in $[J\,m^{-3}]$. Then, the additional driving force is

$$X_\sigma = -\frac{1}{h}\frac{\partial \sigma}{\partial z} = -\frac{1}{h}\frac{\partial \sigma}{\partial c(z)}\frac{\partial c(z)}{\partial z} = -\frac{2\sigma_0}{h}(c(z) - c_0)\frac{\partial c(z)}{\partial z} \tag{12.41}$$

12.2 Calculation of Discontinuous Precipitation Parameters in the Pb–Sn System

The additional part of the free energy dissipation rate due to atomic jumps through the transformation boundary at the interface between α_0 and α phases is as follows:

$$\Psi_{D\sigma} = \frac{bs\tilde{D}2\sigma_0}{\Delta z} \int_0^{\Delta z} (c(z) - c_0) \left(\frac{\partial c(z)}{\partial z}\right)^2 dz = \frac{Lvg_2 BB^2}{8\Delta z} \sinh^3\left(\frac{\Delta z}{L}\right) \quad (12.42)$$

Dissipation processes at the other side of the moving interface, that is, between α_0 and β phases, can be significant and must be taken into account in the general balance of free energy release. In this case, it is possible to introduce two additional terms into the expression of the general rate of free energy dissipation. These terms arise from the grain boundary energy between α_0 and β phases with different structures and compositions of neighboring grains. Let us take into account the energy of grain boundaries between α_0 and β phases in accordance with [35]:

$$\sigma_{\alpha\beta} = \frac{n_s z_s}{2}(1 - c_0)\left[G_{Sn}^{BCT} - 2(1 - c_0)g_2\right] \frac{V_m}{zN_A} \quad (12.43)$$

The first term in the brackets describes the differences between the grain structure and Sn grain boundary. The second one is similar to that in Equation 12.18. Physically, the two mentioned terms mean that formation of the β phase implies both the appearance of a new composition (the second term) and transformation of the lattice structure (the first term).

While determining the corresponding terms of free energy dissipation, one must also bear in mind that each elementary volume starts transforming in the α_0 phase and finishes in the β phase. During the transformation time $t_{tr} = h/v$, each energy change begins from zero and ends in the maximum value (calculated from Equations 12.40 and 12.43). The simplest averaged estimation gives one-half of the maximum value. Thus, the terms Ψ described in Equations 12.40 and 12.43 will be taken as $v \cdot 1/2 \cdot \sigma \cdot 1/h$. Then, the two additional terms entering the expression for the general free energy release will be written in the following form:

$$\Psi_{D\alpha\beta} = \frac{v}{2h}(1 - c_0)^2 \frac{1}{a^2\sqrt{3}} \frac{g_2 V_m}{N_A} \frac{\Delta z^\beta}{\Delta z^{tot}} \quad (12.44)$$

$$\Psi_{DBCT} = \frac{v}{2h}(1 - c_0) \frac{1}{a^2\sqrt{3}} \frac{G_{Sn}^{BCT} V_m}{N_A} \frac{\Delta z^\beta}{\Delta z^{tot}} \quad (12.45)$$

where $\Delta z^{tot} = \Delta z + \Delta z^\beta$ is the total interlamellar distance. It should be noted that the coefficient 1/2 in equations 12.44 and 12.45 correlates with the results from [13], where we read the rate of free energy release amounts to half the energy used at formation of a grain boundary.

12.2.3.1 Optimization Procedure and Calculation Results

Thus, the entropy balance is

$$\Psi = \Psi_{DF}(\Delta z, L) = \Psi_{release}$$
$$\equiv \Psi_{DZ}(\Delta z, L) + \Psi_{D\sigma}(\Delta z, L) + \Psi_{D\alpha\beta}(\Delta z, L) + \Psi_{DBCT}(\Delta z, L) \quad (12.46)$$

This condition relates Δz and L, and turns the dependence Equation 12.46 into a function of one argument Δz.

Since one connection between Δz and L is established, let us use the principle of maximum free energy release to find the optimum of Δz from the condition

$$\frac{d\Psi}{d(\Delta z)} = 0 \qquad (12.47)$$

The process of solving the variation problem (Equations 12.46 and 12.47) will be performed according to the algorithm given in Equation 12.2. For the calculations, we use a set of parameters presented in Equation 12.2. To calculate the triple product $s\tilde{D}h$, we apply the expression for interdiffusion coefficient in Darken's approximations:

$$s\tilde{D}h = s \cdot h \cdot \left(c_{av} D^*_{Pb} + (1 - c_{av}) D^*_{Sn}\right) \left(1 + c_{av}(1 - c_{av}) \frac{g_2}{RT}\right) \qquad (12.48)$$

where c_{av} is the averaged Sn concentration in the grain boundary,

$$hD^*_{Pb} = 7.3 \times 10^{-15} \cdot \exp\left(-\frac{0.41 \times 1.6 \times 10^{-19}}{kT}\right) m^3 s^{-1}$$

$$hD^*_{Pb} = 6.1 \times 10^{-15} \cdot \exp\left(-\frac{0.46 \times 1.6 \times 10^{-19}}{kT}\right) m^3 s^{-1} \qquad (12.49)$$

are the tracer diffusivities of both components in pure Pb, found experimentally in [37]. These diffusion parameters were determined during experimental research on Pb and Sn isotope redistribution in polycrystalline Pb with moving grain boundaries.

Further, we can find $c_{av}(\Delta z, L)$ in Equation 12.48 after solving the optimization problem (Equations 12.46 and 12.47). The triple product is used for the determination of the transformation front velocity also after the optimization parameters Δz^{SOL} and L^{SOL} are found. Therefore, the velocity of the transformation front is derived from the expression for $L^{SOL} = \sqrt{s\tilde{D}h/v}$. Here, the triple product $sh\tilde{D}$ itself is not directly utilized in the optimization procedure. Our model requires only steadiness of $sh\tilde{D}$ for the whole transformation boundary. The actual value of $sh\tilde{D}$ may depend on the averaged concentration behind the transformation front.

First, we analyze the dependence of interlamellar distance and transformation rate on the change of concentration of the supersaturated solution c_0 and using the approximation Equation 12.48. Figure 12.10a presents the calculated values of interlamellar distance taking into account different mechanisms of energy dissipation as compared to the experimental one (line 5). Taking into account energy dissipation along the boundary between α and α_0 phases (taking into account only the term $\Psi_{DZ}(\Delta z^{SOL}, L^{SOL})$ – line 1, the sum of $\Psi_{DZ}(\Delta z^{SOL}, L^{SOL})$ and $\Psi_{D\sigma}(\Delta z^{SOL}, L^{SOL})$ – line 2 indicates a noticeable deviation of the calculated interlamellar distance from the experimentally determined one while treating the alloys with small initial supersaturation c_0. Taking into account the energy dissipation at the α_0/β interface (the term $\Psi_{D\alpha\beta}(\Delta z^{SOL}, L^{SOL})$ – line 3 and additionally the term $\Psi_{DBCT}(\Delta z^{SOL}, L^{SOL})$ – line 4 results in too high values of the interlamellar

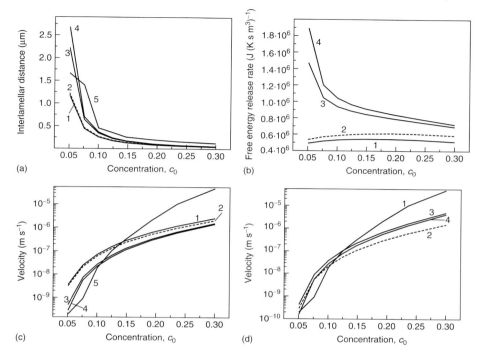

Figure 12.10 Experimental (line 5) and calculated (lines 1–4) concentration dependencies: (a) interlamellar distance; (b) free energy release rate; (c, d) transformation front velocity in dependence on supersaturation c_0 in the Pb–Sn system at 293 K.

distance. This is explained by the fact that total energy dissipation, when taking into account additional terms, increases considerably in case of alloys with small supersaturation, which leads to high interlamellar distances (Figure 12.10b). One can also use concentration-dependent tracer diffusivities in the following form:

$$D^*_{Sn,Pb} = D^*_{Sn,Pb}(0)\, \exp(\alpha c_{av}) \tag{12.50}$$

where $D^*_{Sn,Pb}(0)$ are given in the form of Equation 12.49. Such approximation can be concerned with the concentration dependence of the liquidus line on the Pb–Sn phase diagram [33].

First, we examine the behavior of the velocity of the front moving from c_0 using the approximation (Equation 12.48) for $sh\tilde{D}$. Lines 1 and 2 in Figure 12.10c show the dependence of the transformation front velocity taking into account the energy dissipation at the α_0/α interface (only the term $\Psi_{DZ}(\Delta z^{SOL}, L^{SOL})$ – line 1, the sum of the terms $\Psi_{DZ}(\Delta z^{SOL}, L^{SOL})$ and $\Psi_{D\sigma}(\Delta z^{SOL}, L^{SOL})$ – line 2). Here, the values of the velocity determined from the optimization procedure (i) are overestimated by an order for alloys with small supersaturation, (ii) coincide at average supersaturations, or (iii) are underestimated by an order at large supersaturations. Taking into account dissipation processes at the α_0/β interface (the term $\Psi_{D\alpha\beta}(\Delta z^{SOL}, L^{SOL})$ – line 3 and additional taking into account the term $\Psi_{DBCT}(\Delta z^{SOL}, L^{SOL})$ – line 4 in Figure 12.10b) allows us to obtain the front velocities close to the experimental ones for the alloys with small and average supersaturations treated. At large supersaturations, even in this case the transformation front velocity is underestimated by an order of magnitude.

Now, we consider the influence of concentration dependences (Equation 12.50) on the dependence of the transformation front velocity on the initial supersaturation. Figure 12.10d presents the mentioned concentration dependences of the transformation front velocity as compared to the experimental line 1. Line 2 describes the dependence of the velocity on c_0 taking into account all the energy dissipation modes using the approximation Equation 12.48. In the case when the approximation Equation 12.50 is employed, we obtain lines 3 and 4, getting closer to the experimental values. Here, line 3 was obtained at $n_s = \frac{4}{\sqrt{3a^2}}$, curve 4 at $n_s = \frac{4}{\sqrt{2a^2}}$. As can be seen, if we take into account the tendency of the coefficient of grain boundary diffusion to increase with the growth of the averaged concentration, we observe that the experimental values of the transformation boundary velocity approach the calculated ones.

A special issue is presented by the precipitation coefficient Equation 12.5, since its calculation must be done depending on the concentration of the supersaturated solid solution. From Figure 12.11, it is clear that taking into account only the diffusion part of the energy dissipation along the part between α_0 and α phases results in the increase of the precipitation coefficient W depending on the concentration c_0 in the range from 65% at $c_0 = 5$ at.% to 72% at $c_0 = 30$ at.%. And taking into account the dissipation processes between α_0 and β phases leads to the increase of the precipitation coefficient up to 80 %, which coincides with the experimental value for small and large c_0.

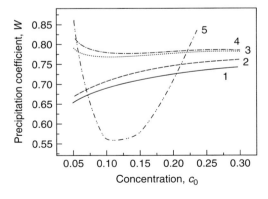

Figure 12.11 Experimental (line 5) and calculated (lines 1–4) concentration dependences of the precipitation coefficient W (Equation 12.5) on the supersaturation c_0 in the Pb–Sn system at 293 K.

Hence, the calculated parameters of discontinuous precipitation which are close to the experimental ones can be obtained on the basis of the described problem involving the extremum of free energy production rate provided the balance both for matter and entropy production is kept. The results of the calculations carried out taking into account additional energy dissipation modes are commensurate with the experimentally determined parameters over a very wide concentration range and reveal a similar concentration dependence at discontinuous precipitation in the Pb–Sn system. Taking into account additional energy dissipation modes, connected with surface energy along the transformation front, leads to obtaining too low values of the boundary velocity and to a very high precipitation coefficient W. This fact is confirmed by the conclusion [3, 14] on the description of the discontinuous precipitation problem in Cahn's model, which is more comparable with experiment provided a very large surface energy between α and β phases is used.

12.3
Model of Diffusion-Induced Grain Boundary Migration (DIGM) Based on the Extremal Principle of Entropy Production by the Example of Cu–Ni Thin Films

The important problem of low-temperature diffusive interaction in binary polycrystalline thin films is a process of solid solution formation as a result of grain boundary migration. The phenomenon of DIGM was experimentally and theoretically investigated from different points of view [38–53]. In the DIGM process, after some incubation period, the grain boundary starts to move inside one of the grains, leaving the solid solution behind. In this process, the second element penetrates along the moving grain boundary from vapor or the neighboring grain of another composition. During DIGM, the free energy of the system decreases but full thermodynamic equilibrium is not reached. Thus the extent of homogenization of the system depends on the thermodynamic stimulus and on the efficiency of possible kinetic mechanisms. The main kinetic parameters are the grain boundary diffusion coefficients and the mobility of the boundary.

DIGM takes place at sufficiently low temperatures when bulk diffusion is frozen. The mobility of the grain boundary depends on the mechanism of grain boundary migration. At high DIGM velocities, the main driving force is the difference in the Gibbs potential between the front and back parts of the migrating boundary (chemical-induced boundary migration). At low velocities, when the concentration step across the migration boundary is absent, the main driving force is the difference in elastic energy in the penetration zone (in the case of a system with different atomic volume of the components). The important part of the driving force can be the energy difference linked with the curvature of the boundary (diffusion-induced recrystallization). Each type of the effective driving force must correspond to the conjugate characteristics, some boundary mobility. This structure-sensitive characteristic depends on the mechanism of migration. The microscopic boundary mobility depends on the mechanism of formation of shift pressure at the opposite sides of the moving boundary. It provides atom (or atomic groups) transition from one grain to the other according to one of the following mechanisms: due to lattice flow, or due to different mobilities of atoms in the grain boundary, or as a result of dislocation climb, or because of the movement of steps and kinks along the boundary. The mobility can be determined using two methods [36, 52–54]. First, it is possible to use the Einstein relation $M = D/kT$, where D is the diffusion coefficient of the solute atom across the grain boundary, which has some intermediate value between the bulk and the grain boundary diffusion coefficients. Second, the boundary mobility can be calculated from the dependence $P(v)$, where P is the solute drag force, and v is the boundary velocity.

Below, we shall try to independently determine (i) the concentration profile of the solid solution formed by DIGM; (ii) the thickness of the formed solid solution layer; (iii) the velocity of the boundary in a steady-state regime in the example of a Cu/Ni/Cu thin film. In our approach, we suppose that the thermodynamic stimulus of the alloying process is determined by the differences in the Gibbs potential before and behind the moving boundary. One part of this stimulus is consumed by free energy release as a result of grain boundary diffusion. The other part of this stimulus provides the boundary shift, that is, the atom transfer across the boundary perpendicular to it. In this case, the dissipation of free energy occurs as the result of friction to boundary migration and the main kinetic parameter for this process is the boundary mobility. We demonstrate that the consideration of the balance of entropy and the principle of maximal free energy release rate unambiguously determine the DIGM parameters in the thin film system viewed on the basis of a minimum of thermodynamic information about the system and kinetic parameters.

12.3.1
Model Description

Let us consider the model of solid solution growth as a result DIGM in the thin film system depicted in Figure 12.12a. In the initial configuration of the sample, the grain boundary preexists in the thin film A with thickness $2\Delta z_0$ between B layers of the moving film. We consider the process in the binary system with full solubility of A and B components. This is illustrated by a model phase diagram in Figure 12.12b. The driving force of the DIGM, as well as of any other process at a constant temperature and pressure, is the decrease in the Gibbs potential. At low temperatures, the homogenization by the usual bulk and grain boundary diffusion would be too slow. Instead, the grain boundary A/A, providing the grain boundary diffusion of the B component, is moved leaving behind the alloyed zone (AB solid solution). In the case of small thickness of the A film ($\sim 0.5/10\,\mu m$), the movement of the planar boundary is possible.

For the determination of the concentration redistribution along the moving boundary, we use Cahn's approach [5], which allows us to determine the concentration interval in the formed solid solution, from some maximal value c^{max} at the joint B/AB films to c^{min} at the middle of the AB film. In case of moving planar

12.3 Model of Diffusion-Induced Grain Boundary Migration

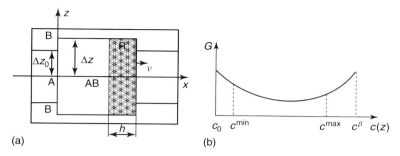

Figure 12.12 (a) Cross section of the sample: $2\Delta z_0$ is the thickness of thin film A, $2\Delta z$ is the thickness of the formed solid solution AB; the transformation front R moves at a constant velocity and coincides with the boundary; (b) a model phase diagram; the initial compositions are $c_0 = 0$ in thin film A and $c^\beta = 1$ in thin film B, the resulting solid solution has a composition between c^{max} and c^{min}.

boundary, the concentration dependence (in the alloyed zone) on the coordinate z along the transformation boundary R, moving with constant velocity, is described by a steady-state diffusion equation based on the balance of component fluxes (without specification of the mechanism of atomic jumps) (for example, [5]):

$$s\tilde{D}\frac{d^2 c(z)}{dz^2} + \frac{c_0 - c(z)}{h} v = 0 \tag{12.51}$$

where

$$c_0 = 0 \quad \text{at} \quad 0 \le z \le \Delta z_0$$
$$c_0 = c^\beta = 1 \quad \text{at} \quad \Delta z_0 \le z \le \Delta z$$

Here, s is the segregation coefficient, equal to 1 in our calculation, $\tilde{D}$ is the coefficient of mutual diffusion in the grain boundary, $c(z)$ is the local concentration in the boundary R, and h is the boundary width.

Then, the solution of Equation 12.51 can be obtained by joining continuously the concentration profiles and their derivatives at the level of the initial contact $z = \Delta z_0$:

$$1 - c^{max} = A \cdot \text{ch}\left(\Delta z/L\right) + B \cdot \text{sh}\left(\Delta z/L\right) \tag{12.52}$$

$$1 - c(\Delta z_0) = A \cdot \text{ch}\left(\Delta z_0/L\right) + B \cdot \text{sh}\left(\Delta z_0/L\right)$$

$$\left.\frac{\partial c^I(z)}{\partial z}\right|_{\Delta z_0+0} = \left.\frac{\partial c^{II}(z)}{\partial z}\right|_{\Delta z_0-0}$$

Then we obtain the next relation for the concentration profile inside the AB solid solution:

$$c^I(z) = 1 - A \cdot \text{ch}(z/L) - B \cdot \text{sh}(z/L) \quad \text{at} \quad \Delta z_0 \le z \le \Delta z \tag{12.53}$$

$$c^{II}(z) = c(\Delta z_0) \frac{\text{ch}(z/L)}{\text{ch}(\Delta z_0/L)} = E \cdot \text{ch}(z/L) \quad \text{at} \quad 0 \le z \le \Delta z_0 \tag{12.54}$$

where

$$A = \frac{\operatorname{sh}(\Delta z_0/L)}{\operatorname{th}(\Delta z/L)}, \qquad B = -\operatorname{sh}(\Delta z_0/L)$$

$$c(\Delta z_0) = \operatorname{ch}(\Delta z_0/L)\left(\operatorname{ch}(\Delta z_0/L) - A\right)$$

$$c^{\max} = 1 - A \cdot \operatorname{ch}(\Delta z/L) - B \cdot \operatorname{sh}(\Delta z/L) \qquad (12.55)$$

$$E = c(\Delta z_0)\frac{1}{\operatorname{ch}(\Delta z_0/L)}$$

Hence, for the unambiguous determination of the solid solution growth parameters in the thin films (with given initial thickness Δz_0), it is necessary to find the thickness of the AB solid solution Δz and the parameter L. The kinetic coefficient L can be determined from the boundary velocity v and the triple product $s\tilde{D}h$,

$$L = \sqrt{s\tilde{D}h/v}$$

Let us analyze the growth of a binary solid solution film with thickness $2\Delta z$ which is larger than the initial thickness $2\Delta z_0$ of the A film by the value $2\Delta z^*$. This thickening is linked with diffusion mixing of A and B components in the alloying zone and must be proved by mass conservation.

12.3.1.1 Mass Conservation and Thermodynamic Description

Let us determine the relation between Δz and Δz_0, using the matter conservation law at components' redistribution in the transformation region R. Thus, at transformation of the elements of the film A with the length dz_0 and concentration c_0 and the film B with the length dz^* and concentration c^β into a section of solid solution of length dz and concentration $c(z)$, the law of mass conservation yields:

$$c(z)dz = c(z)(dz_0 + dz^*) = c_0 dz_0 + c^\beta dz^* \qquad (12.56)$$

If we take into consideration $c_0 = 0$ and $c^\beta = 1$, we obtain the relation

$$dz^* = \frac{c(z) - c_0}{c^\beta - c(z)} = \frac{c(z)}{1 - c(z)}$$

$$dz = \frac{c^\beta - c_0}{c^\beta - c(z)}dz_0 = \frac{1}{1 - c(z)}dz_0 \qquad (12.57)$$

Then

$$\Delta z_0 = \int_0^{\Delta z_0} dz_0 = \int_0^{\Delta z}(1 - c(z))dz$$

$$= \int_0^{\Delta z_0}(1 - c^{II}(z))dz + \int_{\Delta z_0}^{\Delta z}(1 - c^I(z))dz \qquad (12.58)$$

holds, which allows the determination of the dependencies $c(\Delta z_0)$, $c^{\max}$ and the coefficients A, B, and E in Equation 12.55.

To determine independently the main kinetic parameters Δz and L, we use the principle of maximum rate of the free energy release, the balance of the entropy fluxes, and Cahn's solution of the mass transfer equation. Let us assume that the change in the entropy $d_e S$ during the elementary time interval dt as a result of the alloying in the region R (coinciding with the interphase boundary) after shifting by its width is equal to the change in the entropy $d_i S$ as a result of the dissipation process in the same region R. We consider a quasi-stationary process obeying the condition [18]

$$\frac{dS}{dt} = \frac{d_i S}{dt} + \frac{d_e S}{dt} = 0 \tag{12.59}$$

which means that the total entropy change in the moving open system is zero (this condition is valid at a constant transformation front velocity v). In order to pass from the rate of entropy change to the rate of free energy release, we can use the relation (valid at a constant temperature and pressure):

$$\frac{dS_{i,e}}{dt} = -\frac{1}{T}\frac{dG_{i,e}}{dt} \tag{12.60}$$

In this case, the rate of the free energy release in the region R is [18, 19]

$$\Psi \equiv T\frac{d_i S}{dt} = T\int_V \sigma \, dV = -\frac{d_i G}{dt} = \frac{d_e G}{dt} > 0 \tag{12.61}$$

12.3.1.2 Calculation of the Entropy Production Rate due to Grain Boundary Diffusion

For the rate of free energy release as a result of diffusion redistribution of components along the region R, we can put down [16, 18, 19]

$$\Psi_{GB} = \frac{hb}{\Delta z}\int_0^{\Delta z} IX dz = \frac{hb}{\Delta z}\int_0^{\Delta z}(-s\tilde{D})\frac{\partial c(z)}{\partial z}\left(-\frac{\partial \tilde{\mu}(z)}{\partial z}\right) dz \tag{12.62}$$

where I is the generalized flux along the z axis, X is the driving force causing the flux along z axis, and

$$\tilde{\mu}(z) = \mu_A - \mu_B = \frac{\partial g}{\partial c_A}$$

is the generalized chemical potential at a point z of the region R, which is constant over the boundary thickness.

The change of the Gibbs potential can be written as

$$\Delta G(dz) = g(c(z))dz - g(c_0)dz_0 - g(c^\beta)dz^* \tag{12.63}$$

where $g(c(z))dz$ is the Gibbs potential of the formed solution (of the α phase) band of length dz at the point z with the concentration $c(z)$ in the region R, g^β is the Gibbs potential of the B film, and $g(c_0)$ is the Gibbs potential of the A film in front of the transformation boundary. Expanding the Gibbs potential into a Taylor series with respect to $g(c^{\alpha/\beta} = 0.5)$ and retaining terms up to the second order, we obtain

$$\Delta G = \frac{1}{2}\left[(c(z) - c^{\alpha/\beta})^2 - \frac{c^\beta - c(z)}{c^\beta - c_0}(c_0 - c^{\alpha/\beta})^2\right]g''|_{c^{\alpha/\beta}} = f(c(z))g'' \tag{12.64}$$

where $f(c(z)) < 0$ and $g'' = g''|_{\alpha/\beta} > 0$ is the curvature of the Gibbs potential at the point with concentration c^e. Then,

$$\tilde{\mu} = \frac{\partial \Delta G}{\partial c(z)} = \left[(c(z) - c^{\alpha/\beta}) - \frac{(c_0 - c^{\alpha/\beta})^2}{2(c^\beta - c_0)} \right] g'' \qquad (12.65)$$

Finally, for the driving force of grain boundary diffusion we have:

$$X = -\frac{\partial \tilde{\mu}}{\partial z} = -\frac{\partial c(z)}{\partial z} k \qquad (12.66)$$

Now, for the rate of free energy release due to longitudinal grain boundary diffusion (Equation 12.62), using Equations 12.53 and 12.54, we may put down

$$\Psi_{GB} = \frac{d_{GB} G}{dt} = \frac{g'' \tilde{D} h b}{\Delta z} \left(\int_0^{\Delta z_0} \left(\frac{\partial c^{II}(z)}{\partial z} \right)^2 dz + \int_{\Delta z_0}^{\Delta z} \left(\frac{\partial c^{I}(z)}{\partial z} \right)^2 dz \right)$$

$$= \frac{kL\upsilon}{2\Delta z} \left(E^2 \left[\text{sh}\left(\frac{\Delta z_0}{L}\right) \text{ch}\left(\frac{\Delta z_0}{L}\right) - \frac{\Delta z_0}{L} \right] \right.$$

$$+ (A^2 + B^2) \left[\text{sh}\left(\frac{\Delta z}{L}\right) \text{ch}\left(\frac{\Delta z}{L}\right) - \text{sh}\left(\frac{\Delta z_0}{L}\right) \text{ch}\left(\frac{\Delta z_0}{L}\right) \right]$$

$$\left. + (B^2 - A^2) \frac{\Delta z - \Delta z_0}{L} + AB \left[\text{ch}\left(\frac{2\Delta z}{L}\right) - \text{ch}\left(\frac{2\Delta z_0}{L}\right) \right] \right) \qquad (12.67)$$

12.3.1.3 Calculation of the Driving Force

Second, for the rate of free energy production due to the solution formation in the volume R, we can formulate the expression

$$\Psi_e = \frac{d_e G}{dt} = -\frac{\upsilon b}{\Delta z} \int_0^{\Delta z} \Delta G(c(z)) dz$$

$$= -\frac{k\upsilon}{\Delta z} \left(\int_0^{\Delta z_0} \left[(c^{II}(z))^2 - c^{II}(z) \right] dz + \int_{\Delta z_0}^{\Delta z} \left[(c^{I}(z))^2 - c^{I}(z) \right] dz \right)$$

$$= \frac{kL\upsilon}{2\Delta z} \left(\frac{B^2 - A^2}{2} \frac{\Delta z - \Delta z_0}{L} + E \cdot \text{sh}\left(\frac{\Delta z_0}{L}\right) \right.$$

$$- \frac{E^2}{2} \left[\text{sh}\left(\frac{\Delta z_0}{L}\right) \text{ch}\left(\frac{\Delta z_0}{L}\right) + \frac{\Delta z_0}{L} \right] + A \left[\text{sh}\left(\frac{\Delta z}{L}\right) - \text{sh}\left(\frac{\Delta z_0}{L}\right) \right]$$

$$- B \left[\text{ch}\left(\frac{\Delta z}{L}\right) - \text{ch}\left(\frac{\Delta z_0}{L}\right) \right] - \frac{AB}{2} \left[\text{ch}\left(\frac{2\Delta z}{L}\right) - \text{ch}\left(\frac{\Delta z_0}{L}\right) \right]$$

$$\left. - (A^2 + B^2) \frac{1}{2} \left[\text{sh}\left(\frac{\Delta z}{L}\right) \text{ch}\left(\frac{\Delta z}{L}\right) - \text{sh}\left(\frac{\Delta z_0}{L}\right) \text{ch}\left(\frac{\Delta z_0}{L}\right) \right] \right) \qquad (12.68)$$

We presuppose that new grain boundaries do not form in the DIGM process.

The difference between the thermodynamic stimulus and the dissipation energy as a result of grain boundary diffusion can provide the determination of the effective driving force for the movement of the grain boundary. This effective driving force is spent in overcoming the boundary friction with the mobility M. The rate of energy release as a result of the diffusive redistribution of components through (perpendicular to) grain boundary can be written as:

$$\Psi_m = \frac{d_m G}{dt} = \frac{d_e G}{dt} - \frac{d_{GB} G}{dt} = \frac{v^2}{M} \qquad (12.69)$$

Equation 12.69 is written using the balance of entropy (or free energy release at T, $P = $ const). The considered dependences allow us to determine the corresponding value of the parameter L for each value of Δz.

Additional application of the principle of maximum rate of free energy release (entropy production) in the form

$$\frac{d_e G(\Delta z, L)}{dt} \to \max \qquad (12.70)$$

allows us to simultaneously determine both Δz and L; that is, the rate of the steady-state process. This principle means that an open dissipative system that is in a nonequilibrium state evolves to an equilibrium state with maximal possible speed which corresponds the maximal rate of free energy release (entropy production).

12.3.2
Results of Model Calculations for the Cu–Ni System

We apply now the model approach to the Cu/Ni/Cu system with a thickness of the A film of several micrometers at a temperature 888 K. The DIGM in such system for the case of the bulk sample was investigated in [47–49, 51]. We suppose that in such a thin film system it is possible to investigate the movement of the flat grain boundary by DIGM mechanism without the influence the boundary curvature.

12.3.2.1 Determination of the Curvature of the Gibbs Potential

For the determination of g'' in the Cu–Ni system, it is possible to use the subregular model [51, 55–58]:

$$g(c_B) = c_A G_A^0 + c_B G_B^0 + RT(c_A \ln c_A + c_B \ln c_B) + G^E + G^{Mo} \qquad (12.71)$$

where G_i^0 $(i = A, B)$ is the molar energy of each pure element, G^m is the mixing energy, and G^{Mo} is the contribution of the magnetic ordering. The expression for G^m has the form:

$$G^E = c_A c_B \left[{}^0 L_{A,B} + {}^1 L_{A,B}(c_A - c_B)\right] \qquad (12.72)$$

where ${}^0 L_{A,B} = {}^0 L_{Cu,Ni} = 8366.0 + 2.802 \cdot T$, and ${}^1 L_{Cu,Ni} = -4359.6 + 1.812 \cdot T$ J mol^{-1}. The contribution of the magnetic ordering is calculated according to

$$G^{Mo} = RT \ln(\beta + 1) f(\tau) \qquad (12.73)$$

where

$$\tau = \frac{T}{T_c(c_B)}$$

$T_c(c_B)$, β and $f(\tau)$ were described in [51]. Using the polynomial fitting of second order and the method of least-squares, we determine the curvature of the Gibbs potential, which for the whole concentration interval is equal to $g'' = 7400 \, \text{J} \, (\text{mol m}^2)^{-1}$.

12.3.2.2 Diffusion Parameters of the System

In the case of the Cu–Ni system at 888 K, we use the bulk diffusion coefficient before and behind the moving boundary equal to $D_\upsilon = 3.8 \times 10^{-20} \, \text{m}^2 \, \text{s}^{-1}$ [51]. The grain boundary diffusion coefficient is evaluated from the empirical expression:

$$D_{GB} = 10^{-4} \cdot \exp\left(-\frac{8.9 \cdot T_L}{T}\right) = 2.5 \times 10^{-11} \, \text{m}^2 \, \text{s}^{-1} \quad (12.74)$$

where T is the simple average of the melting temperature of Cu and Ni.

12.3.2.3 Grain Boundary Mobility

The calculation of the DIGM parameters is performed using, firstly, the experimentally determined mobility $M = 2 \times 10^{-17} \, \text{m}^4 \, (\text{Js})^{-1}$ from the model [51] in the case when the curvature of the grain boundary is absent; and secondly, by estimating the value of the mobility using the diffusion coefficient $D_\perp$ across (perpendicular to) the grain boundary and the Einstein expression

$$M = \frac{D_\perp a^2 \xi}{kT} \quad (12.75)$$

where $a = 3.52 \, \text{Å}$ is the interatomic distance, and ξ is a geometrical factor equal to $1/4\sqrt{2}$ for the fcc lattice. For the determination of the diffusion coefficient $D_\perp$, we can use the model [59] (see also [60]), which links $D_\perp$ with the bulk diffusion coefficient D_υ and the grain boundary energy ($\gamma = 0.87 \, \text{J m}^{-2}$) for the Cu–Ni system:

$$D_\perp = D_\upsilon \exp\left(\frac{a^2 \gamma}{kT}\right) = 2.5 \times 10^{-16} \, \text{m}^2 \, \text{s}^{-1} \quad (12.76)$$

In this case, the free energy release rate due to transversal grain boundary diffusion ("friction" boundary) has the form:

$$\Psi_m = \frac{\upsilon^2}{M} = \upsilon \frac{\upsilon}{M} = \upsilon \frac{\tilde{D} h k T}{L^2 D_\perp a^2 \xi} \quad (12.77)$$

12.3.2.4 Results of the Model Calculation for the Cu/Ni/Cu-Like System

We solved the variation problem (Equations 12.69–12.70) numerically and found the solutions $\Delta z = \Delta z^{SOL}$ and L^{SOL}. Then we determined the value of υ using the given triple product $sh\tilde{D}$ from the value L^{SOL}. A special numerical procedure for finding the optimal values Δz^{SOL} and L^{SOL} was constructed. To solve this variation problem, it is necessary to maximize the rate of the free energy release $\Psi_e(\Delta z, L)$ taking into consideration the balance of the entropy production (Equation 12.59):

$$\Psi_e(\Delta z, L) = \Psi_{GB}(\Delta z, L) + \Psi_m(\Delta z, L)$$

In this case, it is necessary to find the solution along the line of zero level in Figure 12.13. The dependencies $\Psi(\Delta z, L)$ along this line are presented in

12.3 Model of Diffusion-Induced Grain Boundary Migration

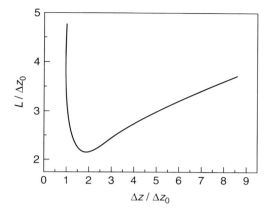

Figure 12.13 Line of entropy balance on the AB film thickness (x axis) and parameter L (y axis).

Figure 12.14. Evidently, along the line of entropy balance the maximal entropy production is reached at some point $\Delta z = \Delta z^{sol}$ and $L = L^{sol}$.

In Figure 12.14, the calculated dependencies of the free energy release rate are shown. Additionally, Table 12.1 gives the calculated DIGM parameter dependencies on mobility for the Cu/Ni/Cu system at the initial thickness of the A film equal to $\Delta z = 10^{-6}$ m. Variant (a) corresponds to a low mobility compared with the experimental determination [51], variant (b) corresponds to a mobility determined in [51] in the case of the maximum value of effective driving force, variant (c) corresponds the mobility calculated by Equation 12.75, and variant (d) corresponds to a high mobility.

From Figure 12.14, we can see that, with an increase of the boundary mobility, the free energy release rate is increased as a result of lateral diffusion along the grain boundary and the free energy release rate is decreased as a result of the transversal grain boundary diffusion ("friction" boundary). Here, the maximum value of $\Psi_{GB}(\Delta z, L)$ is shifted to the right, and the maximum of $\Psi_m(\Delta z, L)$ to

Table 12.1 Calculated DIGM parameters at an initial thickness of the Ni film of 1 μm for various mobilities.

N	1	2	3	4
M (m⁴ (J s)$^{-1}$)	1×10^{-20}	2×10^{-17}	4.2×10^{-16}	1×10^{-15}
L (m)	1×10^{-4}	2.6×10^{-6}	7.7×10^{-7}	5.9×10^{-7}
Δz (m)	4×10^{-6}	3.6×10^{-6}	2.6×10^{-6}	2.5×10^{-6}
Υ, (m s^{-1})	1×10^{-12}	1.7×10^{-9}	1.8×10^{-8}	3.2×10^{-8}
c^{max}	0.75	0.78	0.9	0.92
$c(\Delta z_0)$	0.75	0.67	0.53	0.51
c^{min}	0.75	0.62	0.27	0.18

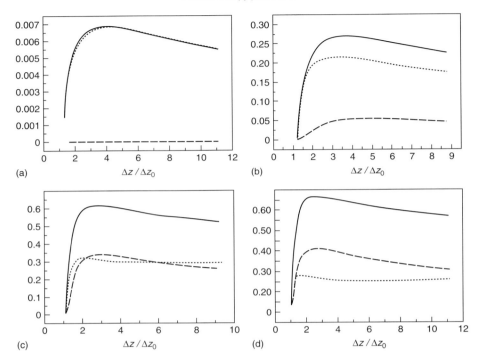

Figure 12.14 Rates of entropy production dependence on thickening of the A film with an initial thickness of 1 μm under different mobilities: (a) $M = 1 \times 10^{-20}$; (b) $M = 2 \times 10^{-17}$; (c) $M = 4.2 \times 10^{-16}$; (d) $M = 1 \times 10^{-15} m^4 \, (Js)^{-1}$. The rates of entropy production: total – solid line, input of grain boundary diffusion – broken line; input of "friction" – dotted line.

the left from the resulting maximum $\Psi_e(\Delta z, L)$. This means that taking into account $\Psi_m(\Delta z, L)$ leads to the reduction of the interlamellar distance in the problems of periodic structures at decomposition or solution formation. From Table 12.1, we can see that the increase of the boundary mobility causes the increase of the boundary velocity in the DIGM process and the decrease of the forming solid solution thickness. In this case, the maximal solubility of the B component goes to 1 and the minimal solubility goes to 0 in the middle of AB film.

Let us consider the influence of the initial thickness of the Ni film on the DIGM parameters (Table 12.2). We also compare the values of velocity obtained experimentally [15] for the high effective driving force $v \approx 10^{-9}$ m s^{-1}, which are shown in Table 12.2. The experimental value corresponds to the velocity in a thick film with an initial thickness 10 μm, which points to the correctness of our model approach.

Table 12.2 Calculated DIGM parameters at mobility $M = 4.2 \times 10^{-16}$ m^4/(J s) by varying the Ni thickness Δz_0.

N	1	2	3	4	5	6
Δz_0 (m)	0.5×10^{-6}	1×10^{-6}	2×10^{-6}	5×10^{-6}	7×10^{-6}	10×10^{-6}
L (m)	7.2×10^{-7}	7.7×10^{-7}	9.7×10^{-7}	1.3×10^{-6}	1.4×10^{-6}	1.5×10^{-6}
Δz (m)	1.5×10^{-6}	2.6×10^{-6}	4.9×10^{-6}	1.1×10^{-5}	1.4×10^{-5}	1.8×10^{-5}
$\Delta z / \Delta z_0$	3	2.6	2.45	2.2	2	1.8
$\Delta z / L$	2.1	3.4	5.05	8.5	10	12
Υ (m s^{-1})	2.2×10^{-12}	1.8×10^{-8}	1.2×10^{-8}	6.7×10^{-9}	5.8×10^{-9}	4.8×10^{-9}
c^{max}	0.82	0.9	0.95	0.98	0.99	0.99
$c(\Delta z_0)$	0.6	0.53	0.51	0.5	0.5	0.5
c^{min}	0.48	0.27	0.13	0.02	0.001	0.001

From the model calculation, we can see that, on increase of the Ni film thickness from 0.5 to 10 µm, the boundary velocity strongly increases; the thickening coefficient $\Delta z / \Delta z_0$ decreases; and the $\Delta z / L$ parameter, on which the kind of concentration profile in the alloying zone depends, increases. Some concentration profiles are shown in Figure 12.15.

We can see that at a small initial Ni film thickness the concentration profile is developing into the whole thickness of the solid solution film. On increasing the initial Ni film thickness, the solid solution inside this film cannot be formed (see Figure 12.15b). In case the boundary mobility depends on the chemical composition, it can cause the curvature to move the boundary during the DIGM and it can be experimentally obtained in a bulk polycrystalline samples [47, 49].

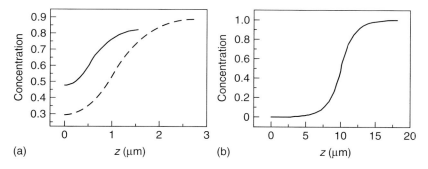

Figure 12.15 Concentration dependences $c(z)$ along the moving grain boundary under the same mobility but different initial thicknesses of Ni film: (a) 0.5 µm – solid line, 1 µm – dotted line; (b) 10 µm.

12.4
Entropy Production as a Regularization Factor in Solving the Inverse Diffusion Problem

Ill-posed inverse diffusion problems [16, 61–63] can be solved by the Tikhonov regularization methods [64]. In this case, an additional condition of smoothness is imposed on the solution. In the general Tikhonov method, the main component can be written as the difference of the solution of the corresponding system of linear algebraic equations. The additional component is written on the basis of the empirical information on the sought-after regularization condition. It can be the squared discrepancy of the first or second derivative. The additional component enters into the general functional with a certain weight α called the regularization parameter. For $\alpha = 0$, the inverse problem is solved without regularization. When solving the inverse problem, the parameter α is chosen in such a way that the deviation of the perturbed initial concentration profile from the profile calculated with the diffusion coefficient determined by solving the inverse problem is comparable with the error of experimental measurements of the initial concentration profile.

We demonstrate that the squared entropy production can be a physically substantiated regularization term for transport problems, for example, diffusion problems [64]. It should be noted that the first and second derivatives of the concentration are used to solve the inverse diffusion problems. They are the quantities that are calculated from the experimental data with large errors and yield ill-posed results (most often the calculated diffusivities are too high and strongly oscillate from one concentration to another). Minimization of the entropy production significantly decreases (up to absolute absence) local diffusion fluxes between neighboring points. Moreover, the larger the weight of the entropy production term entering into the discrepancy functional, the smoother the solution obtained. When α is optimal, a solution of the direct diffusion problem with the diffusion coefficients obtained will yield the concentration profile best fitted to the experimental one.

12.4.1
Description of the Procedure of the Inverse Diffusion Problem Solution for a Binary System

The process of concentration redistribution in a binary system caused by diffusion is described by the second Fick's law [16, 61–63]

$$\frac{\partial c}{\partial t} = \frac{\partial}{\partial x}\left(\tilde{D}(c)\frac{\partial c}{\partial x}\right) \tag{12.78}$$

where $c(x, t)$ is the concentration profile and $\tilde{D}(c)$ is the concentration-dependent diffusion coefficient. A solution of the boundary problem described by Equation 12.78 is a concentration profile calculated at the given moment of time. On the other hand, the concentration profile is determined by X-ray microanalysis

12.4 Entropy Production as a Regularization Factor

of the diffusion zone. These two profiles differ because of errors in measuring the experimental profile. These experimental errors considerably distort the result of inverse problem solution when the concentration-dependent diffusivities are calculated on the basis of the experimental concentration distribution. Using the regularization algorithms in solving the ill-posed inverse problems by Tikhonov's method [65, 66] allows one to determine the concentration-dependent expression for the interdiffusion coefficient with the corresponding limit imposed on smoothness and oscillations; thus, it makes the solution stable with respect to input data perturbations.

Let us consider the discrepancy functional to be minimized in the process of the correct solution of the inverse diffusion problem in the form

$$\Phi(c) = \left[\frac{\partial c}{\partial t} - \frac{\partial}{\partial x}\tilde{D}(c)\frac{\partial c}{\partial x}\right]^2 + \alpha \left|\frac{dS}{dt}\right|^2 \to \min \quad (12.79)$$

Here, the first term characterizes the discrepancy of solving the system of linear algebraic equations for the examined diffusion problem. The second term describes the regularization procedure and is set to be equal to the squared entropy production. The degree of smoothing of the inverse problem solution depends on the regularization parameter α. In our treatment, both the first and second terms are involved in the regularization procedure proceeding from physical interrelations for the described process rather than from the smoothness of the first or second derivative in the finite-difference procedure.

We now express the entropy production during diffusion interaction using standard notions of linear nonequilibrium thermodynamics [16]:

$$\frac{dS}{dt} = IX = \left(-\tilde{D}(c)\frac{\partial c}{\partial x}\right)\left(-\frac{\partial \tilde{\mu}}{\partial x}\right) = \tilde{D}(c)\left(\frac{\partial c}{\partial x}\right)^2 g''(c) \quad (12.80)$$

Here I is the generalized diffusion flux in the binary diffusion pair, X is the driving force of the diffusion flux,

$$\tilde{\mu} = \mu_1 - \mu_2 = \frac{\partial g}{\partial c}$$

is the generalized chemical potential, and $g''(c)$ is the thermodynamic multiplier (taken in units of kT) caused by the energy of mixing of the binary solid solution (in our calculations, it is set equal to zero).

To obtain a linear system of algebraic equations from Equation 12.79, we take advantage of a quadratic dependence of the diffusion coefficient on the concentration:

$$\tilde{D} = a_1 + a_2 c + a_3 c^2 = \sum_{L_1=1}^{N_1+1} a(L_1) c^{L_1-1} \quad (12.81)$$

where $N_1 = 2$, and we also use the parabolic Boltzmann substitution $\xi = x/\sqrt{t}$. We now express the discrepancy functional (Equation 12.79) as

$$\Phi(c) = \sum_{k=1}^{N} \Phi_{1,k}^2 + \alpha \sum_{k=1}^{N} \Phi_{2,k}^2 \quad (12.82)$$

where

$$\Phi_{1,k} = \frac{\xi}{2}\frac{\partial c_k}{\partial \xi} + \sum_{L_1=1}^{N_1+1} a(L_1) \left\{ \frac{\partial^2 c_k}{\partial \xi^2} c_k^{L_1-1} + \left(\frac{\partial c}{\partial \xi}\right)^2 (L_1-1) c_k^{L_1-2} \right\} \quad (12.83)$$

$$\Phi_2 = \frac{dS}{dt} = \sum_{L_1=1}^{N_1+1} a(L_1) \left\{ \left(\frac{\partial c_k}{\partial \xi}\right)^2 c_k^{L_1-1} \right\} \quad (12.84)$$

where k is the serial number of the point on the concentration profile, and N is the number of points on the profile.

The procedure of determining optimal values of the coefficients $a(L1)$ is the following. For different values of the coefficient α, we find a minimum of the discrepancy functional (Equations 12.82–12.84) first with the use of random search for the genetic algorithm [67] and then refining $a(L1)$ values by the Newton–Raphson method of finding the roots for the corresponding system of nonlinear equations or/and by the gradient descent method of minimization of the discrepancy functional [68]. In this case, the system of independent equations for the Newton–Raphson method is formed by differentiation of the discrepancy functional (Equation 12.35) with respect to the parameters $a(L1)$:

$$\frac{\partial \Phi(c)}{\partial a(L_1)} = \sum_{k=1}^{N} \left[\left\{ \frac{\partial^2 c_k}{\partial \xi^2} c_k^{L_1-1} + \left(\frac{\partial c_k}{\partial \xi}\right)^2 (L_1-1) c_k^{L_1-2} \right\} \Phi_{1,k} \right.$$
$$\left. + \alpha \left\{ \left(\frac{\partial c_k}{\partial \xi}\right)^2 c_k^{L_1-1} \right\} \Phi_{2,k} \right] = 0 \quad (12.85)$$

Thus, solving the system of Equations 12.85, we find a set of coefficients $a(L1)$ for the value of the regularization parameter α which minimizes the functional (Equation 12.82).

For each set $a(L1)$, the concentration profile c_k^{calc} is calculated by solving the direct diffusion problem (Equation 12.78). The set of coefficients $a(L1)$ that yields the concentration profile best fitted to the experimental profile c_k^{exp}, that is, whose discrepancy

$$\eta = \frac{1}{N}\sqrt{\sum_{k=1}^{N}(c_k^{exp} - c_k^{calc})^2}$$

is minimum, is considered to be optimal.

12.4.2
Results of Model Calculations

In test examples, we calculated the experimental profile by solving direct diffusion problems (Equation 12.78) with a diffusion coefficient (Equation 12.81) for preset parameters $a(1)^0$–$a(3)^0$. In order to do so, we used the numerical procedure of solving nonlinear parabolic equations described in [68]. After that, the calculated profile was distorted by the addition of a random concentration proportional to the experimental error. The concentration profile c_k^{exp} so obtained was used in the procedure of determining optimal values of coefficients $a(1)-a(3)$, which were then compared with the preset parameters $a(1)^0-a(3)^0$. The parameters $a(1)-a(3)$ were also used to calculate the concentration profile c_k^{calc}, which was compared with the initial profile c_k^{exp}. In the calculations, we used the concentration dependence of the diffusion coefficient (see Figure 12.16a) analogous to that for the Cu–Au, Ni–Pt, or Ni–Pd system [61]. Results of test calculations for the indicated perturbations amplitudes of the initial concentration profile are presented in Table 12.3 and Figure 12.16.

From the table, it can be seen that the suggested procedure for solving the inverse diffusion problem with the use of the regularization term expressed as a squared entropy production is stable enough even when the perturbation amplitude of the concentration reaches 10%. It should be noted that the procedure without regularization yields unstable solutions already when the perturbation amplitude of the initial concentration profile reaches 0.1%. Model calculations demonstrated that the calculated parameters $a(1)-a(3)$ were close in values to the preset parameters $a(1)^0-a(3)^0$ for perturbation amplitudes up to 1%. In this case, the preset and calculated concentration profiles almost coincided (for example, see Figure 12.16b). When the perturbations amplitude of the concentration profile exceeded 1%, the parameters $a(1)-a(3)$ determined by the inverse problem solution were underestimated by several times in comparison with $a(1)^0-a(3)^0$. Moreover, less developed concentration profiles were obtained (see Figure 12.16c and d); however, the solution procedure remained stable.

Table 12.3 Calculated parameters versus the degree of perturbation of the initial concentration profile calculated from the dependence $\tilde{D}(c)$ with parameters $a(1)^0 = 1 \times 10^{-4}$, $a(2)^0 = 6 \times 10^{-4}$, $a(3)^0 = -4 \times 10^{-4}$: α is the regularization parameter, η is the discrepancy of the concentration profile reconstructed from the distorted profile, and $a(1)-a(3)$ are the parameters of the diffusion coefficient calculated by solving the inverse problem.

Error (%)	0	0.1	1
α	1.26×10^{-3}	7.94×10^{-5}	5.02×10^{-3}
η	1.49×10^{-4}	3.39×10^{-4}	3.54×10^{-3}
$a(1)$	9.48×10^{-5}	1.06×10^{-4}	1.25×10^{-4}
$a(2)$	5.29×10^{-4}	5.39×10^{-4}	4.04×10^{-4}
$a(3)$	-4.39×10^{-4}	-3.45×10^{-4}	-3.66×10^{-4}

Error (%)	3	5	10
α	7.94×10^{-1}	5.31×10^{-2}	5.31×10^{-3}
η	1.46×10^{-2}	8.11×10^{-3}	1.99×10^{-2}
$a(1)$	3.14×10^{-5}	5.25×10^{-5}	2.25×10^{-5}
$a(2)$	2.49×10^{-4}	4.33×10^{-4}	7.79×10^{-5}
$a(3)$	-2.73×10^{-4}	-4.57×10^{-4}	-1.11×10^{-4}

12 Optimal Regimes in Cellular Decomposition

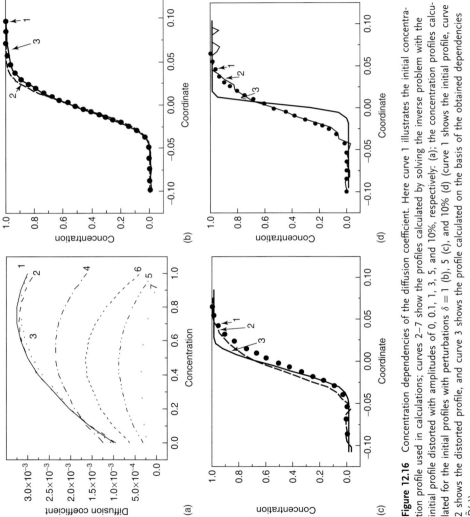

Figure 12.16 Concentration dependencies of the diffusion coefficient. Here curve 1 illustrates the initial concentration profile used in calculations; curves 2–7 show the profiles calculated by solving the inverse problem with the initial profile distorted with amplitudes of 0, 0.1, 1, 3, 5, and 10%, respectively: (a); the concentration profiles calculated for the initial profiles with perturbations $\delta = 1$ (b), 5 (c), and 10% (d) (curve 1 shows the initial profile, curve 2 shows the distorted profile, and curve 3 shows the profile calculated on the basis of the obtained dependencies $\tilde{D}(c)$).

12.5
Conclusions

A model for self-consistent determination of the main parameters of discontinuous precipitation in binary systems at low temperatures is worked out. Here, in addition to the equation of grain boundary diffusion in the moving boundary (which coincides with the transformation region) and bulk diffusion close to the transformation front, two other equations were used: for the balance of entropy fluxes and maximum rate of free energy release. Applying thermodynamic data about the system enables an unambiguous determination of the interlamellar distance, the velocity of the planar transformation front, concentration profile close to the transformation front, residual supersaturation behind the front, and other characteristics of discontinuous precipitation. The calculations demonstrated that energy dissipation in the open system due to bulk diffusion close to the transformation front should be taken into account at supersaturations higher than 20 at.%, though this introduces minor corrections to the results obtained without such a procedure. The found concentration dependences of the discontinuous precipitation parameters are similar to the experimental ones, which allows us to use the proposed formalism in solving the inverse problem, when diffusion parameters at interphase boundaries at low temperatures are determined on the basis of the analysis of discontinuous precipitation parameters.

The model for self-consistent determination of DIGM parameters in a thin film binary system with a planar front of solution formation at low homologous temperatures is developed. Simultaneously, in addition to the equation of grain boundary diffusion in the moving interphase boundary, the following equations were used: the equation for entropy balance and maximum rate of entropy production (maximum rate of free energy release). The model presupposes different ways to estimate the mobility of the boundary. Thermodynamic data provide a self-consistent calculation of thermodynamic stimulus and determination of the driving force of the diffusion process considered. The developed model and subsequent calculations allowed us to find the thickness of the formed solid solution, the velocity of the solution formation front, and the concentration profile in the formed solid solution depending on the initial thickness of the average film. The model calculation for the DIGM process in Cu/Ni/Cu-like thin films system is done at different values of the kinetic parameters and initial conditions. We describe the thin film size effect on the grain boundary velocity, thickening related to solid solution formation, and form of the concentration profile in the forming solid solution. Then, we point at the size effect as the possible reason of the curvature of the moving grain boundary in bulk samples.

The method providing the stability increase for the solution of the inverse diffusion problem in a binary system is presented. It is shown that entropy production may serve as a physically adequate regularization term in the discrepancy functional in the process of solving the inverse diffusion problem. Stabilization of the solution is achieved by the imposed additional condition of minimization of local entropy production with the corresponding weight coefficient: that is, Tikhonov's regularization parameter. Choosing the optimal regularization parameter leads to the solution closest to the experimental one, the initial concentration profiles being considerably distorted.

Model calculations were carried out by the example of a parabolic dependence of the interdiffusion coefficient for a binary system. The efficiency of the method was demonstrated for significant perturbations of the model experimental concentration profile up to 10% with respect to the exact values. The additional smoothing component of the discrepancy functional expressed as squared entropy production can be set to solve any inverse problems of heat and mass transfer. The suggested method, unlike the Matano–Boltzmann method, can be generalized to multicomponent systems with arbitrary sample configurations.

References

1. Kaur, I. and Gust, W. (1989) *Fundamentals of Grain and Interphase Boundary Diffusion*, Ziegler Press, Stuttgart.
2. Larikov, L.N. and Shmatko, O.A. (1976) *Cellular Decomposition of Supersaturated Solid Solutions*, Naukova Dumka, Kyiv (in Russian).
3. Bokstein, B.S., Kopetsky, Ch.V. and Shvindlerman, L.S. (1980) *Thermodynamics and Kinetics of Grain Boundaries*, Metallurgiya, Moscow (in Russian).
4. Chuistov, K.V. (2003) *Aging of Metallic Alloys*, Akademperiodika, Kyiv (in Russian).
5. Cahn, J.W. (1959) *Acta Mettallurgica*, **7**, 18.
6. Lyashenko, Yu.A. (2004) *Technical Physics Letters*, **30**(2), 109. (Translated from (2004) *Pis ma v Zhurnal Tekniheskoi Fiziki*, **30**(3), 54.
7. Gusak, A.M., Bogaturev, O.O., Zaporogetz, T.V. et al. (2004) *Models of Solid State Reactions*, Cherkasy National University, Cherkasy (in Russian).
8. Lyashenko, Yu.A. and Shmatko, O.A. (2003) *Bulletin Technologial University Podillya*, **2**(6), 97 (in Ukrainian).
9. Lyashenko, Yu.A., Gusak, A.M. and Shmatko, O.A. (2005) *Metallofizika i Noveishiye Technologii*, **27**(7), 159 (in Russian).
10. Voronina, N.F., Ziemba, P., Pawlowski, A. and Shmatko, O.A. (1999) *Metallofizika i Noveishiye Technologii*, **18**, 391 (in Russian).
11. Turnbull, D. (1995) *Acta Metallurgica*, **3**, 55.
12. Petermann, J. and Hornbogen, E. (1968) *Zeitschrift fur Metallkde*, **59**, 814.
13. Bogel, A. and Gust, W. (1988) *Zeitschrift fur Metallkunde*, **79**(5), 296.
14. Speich, G.R. (1968) *Transactions of the Metallurgical Society of the AIME*, **242**, 1359.
15. Mullins, W.W. and Sekerka, R.F. (1963) *Journal of Applied Physics*, **34**, 323.
16. Kirkaldy, J.S. and Young, D.J. (1987) *Diffusion in the Condensed State*, The Institute of Metals, London.
17. Sundquist, B.E. (1973) *Metallurgical Transactions*, **4A**, 1919.
18. De Groot, S.R. and Mazur, P. (1962) *Non-equilibrium Thermodynamics*, North-Holland, Amsterdam.
19. Gyarmati, I. (1970) *Non-Equilibrium Thermodynamics*, Springer, Berlin.
20. Glansdorff, P. and Prigogine, I. (1971) *Thermodynamic Theory of Structure, Stability and Fluctuations*, John Wiley & Sons, Ltd, New York.
21. Vehoeven, J.D. and Pearson, D.D. (1984) *Metallurgical Transactions A*, **15**, 1047.

22. Ziegler, H. (1963) in *Progress in Solid Mechanics*, Vol. 4, Ch. 2 (eds I.N. Sneddon and R. Hill), North-Holland, Amsterdam.
23. Bene, R.W. (1987) *Journal of Applied Physics*, **61**, 1826.
24. Gosele, U. and Tu, K.N. (1982) *Journal of Applied Physics*, **53**, 3252.
25. Ivantzov, G.I. (1947) *Doklady Academy Sciences of SSSR*, **58**, 567 (in Russian).
26. Martyushev, L.M., Seleznew, V.D. and Kuznetzova, I.E. (2000) *Journal of Experimental and Theoretical Physics (JETP) (Zhurnal Eksperimental'noi i Teoreticheskoi Fiziki)*, **118**(1), 149 (in Russian).
27. Shapiro, J.M. and Kirkaldy, J.S. (1968) *Acta Metallurgica*, **16**, 579.
28. Podolskiy, S.E. (1996) *Metallofizika i Noveishiye Technologii*, **18**(1), 18 (in Russian).
29. Zhang, L. and Ivey, D.G. (1995) *Canadian Metallurgical Quarterly*, **34**(1), 51.
30. Ghosh, G. (1999) *Metallurgical and Materials Transactions*, **30A**, 1481.
31. Lyakishev, N.P. (ed.) (1999) *Phase Diagrams of Binary Metallic Systems*, Mashinostroenie, Moscow (in Russian).
32. Turnbull, D. and Treaftis, H.N. (1958) *Transactions AIME*, **212**(2), 33.
33. Afanasiev, N.I. and Elsukova, T.F. (1982) *Fizika Metallov i Metallovedeniye*, **53**(2), 354 (in Russian).
34. Koval, Yu.M., Bezuhly, A.M., Didyk, M.I. et al. (2004) *Reports of the National Academy of Sciences of Ukraine*, **2**, 102 (in Ukrainian).
35. Pines, B.Ya. (1961) *Studies by Metallophysics*, Kharkov University, Kharkov (in Russian).
36. Lucke, K. and Detert, K. (1957) *Acta Metallurgica*, **5**, 628.
37. Oberschmidt, J.M., Kim, K.K. and Gupta, D. (1982) *Journal of Applied Physics*, **53**, 5672.
38. Hillert, M. and Purdy, G.R. (1978) *Acta Metallurgica*, **26**, 333.
39. Balluffi, R.W. and Cahn, J.W. (1981) *Acta Metallurgica*, **29**, 493.
40. Shewmon, P.G. (1981) *Acta Metallurgica*, **29**, 1567.
41. Hillert, M. (1972) *Metallurgical Transactions*, **3A**, 2729.
42. Brechet, Y.I.M. and Purdy, G.R. (1989) *Acta Metallurgica*, **37**, 2253.
43. Kajihara, M. and Gust, W. (1998) *Scripta Metallurgica*, **38**, 1621.
44. Rabkin, E. (1994) *Scripta Metallurgica*, **30**, 1413.
45. Cahn, J.W., Fife, P.C. and Penrose, O. (1997) *Acta Metallurgica*, **45**, 4397.
46. Zieba, P. (2001) *Local Characterization of the Chemistry and Kinetics in Discontinuous Solid State Reactions*, IMMS, Cracow.
47. Liu, D., Miller, W.A. and Aust, K.T. (1989) *Acta Metallurgica*, **37**, 3367.
48. den Broeder, F.J.A. and Nakahara, S. (1983) *Scripta Metallurgica*, **17**, 399.
49. Ma, C.Y., Rabkin, E., Gust, W. and Hsu, S.E. (1995) *Acta Metallurgica*, **43**, 3113.
50. Larikov, L.N., Maksimenko, E.A. and Franchuk, V.I. (1993) *Metallofizika i Noveishiye Technologii*, **15**, 44 (in Russian).
51. Moriyama, M. and Kajihara, M. (1998) *ISIJ International*, **38**, 86.
52. Hillert, M. (1999) *Acta Materialia*, **47**, 4481.
53. Mendelev, M.I. and Srolovitz, D.J. (2001) *Acta Materialia*, **49**, 589.
54. Hillert, M. (2004) *Acta Materialia*, **52**, 5289.
55. Inden, G. (1976) *Proceedings of CALPHAD V*, p. 1.
56. Chen, F.-S. and King, A.H. (1988) *Acta Metallurgica*, **36**, 2827.
57. Jansson, A. (1987) *TRITA – MAC 34*, Royal Institute of Technology, Stockholm.
58. Dinsdale, A. (1989) *SGTE data for pure elements*. NPL Report DMA (A), p. 195.
59. Borisov, V.T., Golikov, V.M. and Scherbedinskiy, G.M. (1964) *Fizika Metallov i Metallovedeniye*, **17**, 881 (in Russian).
60. Gupta, D., Vieregge, K. and Gust, W. (1999) *Acta Materialia*, **47**, 5.
61. Borovskii, I.B., Gurov, K.P., Marchukova, I.D. and Ugaste, Yu.E. (1971) *Inter-diffusion Processes in Metals*, Nauka, Moscow (in Russian).
62. Van Loo, F.I.I., Bastin, G.F. and Vrolijk, I.W.G.A. (1987) *Metallurgical Transactions*, **A18**, 801.

63. Dayananda, M.A. (1991) in *Diffusion in Solid Metals and Alloys* (ed. H. Mehrer), Springer Verlag, Berlin, p. 372.
64. Lyashenko, Yu.A. and Shmatko, O.A. (2006) *Izvestiya Vysshikh Uchebnykh Zavedenii, Fizika*, **6**, 79 (in Russian) (2006) *Russian Physics Journal*, **49**(6), 658.
65. Gusak, A.M., Lyashenko, Yu.A., Kornienko, S.V. et al. (1997) *Defect and Diffusion Forum.*, **143–147**, 689.
66. Tikhonov, A.N. and Arsenin, V.Ya. (1986) *Methods of Solving Ill-Posed Problems*, Nauka, Moscow (in Russian).
67. Tikhonov, A.N., Goncharskii, A.V., Stepanov, V.V. and Yagola, A.G. (1990) *Numerical Solutions of Ill-Posed Problems*, Nauka, Moscow (in Russian).
68. Carroll, D.S. *Genetic algorithm GA 170*, http://cuaerospace.com/carroll/ga.html.

Further Reading

Gusak, A.M. and Lyashenko, Yu.A. (1991) *Zavodskaya Laboratoriya*, **4**, 48 (in Russian).

Lyashenko, Yu.A., Derevyanko, L.I. and Handus, T.V. (1999) *Bulletin Cherkasy University*, **9**, 60 (in Russian).

Mokrov, A.P., Akimov, V.K. and Ushakov, O.I. (1977) *Diffusion Processes in Metals*, TPI, Tula (in Russian).

Samarskii, A.A. and Gulin, A.V. (1989) *Numerical Methods*, Nauka, Moscow (in Russian).

Tu, K.N. and Turnbull, D. (1971) *Metallurgical Transactions*, **2**, 2509.

13
Nucleation and Phase Separation in Nanovolumes
Aram S. Shirinyan and Andriy M. Gusak

13.1
Introduction

Nanostructure design is a key aspect of nanotechnology. Nanotechnologies are being considered as the driving force behind a new industrial revolution. Nanotechnology is directed to achieving the ability to build materials and products at an atomic level of high precision and has been one of the most active areas of scientific research in the last decade [1–6]. Metal nanoparticles and nanopowders have a special significance among nanostructured materials, which is due to their wide practical applications such as in transportation, aerospace, sports products, cosmetics, medicine, chemical and food processing, and military applications.

At the same time, nanosystems and, in particular, nanoparticles are mainly studied in the framework of solid state or technological works, such as vacuum evaporation, heterogeneous catalysis, synthesis of very fine powders, nanostructures, nanoelectronics, and so on. Nanoparticles are also seen embedded in structures such as the so-called composite materials or clusters of implanted materials irradiated by particles from nuclear reactors. They are also of interest in other fields such as astrophysics since nanoparticles are present in the dust in free space and in the atmosphere of earth generated by natural events such as volcanic eruptions and forest fires. The toxic effects of nanomaterials that have been brought to light recently show that some types of nanomaterials can be dangerous. Nanotechnology risks in the long term have become a subject of concern insurance companies [7].

In order to develop nanomaterials with the desired structure and properties, the knowledge of basic principles and specific features of self-organization processes at the atomic and/or molecular level is of fundamental importance. However, despite the widely acknowledged importance of nanomaterials, the understanding of the specific features of the evolution of first-order phase transitions in such systems is far from complete. The earliest known use of nanoparticles dates back to the ninth century and the reign of the Abbasid Dynasty. Arab potters used nanoparticles in their glazes so that objects would change color depending on the viewing angle (the so-called polychrome lustre) [8].

Diffusion-Controlled Solid State Reactions. Andriy M. Gusak
Copyright © 2010 WILEY-VCH Verlag GmbH & Co. KGaA, Weinheim
ISBN: 978-3-527-40884-9

One of the well-known physical properties of modern nanomaterials is the variation of the melting temperature with the size of the samples [9–13]. Such kind of behavior is known for a long time and has been verified by recent experimental, theoretical, and computer investigations on melting and superheating of low-dimensional materials [14]. As another example where size-induced phenomena occur, one can mention the increase in solubility of chemical elements as their size decreases. It is also known that a decrease of a system's size can cause the formation of unusual crystalline modifications, amorphization, change of melting temperature, surface tension, difference between temperatures of melting and crystallization, and high ability to form intermetallic compounds. Therefore, the sizes of critical nuclei of the new phases in first-order phase transitions have the same order of magnitude as the whole material. As we shall see in the following, this fact substantially changes the physics and chemistry of the processes in nanomaterials.

Nanomaterials are also called *ultrafine particles* or *dispersed systems (DSs)*. The notion "nanomaterials" is used for systems that contain at least one small (less than 100 nm) dispersed constituent part in one dimension [15, 16]. Those are thin films, amorphous metals, powders, clusters, formations in porous glasses and materials, small metallic particles, clusters, carbon nanotubes, fullerenes, quantum dots, and so on. [15]. Many physical and chemical properties of such DSs depend on their size and this size dependence is of general character. In this respect, it is worth noting that systems with a size in the range of 1–100 nm are in an intermediate state between the solid and the molecular one. When the number of atoms in the system amounts to thousands or above, the properties evolve gradually from molecular to solid ones. For such systems, the ratio of the number of surface to volume atoms is not small. It is, then, obvious that surface effects on cohesive properties of such a system cannot be neglected. When one extrapolates this simple argument to compound materials, one concludes that their behavior and phase diagrams might differ from those of the bulk material [12, 17].

In the present chapter, we analyze first-order phase transitions in nanosystems both from the thermodynamic viewpoint and with respect to specific features of phase diagrams of nanostructures and kinetics of such processes, specifically targeting at metallic nanoalloys. First-order phase transformations are being spoken of as transformations following the nucleation of new phase clusters. Progress in differential scanning calorimetry allows a detailed tracing of the initial stages of phase transformations in solids including formation and growth of new phase nuclei [18, 19]. In the usual treatment of nucleation, it is assumed that the volume of a system is much larger than the volume of the nucleus, so that there is no problem of matter supply. In a DS, the total amount of one of the chemical components may be too small for the synthesis of the critical nucleus or for phase formation during the competition of two and more phases. We call it the *depletion effect*. We shall see that depletion effects become crucial for binary and multicomponent nucleation.

Most practical applications of nanomaterials will require large quantities of a suitable powder product. One of the fundamental problems in nanoscience is the behavior of a nanopowder under changing external conditions such as temperature, pressure, and chemical composition of the environment. These questions are discussed in the last part of the present chapter.

Altogether, one can presume that, from the viewpoint of structural stability, four fundamental problems should be addressed: (i) the effect of size on the properties of nanosystems; (ii) the effect of defects in nanomaterials on their internal structures; (iii) the evolution of nanomaterials at the beginning of their synthesis, (iv) the stability of nanosystems under varying external conditions and fields. All these problems have not been completely solved yet both from theoretical and experimental points of view.

13.2
Physics of Small Particles and Dispersed Systems

There are two main types of unique properties associated with DSs: (i) novel optical, electrical, and magnetic properties due to quantum confinement effects; and (ii) changes in surface and physicochemical properties due to smaller physical dimensions. In our analysis here, we will concentrate on the second type of finite-size effects.

13.2.1
Nano-Thermodynamics

One generally considers that thermodynamics is valid when the number of atoms is "large" [20]. However, what is "large"? Are nanosystems "large" with respect to the thermodynamics definition?

Especially important is this question with respect to phase transformations. In a rigorous theoretical treatment, the phase transition (as a steplike change of the first- or second-order derivatives of Gibbs potential) is defined only in the "thermodynamic limit" $N_0 \to \infty$, $V \to \infty$. N_0 here is the number of atoms and V is the volume of the system. Nevertheless, everybody is applying the notion of phase transformations to rather small (even nano) systems. One should expect that the behavior of various parameters during such "gradual" transformations is smeared by fluctuations so that the transition point should become a transition interval, the solubility line should become a "solubility band," and so on.

We would like to emphasize here three main peculiarities of nanosystems with respect to thermodynamics:

1) large fluctuations,
2) nonnegligible ratio of the number of surface atoms to bulk atoms,
3) considerable depletion of binary or multicomponent nanoparticles at the initial nucleation stage of transformation.

The second issue is the most frequently discussed and used. The importance of issue (3) is much less known and understood.

13.2.2
Production of Dispersed Systems

The synthesis of nanoparticles is not a simple task; however, numerous techniques have been developed for that purpose [14–16]. In all experimental studies, several key aspects are crucial:

1) preparation of the DS,
2) detection of the corresponding phase transition points,
3) determination of DS sizes and their size and space distributions,
4) accurate measurement of temperature, pressure, and composition, and
5) methods to avoid oxidation of small particles.

About 35 years ago, the production of noble-gas clusters and metallic clusters became available as a result of free beam sources [21, 22]. Let us estimate the time scale of 1 nm size nanocluster production by these techniques [23]. There, the typical metal vapor temperature, T, is about 1500 K and the pressure, p, is nearly equal to 500 Pa, and the estimated size of the clusters is $L = 1$ nm. Then, using the kinetic theory of gases, one obtains the incident atomic flux on the cluster, F, which is the number of metal atoms hitting the nanocluster surface per second:

$$F = L^2 \frac{p}{(mkT)^{1/2}} \quad (13.1)$$

where m is the mass of the atoms and k is the Boltzmann constant. Simple algebra yields the value $F = 10^7$ s^{-1} and the corresponding time scale $\tau = F^{-1} = 10^{-7}$ s $= 100$ ns. It means that, in order to create a 1000-atom cluster, one needs about 10 ms.

13.2.3
Anomalous Structures and Phases in DSs and Thermodynamic Estimates

Electronographic analyses of Ta, Nb, W, and Mo films has shown a face-centered cubic (fcc) structure with average sizes of 5–10 nm instead of usual body-centered cubic (bcc) structure [24–26]. Metastable structures of β-W type in Re, W, Mo, and Cr thin films have been found [27]. Thin layers of W, Mo, and Ta form an amorphous phase [24, 28, 29]. The amorphous phase is observed also in small particles of Fe and Cr [15]. One can provide a thermodynamics-based explanation of such behavior.

Our analysis performed here is grounded on the very simple idea of Bublik and Pines [24, 30]. It is clear that the "anomalous" appearance of metastable phases in small systems is related to the change of conditions of the phase equilibrium. In bulk materials, the stable phase (say, phase 1) is the one that has the lowest bulk Gibbs free energy (per unit volume of the system), $g : g_1 < g_2$. The subscripts 1 and 2 refer here to the phases 1 and 2, respectively. In the description of nanosystems, one has to take into account, in addition to the

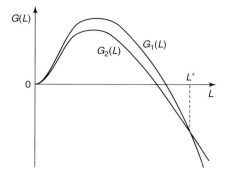

Figure 13.1 Qualitative representation of the size effect on the change of phase equilibrium conditions. L^* is the system's transition size at which the metastable phase 2 becomes stable.

bulk contributions, also surface (and/or interfacial) free energies (per unit area), σ_1 and σ_2. Then the Gibbs free energy G of the transforming system has the form: $G_1 = L^3 g_1 + \sigma_1 L^2$ for phase 1 and $G_2 = L^3 g_2 + \sigma_2 L^2$ for phase 2. Here, the DSs are assumed to be of average size L, and L^3 and L^2 are the estimated volume and surface area (Figure 13.1). Because of the different surface energy contributions of the phases, the equilibrium conditions may be changed so that the metastable (from the usual "bulk" point of view) phase 2 becomes the stable one:

$$G_1 > G_2 \implies g_1 + \sigma_1/L > g_2 + \sigma_2/L$$

For the typical sizes of a nanosystem,

$$L < L^* = (\sigma_1 - \sigma_2)/(g_2 - g_1) \tag{13.2}$$

the metastable phase 2 will be thermodynamically more favorable. Here, the size L^* is the system's transition size: that is, the size of the whole system transforming from phase 1 to phase 2 and vice versa.

Such considerations show that the decrease of the system's size L may lead to a situation in which the phase with a smaller specific surface energy becomes more probable and stable [29]. Thus, size constraints may present the main reason for the formation of a metastable instead of a stable (from macroscopic point of view) phase.

As examples of the influence of size-induced effects, one can mention the films of refractory metals Mo and W, which in bulk have the bcc structure and at small sizes of 50–100 Å have the fcc phase [25]. Electronographic methods show that the DS of rare-earth metals Y, Gd, Tb, Ho, and Tm have an fcc lattice instead of the hexagonal close-packed lattice (as is the case for macroscopic crystals [15]).

13.2.4
Influence of DSs on the Temperature of the Phase Transformation

During thin film structure investigations, one often observes high-temperature phases even at room temperature. For example, in beryllium films, a high-temperature β-phase of Be was found, and similar results were also reported for the β-phase of Co (fcc-structure) [15, 30].

The size-induced melting behavior of nanoparticles and thin films was studied with the help of Vecshinskyy's elegant differential technique [12, 31]. Binary thin films were prepared by subsequent condensation of the components Bi and Sn on amorphous glassy Pt substrates in vacuum so that the thickness of the condensate changed step by step. The temperature gradient created along the substrate enables one to determine the melting temperature and its dependence on sizes:

$$T_m = T_{m,\infty} - a_i \cdot 10^{-5}/(2L) \qquad (13.3)$$

where $T_{m,\infty}$ is melting temperature of a bulk sample, the indices i (Bi, Sn) correspond to the particular chemical element (material constants $a_{Bi} = 4.73$, $a_{Sn} = 2.6$), and L is measured in angstrom.

Beginning from Pawlow's theoretical work [32], performed in 1909, various models have been devised to describe the variation of the melting temperature with the radius of the particle [33–35]. In the case of a nonspherical particle (such as metals embedded in polymers or other materials), it has been argued that the melting point depression might be, depending on its shape, smaller or larger than for a spherical particle [36]. It has also been argued that this effect depends on the chemical environment of the particle, via the surface tension [37].

In the following, we shall give more consideration to nanoparticles. They are not always uniform. In many circumstances, they consist of one core phase, surrounded by another phase, making the shell of the particle. This is the case for elementary particles surrounded by their oxide [37], metal core in the other metal shell [38], and so on. It is also observed that some otherwise metastable phases are stable when they are surrounded by another shell [39]. In some circumstances, core–shell structures seem to appear spontaneously [38], while alloying occurs for other alloys [40]. The experimental data on tin and gold particles, distributed in a wide range of sizes, gave the result similar to Equation 13.3, that is, the melting temperatures decrease with the decrease of the sizes [41, 42]. Polymorphic phase transitions in nanosized Fe particles have also been investigated and, again, a size-induced decrease of the phase transformation temperature was found [43].

Most of the proposed direct experimental verifications and theoretical research of size effects in DSs have been performed so far for pure metals and materials. The situation remains unelucidated for the case of DS compounds and metal alloys and many interesting topics are encountered in this field. Examples of unsolved problems for multicomponent DSs are the problem of solubility of elements, effect of the chemical environment, the influence of magnetic and electrical fields, and state diagrams.

13.2.5
State Diagrams of DSs

During the last decade, phase diagrams of binary nanoalloys were intensively discussed. Phase diagrams are also used in the detailed analysis of nanocrystalline structures on substrates [44–46]. These studies have shown that when the thicknesses of the films of Bi–Pb alloys are 32, 20, and 10 nm, the decrease of the eutectic temperature may reach 5, 10, and 18 K, respectively. For thicknesses less than 50 nm, one observes the lowering of the liquidus and solidus curves: that is, the shift of the phase diagram to lower temperatures.

Alternative approaches for the analysis of state diagrams of DSs are simulation methods. Among them the well-known ones are the following: (i) For investigation of small clusters (up to 10–50 atoms), *ab initio* methods of quantum chemistry (such as the Hartree–Fock method), density functional theory, quantum Monte Carlo calculations, and quantum molecular dynamics; (ii) For big clusters (from 10 to 50 up to 10^6 atoms), methods based on parameters fitted to experimental material properties: that is, methods of semiempirical pair potentials (as the Lennard–Jones one), embedded atom methods and density functional theory, many-body potential methods, second-moment methods and classical Monte Carlo calculations [47–51]. Melting of 38, 55, 57, 135, and 429-atom clusters were studied by calculation of the caloric curve: that is the total energy of the cluster, E, as a function of temperature, T [6, 51]. If a phase transition takes place, then the dependence $E(T)$ has a jump. The derivative of the function $E(T)$ gives the heat capacity dependence $c(T) = (\partial E(T)/\partial T)$ on T. The obtained dependence $c(T)$ exhibits maxima shifted with respect to the corresponding bulk crystalline sample values (nearly by two times) to lower temperatures.

Similar results have been reported in recent theoretical analyses [17, 36], where the authors analyzed a liquidus–solidus phase diagram for a binary system with full mutual solubility of the components. The well-known lens-type shape of a diagram on the temperature–composition surface is shifted to lower temperatures. Thereby, the dependence of the melting temperature on the particle size has been found to obey Equation 13.3:

$$T_m = T_{m,\infty}(1 - \alpha/2L)$$

where the coefficient α varies in the range $\alpha = 0.4/3.3$ nm. One may expect a similar behavior in other first-order phase transformations as well. From the above-mentioned results, one can conclude that the melting temperature of not very small particles (larger than 100 nm) should not essentially depend on size.

13.2.6
Shift of the Solubility Limits in DSs

As an example of experimental work on size-induced solubility change, one can mention the results for Ag–Cu DS [52]. There, thin films of Ag–Cu substitution solution with fcc structures have been investigated (so that the polymorphic transitions are prevented). Ag and Cu were evaporated and condensed in vacuum (10^{-7}–10^{-9} Torr) onto a carbon substrate. Electronographic investigations of such Ag–Cu films demonstrate the increase of solubility: for 27 nm thickness, the solubility of Cu was 6 at.%, and for films of 7 nm it was 15–17 at.%. For macroscopic alloys, the solubility of Cu in Ag is 0.35 at.% [53]. The shift of solubility is also reported for alloys such as Al–Cu, Ag–Ga, In–Au, Fe–Cu, and Au–Sn [46, 53–55].

13.2.6.1 Depletion

In the following, we restrict the discussion to binary systems (containing A and B components) and consider the formation of a two-phase system, where the new phase has a nonzero driving force of transformation.

To study the depletion effect, let us consider the simple model of an isolated binary nanoparticle. Let us also assume that there is no constraint on lattice rearrangement. Then, the process of nucleation of the new phase in the initially homogeneous system is related to the concentration fluctuations. Let C_0 be the molar fraction of species B in the particle before nucleation, C_n is the molar fraction of species B in the new phase (the new phase nucleus will have a concentration different from the parent phase, and $C_n \neq C_0$, N_0, and N_n are the number of atoms in the parent and new phases, respectively.

Let us first examine the consequences of matter conservation inside the nanoparticle. The minimal size N_0^* of such a system in which the single new phase embryo of critical size N_n^* can appear may be found from the condition of conservation of matter: $C_0 \cdot N_0^* = C_n \cdot N_n^*$. If the embryo of the new phase appears, it will need the supply from the region of the parent phase from which it may "draw" B atoms. In the spherical case, the last condition gives the estimate for the number of atoms N and radius R of a particle. The value N_0 and size R should not be less than

$$N_0^* = N_n^* \cdot C_n/C_0$$
$$R^* = (n_1 C_n / n C_0)^{1/3} r_{cr} \qquad (13.4)$$

Here n and n_1 are the atomic density (per unit volume) in the parent and new phases, respectively, and r_{cr} is the radius of the critical nucleus of the new phase. Nucleation and phase transition becomes impossible for particle sizes $R < R^*$ and/or $N_0 < N_0^*$. Therefore, the size R^* does not coincide with the system's critical size L^* discussed in Equation 13.2 and Figure 13.1. Thus the effect of depletion of the parent phase on nucleation and growth in nanovolumes cannot be neglected. Of course, the above-mentioned considerations are not rigorous since the dependence of critical size on particle size was not taken into account.

13.2.7
Concluding Remarks

Thus, the problem of phase transformations and the behavior of DSs is urgent at present and for future science and technology development. Many physicochemical problems remain unsolved so far, especially for multicomponent DSs. Some of these unsolved problems will be discussed in this chapter. In particular, we shall deal with the problem of first-order phase transition evolution in metal DSs. We also show that, in the description of phase transformation in nanomaterials, one must take into account the effective "matter supply region" [56–59].

13.3
Phase Transformations in Nanosystems

As follows from the previous analysis, when dealing with the thermodynamics of nanoparticles it is necessary to specify the size range under discussion. Indeed, in the literature the term *nanoparticle* is used for particles having sizes from 1 nm (a few atoms) up to 500 nm. The behavior of nanoparticles at the margins of the indicated size range is very different. Here we treat cases where the thermodynamic description remains valid. This requirement implies the following:

1) The overall radius of the nanoparticle is relatively large ($R \geq 2$ nm) so that the number of atoms is sufficient large allowing one to apply the thermodynamic approach.
2) The radii of the core and the shell of a nanosystem are also relatively large.
3) The surface of the core is characterized by a single value of the surface tension. This condition is met when the particle is either "rounded" or, conversely, presents the shape of a regular polyhedron with the same kind of facets [60]. The difference between the surface tension and specific surface free energy is neglected.

In the present section, we introduce a thermodynamic model of phase separation in a nanoparticle and analyze how the depletion effects can be taken into account. The last part is devoted to the analysis of the influence of size and depletion on phase diagrams of regular solutions. In the Appendix, we present the rule of parallel tangents construction (not to confuse with common tangent rule) for the extreme points of the phase transition. In the following, the starting single phase is called "old" or *"parent" phase* and the newly formed phase is called "new" one.

13.3.1
Solid–Solid First-Order Phase Transitions

In nanoparticles, the quantity of matter is finite. Hence, one also has to take into account the fact that all stoichiometries are not available because of the above-mentioned depletion of matter [59, 61].

13.3.1.1 Geometry of a Nanoparticle and Nucleation Modes
Let us assume that a small, isolated, initially supersaturated particle of a given alloy is quenched into the two-phase region. Then a phase transition from the single phase to a two-phase state takes place. A single nucleus of a new phase forms inside or at the margin of the particle (Figure 13.2).

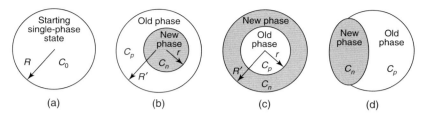

Figure 13.2 Different modes of transformation: (a) initial particle of composition C_0; (b, c, d) the same particle after nucleation in different configurations of "old" and "new" phases, $C_p(r)$ is the concentration of the parent phase, C_n is the concentration of the newly born phase, r is the nucleus size and radius of the parent phase (c), R and R' are the radii of a nanometer-sized isolated particle before and after nucleation, respectively.

The phase separation might lead to either the formation of a core–shell structure (as studied here) or to the disintegration into two new nanoparticles. The condition to obtain the core–shell structure is that the two phases wet each other. Let us look at the evolution of the surface energy when the core–shell structure appears. The initial surface energy of the particle is equal to:

$$A_o = 4\pi R^2 \sigma_p$$

where R is the radius of the particle before nucleation, and σ_p is the surface tension of the free parent phase (Figure 13.2a).

After nucleation, three different situations may be encountered. In the first case (Figure 13.2b), the parent phase is located outside (shell) while the new-born phase is inside (core of the particle). In this case, the total surface energy is given by

$$A_1 = 4\pi \left(R'^2 \sigma_p + r^2 \sigma_{np} \right)$$

where R' is the external radius of the particle, r is the radius of the core material ($r \geq 0$, $r < R$, $r < R'$), and σ_{np} is the specific Gibbs free energy per unit area (interphase tension) of the parent phase–nucleus interface. R' may differ from R when there is a difference in the atomic volumes of components in the parent and new phases.

In the second case (Figure 13.2c), the parent phase is inside (core) while the new phase is outside (shell). In this case, the total surface energy is given by

$$A_2 = 4\pi \left((R')^2 \sigma_n + r^2 \sigma_{np} \right)$$

Here, σ_n is the surface tension of the new free phase, and r is the radius of the parent phase.

In the third case of heterogeneous nucleation at the external interface boundary (Figure 13.2d), the two phases do not show the core–shell structure. After the nucleation, the corresponding value of the surface energy will be:

$$A_3 = \sigma_p S_p + \sigma_n S_n + \sigma_{np} S_{np}$$

Here, S_{np} is the surface area between the nucleus and the parent phase, and S_n and S_p are the free surface areas of the new and parent phases, respectively.

If the atomic densities are practically the same (so that $R' = R$ and $\sigma_n \approx \sigma_p$, the differences in surface energies between the initial single-phase state and the separated two-phase states are given by:

$$\Delta A \approx \sigma_{np} \cdot S_{np} \qquad (13.5)$$

where $S_{np} = 4\pi r^2$ for the cases (b and c) in Figure 13.2. The "thermodynamic and kinetic decoding" of these transformations: (a)→(c), (a)→(b), (a)→(d), (a)→(d)→(c), and so on, will be more fully discussed in the following. Here, only a general thermodynamic consideration of (a)→(b) transition is given.

13.3.1.2 Depletion Effect

When the stoichiometry of the new phase differs from that of the parent phase, the change of composition will lead to the depletion in the parent phase. The mole fractions C_n and C_p of species B in the new and parent phases, irrespective to morphology, are interrelated by the matter conservation condition:

$$C_0 V = C_1 V_1 + C_p(V - V_1) \qquad (13.6)$$

Here $V_1 = 4\pi r^3/3$ is the volume of the nucleus, and the atomic densities are assumed to be equal $n = n_1$. In all models of the present chapter, at any moment and at any size of the growing nucleus, the mole fraction distribution is supposed to be steplike (without transient layers) so that the concentration is treated as uniform within each phase.

13.3.1.3 Regular Solution

Let us choose first the thermodynamic models for new and parent phases. The Gibbs energy (per atom) of the parent phase and of the new phase is assumed to be described by the regular solution theory (Figure 13.3). In the case of a regular solution, the Gibbs free energy (per atom) of formation of the new phase is given by [62, 63]:

$$\Delta g(C) = 0.5Z\{C\varphi_{BB} + (1-C)\varphi_{AA} - 2E_{mix}C(1-C)\} \\ + kT\{C\ln C + (1-C)\ln(1-C)\} + p\{(1-C)\omega_A + C\omega_B\} \qquad (13.7)$$

In this equation, φ_{AA}, φ_{BB}, and φ_{AB} are the interatomic interaction (pair) energies, $E_{mix} = \{0.5(\varphi_{BB} + \varphi_{AA}) - \varphi_{AB}\}$ is the mixing energy, Z is the coordination number, p is the pressure, and ω_A and ω_B are the atomic volumes of A and B atoms, respectively. C is the atomic fraction of the B species.

In the following, for simplicity, we assume that $\varphi_{AA} = \varphi_{BB}$. The parameters used in the theoretical calculations are taken as those typical for metals and given in Table 13.1. We shall not discuss the results and conclusions based on the nonsymmetric approximation of the pair potentials and other values of parameters of the system since they are basically similar. The total number of atoms in the given particle is about $R^3 \cdot n \approx 10^6$ where the classical nucleation theory is appropriate.

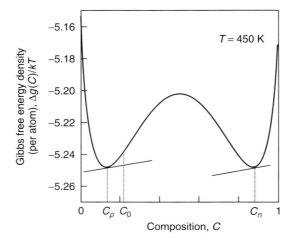

Figure 13.3 Dependence of Gibbs free energy density (per atom) on composition for different temperatures; C_0 is the initial composition. The mole fractions C_n and C_p are linked by the parallel tangents rule; they depend on the nuclear radius r and provide the minimization of ΔG (energy of the system) with respect to concentrations for every given r, R. For the case of full decomposition in the bulk system, these mole fractions C_n, C_p tend to the values determined by the common tangent rule corresponding to stable equilibrium. Temperatures are shown in the plot. Parameters are given in Table 13.1.

Table 13.1 Parameters of the binary nanosystem.

Z	σ, J m^{-2} ($\sigma \equiv \sigma_{np}$)	φ_{AA}, φ_{BB}, J ($\varphi_{BB} = \varphi_{AA}$)	φ_{AB}, J ($\varphi_{AB} = \varphi_{BA}$)	n, m^{-3} ($n = n_1$)	ω_B, m^3 ($\omega_B = \omega_A$)	R, m ($R = R'$)	T_c, K
8	0.15	-8×10^{-21}	-6×10^{-21}	7×10^{28}	1.43×10^{-29}	2×10^{-8}	580

13.3.1.4 Change of Gibbs Free Energy

Now let us take into account all mentioned constraints, that is, depletion Equation 13.6 and the model of regular solution Equation 13.7. Consider case (b) in Figure 13.2 when the depleted parent phase forms the shell enveloping the core, the new phase. Under these conditions, the Gibbs free energy ΔG for the nucleation of the new phase of volume V_1 (radius r) and decomposition can be written as:

$$\Delta G(V_1, C_n) = n_1 V_1 \Delta g(C_n) + n(V' - V_1) \Delta g(C_p) - n V_0 \Delta g(C_0) + 4\pi r^2 \sigma \quad (13.8)$$

where σ (σ_{np}) is the interphase tension, and $V' = V_0 + V_1(n - n_1)/n$ is the volume of the separated particle after the nucleation and/or separation. As pointed out previously, the change of the surface energy may be written as $\Delta A = \sigma_{np} S_{np}$ ($S_{np} = 4\pi r^2$). It is also assumed that the surface tension σ_p of the external surface does not change upon phase separation and $n = n_1$. Equation 13.8 implies that the Gibbs free energy change of the system is a function of two variables: C_n and r (if one takes into account the constraint Equation 13.6 relating C_p to C_n and r).

13.3.1.5 Minimization Procedure

Let us now have a look at equilibrium phase transformation of the nanoparticle. It is known from thermodynamics that the equilibrium is related to the concavity (or convexity) of the thermodynamic potentials [63]. There are two equivalent ways to investigate this.

The first one is the usual method of geometrical thermodynamics [62]. According to it, one plots the Gibbs free energy density as a function of composition taking into account the additional surface energies related to the nucleus surface and the particle surface. Then, one finds the conditions for minimal Gibbs free energy of a given system. Unfortunately, this method is not always correct for nanosystems because of the above-mentioned depletion effect (also Section 13.4).

The second method (used here) is to consider the general thermodynamic equilibrium conditions for the function $\Delta G(r, C_n)$ and write the equations of the first- and second-order derivatives of $\Delta G(r, C_n)$ with respect to their variables [61]. To determine the extreme points of phase transition, one has to solve then the following set of equations:

$$\partial \Delta G(r, C_n)/\partial C_n = 0 \tag{13.9}$$
$$\partial \Delta G(r, C_n)/\partial r = 0 \tag{13.10}$$

The solution of the second equation of this system gives the radii of the phases in the equilibrium states, at constant T, C_0, and R. The solution of the first one leads to the rule of parallel tangents for the extreme points of transformation, at constant r, R, and T.

The driving force for the transformation is generally determined by assuming that the concentration of the parent phase is constant. As shown in the Appendix, this is far from being true for nanoparticles. In the case of limited volume, one must take depletion effects into account. The general peculiarity of nucleation is that the stoichiometry of the nucleus does not coincide either with the initial stoichiometry of the parent phase or with the stoichiometry of the new phase after the transformation, or with the stoichiometry of the parent phase after separation. Sometimes, people wrongly use the rule of common tangent. To avoid confusion, a theorem for extreme points of phase transformation is proven in the Appendix. This theorem states that in the case of a nanosystem the boundaries of the phases are determined by the points at which the slopes of the two free energy density $\Delta g(C)$ curves for new and old phases are equal: that is, have equal (not common) tangents. This is also shown in Figure 13.3.

After some algebra, in the Appendix, Equation 13.A.3 is obtained for the optimal concentration of the depleted parent phase and the optimal concentration of the "new" phase for the chosen case of regular solution model. It reads as follows:

$$4E_{mix}\left(C_n^{opt} - C_p^{opt}\right) = kT \ln \left\{ \frac{C_p^{opt}\left(1 - C_n^{opt}\right)}{C_n^{opt}\left(1 - C_p^{opt}\right)} \right\} \qquad (13.11)$$

Substituting C_n^{opt} from Equation 13.A.5 into this expression and solving it with respect to C_p^{opt}, one obtains the optimal mole fraction in the new phase C_n^{opt} as a function of one "coordinate" r at fixed R and the other parameters of the system ($E_{mix} < 0$). The corresponding optimal solution for the mole fraction in the parent phase C_p^{opt} is determined from Equation 13.A.5. In the following, for simplicity, we write C_n and C_p instead of C_n^{opt} and C_p^{opt}, having in mind the equilibrium concentrations found by the parallel tangent rule Equation 13.A.5. Note that a nanosystem is not "obliged" to choose optimal values, since the fluctuation level for small systems is high. "Optimal" means just most probable but the probability distribution is not that sharp when compared to macrosystems.

After substituting C_n and C_p into Equation 13.8, one obtains the Gibbs free energy $\Delta G = \Delta G(r, C_n)$ of the system as a function of one variable, the radius of the new phase, at constant R and T. Equation 13.8 allows us to find the critical size of the nucleus and other critical parameters of the system. The condition (Equation 13.10) for Equation 13.8 may be rewritten as:

$$4\pi r^2 n \left\{ \gamma \Delta g \left(C_n\right) - \Delta g \left(C_p\right) \right\} + \frac{4\pi}{3} n \left(\left(R'\right)^3 - r^3\right) \left.\frac{\partial \Delta g \left(C\right)}{\partial C}\right|_{C_p} \frac{\partial C_p}{\partial r} + 8\pi r\sigma = 0$$

Here, σ is taken independent of size r, $\gamma = n_1/n$.

The last equation and conservation of matter Equation 13.6 determine the radius of the critical nucleus and the radius of equilibrium (metastable or stable) two-phase configuration of the system:

$$r_0 = -\frac{2\sigma}{n \left\{ \gamma \Delta g \left(C_n\right) - \Delta g \left(C_p\right) - \left(C_n - C_0\right) \left.\frac{\partial \Delta g \left(C\right)}{\partial C}\right|_{C_p} \cdot \frac{\gamma R^3}{R^3 - \gamma r_0^3} \right\}} \qquad (13.12)$$

In the general case, Equation 13.12 is of fourth order with respect to size r_0. Depending on the parameters, the equation has two, one, or zero solutions (we neglect the solution $r_0 = 0$).

In the first case, $\Delta G(r)$ presents one maximum (critical nucleus) and one minimum (separated two-phase state: new phase plus depleted parent phase). In the second one, the maximum and minimum coincide ($\partial \Delta G(r)/\partial r = \partial^2 \Delta G(r)/\partial r^2 = 0$ for $r > 0$). In the last case, nucleation is impossible, and the $\Delta G(r)$ dependence on r is a monotonically increasing function of r.

Strictly speaking, one cannot guarantee that the system kinetically has time to optimize the compositions of both phases before further change of their sizes. Instead, one should consider the surface of Gibbs free energy as a function of two independent arguments and look for saddle points (if any) and local minima of this surface.

13.3.1.6 Probability Factor of the Phase Transformation

In experiment, one generally deals with a large number of particles, so that a statistical approach is required. The nucleation process is also inherently statistical. The crossover of the nucleation barrier is a stochastic process, which can be treated by considering the fluctuations in the system. It is, therefore, mandatory to study the phase separation of nanoparticles by statistical approaches.

Under equilibrium conditions, the probability of concentration fluctuation is given by the theory of thermodynamic fluctuations. The probability factor $f(r)$ is given by the Boltzmann distribution [61]:

$$f(r) = f_0 \exp\left(\frac{-\Delta G(r)}{kT}\right) = \frac{1}{\sum_{r=0}^{r=r_{max}} \exp\left(\frac{-\Delta G(r)}{kT}\right)} \exp\left(\frac{-\Delta G(r)}{kT}\right) \quad (13.13)$$

Here, $f(r)/f_0$ is the fluctuation probability function for particles made of nuclei of size r, $1/f_0$ is the statistical sum, and r_{max} is the maximal possible size of the nucleus when the parent phase is fully depleted by B species: $C_p = 0$. Let us recall that $\Delta G(r)$ and $f(r)$ are functions of T, R, and C_0. In the following analysis, we consider their influences on energy barriers, fluctuation probability, and phase transformation: in particular, we study the effect of

1) changes of temperature T at fixed other parameters;
2) changes of sizes R at fixed other parameters;
3) changes of the initial composition C_0 at fixed other parameters.

13.3.2
Phase Diagram Separation

13.3.2.1 Variation of Temperature T

Let us look at the variations of the dependencies $f(r)$ and $\Delta G(r)$ with T at fixed C_0, R, and other parameters (Table 13.1). Typical temperature-dependent equilibrium fluctuation probabilities $f(r)$ with sizes r and Gibbs free energy dependence $\Delta G(r)$ on sizes r are shown in Figure 13.4 for given sets of the parameters (here, the rule of parallel tangents is used).

Let us start from the single-phase state at high T (Figure 13.2a), and let T decrease. During the decrease of T, the dependencies of $\Delta G(r)$ and $f(r)/f_0$ on r change from the monotonic curves (curves 1 in Figure 13.4) to the nonmonotonic curves with one minimum and one maximum for $r > 0$ (curves 3–5 in Figure 13.4). One can visually compare $\Delta G(r)$ and $f(r)/f_0$. The analysis shows that the minimum of $f(r)/f_0$ corresponds to the maximum − nucleation barrier − of $\Delta G(r)$ (at the so-called critical size of the nucleus). Also, the maximum of $f(r)/f_0$ corresponds to the minimum of $\Delta G(r)$. As a rule, the maximum of $f(r)/f_0$ (minimum of $\Delta G(r)$) corresponds to a two-phase equilibrium state: that is, to the overcritical nucleus of the new phase and ambient parent phase (Figure 13.2b). When T decreases, the minimum of $f(r)$ shifts down and to the left, while the maximum moves to the right and up (Figure 13.4b).

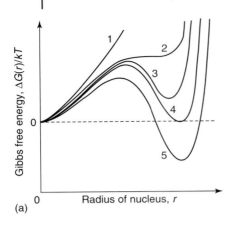 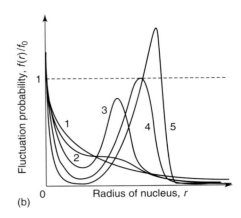

Figure 13.4 Qualitative dependence of Gibbs free energy $\Delta G(r)$ on size (a) and fluctuation probability $f(r)/f_0$ on the radius r of the nucleus (b): (i) for different temperatures T, provided other parameters are fixed; (ii) for different sizes R and fixed other parameters; (iii) for different initial compositions C_0 at fixed other parameters. Explanation is given in the main text. Case 4 represents the separation condition [61].

13.3.2.2 Transition Criterion, Separation Criterion

Of particular importance is the situation where $f(r)/f_0 = 1$ (case 4, Figure 13.4). It corresponds to the separation limit at separation temperature T_{tr}. At T_{tr}, $f(r)/f_0 = 1$, while $\partial f(r)/\partial r = 0$, $\partial^2 f(r)/\partial r^2 < 0$, $r > 0$. According to Equation 13.13, the value of $f(r)$ must be smaller than unity, but the ratio $f(r)/f_0$, which is equal to

$$f(r)/f_0 = \exp\left(\frac{-\Delta G(r)}{kT}\right)$$

may be larger than unity. This criterion will be called *separation criterion* further. This criterion coincides with the transition condition

$$\Delta G(r) = 0, \quad \partial \Delta G(r)/\partial r = 0, \quad \partial^2 \Delta G(r)/\partial r^2 > 0$$

The value of T_{tr} depends on R of the particles at fixed other parameters. In principle, depending on sizes and composition, the transition temperature may vary from a few to hundreds of kelvins.

The analysis of the distribution profile, Equation 13.13 and Equation 13.8, indicates the existence of three characteristic facts: single-phase states (curves 1 in Figure 13.4); two-phase states of the nanoparticle (minimum point on curve 5 in Figure 13.4a and/or maximum point on curve 5 in Figure 13.4b); and metastable states (minimum point for $r > 0$ on curve 3 in Figure 13.4a and/or maximum point on curve 3 in Figure 13.4b). One can obtain all possible states by changing not only the temperature of the particle (at fixed other parameters) but also the sizes of particle R (at fixed other parameters) as well as by changing the initial composition C_0 (at fixed other parameters).

13.3.2.3 Varying R

Let us now look at the influence of R on the solubility at fixed T, C_0, and other parameters. As R increases, the maximum of the fluctuation probability (Equation 13.13) appears and shifts right and up, and the minimum of $f(r)$ shifts down and to the left (Figure 13.4b). By means of the same reasoning as above, one deduces the existence of a separation transition criterion at the size of the nanosystem R_{tr}. This value R_{tr} is a function of the degree of supersaturation (temperature and initial composition).

13.3.2.4 Varying C_0

Let us again fix all parameters of the system and consider the changes of initial composition C_0 at fixed T and R. To investigate this, one should make the above-mentioned thermodynamic analysis using the separation criterion for small particles. As pointed out previously, it is qualitatively similar to the results of temperature or size changes discussed earlier. As the supersaturation C_0 becomes larger, the fluctuation probability, $f(r)$, reveals a maximum which displaces toward large sizes (Figure 13.4b). Furthermore, one can find a very interesting peculiarity: namely, the existence of "critical supersaturation" shown below.

Let us now compare the size-dependent phase diagram with that of the bulk material.

13.3.2.5 Phase Diagram

As a first step, we will look at the equilibrium compositions in the parent phase $C_{p,\infty} \equiv C_p(R \to \infty)$ and in the new phase $C_{n,\infty} \equiv C_n(R \to \infty)$, corresponding to the full separation in the infinite bulk material (at every fixed temperature T). The conditions for optimal concentration $C_{n,\infty}$ and $C_{p,\infty}$ and solubility limits can be found according to the common tangent rule:

$$\Delta g(C_{p,\infty}) + \frac{\partial \Delta g(C)}{\partial C}\bigg|_{C_{p,\infty}} (C_{n,\infty} - C_{p,\infty}) = \Delta g(C_{n,\infty})$$

In the symmetric case ($\varphi_{BB} = \varphi_{AA}$, $\omega_A = \omega_B$):

$$\frac{\partial \Delta g(C)}{\partial C}\bigg|_{C_{p,\infty}} = \frac{\partial \Delta g(C)}{\partial C}\bigg|_{C_{n,\infty}} = 0, \quad \Delta g(C_{p,\infty}) = \Delta g(C_{n,\infty})$$

This leads to the transcendent equation:

$$ZE_{mix}(2C_{p,\infty} - 1) + kT\ln\frac{C_{p,\infty}}{(1 - C_{p,\infty})} = 0, \quad C_{n,\infty} = 1 - C_{p,\infty}$$

For our set of parameters (Table 13.1), this transcendent equation has temperature-dependent roots, $C_{p,\infty}$ and $C_{n,\infty}$. For regular solutions, the critical temperature, T_c, at which the separation is impossible in a bulk material, corresponds to the conditions:

$$\partial \Delta g(C)/\partial C = \partial^2 \Delta g(C)/\partial C^2 = 0, \quad C_{n,\infty} = C_{p,\infty} = 0.5$$

In our case, $T_c = 580$ K. The usual cupola-shaped separation diagram T–C for an infinite matrix is given in Figure 13.5 and pointed by symbols "♦".

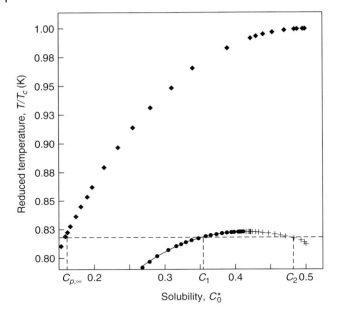

Figure 13.5 Size-dependent state diagram: reduced temperature (T/T_c) versus solubility C_0^* in the interval from 0 to 0.5. Points "♦" show the usual cupola-shaped diagram of a binary system for the case of separation in an infinite system (when $C_0^* = C_{p,\infty}$). Points "●" represent the first stable solution of Equation 13.12 and indicate the cupola-shaped diagram of a small particle at a fixed radius R (the line connecting the experimental points is plotted for visualization of the cupola shape). Points "+" correspond to the second stable solution of the system of equations (Equation 13.12) and continue the size-dependent phase diagram into the high-temperature range. The parameters are given in Table 13.1. Compositions C_p and C_n after the separation are not presented in the figure.

Let us remind ourselves that the usual cupola-shaped equilibrium diagram determines the solubility as well as the equilibrium compositions ($C_{p,\infty}$ and $C_{n,\infty}$), as a result of separation by one line. In fact, in bulk material, the solubility is equal to the equilibrium composition $C_{p,\infty}$. In the following, we will see that in nanosystems the solubility does not coincide with the equilibrium composition after separation (Sections 13.4–13.6). Moreover, one needs to reinterpret the size-dependent separation diagram for nanosystems. Note that in our present analysis we do not treat spinodal decomposition and how it is modified in nanoparticles. This question was discussed, for example, in [64].

13.3.2.6 Size-Dependent Diagram and Solubilities in Multicomponent Nanomaterials

Let us now study phase transitions in a binary nanosystem and plot the corresponding state diagram. Let us fix R and C_0, and let us vary T until T_{tr} is reached. Then we change only the initial composition C_0 at fixed R and again find the new transition temperature T_{tr} (here the rule of parallel tangents is used). The results are shown in Figure 13.5 by symbols "•" and "+". The conclusion of such a procedure is as follows.

According to the rule of parallel tangent construction and our separation criterion for small particle (case 4 in Figure 13.4), one can find the optimal composition C_p ($C_p \equiv C_p(R, T)$) of the parent phase corresponding to the two-phase state condition $f(r)/f_0 = 1$ (or stable $\Delta G(r)$ minimum). Thus, we have three limiting points for the chosen criterion [65]:

1) initial composition as the limit solubility $C_0^*(R, T)$ (further C_0^*) of one component in another (B in A);
2) optimal composition of the depleted ambient parent phase C_p after the separation;
3) optimal composition of the new-born phase C_n as the result of separation.

Let us note that the separation criterion in the regular solution leads to some unexpected result, namely, multiple values for the solubility C_0^* (and respectively multiple equilibrium configurations of the system (Figure 13.2b) at the same sets of initial parameters). So, for simplicity we restrict the comparison with the bulk case within the concentration limits $0 < C_0 < 0.5$ in Figure 13.5.

Let us consider the influence of finite sizes R on the solubility C_0^* change at fixed temperature T (say, $T/T_c = 0.817$) and the cited above parameters (Table 13.1). The solubility C_0^* in a bulk alloy is determined by the point $C_{p,\infty} = 6.4$ at.% in Figure 13.5. In the nanoparticle, we calculate $C_1 = 35$ at.% ($C_{01}^* \equiv C_1 > C_{p,\infty}$) and $C_2 = 48.3$ at.% ($C_{02}^* \equiv C_2$ is the result of the second solution of Equations 13.11–13.12, second critical solubility). This means that the bulk alloy with $C_0 < C_{p,\infty}$ is thermodynamically stable with respect to phase separation. For the chosen temperature $T/T_c = 0.817$ in the $C_{p,\infty} < C_0 < 0.5$ interval, the bulk alloy is unstable, and it will be separated into the new phase of composition $C_{n,\infty} = 1 - C_{p,\infty} = 95.6$ at.% and the parent phase of composition $C_{p,\infty} = 6.4$ at.%. At the same time, the small particle with the same initial concentration C_0 ($C_0 < C_1$ and $C_0 > C_2$) will not be separated, but the particle with composition $C_1 < C_0 < C_2$ will be separated into the new phase and parent phase. Hence, the decrease of the size of a system leads to the increase of the solubility C_0^* (Figure 13.5).

13.3.2.7 Critical Supersaturation

It appears from our analysis that the limit of solubility C_0^* in small particles does not coincide with the equilibrium composition C_p after the separation. This difference between the limiting mean mole fraction of component B in an initially saturated alloy (or solubility – concentration corresponding to the separation criterion) and optimal (or equilibrium) concentration in the parent phase after the separation was called *critical supersaturation* [65, 66]. Here, the difference $\Delta C^* = C_0^* - C_p$ is the "critical supersaturation." The effect of critical supersaturation is that the separation is possible only (at some fixed temperature and size) if the supersaturation $\Delta C = C_0 - C_p$ is larger than ΔC^*. If the supersaturation $\Delta C < \Delta C^*$, then nucleation and separation are impossible. It is also seen that the shape of phase diagram is changed in nanosystems as well.

13.3.2.8 Concluding Remarks

The classical nucleation theory uses macroscopic arguments to estimate the Gibbs free energy required to form a new phase. Owing to the competition between bulk driving force and surface terms, the Gibbs free energy required to form a nucleus of a new phase goes through a maximum, the so-called nucleation barrier. This maximum is reached at a size called the *critical nucleus size*. As was pointed out earlier, the modification of classical nucleation theory for the case of nucleation in binary and multicomponent nanoparticles takes into account the depletion effect. The depletion results in the existence of system's phase transition temperature T_{tr} and size R_{tr} (which is not a critical nucleus size and not the system's critical size R^*). For sizes of the system smaller than a size R_{tr}, the Gibbs free energy of the system is monotonically increasing. For sizes larger than this transition size, the Gibbs free energy of the system presents the classical form with one maximum defining the nucleation barrier and a minimum corresponding to decomposition (two-phase state). Therefore, there exist intermediate situations with metastable minimums of Gibbs energy higher than the one at the initial state. In fact, the condition at which the Gibbs free energy dependence on size of a new phase becomes nonmonotonic with maximum and zero second minimum is taken here as the phase transition criterion. Theoretically speaking, the corresponding value of the probability of transforming is expected to be equal to unity. It was found also that the decrease of the nanosystem's size would increase the nucleation barrier.

13.4
Diagram Method of Phase Transition Analysis in Nanosystems

For a simple illustration of the presented results for binary DSs and visualization, let us use the diagram method of phase transition analysis in nanosystems commonly accepted by experimenters for the interpretation of phase transitions. By doing this, we show that phase diagrams in nanosystems are not only *shifted* but are also *split* [61, 65, 66], implying the reconsideration of such basic concepts as phase diagram, solubility curve, and so on. This is the topic of the present section, and it is aimed at describing the fundamental differences between phase diagrams for bulk and nanomaterials related to the non-negligible depletion of nanosystems. Here, we restrict ourselves to the cases of melting and freezing of compound systems, associated with the corresponding change of compositions. First, we remind ourselves the usual reasoning about size effects, based on surface energy input to effective Gibbs energy per atom of the nanoparticle. Then we demonstrate the effect of *splitting*. Finally, we propose a possible way of reformulation of some basic concepts of phase transformations in binary and multicomponent nanosystems.

13.4.1
Gibbs's Method of Geometrical Thermodynamics

As discussed above, size effects in phase transformations are well known in physics and chemistry first of all as effects of the *shift* of phase equilibrium in small particles depending on their size. The corresponding explanation is generally related to the additional energy of the external surface, shifting the Gibbs free energy *per atom* (and hence, shifting the phase equilibrium) by a value inversely proportional to the particle size. This is equivalent to treating the additional energy under a curved surface due to Laplace tension.

Let us now briefly discuss the effect of depletion on such a *shift* from the thermodynamic point of view [61, 67, 68]. Let us first look at the problem of a liquid–solid transition in bulk materials. Figure 13.6 determines the "cigar"-type solubility behavior (lens-type liquidus/solidus diagram like in Au–Ag, Cu–Ni, Ge–Si, or Nb–W systems discussed qualitatively in Section 13.6). The driving force of transformation and solubility limits are usually determined in the framework of Gibbs geometrical thermodynamics [62, 63] by assuming a constant concentration of the parent phase. That is, classical thermodynamics states that, at fixed temperature T and average composition x between the points $C_S(\infty)$ and $C_L(\infty)$ of common tangent, the material is a mixture of solid and liquid phases, each with compositions $C_S(\infty)$ and $C_L(\infty)$ (Figure 13.6). It is worth noting that this theory says nothing about the dimensions of the solid and liquid particles (if any) present

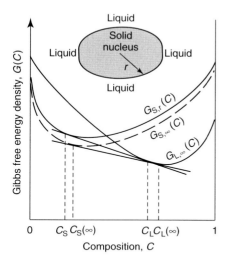

Figure 13.6 Effect of size on equilibrium state and solubility limits found by the common tangent method. Curves $G_{S,\infty}(C)$ and $G_{L,\infty}(C)$ characterize the energy density dependence on composition for solid and liquid phases in bulk form, respectively. $G_{S,r}(C)$ is the Gibbs free energy of the solid nucleus shifted with respect to $G_{S,\infty}(C)$ due to Laplace pressure. C_S and C_L are new solubility boundaries. Explanation is given in the text. A schematic representation of configuration of the system is shown at the top of the figure.

in the material. In other words, according to Gibbs method, the common tangent is being constructed for the two $G(C)$-curves for two coexisting phases, where $G(C)$ is the Gibbs potential per atom and C is the atomic concentration. The tie line or the so-called conode (line segment between two tangency points of the common tangent) connects equilibrium compositions of two phases, corresponding to solubilities. The necessary condition for such construction is that the formation of a new phase inside an old phase does not change the $G(C)$-curves for both phases. It is true only if both new and old phases have macroscopic sizes. If a nanoparticle (nucleus) is formed inside the bulk old phase, one should shift the $G(C)$-curve for the new phase by the product of Laplace pressure of newly formed interface and atomic volume. However, if the old phase is also of nanometric size, the above-mentioned construction becomes totally invalid. At least, just an additional shift of the $G(C)$-curve for the old phase and the corresponding shift of the $G(C)$-curve for the new phase due to Laplace pressure of the external surface are not enough to describe solubility and even the phase equilibrium itself. As shown above, the general peculiarity of nucleation in nanosystems is that the *initial* stoichiometry of the parent phase does not coincide with the stoichiometry of the parent phase *after* the phase transition [59, 61, 69]. So, we cannot use quantitatively the analysis based on the usual method of geometrical thermodynamics.

13.4.2
Nucleation of an Intermediate Phase

The interrelation between the effects of size, nucleation, phase transition, and depletion in first-order phase transitions has been studied elsewhere for the cases of ideal and regular solutions and parabolic approximations [57, 59, 61]. To generalize these results, let us consider other thermodynamic models for the new and parent phases. We will consider the formation of a spherical nucleus of an intermediate phase inside a spherical particle of the supersaturated solid solution at an initial concentration C_0, when the composition of a new phase is assumed fixed and known.

13.4.2.1 Phase Transition Criterion
The condition that the Gibbs free energy of the total system for a new (two-phase) configuration (Figure 13.3b–d) is smaller than for the starting (single-phase) one (Figure 13.3a) is taken as the phase transition criterion.

13.4.2.2 Model of Intermediate Phase
Without loss of generality, let us choose the model of the new phase as a "line" (strictly stoichiometric) intermediate phase with composition $C_n = C_1 = 0.5$, and exclude the elastic contributions to the Gibbs energy. The parent phase in the vicinity of the phase transition points will be described by the ideal solution law and will be also denoted as the α-phase. In fact, the model of regular solution would be more reasonable since the existence of intermediate phases usually correlates with negative mixing energy. Yet, for simplicity, below we restrict ourselves to the

case of negligible mixing energy. Hence, for the formation of a two-phase system, where the new phase has a nonzero driving force of transformation, one can write:

$$\Delta g_1(T, C) = \Delta g_1 + \alpha kT, \quad (C = C_1) \tag{13.14}$$

$$\Delta g_0(T, C) = \Delta g_0 + kT\{C\ln C + (1 - C)\ln(1 - C)\} \tag{13.15}$$

Here, $\Delta g_0(T, C)$ is the Gibbs free energy (per atom) of the parent supersaturated phase, $\Delta g_1(T, C)$ is the Gibbs free energy of formation of the new phase 1, and $\alpha > 0$ is a dimensionless parameter determining the temperature-dependent behavior of the driving force $\Delta g_1(T, C)$ for the new phase. One can say that the new phase has infinitely a thin composition dependence. The linear approximation for $\Delta g_1(T)$ corresponds to experimental situation [70, 71] and is commonly used for determining phase diagrams [17].

To guarantee the decrease of the driving force with increasing temperature, we have used α, which leads to metastability at low temperatures and corresponds to the case of common tangent at certain temperature T^*:

$$\Delta g_0(T^*, C) + \left.\frac{\partial \Delta g_0(T^*, C)}{\partial C}\right|_{C_0} (C_1 - C_0) = \Delta g_1(T^*, C_1)$$

13.4.2.3 Separation in a Macroscopic Sample: Equilibrium State Diagram

Let us take typical values for metallic materials:

$$n = 7 \times 10^{28} \text{ m}^3, \quad \sigma = 0.15 \text{ Jm}^{-2}, \quad \Delta g_1 - \Delta g_0 = -3 \times 10^{-20} \text{ J},$$
$$C_1 = 0.5, \quad \alpha = 2.4$$

for the chosen set of parameters, $T^* \approx 710$ K. First of all, we will find the equilibrium composition in the parent phase C_p^∞ corresponding to the full separation in an infinite matrix (at every fixed temperature T). The second equilibrium concentration is equal to C_1 (which is known). The condition for optimal concentration C_p^∞ and solubility limits C_0 can be found according to the common tangent rule for a conode:

$$\Delta g_0(T, C_p^\infty) + \left.\frac{\partial \Delta g_0(T, C)}{\partial C}\right|_{C_p^\infty} (C_1 - C_p^\infty) = \Delta g_1(T, C_1) \tag{13.16}$$

This is a transcendent equation, which for the case $C_1 = 0.5$ (which we will consider below), after easy algebra this transcendent equation, may be rewritten as a quadratic equation and it has the root:

$$C_p^\infty = C_1\left(1 - \sqrt{1 - 4\exp\left(2\left(\frac{\Delta g_1 - \Delta g_0}{kT} + \alpha\right)\right)}\right)$$

The second root

$$C_p^\infty = C_1\left(1 + \sqrt{1 - 4\exp\left(2\left(\frac{\Delta g_1 - \Delta g_0}{kT} + \alpha\right)\right)}\right)$$

appears because of the symmetry of the problem with respect to $C_1 = 0.5$ and corresponds to an initial concentrations $C_0 > 0.5$; so we do not consider this one

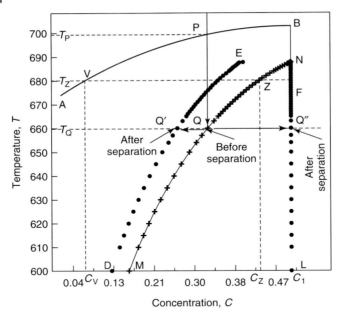

Figure 13.7 Size-dependent state diagram: temperature–composition. Continuous line AVHPNF: cupola-shaped diagram of a binary system for the case of separation in an infinite system, which is found analytically when the new phase is strictly stoichiometric; MQZNL: cupola-shaped diagram of small particle at fixed radius R (line connecting the experimental points "+" is plotted for visualization of cupola shape); the set of points DQ'ENL determines the result of separation: C_p^{opt} (points DQ'E) and $C_1 = 0.5$ (points NL) in a small particle. The parameters are $R = 10^{-8}$ m, $n = 7 \times 10^{28}$ m^{-3}, $\sigma = 0.15$ Jm^{-2}, $\Delta g_1 - \Delta g_0 = -3 \times 10^{-20}$ J, $C_1 = 0.5$, and $\alpha = 2.4$.

further. The cupola-shaped separation diagram T–C for an infinite matrix shifted toward strong stoichiometric phase is presented in Figure 13.7 (case AVPBNF).

13.4.2.4 Separation in DSs: Size-Dependent Phase Diagram

Let us now specify the size of our binary system as $R = 10^{-8}$ m, and look at the influence of the system's size on the change of the T–C state diagram. Assume that the radius R of the nanoparticle, pressure, and temperature remain constant and the nucleus is formed inside the nanoparticle (Figure 13.2b). The expression for the change of Gibbs free energy after nucleation of phase 1 nucleus is

$$\Delta G = nV_1 \Delta g_1(C_1) + n(V - V_1)\Delta g_0(C_P) - nV\Delta g_0(C_0) + 4\pi r^2 \sigma \quad (13.17)$$

Here, again, $V_1 = 4\pi r^3/3$ is the volume of the nucleus, and $n = n_1$ and the compositions obey the rule Equation 13.6.

We repeat the analysis of variation of C_0 for the definition of the extremes of the $\Delta G(r, R)$-function in Equation 13.17 by direct calculation of ΔG for all reasonable sizes of the nucleus (with small step). At that, the transition (separation) criterion (case 4 in Figure 13.4) $\Delta G(r) = 0$, $\partial \Delta G(r)/\partial r = 0$, and $\partial^2 \Delta G(r)/\partial r^2 > 0$ $(r \neq 0)$

hold. According to this procedure, one can find the optimal composition C_p^{opt} of the parent phase corresponding to a stable $\Delta G(r)$ minimum. Thus, we have two limiting points (the third one, $C_1 = 0.5$, is determined from the initial condition) for the chosen criterion: initial composition as the limit solubility, C_0^*, of one component in another (B in A) and optimal composition of the depleted ambient parent phase C_p^{opt} as the result of separation. The size-dependent separation diagram is presented in Figure 13.7 (case MQZNL).

13.4.2.5 Influence of Size on Limiting Solubility

Consider the influence of sizes on the solubility C_0^* change at the fixed temperature T_V ($T_V = T_Z$ in Figure 13.7) and cited above parameters. The solubility C_0^* in a bulk alloy is determined by the point C_V in Figure 13.7, and in small particles by the point C_Z ($C_Z > C_V$). For example, at $T_V = 680$ K the solubility in the bulk alloy will be $C_V = 6.4$ at.% and in a small particle it will be $C_Z = 43$ at.%. In other words, the bulk alloy in the concentration interval $C_0 < C_V$ will be thermodynamically stable with respect to phase separation. Within the $C_V < C_0 < C_1$-interval, the bulk alloy is unstable, and will be separated into a new phase of composition C_1 and the parent phase of composition C_V. A small particle with the same initial concentration $C_0 < C_Z$ will not be separated, but the particle with composition $C_Z < C_0 < C_1$ will be separated into a new phase of composition C_1 and the parent phase of a composition determined by the corresponding point of series DQ'E in Figure 13.7. Hence, the decrease of the size of the particle yields the increase of the solubility of the components.

Notice that, in contrast to the analysis for cupola-shaped separation diagram of bulk alloys, one needs to interpret the size-dependent separation diagram for small particle differently. This is clear from the following reasons. Indeed, the usual cupola-shaped equilibrium diagram determines the solubility as well as equilibrium compositions as a result of separation by one line AVPBNF. For a small particle, the equilibrium diagram becomes doubled (and shifted and size dependent). So, instead of one line, one needs to deal with two lines, namely, line MQZNL of solubility C_0^* and line DENL of separation results: C_p^{opt} and C_1. It appears from depletion effect (splitting, $C_0^* \neq C_p^{opt}$). We define the "critical supersaturation" $\Delta C^* = C_0^* - C_p^{opt}$ (QQ' in Figure 13.7). The critical supersaturation ΔC^* is the difference between the limit of solubility (concentration corresponding to the separation criterion) and optimal (or equilibrium) concentration in the parent phase *after* the separation. At low temperatures, the curves MQZN and DEN shift to low compositions and gradually fuse into one curve (MD < QQ' < NE) as a result of the large bulk driving force of transformation. Thus, the decrease of temperature leads to the decrease of the critical supersaturation ΔC^*.

13.4.2.6 Influence of Size of an Isolated Particle on the Phase Transition Temperature

We consider the influence of sizes on the separation temperature. We start from the single-phase particle at high T (PQ in Figure 13.4) and then decrease T at fixed R and C_0. As compared with the bulk phase transition temperature T_P, in a nanoparticle it is attained at $T_Q < T_P$ for the above chosen set of parameters at $C_0 = 0.33$, $T_Q = 660$ K, and $T_P = 699$ K. Thus, the decrease of size leads to the decrease of temperature of the phase transition. Under separation temperature T_{tr}, we will again understand the temperature at which the Gibbs free energy dependence on radius of a new phase becomes nonmonotonic with maximum and zero second minimum: that is, which corresponds to the separation criterion (Figure 13.8).

13.4.2.7 Concluding Remarks

When describing the separation of a supersaturated solution in a small Particle, one must distinguish between the solubility limit (maximal concentration of impurity before separation) and the equilibrium concentration of the depleted parent phase after separation, the difference being called a *critical supersaturation*. Critical supersaturation is the thermodynamic characteristic depending on temperature and size. So, one can call it a size-dependent "critical supersaturation." The solubility in a small particles increases even without accounting for interactions with the boundaries of the particle (if the density change is neglected).

The physical reason of such peculiarity consists of two factors. The first one is the conservation law effect; the second one is that the separation in a small particle may start only by nucleus formation, the volume of the nucleus being not

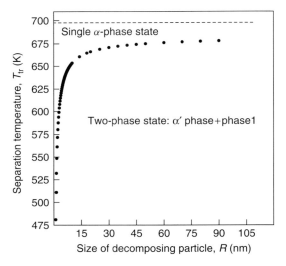

Figure 13.8 Separation temperature T_{tr} dependence versus size of decomposing particle [65]. The horizontal broken line characterizes the separation temperature in an infinite matrix ($T_\infty = 698$ K). Points "●" show the results of Gibbs free energy analysis in a small particle at separation condition: $\Delta G(r) = 0$, $\partial \Delta G(r)/\partial r = 0$, $\partial^2 \Delta G(r)/\partial r^2 > 0$ ($r \neq 0$). The solid line is the approximation function $T_{tr} = T_\infty \cdot (A + B/R)$, where $A = 0.9656$, and $B = -2.894 \times 10^{-10}$ m. Explanations and parameters are given in the main text. For example, at $R = 7 \times 10^{-9}$ m (and the same set of parameters), it yields $T_{tr} = 644$ K.

small compared to the total system's volume. The model gives the decrease of transformation temperature with the decrease in size.

Note that the equilibrium condition in a small particle means equal depths of two pits (in Gibbs free energy dependence on radius r), which are separated by a thermodynamic barrier. In the case of a small particle, this barrier is of the order of the nucleation barrier and may be not high ($<50kT$). It means that the phase equilibrium in an ensemble of small particles will correspond to a statistical distribution (Figure 13.4) at which one fraction of particles will be in the single-phase state and the other fraction in the two-phase state. The corresponding kinetic analysis is discussed in Section 13.7.

13.5
Competitive Nucleation and Growth of Two Intermediate Phases: Binary Systems

As was shown in the previous section, confined volumes of the decomposing alloy can change both the rate and the very result of decomposition. If two phases have a driving force to nucleate, the competition between them is inevitable: first of all, competition for the necessary constructing material. One can expect interesting possibilities if phase 2 has both a smaller bulk driving force of transformation and interface energy [58, 59]. One example of competitive formation of new phases in nanovolumes is the precipitation of coherent precipitates of the metastable Al_3Ni ordered phase instead of the stable Al_1Ni_1 in a supersaturated solid solution of AlNi [72, 73]. In the mentioned case, phase 1 is the Al_1Ni_1 and phase 2 is Al_3Ni.

The question of appearance of different crystalline modifications in precipitating supersaturated solutions or crystallization from the melts was formulated by Ostwald at the end of the nineteenth century based on empirical research. According to Ostwald's rule, instead of the thermodynamically most stable phase modifications, the nearest possible metastable ones are initially formed [74–76]. In other words, the phase with the smallest free energy difference to the ambient phase has the highest chance to be initially formed [76]. Stranski and Totomanov introduced a theoretical concept for both the fulfillment and the exceptions of Ostwalds' rule when applied to problems of phase formation. According to Stranski and Totomanov, under the competition in the formation of critical clusters of different modifications, which is possible under the given thermodynamic conditions, the maximal nucleation rate determines the dominant appearance of the corresponding structure. As a consequence of this kinetic rule, the maximal value of the nucleation barrier is considered as being responsible for the competition between the different phases to be formed. Obviously, this rule does not work if a system at the stable equilibrium state is under the action of an external force. The latter case is described by the le Chatelier–Brown principle. In the simplest case of polymorphic transformation, the description of phase competition can be performed employing Bublik and Pines's idea (Section 13.2). In the case of a phase transition with composition redistribution, one must take into account the depletion effect, which complicates the analysis (Sections 13.3–13.4).

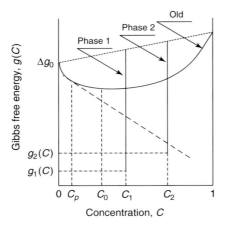

Figure 13.9 $g(c)$: Gibbs free energy (per atom) as a function of composition for "old" and "new" phases; Δg_0 is the isothermal Gibbs free energy density of the α-alloy from pure solid components. C_0 is the initial composition of the parent phase, C_P is the composition of the depleted parent phase (in this diagram the components are implicitly supposed to have the same structure), and C_1 and C_2 are the compositions of the corresponding new phases.

Here we consider the decomposition of a supersaturated binary alloy α (containing A and B components) leading to the formation (when it is not suppressed) of a two-phase system $\alpha' + 1$ or $\alpha'' + 2$, where 1 and 2 are intermediate phases both having a nonzero driving force of transformation (Figure 13.9).

The precipitation develops in nanometric volumes, which can be realized either in nanometric isolated particles or in small spherical regions (of radius R) around nucleation sites in case of simultaneous nucleation at many sites such as in a highly imperfect supersaturated alloy or fast multiple homogeneous nucleation in bulk metallic glasses. Even if we assume that the formation of a new phase has a symmetric form, in principle, different possibilities must be considered. They are represented in Figure 13.10: $\alpha \to \alpha' + 1$, $\alpha \to \alpha'' + 2$, $\alpha \to \alpha''' + 1 + 2$. The detailed analysis of competitive nucleation is presented in [58, 59, 66].

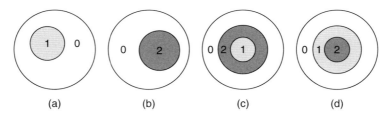

Figure 13.10 Possible modes of nucleation and growth of new phases at the center of a small particle or a region of supply: (a) nucleation of phase 1; (b) nucleation of phase 2; (c), (d) competitive formation of phase 1 and phase 2.

13.5 Competitive Nucleation and Growth of Two Intermediate Phases: Binary Systems

In this section, at every fixed C_p, the Gibbs energy is minimized. Naturally, the following question remains: is such quasi-equilibrium process kinetically possible? The answer to this question can be given by solving the kinetic equations of the Fokker–Planck type or by modeling the corresponding kinetics in an alternative way (Section 13.7). But it must be recalled that in the present section the kinetic aspect is not considered.

To catch the main idea, let the new intermediate phases be line compounds with fixed compositions, C_1 for phase 1 and C_2 for phase 2. The conservation law (at $n_1 = n = n_2$) leads to:

$$C_0 V = C_1 V_1 + C_2 V_2 + C_P(V - V_1 - V_2). \tag{13.18}$$

The expression for the driving force is:

$$\Delta G = nV_1 \Delta g_1 + nV_2 \Delta g_2 + n(V - V_1 - V_2)\Delta g_0(C_P) - nV\Delta g_0(C_0) + \Delta Gs \tag{13.19}$$

According to the configurations represented in Figure 13.10, it follows that

(a) $V_1 = 4\pi r_1^3/3,\quad V_2 = 0,\quad \Delta Gs = 4\pi r_1^2 \sigma_{10}$
(b) $V_1 = 0,\quad V_2 = 4\pi r_2^3/3,\quad \Delta Gs = 4\pi r_2^2 \sigma_{20}$
(c) $V_1 = 4\pi r_1^3/3,\quad V_2 = 4\pi(r_2^3 - r_1^3)/3,\quad \Delta Gs = 4\pi r_1^2 \sigma_{21} + 4\pi r_2^2 \sigma_{20}$
(d) $V_1 = 4\pi(r_1^3 - r_2^3)/3,\quad V_2 = 4\pi r_2^3/3,\quad \Delta Gs = 4\pi r_2^2 \sigma_{21} + 4\pi r_1^2 \sigma_{10}$

$$\tag{13.20}$$

In Equations 13.18–13.20, V_1 and V_2 are the nuclei volumes; V is the volume of the nanoparticle; n is the number of atoms in unit volume; σ_{10} (σ_{20}) is the specific surface energy between phase 1 (2) and parent phase; and $\sigma_{12} = \sigma_{21}$ is the specific surface energy between the new phases.

The parent supersaturated phase is supposed to behave like an ideal solution (Equation 13.15):

$$\Delta g_0(T, C) = \Delta g_0 + kT\{C\ln C + (1 - C)\ln(1 - C)\} \tag{13.21}$$

Here, Δg_0, Δg_1, and Δg_2 are the isothermal Gibbs free energy densities of formation of the corresponding alloys from the pure solid components:

$$\Delta g_0(C) = g_0(C) - Cg_{0B} - (1 - C)g_{0A}, \quad \Delta g_1 = g_1(C_1) - C_1 g_{0B} - (1 - C_1)g_{0A},$$
$$\Delta g_2 = g_2(C_2) - C_2 g_{0B} - (1 - C_2)g_{0A}$$

Therefore, ΔG can be treated as a function of two independent parameters: the size of the new phases and the composition C_P.

The numerical analysis shows that the minimization of ΔG at fixed C_P gives $r_1 = 0$ or $r_2 = 0$. It means that the three-phase configurations $\alpha \rightarrow \alpha''' + 1 + 2$ (cases (c) and (d) in Figure 13.10) appeared to be less favorable than two-phase configurations $\alpha' + 1$ (Figure 13.10a) or $\alpha'' + 2$ (Figure 13.10b). Therefore, dependencies of ΔG are presented only for the regimes described by Equations 13.20a and 13.20b. The results can be gathered into the following three cases:

Case 1

When the interface energies are such that $\sigma_{10} < \sigma_{20}$ with bulk driving force for phase 1 being greater than for phase 2, phase 1 has a lower nucleation barrier and its formation is more favorable in general. That is the usual case of total suppression of the metastable phase 2 by stable phase 1. For example, for $C_1 > C_2$ with the set of parameters $n = 7 \times 10^{28}$ m^{-3}, $C_0 = 0.03$, $kT = 7 \times 10^{-21}$ J, $R = 1.3 \times 10^{-8}$ m, $\sigma_{10} = 0.25$ Jm^{-2}, $\Delta g_1 - \Delta g_0 = -2.2 \times 10^{-20}$ J, $C_1 = 0.6$, $\sigma_{20} = 0.3$ Jm^{-2}, $\Delta g_2 - \Delta g_0 = -1.8 \times 10^{-20}$ J, and $C_2 = 0.5$, the results are shown in Figure 13.11. The system will stop in the absolute minimum state (decomposition into $\alpha' + 1$ at point B).

Case 2

When $\sigma_{20} \ll \sigma_{10}$ with bulk driving force for phase 1 being larger than for phase 2, that is when

$$\left|\Delta g_1 - \Delta g_0 (C_0)\right| > \left|\Delta g_2 - \Delta g_0 (C_0)\right| - \left.\frac{\partial \Delta g_0 (C)}{\partial C}\right|_{C_0} \cdot |C_2 - C_1|$$

nucleation of phase 1 in the nanoparticle is more difficult. It appears that phase 2 will nucleate first (suppressing the formation of phase 1) and the system will stop in a state corresponding to Figure 13.10b ($\alpha'' + 2$). In this case, a small volume

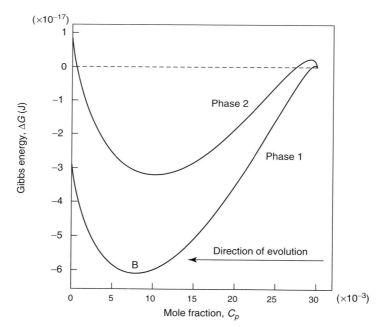

Figure 13.11 Gibbs free energy dependencies on the depletion of the parent phase. Note that the evolution proceeds with decreasing mole fraction C_p in the parent phase (which corresponds to the direction from the right to the left).

helps phase 2 to suppress the precipitation of phase 1. Moreover, it is noteworthy that the absolute minimum of Gibbs energy for phase 2 is reached during the evolution of reaction coordinate before the corresponding value for phase 1.

With the following values of the parameters,

$$C_1 < C_2, \quad n = 6 \times 10^{28} \text{ m}^{-3}, \quad C_0 = 0.02,$$
$$kT = 10^{-20} \text{ J, and } R = 10^{-8} \text{ m}$$

for phase 1:

$$\sigma_{10} = 2.02 \text{ Jm}^{-2}, \quad \Delta g_1 - \Delta g_0 = -4 \times 10^{-20} \text{ J}, \quad C_1 = 0.3$$

and for phase 2, correspondingly,

$$\sigma_{20} = 0.02 \text{ Jm}^{-2}, \quad \Delta g_2 - \Delta g_0 = -3 \times 10^{-20} \text{ J}, \quad C_2 = 0.6 \quad \sigma_{21} = 2 \text{ Jm}^{-2}$$

are taken. The results are presented in Figure 13.12. All cases including the impossibility of phase 1 formation (owing to too small volumes) even in the absence of phase 2 must also be considered. Actually, they have been discussed in Section 13.3 for the case of nucleation of a single phase.

Case 3 or Crossover Regime

When the conditions $\sigma_{20} < \sigma_{10}$ with a driving force for phase 1 being larger than for phase 2 are satisfied, that is,

$$\left|\Delta g_1 - \Delta g_0 (C_0)\right| > \left|\Delta g_2 - \Delta g_0 (C_0)\right| - \left.\frac{\partial \Delta g_0 (C)}{\partial C}\right|_{C_0} |C_2 - C_1|$$

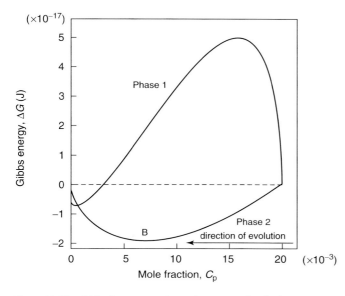

Figure 13.12 Gibbs free energy dependencies on the depletion of the parent phase. The parameters are given in the main text.

the following other situation can be set up. Nucleation remains easier for phase 2 but with a minimum of Gibbs energy for phase 1 being deeper than for phase 2 (but this is for a greater depletion, smaller C_p). Then, a crossover from regime (a) to regime (b) becomes possible (Figure 13.10).

At the first stage of the evolution path of the system, phase 2 should nucleate and grow since the nucleation barrier for phase 2 is smaller than for phase 1. But beyond the crossover point, phase 1 becomes more favorable, so that further depletion beyond this point favors transformation $\alpha'' + 2 \rightarrow \alpha' + 1$. However, this transformation (second stage of evolution) obviously means one more Gibbs energy barrier ΔG_{12} due to the required formation of the interface between phase 1 and phase 2. If this additional barrier ΔG_{12} is small enough, so that the sum of values ΔG_{12} and nucleation barrier for phase 2 will be less than nucleation barrier for phase 1, then the evolution path of the formation of the phase 1 via phase 2 becomes more favorable and phase 2 helps to form the phase 1. On the contrary, the sum of ΔG_{12} and the value of the nucleation barrier for phase 2 will be larger than nucleation barrier for phase 1, ΔG_{12} becomes sufficiently high to hinder the transformation. In this case, the system can reach the metastable state and is able to remain there for a long time. The last case is similar to the case 2 discussed previously.

Let us take, for example,

$$n = 8.5 \times 10^{28} \text{ m}^{-3}, \quad C_0 = 0.04, \quad kT = 8 \times 10^{-21} \text{ J}, \quad \text{and} \quad R = 1.2 \times 10^{-8} \text{ m}$$

for the parent phase and

$$\sigma_{10} = 1.2 \text{ Jm}^{-2} \quad \Delta g_1 - \Delta g_0 = -2.3 \times 10^{-20} \text{ J} \quad \text{and} \quad C_1 = 0.4,$$
$$\sigma_{20} = 0.5 \text{ Jm}^{-2}, \quad \Delta g_2 - \Delta g_0 = -2 \times 10^{-20} \text{ J}, \quad C_2 = 0.5$$

for both new phases. Here, $\sigma_{21} = 0.7 \text{ Jm}^{-2}$, and, in addition, $\sigma_{21} + \sigma_{20} = \sigma_{10}$ holds. As one can see in Figure 13.13, there exists the possibility when first a metastable phase 2 is formed with a lower energy barrier KP < MN but at point S (before the minimum is reached) phase 1 becomes more advantageous.

It is now of interest to study, through a practical example, whether or not a metastable state, at the sample level, can be attained owing to a very high nucleation frequency, and how such a metastable state may compete with the formation of the stable phase.

13.5.1
Application to the Aluminum–Lithium system

The decomposition of a supersaturated solution of Li in Al usually leads to dispersion of coherent precipitates of the metastable phase Al_3Ni within the disordered matrix which hinders the subsequent transformation into the stable phase AlNi [72, 73]. From a qualitative point of view, this behavior takes place mainly because of the coherence between the metastable phase and the matrix yielding a small interface energy. We studied this practical example of the Al–Li alloy using the data from experiments [70–73, 77–79]. Experiments show that Al(Li) alloys with more than 5

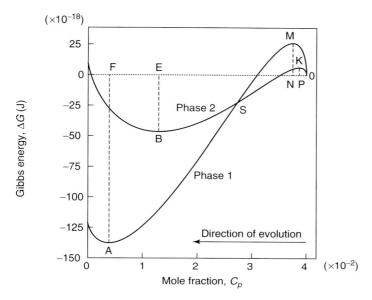

Figure 13.13 Gibbs energy dependence on the depletion of the matrix in the crossover regime. Point S is the point of crossover of energy advantage.

at.% of Li separate in the process of annealing [77–79]. In the annealed state, they are two-phase alloys and contain spherical Al_3Li_1 phase particles with the average size of 2–4 nm. In most investigations, the authors report that the Al_3Li_1 phase is coherent to the matrix (the lattice mismatch is near 0.1%), and Al_1Li_1 phase is a noncoherent one. Thus, most probably the metastable phase Al_3Li_1 appears homogeneously, and phase Al_1Li_1 heterogeneously. Experimentally, the Al_3Li_1 phase is found in the interval of temperatures 20–400 °C, and the Al_1Li_1 phase in the 230–570 °C interval [73].

For industrial applications, Al_1Li_1 is an unwanted phase because of the deterioration of mechanical and corrosion properties of the Al(Li) alloy. The characteristic feature of Al_1Li_1 phase formation is that around the Al_1Li_1 phase regions there is no metastable Al_3Li_1 phase. It is also found experimentally that the elastic modulus increase is well related to the presence of the intermediate metastable Al_3Li_1 phase [70, 77–79]. It would be interesting to define whether it is possible to stabilize the metastable Al_3Li_1 phase because of small particles or some other way. The basic results of this section were published first in [58, 59, 66].

Our analysis leads to the main conclusion that there exists the possibility of formation and total stabilization of the metastable Al_3Li_1 phase instead of the stable Al_1Li_1 phase in small particles of Al −5/50 at.% Li alloys. This suggests that too high an interface energy σ_{21} (between Al_3Li_1 and Al_1Li_1) may hinder the subsequent transformation from the metastable phase Al_3Li_1 to the stable phase Al_1Li_1 in particles of nanometric scale ($R = 2$ nm).

13.5.2
Concluding Remarks

The competitive nucleation and growth of two intermediate phases in volumes of nanometric size is investigated, taking into account the mole fraction depletion of the parent phase. The problem has been solved in the framework of the classical method by representation of the thermodynamic potential in the mole fraction–size space. It is shown that, depending on the particle size and thermodynamic parameters of parent and intermediate phases, there exist the following possible situations: (i) total inhibition of separation, (ii) formation and total stabilization of the metastable phase instead of a stable one, (iii) relative stabilization of the metastable phase with the temporary delay of its transformation into the stable phase, (iv) formation and growth of a stable phase when the metastable phase does not appear at all, (v) formation and growth of the stable phase via the metastable phase. This was applied to the coherent precipitation of metastable Al_3Li ordered phase in supersaturated solid solution Al(Li).

13.6
Phase Diagram Versus Diagram of Solubility: What is the Difference for Nanosystems?

According to classical statistical physics, phase transitions can be rigorously described only in the thermodynamic limit of an infinite system. Nevertheless, "something" happens with nanosystems with the change of temperature, concentration, and size. People treat this "something" using the familiar word "phase transformations." Indeed, as we see below, the usual language of phase transition theory becomes invalid in this case [61, 65, 80]. Thus, we need some new language, and a version of such a language is presented below [81].

The previous thermodynamic analysis clearly showed that after the transition one can find the optimal compositions corresponding to the phase transition. Actually, we always have three characteristic points (corresponding to one conode):

1) *initial* composition C_0 as the limit solubility of one component in another;
2) composition C_p of the depleted ambient parent phase *after* the phase transition;
3) composition C_n of the new-born phase as a *result* of the phase transition.

These compositions are different because of the above-mentioned depletion and finite size of the system. When a nanoparticle separates into two different phases or when the liquid nanoparticle solidifies or when the solid nanoparticle melts, the equilibrium phase diagram appears to be *shifted* (which is familiar) and *split* (which is new), as compared with that of the bulk material. It is also size dependent [80].

Under experimental conditions, one generally deals with a change of temperature, T [12, 69]. It is therefore mandatory to discuss the phase transition in nanoparticles and definitions from this point of view. Qualitatively, the shift of a phase diagram

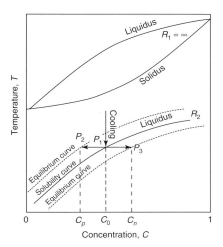

Figure 13.14 Freezing and melting as examples of liquid–solid transition in bulk material ($R_1 = \infty$). Representation of the T–C diagram of a small particle at fixed dimension $R_2 < R_1$ (solubility curve is shown for liquidus). Point P_1 indicates the initial composition C_0 before nucleation, point P_2 characterizes equilibrium composition C_p after the transition, and P_3 shows the optimal mole fraction in the new phase C_n. The conode links the points P_1, P_2, and P_3 corresponding to states with equal Gibbs free energy value and to the lever rule for starting phase and new two-phase (solid–liquid) equilibrium.

of solid–liquid transition and depletion effect are shown (only for liquidus for simplicity) in Figure 13.14.

Let us briefly discuss Figure 13.14 (case $R_2 < R_1$). Here, we start from the liquid particle at high temperature, T, and then decrease T at fixed R and C_0. Since the radius of the nanoparticle is small, the liquidus line is shifted as compared to the bulk one. It is attained at the point $P_1(C_0, T)$. When going to lower T, a solid embryo is assumed to be formed inside the nanoparticle. Thus, starting from the point P_1, the two-phase solid–liquid configuration of the nanoparticle (Figure 13.2b–d) has a minimum of Gibbs free energy lower than that at the initial single-phase liquid state. That is the transition criterion. This event indicates the appearance of the solid part in a nanoparticle, that is, nucleation. In the usual phase diagram methods, the compositions of liquid and solid phases are given by the compositions, at fixed T, of the liquidus and solidus curves. However, the amount of matter in the nanoparticle is limited. So, the corresponding stoichiometry of the solid embryo cannot be attained. Gibbs free energy calculations show that the stoichiometry of the new-born solid phase C_n is determined by the corresponding point P_3 of the equilibrium curve (Figure 13.14). At the same time, the liquid part of the same nanoparticle (i.e., depleted parent phase after phase transition) will have the composition C_p determined by corresponding point P_2 of the equilibrium curve (Figure 13.14). In other words, the conode $P_2 P_3$ does not have the ends on the liquidus and solidus lines (Figure 13.15). The values C_p, C_n, and C_0 are different

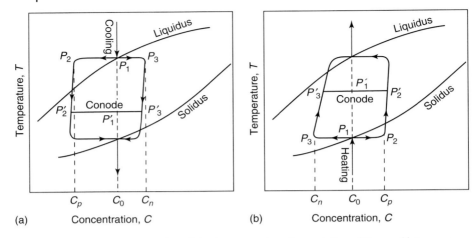

Figure 13.15 Representation of the freezing process for the liquid nanoparticle (a) and melting of the solid nanoparticle (b) at fixed size R and initial composition C_0. Fragments of solidus and liquidus curves of the same particle (a, b). The melting and freezing loops between solidus and liquidus show the evolution of the equilibrium compositions C_n and C_p of the corresponding solid and liquid parts in the transforming nanoparticle. The conodes link the points P'_1, P'_2, and P'_3 corresponding to the lever rule for mass conservation.

because of the depletion effect. When T decreases further, one obtains the loop-like split path (called *below just a loop*) (Figure 13.15).

Consider now the melting of an initially solid nanoparticle [81]. Let us start from the solid particle and go from low to high T (Figure 13.15b). Then melting will be indicated by the appearance of a liquid (two-phase solid–liquid equilibrium). This is the solidus temperature and solidus composition. If one continues with the increase of T, then the value of T at which the same two-phase nanoparticle transforms to complete liquid is taken as the liquidus temperature (end of loop at the up curve, at liquidus in Figure 13.15b). The temperature interval between these two events defines the range of T over which the solid and liquid parts in a given nanoparticle coexist in equilibrium (Figure 13.2b–d). It is worth noting that this loop is valid for a given dimension of the nanoparticle (in fact, a given total number of atoms) and for a given overall composition, C_0. Furthermore, at different initial compositions C_0, the melting and/or freezing loops are different. If one starts from a pure liquid particle and decreases T, then freezing begins (Figure 13.15a). So, the corresponding liquidus temperature (and composition in the starting phase) indicates the presence of a solid part in a nanoparticle. The solidus temperature shows the transition of the same two-phase particle to complete with the solid particle (end of loop at the down curve, at solidus in Figure 13.15a). Also, the loops of melting and freezing processes (Figure 13.15) at the same C_0 and R may be different due to possible differences of the nucleation mechanisms (Figure 13.2) and different energy barrier dependencies on compositions and sizes. The melting loop is symmetrical to the freezing one only if the configuration of the system is the same as during the cooling process. One can see that the solidus and liquidus lines indicate only the start and the end of melting and freezing but not the intermediate states of two-phase equilibrium. These intermediate states are shown by loops in Figure 13.15 and characterize the evolution paths of compositions C_0, C_n, and C_p during the temperature variation. Finally, we have drawn the lines of coexistence (tie lines) at some intermediate T between the solidus and liquidus temperatures in order to show that the lever rule for mass conservation does not work for the liquidus and solidus curves, but it does work for points on the loops: P'_1, P'_2, and P'_3.

While the theory presented here has a rather general character, one can report some specific applications of experimental interest for the theoretical approach developed. The origin of the presented results derives from the variation in energy with size and composition. So, one might expect the confirmation of the experimental results by the calculated ones in case of the size- and composition-dependent material properties. In this respect, isolated nanoparticles of Pb–Bi alloys have a loop-like split diagram and size-induced melting behavior, observed by hot-stage transmission electron microscopy [69]. Loops similar to the presented ones are obtained theoretically for separation of the solid nanoparticle [80]. Our recent analysis for Cu–Ni and Au–Cu binary nanoparticles shows the quantitative predictions [82].

From the previous reasoning, it turns out that some difficulties appear in explanations of state diagrams of nanosystems as well as of such notions as "phase diagram," "solubility," "solidus," and "liquidus." Hence one needs to review them. So, first we should remember the basic well-known notions.

13.6.1
Some General Definitions

Before going further let us remember the general definitions of such notions as "solubility," "solidus," and "liquidus."

13.6.1.1 What are the "solidus" and "liquidus"?
The "liquidus" and "solidus" lines are defined from the phase diagram. The liquidus curve is "in a temperature–concentration diagram, the line connecting the temperatures at which fusion is just completed for various compositions" [83]. Similarly, the solidus curve is the "curve representing the equilibrium between the solid phase and the liquid phase in a condensed system of two components. The points on the solidus curve are obtained by plotting the temperature at which the last of the liquid phase solidifies against the composition, usually in terms of the percentage composition of one of the two components" [83].

13.6.1.2 What is the "Limit of Solubility"?
It is defined in terms of a solution which is a "homogeneous mixture of two or more substances in relative amounts that can be varied continuously up to what is called the *limit of solubility*" [84].

Summarizing the definitions, one can say that, from the common point of view, the solubility (or solubility limits) and equilibrium compositions after the transition in bulk material coincide [63, 83, 84]. They are given by *solidus* and *liquidus*. In nanosystems, this is far from being true. Thus, the notions *solidus* and *liquidus* have to be reexamined when dealing with nanoparticles.

We want to outline here the definition of *solubility diagram* and separate it from the definition of *phase diagram* which is now transformed into the *nanophase diagram*. Strictly speaking, the phase diagram is split, so the definitions should be split as well [81].

13.6.2
Nanosized Solubility Diagram

13.6.2.1 Solubility Limit

Under the solubility or solubility limits, we shall understand "the limiting compositions at which the starting (single phase) state remains without transition into another (two- or multiphase) state." Varying C_0 the solubilities, for different T, we form the solubility curve (the liquidus and the solidus). This notion is applied to melting, freezing, and phase separation as well as to any other first-order phase transition. Let us now introduce the solubility curve notions, namely, the "liquidus" notion related to initially liquid phase and the "solidus" notion with respect to a starting solid phase.

13.6.2.2 Liquidus

Liquidus is the solubility curve for liquid particle. So in our Interpretation, the liquidus curve is "in a temperature–concentration diagram, the line connecting the temperatures at which freezing is just started for various compositions of a starting liquid phase."

13.6.2.3 Solidus

Solidus is the solubility curve for a solid particle. Hence the solidus curve is the "curve representing in a temperature–concentration diagram, the line connecting the temperatures at which fusion is just started for various compositions of a starting solid phase."

13.6.2.4 Nanosized Solubility Diagram

When one deals with the problem of solubility in a nanosystem, one must determine the starting sizes of a given system as well as initial and final configurations (Figure 13.16).

For example, if we want to find the liquidus, we must start only from the initial fully liquid particle as a single phase state, and calculate or make an experiment on the liquid–solid transition. The solubility limit in this case is the limit composition of one of the components at which the liquid–solid transition *starts*. Then, plotting the corresponding points at the $T-C$-diagram one obtains the curve which is the "diagram of solubility in the liquid substance" (liquidus). The same reasoning is

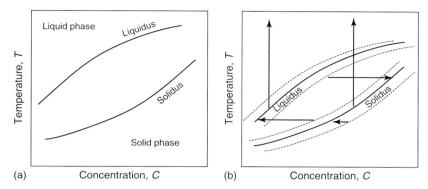

Figure 13.16 Qualitative T–C size-dependent diagrams: (a) solubility diagram; (b) phase diagram. The effect of size increase on the phase diagram of the transforming system is indicated by vertical and horizontal arrows.

applicable to the solidus curve. Therefore, one can define the solubility diagram as follows: the "solubility diagram is the temperature–composition diagram at a fixed quantity of matter of a nanosystem obtained by plotting the solubility curves." Qualitatively, it may be presented as Figure 13.15 but only without the melting and freezing loops. Solubility curves defined by such a procedure will not explain the usual equilibrium conditions (Figure 13.16). The lines of solidus and liquidus may even intersect each other depending on the mechanism of nucleation during the processes of melting and freezing [82]. It is worth noting that the notion "solubility diagram" may be applied both to R–C and T–R diagrams (see further).

13.6.3
Nanosized Phase Diagram

For any fixed T and initial composition C_0, the equilibrium compositions are C_n and C_p after the transition. They do not correspond to the ones given by the classical reasoning on the phase diagram because of the mentioned depletion. For our melting (or freezing) example, when the starting C_0 changes to C_0', at fixed size R, a new composition loop appears. If one plots all of them on one size-dependent T–C-diagram, one will obtain a large number of loops (connecting solidus and liquidus). It is confusing. In Figure 13.15, we showed only two loops for the same fixed C_0 and R.

13.6.3.1 Three Types of Diagrams
In the case of a nanosystem, the size R becomes an external parameter such as temperature or composition. So, in contrast to the bulk case, when analyzing nanosystems one must use three-dimensional diagrams T–C–R, which may be reduced to three types of two-dimensional ones: (a) T–C diagram at fixed R; (b) T–R diagram at fixed C_0; (c) R–C diagram at fixed T (Figure 13.17). It is clear that only the first one has a bulk analogy and may be compared with the T–C-diagram

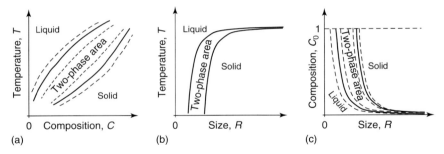

Figure 13.17 Three types of state diagrams of a nanosystem: (a) T–C diagram at fixed R; (b) T–R diagram at fixed C_0; (c) R–C diagram at fixed T.

at infinite size $R_1 = \infty$. The other two diagrams represent the characteristic feature of the nanosystems and have no bulk analogy. Note that one can use the "solubility diagram" notion for R–C and T–R diagrams as well. The analysis of T–R and R–C-diagrams and their peculiarities have been discussed in the previous work [57]. Here we discuss only the T–C diagram.

13.6.3.2 T–C Diagram at Fixed R

Despite the above-mentioned problems, from our point of view, it is convenient to discuss the T–C "phase diagram" at fixed R (or "nanophase diagram") as follows. Namely, under the notation of "phase diagram" we shall understand the "diagram at which the temperature–composition boundaries of the phases of the system found at fixed quantity of matter of the system by transition criterion and plotted as functions of composition on T–C coordinate axes" [81]. From our point of view, the most essential (informative) values of temperatures and compositions correspond to the solubilities and their (corresponding to transition criterion) equilibrium compositions after the transition (Figure 13.14). By varying C_0, one can find that the solubilities gather into a solubility curve, and the corresponding equilibrium compositions after the transition gather into two equilibrium curves. Such defined T–C nanophase diagram indicates the phase fields (single solid-phase state, single liquid phase state, and two-phase states) separated by solubility curves for different initial compositions [83]. May the lever rule of mass conservation be used in such nanophase diagram? Yes, but only for splitting of solubility curve into two equilibrium curves (in bulk case the solubility curve coincides with one of the equilibrium curves). For example, the points P_1, P_2, and P_3 in Figures 13.14–13.15 correspond to the lever rule: $(N - N_n) \cdot |P_1 P_2| = N_n \cdot |P_3 P_1|$, where N_n is the number of atoms in the new phase, the composition interval $|P_1 P_2| = \Delta C = |C_0 - C_p|$, and $|P_3 P_1| = |C_n - C_p|$.

It follows that, when one deals with the phase diagram, one must determine and plot the solubility curves in one T–C nanophase diagram (in the authors' definitions) as well as the final equilibrium compositions, the boundaries of the phases of the already transformed system found by the transition criterion

(Figure 13.16). The $T–C$ nanophase diagram must combine all the mentioned plots and, obviously, it becomes much more complicated in the case of a nanosystem. For example, for solid–solid transition (separation) the $T–C$ nanophase diagram may consist of three curves [57, 81] instead of one; for our example of liquid–solid transition, the $T–C$ nanostate diagram may show six curves (liquidus plus its two equilibrium curves and solidus plus corresponding two equilibrium curves, Figures 13.7, 13.14, 13.16b, and 13.17a). On the other hand, one can define the notion of phase diagram as follows: "the temperature–composition boundaries of the phases of the system found at fixed R and C_0 and plotted as functions of composition." If we fix R and C_0, then we can plot the diagram which will be just a loop (Figure 13.15, but without liquidus and solidus). This loop will give the equilibrium compositions C_n and C_p for all the temperatures (but not for all different values C_0 as well as R). Because of this, such definition looks unreasonable.

13.6.3.3 Varying R

As the size of the particle increases, the solubility and equilibrium curves merge into the usual bulk curves (Figure 13.16b). In the infinite case, one obtains the usual state diagram in which the solubility limits coincide with the equilibrium compositions. It means that the solubility diagram and the phase diagram coincide in the bulk case.

13.6.3.4 Concluding Remarks

Thus, in the present section we used these theoretical and experimental results to modify the notions of "solubility," "solidus," "liquidus," and "vaporous" and outline the new notions of "solubility diagram" and "nanophase diagram." It is worth noting that the introduction of these new notions does not modify the classical thermodynamics and the transition criterion. If one extrapolates the same arguments to boiling and to binary (or multicomponent) liquid and its vapor in a container of fixed nanovolume, one concludes that the solubility and equilibrium curves might be explained in a similar way. It will be the future aim to analyze the liquid–vapor transition in nanovolumes in the framework of the phase diagram approach and the solubility presented here.

13.7
Some Further Developments

13.7.1
Solubility Diagram of the Cu–Ni Nanosystem

We used the mentioned results and the notion of size-induced "solubility diagram" to discuss the particular case of the Cu–Ni nanosystem [82]. Here, we restricted ourselves to the thermodynamic study of melting and freezing. Our additional analysis on the example of the Cu–Ni system demonstrated that for a nanosystem there may exist several lines of solidus (and liquidus), which may be gathered into one curve. The following effects are predicted in [82]: (i) The effective width of the diagram decreases: the two-phase field narrows and even pinches off the solidus and liquidus branches to a line for compositions about 0 and 1; (ii) There exists the possibility of the overlap and intersection of solidus and liquidus at solubility diagram; (iii) The shapes of the solubility curves on the diagram change; (iv) The solubility diagram shifts down to lower temperatures as compared to bulk state diagram (in other words, it is size dependent).

13.7.2
Size-Induced Hysteresis in the Process of Temperature Cycling of a Nanopowder

While the previous sections are devoted to a thermodynamic analysis, here we deal with a powder of nanosized particles subjected to temperature cycling and discuss the kinetics of first-order phase transitions in such a system. When one extrapolates the argument of the limited amount of atoms in a volume to nanomaterials under various external conditions, one concludes that their kinetic behavior should differ from that of the bulk material.

Experimental results on the effect of size on the kinetic behavior of phase transitions in alloys have been obtained in several works [14, 23, 85]. As an example of size-induced effects and, in particular, size-induced hysteresis phenomena in kinetics, one can mention capillary condensation (phase transitions in pores or capillaries; for some recent work in this direction, see [69, 86, 87]). Controllable, continuous, and reversible coexistence of different crystalline and disordered phases in gallium nanoparticles under electron beam excitation has also been demonstrated [88]. Moreover, recent theoretical, experimental, and numerical Monte Carlo results on first-order phase transitions in nanovolumes demonstrate hysteresis phenomena [59, 65, 86, 89–92].

Having in mind Equation 13.3, such hysteresis cannot be explained by thermodynamics alone but has to involve a description of the kinetics of first-order phase transformation as well. In order to develop a theoretical description, we applied the previous thermodynamic approach to the study of the kinetic "decoding" of back and forth transitions during temperature cycling of nanopowders and presented a numerical analysis within the framework of the standard kinetic equation approach [93–95]. As a continuation of our approximation in Section 13.4, we arbitrarily assumed that the new thermodynamically advantageous phase has strict stoichiometry $C_1 = 0.5$. The composition in each initially supersaturated solid particle is equal to C_0 (so that the new phase nucleus has a composition different from that of the parent phase, $C_1 \neq C_0$).

We introduced finite rates of temperature changes and considered the temperature cycling of a nanopowder. First, we start from the single-phase state at high T and let T decrease (cooling). Then we stop the temperature change at some point, when the alloy is quenched into the two-phase region, and make the process direction opposite, that is increase T (heating) at the same rate. One cycle

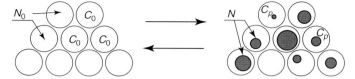

Figure 13.18 Nanopowder under temperature cycling: that is, particles of initial composition C_0 at high temperature (a) and the same particles after the phase transition at low temperature (b). Each nanosized particle is described as isolated and consisting of N_0 atoms, each nucleus as a new phase inside the particle and consisting of N atoms. C_p is the concentration of the ambient parent phase after the nucleation.

refers to the complete forward and backward changes of T from some initial point back to the same point, say from $T = 1100$ K to $T = 800$ K back to $T = 1100$ K. In the case considered here, the temperature T changes with time as a linear function at constant rate, $|dT/dt| = $ const, for cooling and heating. The Gibbs free energy Equation 13.8 or 13.17 may be expressed as a function of time (via the time-dependent temperature): $\Delta G(N, T) \equiv \Delta G(N, t)$. Here, N is the number of atoms inside the newly born nucleus.

We considered the nonequilibrium size distribution function $f(N, t)$, the number of new phase droplets consisting of N structural units at time t. The evolution of the ensemble of clusters formed by nucleation and growth processes is described by the kinetic equation of the Fokker–Plank type [94, 95]. Here we present the case of particles of equal size. The main task of such a kinetic model is to describe the volume fraction ρ of the new phase 1 during the temperature cycling of the isolated nanoparticle ensemble (Figure 13.18). The volume fraction of the new phase ρ is obtained as a function of T and N_0. In this section, we report the obtained kinetic result: size-induced hysteresis.

Our model shows that the width and shapes of hysteresis loops depend on the finite (non-negligible) depletion of the parent phase in nanoparticles of the powder; finite rates of change of the external parameters (temperature changes during annealing and heating); the finite size of the system; finite rates of the transfer of atoms across the parent phase–nucleus interface; and bulk diffusion (existence of free energy barrier for diffusion) [94, 95].

Let us examine the influence of different values of N_0 (effect of different sizes of nanoparticles). The corresponding result, at fixed rate of temperature change, V, and other parameters fixed, is presented in Figure 13.19. We see that larger the system size, the greater is the effective width of the hysteresis loop under the same cycling conditions. The less the size of particles, the less is the effective width of the hysteresis loop. In general, both thermodynamic and kinetic factors result in the existence of hysteresis. Here, one needs to differentiate two cases: cases of slow rates (when the rate of change of the external parameters is slower than the rate of nucleation/growth processes) and cases of high rates [95].

One may expect that the "pure" hysteresis loop presented here will be different from the real experimental one. This is due to the existence of irregularities, defects of any kind, particle size distribution, and diffusion between the particles of a

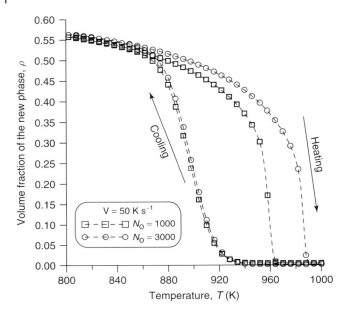

Figure 13.19 Effect of size on hysteresis: new phase volume fraction loops for the fixed rate $V = 50$ K s^{-1} of temperature cycling between 1100 and 800 K and different number of atoms in the particles: $N_0 = 1000$, $N_0 = 3000$.

powder in real experiments which will lead to a shift of the separation and/or nucleation points, barriers, and so on. Furthermore, the previous reasoning is based on the assumption that the nanopowder consists of particles of equal size. Experimental studies usually show that the highly monodisperse particles in a powder have a standard deviation of less than 5–10% [91, 96, 97]. The effect of scatter in particle sizes has also been taken into account and did not lead to any fundamental changes [95].

Recently, the structural transitions without diffusion redistribution (polymorphic transitions) have been analyzed. We presented a general thermodynamic and kinetic analysis for a polymorphic phase transitions in a nanopowder. The corresponding model describes a size-induced change of critical temperatures of phase transitions in small, isolated nanoparticles, and application to the γ-Fe to α-Fe polymorphic phase transformations gives the qualitative similar size-induced results [93, 98, 99].

13.7.2.1 Concluding Remarks

The kinetics of nucleation/growth and separation processes in an ensemble of nanoparticles under time-dependent temperature is presented. Grounding upon our numerical studies, we conclude that under temperature cycling one should observe a hysteresis behavior. Such hysteresis is conditioned by the finite size and depletion effects and is described on the basis of the kinetic equation approach. The model shows that the width of hysteresis loop depends on (i) thermodynamic constraints at the fixed rate of temperature change and other parameters and (ii) kinetic constraints. In particular, as the (a) size of a system, (b) the rate of temperature change, (c) the interphase tension, and (d) the energy barrier for diffusion decrease, the hysteresis loop narrows showing the tendency to disappear. The given approach allows us to differentiate the influence of thermodynamic constraints on hysteresis (related to the Gibbs free energy dependence on size and nucleation barrier) from the influence of kinetic constraints on hysteresis (related to the activation energy for the diffusion across the parent phase–nucleus interface). At high T the hysteresis is related to thermodynamic control of the process, whereas at low T the hysteresis is related to kinetic control of the process.

Besides, so-called bulk metallic glasses often devitrify with a very high nucleation rate, yielding a dispersion of nanocrystals in the amorphous matrix at the first crystallization stage [100–102]. It means that precipitation develops in nanometric volumes, which can be realized in nanometric spherical regions (of radius R) around nucleation sites. One should expect a similar hysteresis behavior in bulk metallic glasses as a result of the depletion and size effects shown in the present paper. The corresponding analysis will be made elsewhere.

It is also worth noting that the shown hysteresis is very similar to magnetic (such as paramagnetic–ferromagnetic) phase transitions under an imposed magnetic field. Such behavior should allow one to use the size-induced hysteresis effect in a way similar to that of magnetic materials for a variety of applications such as high-tech DVD and CD technologies, coatings, shape memory systems, and many other important applications.

13.A
Appendix: The Rule of Parallel Tangent Construction for Optimal Points of Phase Transitions

Equation 13.8 implies that the Gibbs free energy change of the system is a function of two variables: C_n and r. In order to consider the general thermodynamic equilibrium conditions for the function $\Delta G(r, C_n)$, one must write the equations of the first and second derivatives of $\Delta G(r, C_n)$ with respect to the variables. To determine the extreme points of phase transition, one has to solve the following equations:

$$\partial \Delta G(r, C_n)/\partial C_n = 0 \quad (13.A.1)$$
$$\partial \Delta G(r, C_n)/\partial r = 0 \quad (13.A.2)$$

Theorem

The equilibrium concentrations C_n^{opt} in the new phase and in the ambient parent phase C_p^{opt} are determined by the rule of the parallel (not common) tangents drawn from lines of concentration dependencies of Gibbs free energy densities on concentrations for "new" and "old" phases [61].

Proof

Let us find the equilibrium concentrations from the above set of equations at other fixed parameters. Considering the possible set of concentrations C_n in the nucleus, let us find the optimal one from Equation 13.A.1.

According to Equation 13.6, one should find the concentration in the parent phase C_p as a function of concentration C_n and the volume V_n ($V_n = V_1$) at fixed other parameters. Analyzing Equation 13.8, with regard to condition Equation 13.A.1, one easily obtains the relation:

$$n_1 V_n \left. \frac{\partial \Delta g(C)}{\partial C} \right|_{C=C_n} = -n(V' - V_n) \left. \frac{\partial \Delta g(C)}{\partial C} \right|_{C=C_p} \frac{\partial C_p}{\partial C_n} \quad (13.A.3)$$

Applying Equation 13.6, one obtains:

$$\frac{\partial C_p}{\partial C_n} = -\frac{n_1 V_n}{n V_0 - n_1 V_n} = -\frac{n_1 V_n}{n(V' - V_n)}$$

Substituting latter equation into Equation 13.A.3, one obtains the rule of parallel tangent construction for optimal concentrations in the 'new' phase and in the "old" parent phase, which was to be proved:

$$\left. \frac{\partial \Delta g(C)}{\partial C} \right|_{C_n^{opt}} = \left. \frac{\partial \Delta g(C)}{\partial C} \right|_{C_p^{opt}} \quad (13.A.4)$$

Then the expressions obtained for the optimal mole fraction in the parent phase will be:

$$C_p^{opt} = \frac{C_0 n V - C_n^{opt} n_1 V_n}{n V - n_1 V_n} \quad (13.A.5)$$

or

$$C_p^{opt} = C_0 + \frac{n_1 r^3}{n R^3 - n_1 r^3} \left(C_o - C_n^{opt}(r|R) \right)$$

13.A.1
Resume

The rule of parallel tangents gives the optimal path of evolution of a system. This rule is a consequence of thermodynamic approach when Gibbs energy Equation 13.8 is minimized: that is, $\partial \Delta G / \partial C_n = 0$. Solving this variation problem, an optimal depletion for fixed nucleus size is obtained with the help of the condition $\partial^2 \Delta G / \partial C_n^2 > 0$. Of course, the condition of fixed size can be kinetically unattainable, and then one should switch to kinetic analysis of transformation. This rule is valid for a critical nucleus as well as for an equilibrium, two-phase state (supercritical nucleus of optimal concentration and separated parent phase being in the state of minimum of Gibbs free energy) because it is determined from the extremum condition for Gibbs free energy change of the system. One can show that in the extreme case of separation (not nucleation) in an infinite matrix, the equal tangents become common and the driving force will be determined by the well-known rule of common tangents.

In the limiting case ($R \to \infty$), the rule of parallel tangents allows us to find the concentrations in the critical nucleus and matrix correspondingly at once on the phase diagram $\Delta g(C) - C$. Here, the composition of the parent phase stays unchanged C_0. In the case of a composition-dependent surface energy, the contribution before drawing the tangents for critical nucleus and matrix or for equilibrium two-phase state ("new" phase plus parent phase) on the diagram $\Delta g(C) - C$ one must solve Equation 13.A.2 first, and then solve the Equation 13.A.1. Here, the rule Equation 13.A.4 of parallel tangents will be substituted by composition-dependent surface energy contribution:

$$\left.\frac{\partial \Delta g(C)}{\partial C}\right|_{C_n^{opt}} + \frac{1}{3nr}\left.\frac{\partial \sigma}{\partial C}\right|_{C_n^{opt}} = \left.\frac{\partial \Delta g(C)}{\partial C}\right|_{C_p^{opt}} \tag{13.A.6}$$

In the general case of size-composition-dependent surface energy contribution for equilibrium two-phase state, one must solve the above given system of equations with complementary parameters and terms $\partial \sigma / \partial r$ and $\partial \sigma / \partial C$. Furthermore, the rule may be applied to a multicomponent system as well when a new phase is not determined by strong stoichiometric composition: that is, there exists the solubility interval on the diagram Gibbs free energy density–concentration ($\Delta g(C) - C$). As was mentioned before in the presented case, Equation 13.A.4 is applied to nucleation and separation of nanoparticles in which the composition of the new phase is a function of size.

References

1. Buffat, Ph. and Borel, J.P. (**1976**) *Physical Review A*, **13**, 2287.
2. Couchman, P.R. and Jesser, W.A. (**1977**) *Nature*, **269**, 481.
3. Gleiter, H. (**1995**) *Nano-structural Materials*, **6**, 3.
4. (**1999**) in *Nano-technology* (ed. G. Timp), Springer-Verlag, New York.
5. (**2004**) in *Springer Handbook of Nano-technology* (ed. B. Bhushan), Springer-Verlag, Berlin.
6. Petrov, Y.I. (**1982**) *Physics of Small Particles*, Science, Moscow (in Russian).
7. Lauterwasser, C. (**2005**) *Small Sizes that Matter: Opportunities and Risks of Nano-technologies - Report in Co-operation with the OECD International Futures Programme*, Allianz Center for Technology.
8. Electronic resource – CNRS Institut des Sciences Chimiques Seine-Amount (**2004**) *The oldest known nanotechnology dates back to the 9th century*, New Materials International, 22 May. http://www.newmaterials.com/customisation/news/nanotechnology/the_oldest_known_nanotechnology_dates_back_to_the_9th_century.asp
9. Robbins, M.O., Grest, G.S. and Kremer, K. (**1990**) *Physical Review B*, **42**, 5579.
10. Nagaev, E.L. (**1992**) *Uspehi Fizicheskih Nauk (Progress in Physics Sciences)*, **162**, 50 (in Russian).
11. Liang, L.H., Liu, D. and Jiang, Q. (**2003**) *Nano-technology*, **14**, 438.
12. Gladgkikh, M.T., Chigik, S.P., Larin, V.N. et al. (**1988**) *Doklady Academii Nauk SSSR (Reports of the Academy of Sciences of USSR)*, **300**, 588 (in Russian).

13. Cooper, S.J., Nicholson, C.E. and Liu, J. (2008) *Journal of Chemical Physics*, **129**, 124715.
14. Mei, Q.S. and Lu, K. (2007) *Progress in Materials Science*, **52**, 1175.
15. Morohov, I.D., Petinov, V.I., Trusov, L.I. and Petrunin, V.F. (1981) *Uspekhi Fizicheskih Nauk (Progress in Physics Sciences)*, **133**, 653 (in Russian).
16. Skorohod, V.V., Uvarova, I.V. and Ragulya, A.V. (2001) *Physical-chemistry Kinetics in Nano-structural Systems*, Kiev, Academperiodika (in Russian).
17. Wautelet, M. (2000) *Nano-technology*, **11**, 6.
18. Barmak, K., Michaelsen, C. and Lucadamo, G. (1997) *Journal of Material Research*, **12**, 133.
19. Barmak, K., Michaelsen, C., Vivekanand, S. and Ma, F. (1998) *Philosophical Magazine A*, **77**, 167.
20. Wautelet, M. and Shirinyan, A. (2009) *Pure and Applied Chemistry*, **81**, 1921.
21. Raoult, B. and Farges, J. (1973) *Review of Science Instruments*, **44**, 430.
22. Sattler, K., Muhlback, J. and Recknagel, E. (1980) *Physical Review Letters*, **45**, 821.
23. Baletto, F. and Ferrando, R. (2005) *Reviews of Modern Physics*, **77**, 371.
24. Comnik, Y.F. (1979) *Physics of metal films. Size and structural effects*, Atomizdat, Moskow (in Russian).
25. Denbigh, P.N. and Marcus, R.B. (1966) *Journal Applied Physics*, **37**, 4325.
26. Hutchinson, T.E. (1963) *Applied Physics Letters*, **3**, 51.
27. Bond, W.L., Cooper, A.S., Andres, K. et al. (1965) *Physical Review Letters*, **15**, 260.
28. Chopra, K.L. (1969) *Solid Status Solidi*, **32**, 489.
29. Nepiyko, S.A. (1985) *Physical Properties of Small Metal Particles*, Kiev, Naukova Dumka (in Russian).
30. Bublik, A.I. and Pines, B.Y. (1952) *Dokladu Academiji nayk USSR (Reports of the Academy of Sciences of USSR)*, **87**, 215 (in Russian).
31. Palatnik, L.S. and Komnik, Y.F. (1960) *Fizika Metallov Metallovedenie (Journal of Metals Physics and Metals Science)*, **9**, 374. in Russian).
32. Pawlow, P. (1909) *Zeitschrift für Physikalische Chemie*, **65**, 1.
33. Couchman, P.R. and Ryan, C.L. (1978) *Philosophical Magazine A*, **37**, 369.
34. Nanda, K.K., Sahu, S.N. and Behera, S.N. (2002) *Physical Review A*, **66**, 013208.
35. Sun, C.Q., Tay, B.K., Zeng, X.T. et al. (2002) *Journal of Physics: Condensed Matter*, **14**, 7781.
36. Wautelet, M., Dauchot, J.P. and Hecq, M. (2003) *Journal of Physics: Condensed Matter*, **15**, 3651.
37. Allen, G.L., Bayles, R.A., Gile, W.W. and Jesser, W.A. (1986) *Thin Solid Films*, **144**, 297.
38. Shibata, T., Bunker, B.A., Zhang, Z. et al. (2002) *Journal of American Chemical Society*, **124**, 11989.
39. Ramos de Debiaggi, S., Campillo, J.M. and Caro, A. (1999) *Journal of Material Research*, **14**, 2849.
40. Yasuda, H. and Mori, H. (1992) *Physical Review Letters*, **69**, 3747.
41. Wronski, C.R.M. (1967) *British Journal of Applied Physics*, **18**, 1731.
42. Blackman, M. and Sambles, J.R. (1970) *Nature*, **226**, 938.
43. Liu, X., Wang, C.P., Jiang, J.Z. et al. (2005) *International Journal of Modern Physics B*, **19**, 2645.
44. Yeadon, M., Ghaly, M., Yang, J.C. et al. (1998) *Applied Physics Letters*, **73**, 3208.
45. Edelstein, A.S., Harris, V.G., Rolison, D.R. et al. (1999) *Applied Physics Letters*, **74**, 3161.
46. Gladgkikh, N.T., Bogatyrenko, S.I., Kryshtal, A.P. and Anton, R. (2003) *Applied Surface Science*, **119**, 338.
47. Mitas, L., Grossman, J.C., Stich, I. and Tobik, J. (2000) *Physical Review Letters*, **89**, 1479.
48. Car, R. and Parinello, M. (1985) *Physical Review Letters*, **55**, 2471.
49. Gupra, R.P. (1981) *Physical Review B*, **23**, 6265.
50. Sutton, A.P. and Chen, J. (1990) *Phylosophical Magazine Letters*, **61**, 139.
51. Franz, D.D. (2001) *Journal of Chemical Physics*, **115**, 6136.
52. Chigik, S.P., Gladgkikh, M.T., Grigorijeva, L.K. et al. (1985) *Izvestiya*

Academiji nayk SSSR –Metals (Proceedings of the Academy of Sciences of USSR), **2**, 175 (in Russian).
53. Hansen, M. and Anderko, K. (**1962**) *Structures of Binary Alloys*, Metalurgizdat, Moskow (in Russian).
54. Palatnik, L.S., Fuks, M.Y. and Kosevich, B.M. (**1972**) *Mechanism of Formation and the Substructure of Condensed Films*, Nauka, Moskow.
55. Yasuda, H., Mitsuishi, K. and Mori, H. (**2001**) *Physical Review B*, **64**, 094101.
56. Gusak, A.M. and Shirinyan, A.S. (**1999**) *Metal Physics and Advanced Technologies*, **18**, 659.
57. Shirinyan, A.S. and Gusak, A.M. (**1999**) *Ukrainian Journal of Physics*, **44**, 883 (in Russian).
58. Gusak, A.M. and Shirinyan, A.S. (**2000**) *Metallofizika i Noveishie Tekhnologii (Metal Physics and Advanced Technologies)*, **22**, 57 (in Russian).
59. Shirinyan, A.S., Gusak, A.M. and Desre, P.J. (**2000**) *Journal of Metastable and Nano-crystalline Materials*, **7**, 17.
60. Sheng, H.W., Xu, J., Yu, L.G. et al. (**1996**) *Journal of Material Research*, **11**, 2841.
61. Shirinyan, A. and Wautelet, M. (**2004**) *Nano-technology*, **15**, 1720.
62. Christian, J.W. (**1965**) *Theory of Transformation in Metals and Alloys*, Pergamon Press, New York.
63. Seitz, F. (**1940**) *The Modern Theory of Solids*, McGraw-Hill, New York and London.
64. Trusov, L.I., Petrunin, B.F. and Kats, E.I. (**1978**) *Fizika Metallov Metallovedenie (Journal of Metals Physics and Metals Science)*, **47**, 1229 (in Russian).
65. Shirinyan, A. and Gusak, A. (**2004**) *Philosophical Magazine A*, **84**, 579.
66. Gusak, A.M., Bogatyrev, A.O., Kovalchuk, A.O. et al. (**2004**) *Uspehi Fiziki Metalov (Journal of Progress in Physics of Metals)*, **5**, 433.
67. Rusanov, A.I. (**1967**) *Phase Equilibrium and Suface Phenomena*, Chemistry, Leningrad (in Russian).
68. Ulbricht, H., Schmelzer, J., Mahnke, R. and Schweitzer, F. (**1988**) *Thermodynamics of Finite Systems and Kinetics of First-order Phase Transitions*, BSB Teubner, Leipzig.
69. Jesser, W.A., Shneck, R.Z. and Gille, W.W. (**2004**) *Physical Review B*, **69**, 144121.
70. Saunders, N. (**1998**) in System Al-Li, in *Thermochemical Database for Light Metal Alloys*, vol. 2 (eds I. Ansara, A.T. Dinsdate and M.H. Rand), Office for Official Publications of the European Communities Luxembourg, pp. 40–43.
71. Chen, S.W., Jan, C.H., Lin, J.C. and Chang, A. (**1989**) *Metallurgical Transactions A*, **20A**, 2247.
72. Berezina, A.L., Volkov, V.A., Golub, G.V. et al. (**1991**) *Metalofizika (Journal of Metal Physics)*, **13**, 54.
73. Koval, Yu.N. (ed.) (**1992**) *Aluminium-lithium Alloys. Structure and Properties*, Kiev, Naukova Dumka (in Russian).
74. Ostwald, W. (**1897**) *Zeitschrift für Physikalische Chemie*, **22**, 282.
75. Schmelzer, J., Moller, J. and Gutzow, I. (**1998**) in *Zeitschrift fur Physikalische Chemie*, vol. 204 (ed. R. Oldenbourg), Verlag, Munchen, p. 171.
76. Milev, A.S. and Gutzow, I.S. (**1996/97**) *Bulgarian Chemical Communications*, **29**, 597.
77. Noble, B. and Thomson, G.E. (**1971**) *Metal Science Journal*, **5**, 114.
78. Williams, D.B. and Edihgton, J.W. (**1975**) *Metal Science Journal*, **9**, 529.
79. Manabu, T., Tzutomi, M. and Tarashisy, H. (**1970**) *Journal of the Japan Institute of Metals*, **34**, 919.
80. Shirinyan, A. and Wautelet, M. (**2006**) *Materials Science and Engineering C*, **26**, 735.
81. Shirinyan, A., Gusak, A. and Wautelet, M. (**2005**) *Acta Materialia*, **53**, 5025.
82. Shirinyan, A., Wautelet, M. and Belogorodsky, Y. (**2006**) *Journal of Physics: Condensed Matter*, **18**, 2537.
83. D. Van Nostrand Company, Inc. (**1968**) *Van Nostrand's Scientific Encyclopedia*, 4th edn, D. Van Nostrand Company, Inc., Princeton, New Jersey.
84. Encyclopedia Britannica, Inc. (**1998**) *The New Encyclopedia Britannica, Vol. 9, Micropedia. Ready Reference*, 15th

edn, Encyclopedia Britannica, Inc., Chicago.
85. Ludwig, F.P. and Schmelzer, J. (**1996**) *Journal of Colloid and Interface Science*, **181**, 503.
86. Neimark, A., Ravikovitch, P.I. and Vishnyakov, A. (**2002**) *Physical Review E*, **65**, 031505:1.
87. Gelb, L.D., Gubbins, K.E., Radhakrishnan, R. and Sliwinska-Bartkowiak, M. (**1999**) *Report Progress Physics*, **62**, 1573.
88. Pochon, S., MacDonald, K.F., Knize, R.J. and Zheludev, N.I. (**2004**) *Physical Review Letters*, **92**, 145702.
89. Chen, C.C., Herhold, A.B., Johnson, C.S. and Alivisatos, A.P. (**1997**) *Science*, **276**, 398.
90. Kamenev, K., Balakrishnan, G., Lees, M.R. and Paul, D.Mc.K. (**1997**) *Physical Review B*, **56**, 2285.
91. Qadri, S.B., Skelton, E.F., Hsu, D. et al. (**1999**) *Physical Review B*, **60**, 9191.
92. Baletto, F., Mottet, C. and Ferrando, R. (**2003**) *Physical Review Letters*, **90**, 135504.
93. Shirinyan, A. and Belogorodsky, Y. (**2009**) *Journal of Phase Transition*, **82**, 551.
94. Shirinyan, A. and Pasichnyy, M. (**2005**) *Defect and Diffusion Forum*, **237-240**, 1252.
95. Shirinyan, A. and Pasichnyy, M. (**2005**) *Nano-technology*, **16**, 1724.
96. Carpenter, D.T., Codner, J.R., Barmak, K. and Rickman, J.M. (**1999**) *Materials Letters*, **41**, 296.
97. Jacobs, K., Zaziski, D., Scher, E.C. et al. (**2001**) *Science*, **293**, 1803.
98. Shirinyan, A. and Belogorodskyy, Y. (**2008**) *Metallofizika i Noveishie Tekhnologii (Journal of Metal Physics and Advanced Technologies)*, **30**, 1641 (in Russian).
99. Shirinyan, A., Belogorodskyy, Y. and Schmelzer, J.W.P. (**2009**) *Acta Materialia*, **57**, 5771.
100. Inoue, A., Zhang, T. and Masumoto, T. (**1990**) *Materials Transaction JIM*, **31**, 425.
101. Paker, A. and Johnson, W.L. (**1993**) *Applied Physics Letters*, **63**, 2342.
102. Calin, M. and Koster, U. (**1997**) *Materials Science Forum*, **269-272**, 749.

Index

b
balance equations 24, 56, 58, 231, 263, 291, 305, 311, 313, 328, 337, 344, 387

c
concentration gradient 3–5, 7, 14, 16, 21, 22, 31, 39, 55, 62, 64, 67–74, 77–79, 81, 82, 91, 92, 100–104, 110, 112, 114–116, 120, 127, 128, 163, 176, 178, 179, 182, 191–194, 197, 219, 229, 238, 244, 263, 272, 321, 325, 329, 334, 343, 346, 369
concentration triangle 307, 312, 321, 324, 325, 328, 334, 336, 337, 346, 350, 352
critical nuclei 2, 44, 45, 47, 48, 52, 57, 79, 260, 268, 293, 320, 426

d
depletion 14, 27, 175, 183, 189, 285, 426, 427, 432, 433, 435–437, 444–446, 449, 451, 454–460, 467–470
diffusion couple 7, 40, 42, 45, 58, 61, 100, 112, 121, 127, 163, 164, 166, 170–172, 179, 217, 269, 275, 304, 321, 324–326, 328, 329, 342, 343, 350
diffusion zone 4, 13, 17, 30, 34, 38, 42, 52, 54, 62, 63, 68, 74, 81, 99–101, 103, 106–110, 112, 122, 127, 131, 164, 179, 239, 259, 260, 270, 271, 275, 289–292, 295–298, 302, 321, 326, 333, 336, 337, 339, 340, 342, 343, 345, 356, 368, 369, 376, 377
driving force 3–5, 31, 40, 53, 70, 75, 76, 81–85, 87–90, 92, 94, 95, 100–107, 112–115, 120, 121, 127, 129, 135, 136, 138, 139, 143, 150, 174, 179, 181, 191, 192, 194, 218, 220, 223, 224, 228, 242, 243, 297, 322, 335, 346, 371–375, 377–379, 383, 384, 389–392, 395, 400, 405, 406, 409, 410, 413, 414, 417, 421, 425, 432, 437, 444, 445, 447, 449, 451–455, 471

e
effective diffusivities 291
electromigration 7, 136, 174, 175, 177, 180, 183, 186, 191, 245, 246, 250, 263, 273–275, 280
entropy production 362, 363, 365–367, 382, 383, 387, 389, 397, 405, 409, 412–414, 416, 417, 419, 421, 422
extremum principles 367

f
failure mechanism 6
Fick's law 146, 346, 347, 416
Frenkel voiding 190

g
Gibbs phase rule 54, 335
Gibbs–Thomson effect 6, 31, 190–192, 194, 216, 218–222, 225, 235, 228, 234, 238, 242

h
hysteresis 466–469

i
incubation time 44, 48, 52, 53, 55, 58, 100, 121, 127–129, 132, 271, 294, 327
inter-diffusion 314
interaction energies 121–123, 131, 230, 239
interdiffusion 7, 8, 12–14, 16, 17, 19, 20, 22, 23, 37–40, 43, 50–52, 54–56, 61, 63, 65, 70, 81, 189, 190, 194, 215–217, 229, 260, 261, 263, 272–275, 299, 301, 302, 305, 310–312, 314–321, 368, 369, 371, 381, 382, 385, 396, 402, 417, 422

Diffusion-Controlled Solid State Reactions. Andriy M. Gusak
Copyright © 2010 WILEY-VCH Verlag GmbH & Co. KGaA, Weinheim
ISBN: 978-3-527-40884-9

Index

intermediate phase 3, 16, 17, 23, 37, 38, 40, 49–53, 61–63, 65, 69–71, 74, 81–83, 85, 89, 91, 100, 109, 112, 121, 122, 126, 127, 132, 143, 275, 278, 289, 290, 293, 297, 300, 301, 303–305, 308, 309, 311, 321, 324–329, 349, 360, 369, 371, 376, 446
interphase boundary 115, 299–302, 334, 336–338, 342, 343, 348, 381, 384, 386–388, 391, 400, 421
IPB 112–114, 121

k

Kirkendall effect 6, 13, 27, 29–31, 57, 164, 186, 190, 208, 216, 241, 261, 345, 348, 356
Kirkendall shift 6, 8, 135, 184, 190, 209, 216, 218, 221, 230

m

MC 238
metastable phase 50, 51, 54, 55, 63, 65, 456–458
Monte Carlo 100, 207, 238, 431
morphological 131, 132
morphology 6, 7, 121, 136, 143, 144, 189, 297, 307, 333, 337–339, 342, 373, 381, 386
multilayers 61, 100, 120, 132

n

nanoparticle 11, 30, 189, 215, 233, 244, 425, 428, 430, 433, 439, 444, 458, 461, 466–468, 471
nanoshell 6, 8, 30, 31, 58, 190–194, 196, 201, 202, 204, 205, 207–212, 214–218, 225–229, 231, 234, 237–239, 241–243
nanovoids 6, 189, 191, 233, 243
nucleation 2–5, 22, 37, 38, 41, 44, 50, 51, 55, 61–64, 66–74, 77–89, 91–97, 99–107, 109–113, 121, 122, 124–130, 132, 138, 143, 157, 190, 218, 225, 228, 233, 260, 267, 275, 289, 296, 297, 321, 326–328, 364, 368, 369, 375, 426, 427, 432, 433–439, 444, 446, 448, 451, 452, 454–456, 458, 460, 463, 467–469, 471

o

Onsager's coefficients 362, 367, 371

p

phase competition 2, 3, 37, 55, 61, 228, 259, 268, 360, 369, 451
phase diagram 45, 52, 290, 291, 296–298, 300–303, 311, 333, 335, 344, 350, 369, 384, 385, 393, 403, 406, 431, 441, 442, 444, 458, 459, 462–465, 471

phase growth 3, 17, 22, 38, 41, 45, 54, 95, 113, 121, 122, 145, 259, 262, 268, 271, 273, 275, 278–280, 282, 284, 286, 289, 290, 293–295, 297, 333, 334, 368, 371, 379
phase separation 180, 181, 186, 356, 433, 434, 437, 439, 462
phase suppression 47, 85, 91, 95, 268
pseudobinary approximation 342

r

reaction regime 289
reactive diffusion 1–3, 37, 38, 42, 73, 95, 99–101, 112, 114, 121, 122, 124, 131, 135, 189, 278, 302, 305, 320, 333, 381
regular solution 103, 382, 384, 391, 392, 435–437, 443, 446

s

segregation 6, 30, 51, 138, 179, 180, 183, 184, 186, 191–194, 197, 198, 202, 205, 206, 208, 209, 211, 214–216, 234, 235, 239, 241, 242, 244, 334, 381, 383, 385, 396, 407
shrinkage 6, 30, 32, 152, 160, 161, 190–194, 196, 197, 199, 200, 202–205, 207–209, 214–216, 228, 229, 231, 233–238, 240–244
size-induced phenomena 426
solid-state reactions 39, 81, 99, 100, 122, 132, 289, 290, 289, 295, 296, 334, 400
solubility limit 290, 450, 462
SSRs 6–9
supersaturation 27, 33, 53, 76, 115, 117, 126, 136, 137, 140, 147, 174, 175, 206, 218, 234, 238, 242, 243, 245, 285, 381, 383, 391, 402–404, 421, 441, 443, 449
surface curvature 247
surface energy 30, 48, 66, 122, 127, 135, 138, 190, 194, 251, 255, 256, 345, 348, 390, 398, 405, 429, 434, 436, 444, 453, 471

t

temperature cycling 466–468
ternary system 290, 291, 297, 303, 318, 320, 322, 336, 337, 349, 354
Tikhonov regularization 382, 416
transformation front 289, 290, 298, 337, 344, 381, 383–385, 387–389, 391–396, 398–400, 402–404, 407, 409, 421
transition criterion 441, 444, 446, 459, 464, 465
two-phase zone 304, 305, 307, 309, 316, 317, 320, 321, 335–345, 348, 349, 354–356

v

void velocity 247, 250, 251